DESIGNING AND PROGRAMMING MODERN COMPUTERS AND SYSTEMS

Volume I

LSI MODULAR COMPUTER SYSTEMS

Svetlana P. Kartashev, ed.
University of Nebraska

Steven I. Kartashev, ed.
Dynamic Computer Architecture, Inc.

PRENTICE-HALL, INC., Englewood Cliffs, New Jersey 07632

Library of Congress Cataloging in Publication Data
Main entry under title:

Designing and programming modern computers and systems.

   Bibliography: v. 1, p.
   Includes index.
   Contents: v. 1. LSI modular computer systems.
   1. Computer engineering. 2. Computer architecture.
I. Kartashev, Svetlana. II. Kartashev, Steven I.
TK7885.D474   001.64′4   81-21078
ISBN 0-13-201343-6 (v. 1)   AACR2

Editorial Production Supervision
and Interior Design: *Lynn S. Frankel*
Cover Design: *Frederick Charles, Ltd.*
Cover Photograph: *With grateful appreciation
to Hughes Aircraft Company for their kind permission
to reproduce an LSI wafer fabricated using
P-channel MOS technology.*
Manufacturing Buyer: *Gordon Osbourne*

© 1982 by Prentice-Hall, Inc., Englewood Cliffs, N.J. 07632

All rights reserved. No part of this book
may be reproduced in any form or by any means
without permission in writing from the publisher.

Printed in the United States of America

10 9 8 7 6 5 4 3 2 1

ISBN 0-13-201343-6

Prentice-Hall International, Inc., *London*
Prentice-Hall of Australia Pty. Limited, *Sydney*
Prentice-Hall of Canada, Ltd., *Toronto*
Prentice-Hall of India Private Limited, *New Delhi*
Prentice-Hall of Japan, Inc., *Tokyo*
Prentice-Hall of Southeast Asia Pte. Ltd., *Singapore*
Whitehall Books Limited, *Wellington, New Zealand*

CONTENTS

EDITORS' INTRODUCTION xv

CHAPTER I

Historic Progress in Architectures for Computers and Systems

Svetlana P. Kartashev
Steven I. Kartashev
Charles R. Vick

PREVIEW 1

THE MICROOPERATION APPROACH AND ITS USE IN DESCRIBING COMPUTERS AND SYSTEMS 3

1. GENERAL CONCEPTS   3
2. MICROOPERATIONS AND MICROCOMMANDS   4
   - 2.1   Microcommands   *4*
   - 2.2   Microoperations   *6*
3. MICROOPERATION DELAY   7
4. TIMING IN MICROOPERATION EXECUTION   8
   - 4.1   The Control Circuit Precedes the Execution Circuit   *8*
   - 4.2   The Control Circuit Follows the Execution Circuit   *9*
5. SELECTION OF A CLOCK PERIOD T_0 10
6. INSTRUCTION MICROPROGRAMS   11
   - 6.1   Microprogram Graphs   *11*
   - 6.2   Conditional and Unconditional Transitions   *12*

7.	MICROOPERATION HARDWARE DIAGRAMS	18
8.	THE USE OF THE MICROOPERATION APPROACH FOR THE DESCRIPTION OF COMPUTER ARCHITECTURE	21
	8.1 Conditional Branch Instructions *21*	
	8.2 The Floating-Point Multiplication Instruction *25*	

TRADITIONAL COMPUTER ARCHITECTURES 30

9.	PECULIARITIES OF THE VON NEUMANN ARCHITECTURE	31
	9.1 Principle of Addresses *31*	
	9.2 Fixed Instruction Microprograms *32*	
	9.3 Weaknesses of von Neumann Architectures *32*	
	9.4 Improving on von Neumann's Architecture *33*	
	9.4.1 Reduction in the Number of Addresses 33	
	9.4.2 Increase in the Number of Processor Registers 34	
10.	THE ARCHITECTURE OF MICROPROGRAMMED COMPUTERS	35
	10.1 Basic Principles of Microprogrammed Architecture *35*	
	10.2 Reconfiguration Fields *36*	
	10.3 Microinstruction Formats *40*	
	10.3.1 Microinstruction Classes 40	
	10.3.2 Phased Execution of Microinstructions 41	
	10.4 Block Diagram of a Microprogrammed Architecture *43*	
	10.5 Two-Level Control Memory *45*	
	10.5.1 One Microinstruction Cycle 46	
	10.5.2 Overlapped Execution of Phases 49	
	10.5.3 Problems of Conditional Branches 51	
	10.5.4 One-Phase Microinstruction Cycle 52	
	10.5.5 Memory Economy Accomplished with the Two-Level Control Memory 52	
11.	SUMMARY	56

PARALLEL SYSTEMS 57

12.	MULTICOMPUTER SYSTEMS	57
13.	MULTIPROCESSOR SYSTEMS	59
14.	ARRAY SYSTEMS	61
15.	PIPELINE SYSTEMS	63

15.1 Pipeline Computation *63*
15.2 Organization of a Pipeline System *67*
15.3 Problems of Pipeline Computation *68*

THE IMPACT OF TECHNOLOGICAL PROGRESS ON THE ARCHITECTURE OF COMPUTERS AND SYSTEMS 70

16. **LOWERING THE LOWER BOUNDS OF COMPUTERIZATION** 71
17. **RAISING THE UPPER BOUND OF COMPUTERIZATION** 71
    17.1 Requirements for Supersystems *72*
    17.2 New Architectures for Supersystems *72*
    17.3 Effect of LSI Technology on Supersystem Architectures *73*
18. **PROBLEMS IN DESIGNING A MODULAR COMPUTER** 74
19. **BIT-SLICED MODULAR COMPUTERS** 76
    19.1 Benefits *76*
    19.2 Weaknesses *77*
20. **MODULAR COMPUTERS FROM OFF-THE-SHELF MODULES** 78
21. **LSI TECHNOLOGY AND TRADITIONAL ARCHITECTURES** 79
22. **EVOLUTION OF COMPUTER DESIGN** 80
23. **ADAPTABLE ARCHITECTURES—PAST AND PRESENT** 81
    23.1 Microprogrammable Architectures *81*
    23.2 Reconfigurable Architectures *82*
    23.3 Dynamic Architectures *83*
24. **MODERN PROGRESS IN ADAPTABLE ARCHITECTURES** 83
    24.1 Reconfigurable Supernetworks *84*
    24.2 Problems of Interconnections in Reconfigurable Supernetworks *85*
    24.3 One-Stage Networks *88*
    24.4 Multistage Networks *88*
    24.5 Assessment of Supernetworks *89*
25. **DYNAMIC ARCHITECTURES** 90
    25.1 Execution of Additional Program Streams on the Same Hardware Resources *91*
    25.2 Speed-Up of Operations *91*
    25.3 Speed-Up of Between-Computer Communications *91*
    25.4 Assignment of Hardware Resources *92*
26. **EVOLUTION OF ADAPTABLE ARCHITECTURES** 92
    CONCLUSIONS 93
    REFERENCES 94

CHAPTER II
Reconfigurable Parallel Array Systems
Hubert H. Love

PREVIEW 99

1. INTRODUCTION 101
    1.1 Classification of Parallel Architectures *101*
    1.2 Parallel Array Systems *102*
    1.3 Applications of Parallel Array Systems *103*
    1.4 Reconfigurability in Parallel Systems *105*
    1.5 An Example of Array Operation *106*
    1.6 Applications for Reconfigurable Parallel Array Systems *107*
    1.7 References *108*

2. BACKGROUND, GENERAL BENEFITS, AND APPLICABILITY OF RECONFIGURABLE PARALLEL ARRAY SYSTEMS 108
    2.1 Background *108*
    2.2 Benefits in Building and Using Parallel Systems *109*
        2.2.1 *Design and Fabrication* *109*
        2.2.2 *Performance* *111*
    2.3 Determining the Suitability of the Application *111*
        2.3.1 *Parallelism in the Algorithms* *111*
        2.3.2 *Complexity of the Process* *112*
        2.3.3 *Input Considerations* *112*
        2.3.4 *Output Considerations* *113*
        2.3.5 *Other Parallel Attributes* *114*
        2.3.6 *Communication Channel Considerations* *115*
        2.3.7 *Avoiding Serial Operations* *115*
        2.3.8 *Another Pitfall to Avoid* *115*

3. SURVEY OF PROCESSOR DESIGNS 116
    3.1 Parallel Network Processors *116*
        3.1.1 *General Description* *116*
        3.1.2 *The ILLIAC IV* *122*
        3.1.3 *PEPE, The Parallel Element Processing Ensemble* *130*
    3.2 Associative Processors *135*
        3.2.1 *General Description* *135*
        3.2.2 *The STARAN* *143*
        3.2.3 *The Associative Linear Array Processor (ALAP)* *155*
        3.2.4 *Comparison of STARAN and ALAP* *161*

3.3　Distributed-Control Arrays: The Holland Machine　*161*

    3.3.1　Organization of the Machine　162
    3.3.2　Storage of Programs and Operands　163
    3.3.3　Path-Building　164
    3.3.4　Storage and Auxiliary Register Functions　165
    3.3.5　Other Features and Conventions　166
    3.3.6　Execution of Instructions　166
    3.3.7　Construction of a Holland Machine　167
    3.3.8　Applying and Programming the Holland Machine　167

3.4　References on Parallel Computer Architectures　*169*

4. DESIGNING A RECONFIGURABLE PARALLEL ARRAY PROCESSING SYSTEM: A CASE STUDY　169

4.1　A Note on the General Approach　*169*
4.2　The Application　*170*
4.3　The Processing Concept　*173*
4.4　The Project Objectives　*175*
4.5　The Initial Task: Developing the Algorithms　*176*
4.6　The Signal-Sorting Algorithms　*177*

    4.6.1　Pulse-Train Maintenance　177
    4.6.2　Pulse-Train Acquisition　179
    4.6.3　Example of Pulse-Train Startup　180

4.7　The Next Task: Determination of Suitability for Parallel Processing　*183*

    4.7.1　Satisfaction of Four Parallel-Processing Criteria　184
    4.7.2　Problems with Two Parallel-Processing Criteria　185

4.8　Preliminary Analysis of Three Candidate Systems　*185*

    4.8.1　Fast Sequential Processors　185
    4.8.2　Multiple-Processor Systems　187
    4.8.3　Associative Processors　190

4.9　Comparison of Results: Deciding on the System Approach　*192*

    4.9.1　Pipeline Processor Ruled Out　192
    4.9.2　Multiple Arrays of PEs as a Last Resort　192
    4.9.3　Improving PEPE Performance　193
    4.9.4　Improving ALAP Performance　193

4.10　Detailed Programming Analysis　*195*
4.11　The Use of Faster Gates　*198*
4.12　Improving Algorithm and Program Efficiency　*200*
4.13　Improving ALAP Array Performance　*200*
4.14　Improving System Performance　*204*
4.15　The First-Pass ALAP System　*216*
4.16　Assessment and Explanation of Risks　*216*

5. PROGRAMMING RECONFIGURABLE PARALLEL ARRAY PROCESSORS　218

5.1　Programming Languages　*218*
　　　5.2　Software Programming Tools　*220*
　　　　　5.2.1　Source-File Editors, File Managers, User Interfaces　221
　　　　　5.2.2　Symbolic Assembler　221
　　　　　5.2.3　Linkage Editor　221
　　　　　5.2.4　Functional Simulator　221
　　　5.3　Programming Techniques for Array Processors　*222*
　　　　　5.3.1　General　222
　　　　　5.3.2　Programming the ALAP　223
　6.　**FUTURE TRENDS IN RECONFIGURABLE ARRAY DESIGNS**　**233**
　　　6.1　The Emerging VLSI Technology　*234*
　　　6.2　Future Applications for Array Processors　*238*
　　　6.3　Limitations in the Performance of Present Designs　*239*
　　　　　6.3.1　Throughput Limitations　239
　　　　　6.3.2　Limitations on Fault-Tolerant Capability　240
　　　　　6.3.3　Limitations with Regard to VLSI Implementation　240
　　　6.4　Recommendations, Predictions, and Suggestions　*241*
　7.　**GENERAL REFERENCES**　**241**
　　　7.1　Reference Books　*242*
　　　7.2　Reference Papers　*242*

CHAPTER III
Designing and Programming Supersystems with Dynamic Architectures

Steven I. Kartashev
Svetlana P. Kartashev

PREVIEW　245

ANALYSIS OF TRADITIONAL AND NEW TECHNIQUES FOR INCREASING THROUGHPUT IN SUPERSYSTEMS　247

1. **INTRODUCTION**　**247**
2. **TRADITIONAL TECHNIQUES FOR AUGMENTING SYSTEM THROUGHPUT**　**249**
　　　2.1　The Use of High-Speed Components　*250*
　　　2.2　Modular Expansion　*250*

          2.2.1   Technological Restrictions *250*
          2.2.2   Increase in Delays or Complexities Introduced by
                  Interconnection Logic *251*
    2.3   The Use of Dedicated Architectures *253*
    2.4   Utilization of the Concurrency Present in Algorithms *254*
    2.5   Computation of an Algorithm in Subsystems with Different Types of
          Architectures *256*
    2.6   Optimization In Data Exchanges *257*
    2.7   Section Summary *259*

3. **NEW SOURCE OF AUGMENTING SYSTEM THROUGHPUT**   **259**

    3.1   Adaptation of Hardware Resources to Instruction and Data
          Parallelism *260*
    3.2   Reconfiguration of Hardware Resources into Different Types of
          Architecture *262*
          3.2.1   Multicomputer Computations *262*
          3.2.2   Multiprocessor Computations *263*
          3.2.3   Array Computations *264*
          3.2.4   Pipeline Computations *267*
          3.2.5   Mixed Computations *270*
    3.3   Section Summary *271*

DESIGNING SYSTEMS WITH DYNAMIC ARCHITECTURES 272

4. **INTRODUCTION**   **272**
5. **DC-GROUP WITH MINIMAL COMPLEXITY**   **273**

    5.1   DC-Group Resources *273*
    5.2   Control Organization in DC-Group Computers *280*
          5.2.1   New Requirements for Control Organization *280*
          5.2.2   Modular Control Organization: Basic Concepts *281*
          5.2.3   Hardware Realization of the Modular Control Device *284*
          5.2.4   Modular Control Organization and LSI Technology *286*
    5.3   Description of a Universal Module *287*
          5.3.1   Execution Portion *288*
          5.3.2   Organization of Instruction Sequencing and Variable Time
                  Intervals *291*
          5.3.3   Generation of Microcommands *292*
    5.4   Memory Management *293*
          5.4.1   PS Exchange *293*
          5.4.2   Addressing *297*
          5.4.3   The Instruction "Jump $ME_i \rightarrow ME_j(A_d)$" *298*
    5.5   Reconfigurable Paths for the Processor Signals *300*

            5.5.1   Reconfigurable Busses for the End-Around Carry   *301*
            5.5.2   Reconfigurable Bus for the Equality Signal   *306*

    5.6   Organization of Architectural Reconfigurations in System   *308*

            5.6.1   Task Synchronization and Priority Analysis   *309*
            5.6.2   Priority Analysis   *309*
            5.6.3   Storage of Variable Control Codes   *311*
            5.6.4   Architectural Switch to a New State   *314*

6. DC-GROUP WITH MINIMAL DELAY   **317**

    6.1   Resource Diagram   *318*
    6.2   Organization of Communications Between PE and ME   *318*

            6.2.1   Address Path   *318*
            6.2.2   The h-bit Data Path   *320*

    6.3   Communications Inside One Computer   *321*

            6.3.1   Instruction Path   *321*
            6.3.2   Data Path   *322*

    6.4   Communications Between Computers   *325*

            6.4.1   Information Exchange Between the Processor and the Memory of Two Different Computers   *325*
            6.4.2   Parallel Byte Exchange Between the Processor and the Memory of Two Different Computers   *326*
            6.4.3   Information Exchange Between the Processors of Two Different Computers   *331*
            6.4.4   Parallel Byte Exchanges Between Two Processors   *332*
            6.4.5   Parallel Shifts Inside One Computer   *334*
            6.4.6   Information Exchanges Between the Memories of Two Computers   *335*

    6.5   Reconfigurable Paths for Processor Signals   *336*
    6.6   Organization of Architectural Reconfigurations in a System   *336*

PERFORMANCE OF DYNAMIC ARCHITECTURES   338

7. INTRODUCTION   **338**
8. TIME REQUIRED FOR ARCHITECTURAL RECONFIGURATION   **339**
9. DELAYS INTRODUCED BY RECONFIGURABLE BUSSES   **340**
10. MODULAR EXPANSION   **341**
11. COST OF REALIZATION   **341**
12. PERFORMANCE EVALUATION   **342**

    12.1   Assessment of the Minimal Delay Memory–Processor Bus   *342*

            12.1.1   Time Required for Architectural Reconfiguration   *343*

12.1.2 Delays Introduced by Reconfigurable Interconnections   344
12.1.3 Time Required for Information Exchanges Between Independent Computers   344
12.1.4 Modular Expansion   344
12.1.5 Cost of Realization   345

12.2 Assessment of the Minimal Complexity Memory–Processor Bus   345

12.2.1 Time Required for Architectural Reconfiguration   345
12.2.2 Delays Introduced by Reconfigurable Interconnections   345
12.2.3 Time Required for Information Exchanges Between Independent Computers   345
12.2.4 Modular Expansion   345
12.2.5 Cost of Realization   346

12.3 Assessment of Processor Busses   346

12.3.1 Minimal Delay Processor Bus   346
12.3.2 Minimal Interconnection Processor Bus   347

13. SUMMARY   348

PROGRAMMING SYSTEMS WITH DYNAMIC ARCHITECTURE   348

14. INTRODUCTION   348
15. FINDING THE HARDWARE RESOURCES REQUIRED FOR A USER'S PROGRAM   350

15.1 Bit Sizes of Computed Variables   350

15.1.1 Algorithm for Constructing a Program Graph (Algorithm 1)   351
15.1.2 Finding the Number of Iterations in the Loop   357
15.1.3 Two Types of Bit Sizes   360
15.1.4 Bit Sizes of Integers   361
15.1.5 Bit Sizes of Real and Floating-Point Variables   364
15.1.6 Bit Sizes of Variables Computed in Graph Nodes   365
15.1.7 Array Dimensions of Variables Computed in Graph Nodes   367

15.2 Diagram of the Hardware Resources   367

15.2.1 Bit Size Diagram of a Program Graph   367
15.2.2 Alignment of the Bit Size Diagram   369
15.2.3 Finding the Time for Computing Each Task   370
15.2.4 P-resource Diagram   373

16. ASSIGNMENT OF THE SYSTEM RESOURCES AMONG PROGRAMS   373

16.1 CE Resource Diagram   374
16.2 DC-Group Flow Chart   377
16.3 ME Resource Diagram   379

17.	SUMMARY	382
	CONCLUSIONS	382
	REFERENCES	384

CHAPTER IV
Verification of Complex Programs and Microprograms

R. Negrini
M. G. Sami
R. Stefanelli

PREVIEW 387

1.	INTRODUCTION	389
2.	DEFINITION OF THE PROBLEM	394
3.	PROGRAM GRAPH MODELS AND REPRESENTATION OF THE RELATIONS AMONG VARIABLES	399
	3.1 Program Graph Models *399*	
	3.2 Extension to Microprograms *409*	
	3.3 Modeling Timing and Synchronization Characteristics *415*	
4.	TERMINATION OF SERIES-PARALLEL CONTROL GRAPHS	420
	4.1 Simple Loop *424*	
	4.2 Complex Series-Parallel Control Graphs *432*	
5.	FORMAL EQUIVALENCE BETWEEN C-GRAPHS AND TRANSITION GRAPHS	444
6.	REDUCTION PROCEDURE	447
7.	CONCLUSIONS	451
	REFERENCES	453

CHAPTER V
Requirements Engineering for Modular Computer Systems

David F. Palmer

PREVIEW 455

1.	INTRODUCTION	457

    1.1 Motivation *457*
    1.2 Scope *459*
    1.3 Overview of Approach *462*
2. **DETAILS AND EXAMPLES** 463
    2.1 Analysis *463*
    2.2 Partitioning *481*
    2.3 Allocation *490*
    2.4 Synthesis *494*
3. **CONCLUDING REMARKS** 497
    REFERENCES 498

CHAPTER VI
Design and Diagnosis of Reconfigurable Modular Digital Systems

Stephen Y. H. Su
Yu-I Hsieh

PREVIEW 501

1. **INTRODUCTION** 502
2. **REDUNDANCY TECHNIQUES FOR FAULT-TOLERANT DESIGN** 506
    2.1 Hardware Redundancy *506*
        2.1.1 *Static Redundancy* *507*
        2.1.2 *Dynamic Redundancy* *510*
        2.1.3 *Hybrid Redundancy* *512*
        2.1.4 *N-Modular Redundancy/Bipurge* *513*
        2.1.5 *N-Modular Redundancy/Unipurge* *515*
        2.1.6 *TMR/Spares* *517*
        2.1.7 *A Scheme for Multiple Fault-Tolerance* *518*
    2.2 Software Fault-Tolerance *524*
        2.2.1 *Software Redundant Architecture* *524*
        2.2.2 *Software Error Diagnosis* *526*
        2.2.3 *Software Error Recovery* *527*
        2.2.4 *Time Redundancy* *530*
3. **RELIABILITY MODELS OF RECONFIGURABLE FAULT-TOLERANT SYSTEMS** 530
    3.1 System Reliability of Fault-Tolerant Computing Systems *532*
    3.2 Other Reliability Related Measures *536*
    3.3 Diagnostic Performance Measure *542*

4. DIAGNOSIS OF MODULAR SYSTEMS 545
    4.1 t-Fault Diagnosable Systems for Permanent Faults   *545*
    4.2 t/s (t-out-of-s) Diagnosability   *551*
    4.3 t_i-Fault Diagnosable Systems for Intermittent Faults   *558*
    4.4   Diagnostic Reconfiguration   *559*
    4.5   Concurrent Computation and Diagnosis   *562*

5. SOME UNSOLVED PROBLEMS 565
REFERENCES 568

GLOSSARY OF DEFINITIONS USED IN THE BOOK 571

BIBLIOGRAPHY OF BASIC TOPICAL REFERENCES 615

DESCRIPTION OF SYSTEMS 633

INDEX 635

EDITORS' INTRODUCTION

During the past few years there has been a new direction in research dealing with LSI modular computers, systems, and networks. These are computers of the fourth generation. As a rule, they are distinguished by the following characteristics:

- *their architectures are modular;*
- *they support distributive processing; and*
- *they provide for software-controlled reconfiguration of between-module interconnections.*

Many computer manufacturers are beginning to produce complex modular computers, multimicrosystems assembled from microcomputers and microprocessors, data flow systems, etc. The computer industry will soon begin production of essentially new computer systems with adaptable architectures. Such systems will be capable of forming an available hardware resource into different numbers of computers that may change their type of architecture, transforming, say, a multicomputer system into a pipeline or array system and vice versa. They may automatically form a new instruction set that reflects the computational specificities of the program being computed, and that contains dedicated instructions specific to the given computation.

Computers and systems of the fourth generation introduce essential changes into traditional techniques of designing and programming. Thus they require that computer designers and programmers master new knowledge in the area. Relevant information is scattered in numerous computer journals and technical reports, however, and locating it can be complicated and frustrating. This makes it far more difficult to organize a continuous education for thousands of computer specialists.

Therefore, the objective of this new series of books, initiated with this pilot, is to eliminate this gap between the requirements of modern computer technology and the available literature.

DESIGNING AND PROGRAMMING MODERN COMPUTERS AND SYSTEMS: ABOUT THE SERIES

This series will present comprehensive and systematic information contributed by multiple authors of the modern state-of-the-art in designing and programming modern computer systems and networks.

Its purpose is to become an encyclopedia of modern knowledge in computers and systems of the fourth generation and to be useful to the following categories of readers:

- *Computer designers* involved in development of new computer projects.
- *Programmers* who begin working with LSI modular computers and systems.
- *Project and industrial managers* becoming acquainted with modern progress in computers and systems.
- *Computer scientists and professors* willing to have a more complete orientation in a particular area and/or to build a new course in this area for undergraduate and graduate education.
- *University and college students* who receive the educational background necessary for successful work in the industry or in modern computer research.

In addition, we would like to emphasize the usefulness of this series for various types of *specialists working with microcomputers and microprocessors*.

Many of them have come to the computer profession from other areas. Therefore, for them it becomes a matter of importance to acquire a systematic and thorough knowledge of complex modular systems assembled from microprocessors. This will expand their professional opportunities since they will be able to design and program more complex multimicrosystems, modular computers and networks—the next milestone in computer evolution initiated by LSI technology with the advent of microprocessors.

LSI MODULAR COMPUTER SYSTEMS: ABOUT THE BOOK

The first volume of the series is called "LSI Modular Computer Systems." It is written by many contributors who have succeeded in preparing a final text structured into a consistent and well-defined logical sequence of covered topics as if it were written by a single author. Since each chapter is written by highly skilled specialists in its area, it gives an in-depth treatment of the subject matter. Therefore, broadening the spectrum of book chapters achieved via multi-authorship did not sacrifice the depth and logical structure within each source presentation.

2.1 Basic Features

The book provides for the comprehensive coverage of the following major topics:

- *Detailed description of historic progress in computers and systems* that explains and justifies major distinctions among past, present, and future computer systems. The following major architectures are analyzed: von Neumann computers, microprogrammed computers, parallel computer systems, and LSI modular computer systems (Chapters I, II, III).
- *Introduction of a new microoperation* approach in describing a computer architecture, one that incorporates a *timing description* of data processing performed by the hardware as a sequence of activated microoperations and a *diagram description* of computer architecture that gives a hardware diagram of each microoperation (Chapter I).
- *Complete background, general benefits, and applicability of reconfigurable array parallel systems.* Description of major array systems either manufactured (Solomon 1, Illiac IV, Pepe, Staran, ALAP) or famous paper designs that caused a profound influence on further architectural thinking (Holland machine) (Chapter II).
- *The in-depth study of the complex reconfigurable array system (ALAP) designed at the Hughes Aircraft Company.* This material is unique inasmuch as, for the first time in the literature, all the hardware and software design of a complex parallel system manufactured by an industry has been described in such detail. Given are all the contradictory and painful trade-offs, problems, and successful and unsuccessful tries that were made by the designing team of which the author was a principal architect. This material is addressed to the working engineer and programmer who wishes to gain an understanding of the nature, applicability, designing, and programming of complex reconfigurable parallel systems (Chapter II).
- *Comprehensive and systematic description of designing and programming of modern computer architectures that perform microprogrammable, reconfigurable, and dynamic adaptations to the algorithm.* For the first time, students of computer architecture are presented with a fundamental and in-depth treatise into the basic characteristics of adaptable architectures which proved to be effective for the modular computer systems (Chapters I, II, III).

- *A systematic study into designing and programming dynamic architectures* that are feasible for distributed Supersystems designated for complex real-time algorithms (Chapter III).
- *Analysis of programs run on modular computer systems* aimed at finding out whether or not programs contain infinite loops. The procedures developed are simple and straightforward and thus can be applied to complex programs and microprograms. The techniques presented are based on an elegant mathematical idea and the authors effectively present their material by illustrating each step of the techniques with numerous examples and illustrations (Chapter IV).
- *Detailed techniques of software design for modular computer systems* aimed at mapping a complex application algorithm onto a modular computer system and finding the configuration of its hardware resource units that is optimal from the viewpoint of performance requirements for this algorithm. These techniques are useful for finding configurations of modular computer systems proceeding from the properties of application algorithms they are designed to compute (Chapter V).
- *Comprehensive treatment of hardware and software fault-tolerance* for the modular computer systems. Description of all major hardware and software techniques that increase reliability of computers and systems.

2.2 Education Qualifications of the Book "LSI Modular Computer Systems"

All contributors to this volume have made painstaking efforts to work as a team in presenting those topics in which they are most skilled.

Prerequisite Chart of Major Topics. The result of this teamwork evolves in the Prerequisite Chart of Major Topics covered in the book (see pages xx–xxi). Using this diagram the reader can see the logical sequence of the topics covered and how they can be most effectively learned.

Suggested Areas of Courses That Can Adopt This Book as a Text. Because of the vast diversity of topics and a comprehensive study of each topic, the book can be adopted for the following courses:

1. *A college text on computer hardware courses* on the graduate and undergraduate level: organization and architecture (Chapters I, II, III), fault-tolerance (Chapter VI), logical design (Chapters I, II, III, VI), parallel systems (Chapters I, II, III, V).
2. *A college text on software courses of the modular computer systems* covering such important and advanced areas as: software design (Chapter V), operating systems and resource assignments for modular computer systems (Chapters III, V), programming reconfigurable

arrays (Chapter II), termination analysis of complex programs run on modular computer systems (Chapter IV), and software fault-tolerance and reliability (Chapters IV and VI).

2.3 Teaching and Learning Aids in the Book

Cross References. A unique feature of this book is the use of cross references organized as follows: if some topic in a particular chapter is incidental and there is another chapter in the book where this topic is treated in greater detail or using another angle, then marginal sidenotes are given with references to other chapters (cross references). We feel this style of cross referencing is a valuable learning and teaching tool. It allows readers to study the multifaceted material while enriching themselves with each author's expertise in presenting one or several facets of the same subject. A college professor will find quite helpful both the cross references and prerequisite topical chart in preparing a course that is based on the material in this book and that fulfills the specific curriculum needs of his or her university.

Glossary of Definitions. The book contains an appendix which is a glossary of definitions used in all chapters. This will be useful to a student who, in reading a chapter, has forgotten the meaning of some term encountered.

Bibliography. Each chapter of the text is presented with a detailed bibliography. In addition, the book contains over 300 references partitioned into topics of discussion. Readers can then explore those topics that interest them most.

2.4 Editors' Acknowledgment

We would like to thank all book contributors for harmonious teamwork and the hard individual efforts each of them has displayed. The authors made a concerted effort to have their messages be clear and understandable in the overall perspective of the book's subject matter.

We are hopeful that the book will be useful for the reader for whom it is designated.

Svetlana Kartashev
Steven Kartashev
Editors

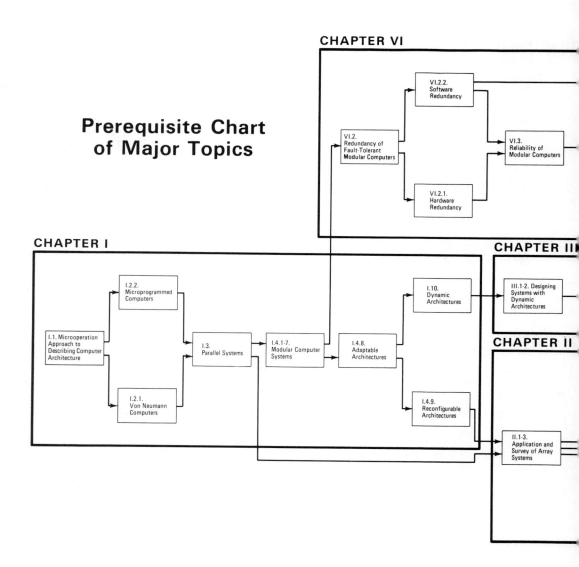

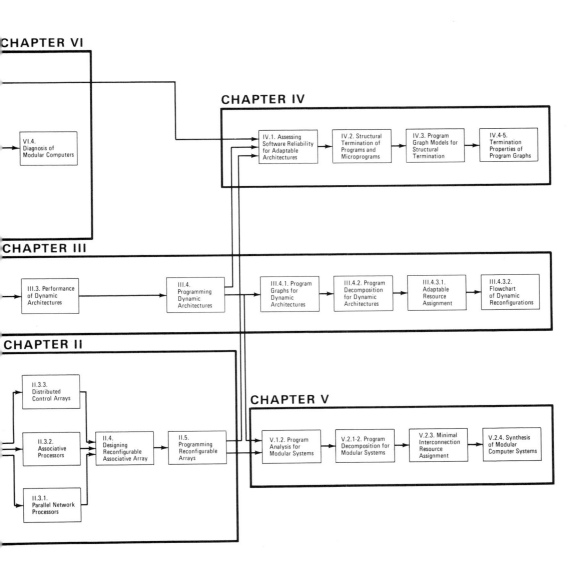

ABOUT THE AUTHORS

SVETLANA P. KARTASHEV Svetlana Kartashev is a professor of Computer Science at the University of Nebraska in Lincoln. She received her B.S. and M.S. degree in computer science from Kiev Polytechnical Institute and her Ph.D. in computer science from the Institute of Cybernetics in Kiev, USSR.

A resident of the United States since 1969, she became an American citizen in 1979. Dr. Kartashev is currently involved in designing and programming powerful multi-computer systems with dynamic architectures. Her research interests also include modular networks with reconfigurable interconnections.

Dr. Kartashev has authored more than 50 technical papers on computers, which have been published in leading, national and international computer journals.

STEVEN I. KARTASHEV Dr. Kartashev received his B.S., M.S., and Ph.D. degrees, all in computer science, at the Institute of Cybernetics in Kiev, USSR. Since 1970, he has worked in the United States, and he became an American citizen in 1979. He is currently the President of Dynamic Computer Architecture, Inc. in Lincoln, Nebraska.

Dr. Kartashev is involved in designing complex parallel systems and supersystems with new types of architectures. He is the principal investigator of the federal project aimed at constructing a complex parallel system with dynamic architecture for Ballistic Missile Defense Applications. Dr. Kartashev has authored approximately 60 technical papers on computers, which have been published in leading journals all over the world.

CHARLES R. VICK Charles Vick holds a Ph.D. in electrical engineering and computer science from Auburn University. Currently, he is the Vice President and technical director of System Control Technology, a company deeply involved in programming complex parallel systems for military real time applications.

For more than 12 years, he was the director of data processing research and development for the U.S. Army Ballistic Missile Defense Advanced Technology Center in Huntsville, Alabama. During this time, he sponsored and supervised the research and development of the most advanced computer systems that were produced in the United States in the 1970s. One such system was PEPE, a complex reconfigurable array system, constructed under his leadership.

A fellow of IEEE, Dr. Vick serves as an associate editor on the editorial board for the IEEE Transaction on Software Engineering. In 1979, he founded and was the first General Chairman of the International Conference on Distributed Computing Systems.

ABOUT THE AUTHORS

HUBERT H. LOVE Hubert Love is a project engineer in the Radar Systems group of Hughes Aircraft Company, where he is engaged in the design of radar data processing systems.

Having received his B.S. in Mathematics from the University of Arizona and his M.S. in Mathematics from the University of Delaware, Mr. Love has worked in industry as a programmer and systems engineer since 1958. Much of his experience has been in the design of architectures, languages, and software for highly-parallel computers. Among the results of these activities are one of the first relational data base designs, one of the first languages for the construction of knowledge-based systems, and the system design for an associative processor constructed at Hughes Aircraft Company.

Using his unique and lengthy experience, he provides readers with the enormity of first-hand details on how to design and program a complex reconfigurable array system.

MARIAGIOVANNA SAMI Ms. Sami is currently a professor of digital computers at the Politecnico di Milano in Italy, where she received her Dr. Ing degree in electrical engineering and her Libera Docenza degree in computers and switching theory.

She has worked in the areas of computer-aided design, program models and characteristics, and computer architecture. She was Program Chairperson for Euromicro Symposium 1980 on microprocessing and microprogramming. Currently, she is the editor of Euromicro Journal, and is a member of IEEE, AEI, and Euromicro.

RENATO STEFANELLI With his Dr. Ing and Libera Docenza degrees in electrical engineering and computer science, respectively, Mr. Stefanelli presently teaches both disciplines at the Politecnico di Milano in Italy.

He is involved with research in fast parallel multiplier and divider circuits, pattern recognition, and microprogramming in computer arithmetic, and is a member of the Associazione Italiana Calcolo Automatico.

ROBERTO NEGRINI Mr. Negrini received his electrical engineering degree from the Politecnico di Milano in Italy. He is currently Assistant Professor at the Institute di Elettrotecnica ed Elettronica del Politecnico di Milano, and his research involves reliability and testing of fault-tolerance multimicroprocessor systems and computer architecture.

DAVID F. PALMER With a background in radar, control systems, multilevel verification of programs, and real time simulation for embedded processor testing, Dr. Palmer is presently manager of the Distributed Data Processing (DDP) Program at the General Research Corporation. Since 1976 he has lead the DDP Program through studies of specific design issues, overall methodology of DDP design, and techniques for designing in fault tolerance, flexibility to

changes, and other so-called operational requirements.

Dr. Palmer holds a Ph.D. from Duke University, an M.S. from the University of New Mexico, and a B.S. from the University of Washington, all in electrical engineering. He has taught courses in electrical engineering and computer science at the University of California at Santa Barbara for several years. He is a member of IFIP Working Group 10.3 on software/hardware interrelation, and he was an IEEE Distinguished National Lecturer for 1979–80.

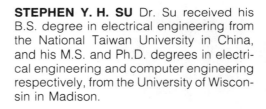

STEPHEN Y. H. SU Dr. Su received his B.S. degree in electrical engineering from the National Taiwan University in China, and his M.S. and Ph.D. degrees in electrical engineering and computer engineering respectively, from the University of Wisconsin in Madison.

He is currently a professor of computer science at the State University of New York at Binghamton. He has taught courses, performed research, and published over 45 papers in the areas of design automation, fault-tolerant computing, software engineering, computer architecture, and multiple-valued logic. His industrial experience has been obtained through summer work for IBM on array logic, for UNIVAC on design automation, for Bell Laboratories on digital simulation and fault diagnosis, and for Fabri-Tek, Inc. on logic design. He has also worked in the computer software area for the Biomedical Computing Division of the University of Wisconsin, and has served as a consultant for various industrial firms on software engineering, design automation, fault-tolerant design, computer architecture, and logic design.

YU-I HSIEH Yu-I Hsieh is a Ph.D. candidate in the computer science program at the State University of New York in Binghamton. His research interests include microprocessor design and Chinese Beta processing system design.

Mr. Hsieh received his B.S. degree in electronic communications engineering from National Chiao-tung University in Taiwan. He holds an M.S. degree in electrical engineering from National Taiwan University.

He has written six papers on design automation and microprogramming optimization.

CALL FOR CHAPTERS FOR VOLUMES II AND III OF THE SERIES, "DESIGNING AND PROGRAMMING MODERN COMPUTERS AND SYSTEMS"

Chapters are solicited for volumes II and III of the series.

Three types of chapters will be under consideration: (a) *industrial chapters* that describe actual design and programming experiences; (b) *research chapters* that are dedicated to the solution of a salient research problem(s), and (c) *tutorial chapters* that describe designing and/or programming methodologies established in the area.

Each chapter must be provided with comprehensive information that puts individual efforts of the author(s) into a proper perspective of an existing state-of-the-art. The detailed presentation must contain numerous examples and illustrations in order to comply with the educational standards maintained by Prentice-Hall.

Volume II will be dealing with the *Supersystem* issue. The following topics are of interest:

1. Description of actual Supersystem projects (applications, major architectural concepts, operating systems, memory management, conflict-free memory access, programming, etc.).
2. Salient scientific problems (design and programming methodologies, novel architectural ideas, improvements of interconnection networks, file allocation techniques, improving data bases, algorithms for fast and concurrent reconfigurations of Supersystem architecture, etc.).

Volume III will be dealing with *Multimicrosystems*. The following topics are solicited:

1. Description of industrial or otherwise implemented multimicrosystems, which are understood as complex modular networks assembled from modules (applications, distributed operating systems, communication between nodes, program distribution among nodes, distributed software-compilers, assemblers, used techniques for allocation, description of network nodes, and used reconfiguration techniques).
2. Outstanding research problems (data flow techniques used in programming, algorithms of fast network reconfigurations, minimization in the communication times between network nodes, basic methodologies for distributed operating systems, techniques for program decompositions, etc.).

A typical chapter format is 100 to 200 pages. *Six copies* should be submitted to the following address:

Drs. S. P. Kartashev and S. I. Kartashev
Dynamic Computer Architecture, Inc.
1210 Carlos Drive
Lincoln, Nebraska 68505

The authors of accepted chapters will be paid honorariums in accordance with the policy maintained by Prentice-Hall.

INFORMATION FOR PROSPECTIVE AUTHORS

Actions	*Deadlines*
Manuscript Submissions	August, 1982
Refereeing Process	October, 1982
Revisions of Accepted Chapters	December, 1982
Notification of Acceptance	January, 1983
Tentative Publication of Volumes II and III	December, 1983

Chapter I

Historic Progress in Architectures for Computers and Systems

Svetlana P. Kartashev
Steven I. Kartashev
Charles R. Vick

PREVIEW

At every stage of their development, large computer architectures have displayed the efforts of their designers to achieve the maximum processing speed attainable. Computer designers have traveled a historic path in their quest, with milestones such as von Neumann computers, microprogrammed computers, and parallel systems (multicomputers, multiprocessors, arrays, and pipelines). The current state-of-the-art features complex parallel systems that are able to assume a number of different architectural states and partition their resources into a variable number of processors and computers.

The great diversity of types of architectures makes discovering their general features very difficult. The problem is not helped by the fact that there does not presently exist a universal technique for describing a computer's architecture. Existing techniques of architectural description depend on traditions accepted in various computer companies and reflect the natural preferences of authors to use the mnemonics or languages they are most familiar with. Thus, any student of a particular architecture is confronted with the following task: Prior to understanding how a given computer works, he or she must study a language adopted by a particular author or a documentation technique used by the company that produces this architecture. Therefore, it is important to outline an approach for describing a computer's architecture that makes it possible to study different architectures independently of their complexity or their technology. One such approach is the so-called *microoperation description of the architecture* in which any architecture is described by the set of elementary actions it realizes, and all computations are reduced to the activation of a sequence of these actions.

This chapter shows how the microoperation approach can effectively explain the distinctions among two traditional architectures—von Neumann and microprogrammed. A microprogrammed architecture can be conceived of as an improvement on a von Neumann architecture since it reduces the time required for memory access of the von Neumann architecture.

We proceed by describing the evolution of these architectures in the 1960s and 1970s: while the 1960s was the decade that saw the development of basic architectural concepts for parallel systems (multicomputers, multiprocessors, arrays, and pipelines) the 1970s produced new computer architectures (modular computer systems) that are a direct outgrowth of LSI technological advances.

We show that the major distinctions among multicomputer, multiprocessor, array, and pipeline systems are in the way they interconnect their available processor and memory resources, although each individual processor continues to be either von Neumann or microprogrammed.† Therefore, the contributions of parallel systems to faster and more efficient computations are that they allow implementation of instruction parallelism and data parallelism in addition to the microoperation parallelism that was implemented in the original von Neumann and microprogrammed computers.

Recent advances in computer and systems technologies have included the appearance of LSI modules with high throughputs. The LSI modules have become the building blocks of fourth-generation computers. These fourth-generation computers (sometimes called *modular computer systems*) are distinguished by the following characteristics:

1. Their architectures are modular.
2. They support distributed processing.
3. They provide for software-controlled reconfiguration of between-module interconnections.

To be cost-effective these systems require the creation of new architectures that significantly alter traditional approaches to computer design

---

†We are not discussing associative processors here. They are treated very thoroughly in Chap. II of this book. (Ed.)

and programming. They are thus forcing serious changes to be made in traditional views of information processing. This chapter concludes with an analysis of the effects of LSI technology on computation and computer organization.

THE MICROOPERATION APPROACH AND ITS USE IN DESCRIBING COMPUTERS AND SYSTEMS

1. GENERAL CONCEPTS

All processing in any computer can be reduced to the execution of sequences of *microoperations*. A microoperation can be described as the smallest operation executed in one clock period. The average computer contains hardware circuits for 100 to 150 different microoperations. Each machine instruction activates a unique *microprogram*, which is a sequence or set of microoperations that realize the instruction. If the instruction set needs to be supplemented with a new machine instruction, the computer must already contain all the microoperations to be activated by the microprogram of this instruction. Thus, the instructions in the instruction set are essentially limited by the microoperations in the microoperation set of a given computer's architecture.

Since each program is executed in a computer as a sequence of machine instructions and each instruction is interpreted in the computer as a sequence of microoperations, the microoperation set may be used to characterize any architecture regardless of its type, hardware complexity, or technological realization.

The microoperation set is a qualitative characteristic of a computer's architecture that is independent of its technology of fabrication, the bit size of its processor(s), the size of primary memory, the number of I/O devices, etc. These parameters, on the other hand, are quantitative characteristics that specify the speed of information processing, the size of information arrays that can be stored in computer memories, the form in which information is received or sent to I/O devices, etc.

The following analogy may prove appropriate. The English alphabet forms a set made of 26 different letters. Any meaningful concept can be described with the use of these letters. Each letter is an elementary carrier of information, and each microoperation is likewise an elementary operation in the architecture. Letters are assembled into words, and microoperations are assembled into microprograms which carry out machine instructions. Words are assembled into sentences and paragraphs, and machine instructions are assembled into routines and programs. Therefore, in order to write a meaningful concept one has to have a set of letters, and in order to perform processing one has to have a set of microoperations. If several letters are withdrawn from the alphabet, some words cannot be written. Likewise, if several microoperations are withdrawn from an architecture, some microprograms cannot be constructed. The set of letters is thus a qualitative characteristic of a given carrier of information. Its quantitative characteristics are the size of a line (some analogy with processor size), the number of lines per page and the number of pages (size of a memory), etc.

2. MICROOPERATIONS AND MICROCOMMANDS

A computer uses two kinds of information for its computations: data and instructions. Both come packaged in *information words*, or simply words. Each word in the computer is either stored in a memory element or transformed by logical circuits. Thus, all computer components are divided into two categories: memory elements and logical circuits.

There are two types of memory elements: registers and memory cells. The difference between them is that registers may be connected to each other for data transfer while memory cells are always disconnected from each other. Each cell is connected to one or to several registers.

Logical circuits in a computer may be either control circuits or execution circuits. Control circuits are designed to cause a word to transfer between two memory elements: register to register; register to memory cell; memory cell to register, etc. An execution circuit performs a functional transformation of a word while it is being transferred between memory elements. All logical circuits can be implemented with the three primitive logical components: AND, OR, and NOT. There are other primitive sets such as (AND, NOT (NAND)), (OR, NOT(NOR)), etc.

2.1 Microcommands

To transfer a word from one register to another the registers should be connected by a data path. To make this data transfer controllable (i.e., performed at the moment of time specified by the control unit) the data path between the registers is interrupted by a control circuit. A simple control circuit is a collection of two-input AND gates. Each gate has one of its inputs connected to the information output (0 or 1) of one bit of the source register that stores the data word, and the second input (control input) connected to the control unit (Fig. I.1). The output of this gate is connected to the information input (0 or 1) of one bit of the destination register that is to receive this word. To activate this word transfer, the control unit sends a signal through the control line that connects all the control inputs of the AND gates together, and thus causes a word to be transferred from the source register to the destination register. Henceforth, a signal sent by the control unit to activate

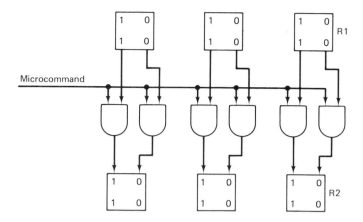

FIGURE I.1 Description of a simple control circuit

the transfer of a word from a source register to destination register will be called a *microcommand*.

Example 1. Suppose a 3-bit word stored in source register $R1$ is to be transferred to destination register $R2$. The data path between these two registers is intercepted by a control circuit containing six gates, so that each bit of $R1$ is connected with one bit of $R2$ via two gates devoted to the transfer of 1's and 0's, respectively (Fig. I.1). At a moment of time selected by the control unit, the control unit issues a microcommand that activates the control input of each gate and thus allows a word to travel between these two registers.

Note that the control circuit of Fig. I.1 can be reduced from six to three control gates (in general, from $2n$ to n gates) provided that destination register register $R2$ is cleared each time before a new word is written into it. For this case $R2$ is usually clocked and is provided with a separate clear input activated by the same microcommand that enables word transfer between registers. Since there are different types of flip-flops that may be assembled into registers, control circuits for them may differ from the one shown in Fig. I.1. ∎

In order to simplify the notation, the control circuit between two registers will be replaced by a solid circle, and the microcommand that activates this circuit will be shown with an arrow. Then the register-to-register transfer of Fig. I.1 is reduced to Fig. I.2. All microcommands that activate register-to-register transfers may be numbered and so simplify the description of transfers. It is sufficient to give the number of a microcommand to indicate the transfer intended.†

---

†This approach of representing microoperations is very useful for structural termination analysis of microprograms aimed at finding out if they contain infinite loops or not. (See Chap. IV, Sec. 3.) (Ed.)

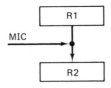

FIGURE I.2 Symbolic notation of the control circuit in Figure I.1

Example 2. Figure I.3 shows three registers ($R1$, $R2$, and $R3$) connected to each other via five control circuits. To transfer a word from $R1$ to $R3$, it is sufficient for the control unit to generate microcommand 3 which activates control circuit 3, etc. ■

2.2 Microoperations

The most elementary processing done in a computer consists of executing a *microoperation* during a word transfer between registers. To execute a microoperation, a data path between registers must contain not only the control circuit, but also an execution circuit. An execution circuit implements a boolean function that transforms input data word(s) stored in one or several source registers into an output data word broadcast to one or several destination registers. Since a microoperation includes one control circuit, it is activated by the microcommand that opens this control circuit. Each microoperation will be assigned a number that identifies the microcommand that activates it.

Note that each register-to-register transfer may also be conceived as a microoperation without an execution circuit. The data word merely passes through the control circuit.

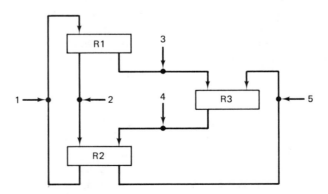

FIGURE I.3 Registers R1, R2 and R3 connected via control circuits 1 through 5

Example 3. Suppose that two data words, A and B, stored in registers $R1$ and $R2$, are to be added modulo 2 and the result sent to $R3$. This microoperation will be written symbolically as $R1 \oplus R2 \rightarrow R3$. Let us assign this operation the number 3.† To implement this microoperation it is necessary that each bit of registers $R1$, $R2$, and $R3$ conform to the boolean function of mod 2 addition, $Z_i = A_i \cdot \overline{B_i} \lor \overline{A_i} \cdot B_i$, where A_i and B_i are values values stored in registers $R1$ and $R2$, and Z_i is the functional output sent to register $R3$ (Fig. I.4). This microoperation also includes the control circuit. ■

3. MICROOPERATION DELAY

Any computer component (flip-flop or logical element) requires time to switch; i.e., time to generate an output after it receives an input(s). This time is called the *delay* of the component. Delay is a standard component parameter that can be found in any circuit catalog. Since every microoperation is a circuit assembled from logical components, each microoperation also introduces a delay (called the *microoperation delay*) in the generation of the output word broadcast to the destination register. Since the microoperation includes both control and execution circuits, its delay MID = COD + EXD, where COD and EXD are the delays introduced by control and execution circuits, respectively. To find MID one has to find the delay of the longest path of consecutive logical components the signal must pass through before it reaches the destination register. Therefore, to find the microoperation delay, we have to find all the logical paths in the circuit, find delay of each path, PD, and then select the maximal delay: MID = max(PD_i). Microoperation delay can be expressed in nanoseconds (or microseconds) or in the number of primitive logical components (AND, OR, NOT) in the logical path with maximal delay. For the latter case assume that each component's delay is t_d; then a logical path containing k components is delayed by $k \cdot t_d$.

Example 4. Find the delay introduced by microoperation 3 of Example 3. For this microoperation one can distinguish two separate logical paths for signal propagation (Fig. I.4): Path 1 including gate 1 (AND), gate 3 (OR), and gate 5 (AND); and path 2 including gate 2 (AND), gate 3 (OR), invertor 4 (NOT), and gate 6 (AND). Two other paths (1, 3, 4, 6 and 2, 3, 5) are equivalent to these two. If one assumes that all components introduce the same delay t_d, then the delay of path 1, $PD_1 = 3t_d$, the delay of path 2, $PD_2 = 4t_d$, and the microoperation delay MID = max($3t_d$, $4t_d$) = $4t_d$. ■

†From now on all microoperations introduced will be listed in the Microoperation Table. (See page 19.)

8 CHAPTER I

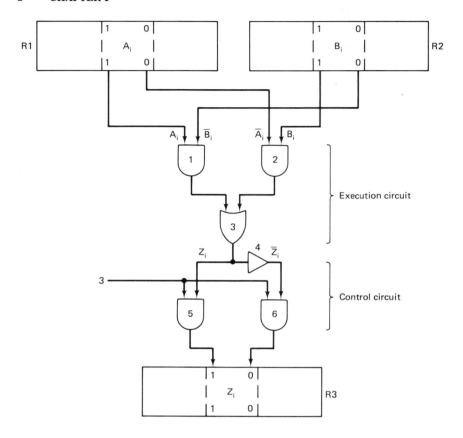

FIGURE I.4 Implementation of the module 2 addition microoperation

4. TIMING IN MICROOPERATION EXECUTION

As a rule, a microoperation is executed during one clock period, T_o, which is the time between the fronts of two consecutive synchronization pulses, τ. (Fig. I.5). The meaning of T_o depends on the ordering of the control and execution circuits in the microoperation circuit. Consider two possible options for their interconnection.

4.1 The Control Circuit Precedes the Execution Circuit

For this case, the information inputs of the control circuit are connected to the source register(s) (Fig. I.6(a)). The word stored in a source register is blocked from propagating through the microoperation circuit unless the microcommand issued by the control unit reaches the control circuit.

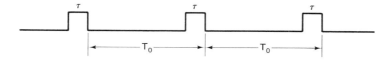

FIGURE I.5 Clock periods, T_0

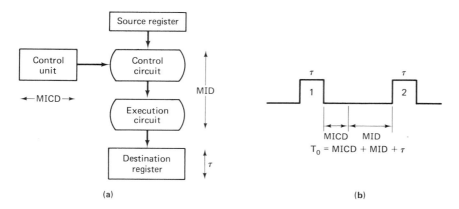

FIGURE I.6 A clock period time for the case control circuit precedes executional circuit

Since the microcommand is also delayed when it passes through the logical circuits in the control unit, the duration of clock period, T_0, is

$$T_o = \text{MICD} + \text{MID} + \tau \tag{1}$$

where MICD is the microcommand delay, MID is the microoperation delay, and τ is the duration of synchronization pulse (Fig. I. 6(b)).

Let us specify the meanings of these values. The control unit may generate a microcommand as an output of its control sequencer or as a signal that is a decoding of either the opcode or another control field in the machine instruction. For both cases microcommand generation begins only after a new word is written to the sequencer or the instruction register; i.e., after pulse τ. Therefore, the time MICD includes time for decoding a word stored in the sequencer or instruction register. Techniques for finding the MID delay were introduced in the previous section.

4.2 The Control Circuit Follows the Execution Circuit

For this case, when an input word is written to a source register (after pulse 1, Fig. I.7(b)) it propagates through the execution circuit concurrently with

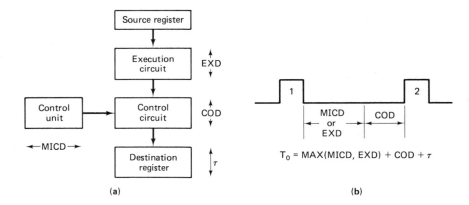

FIGURE I.7 A clock period time for the case executional circuit precedes control circuit

the microcommand propagating through the logic circuits of the control unit (Fig. I.7(a)). Thus, the time T_0 is specified by

$$T_0 = \max (MICD, EXD) + COD + \tau \qquad (2)$$

where EXD and COD are delays introduced by the execution and control circuits, and max (MICD, EXD) is the maximum of the two delays, MICD and EXD.

Note that for many cases, the execution circuit of the microoperation contains a level of AND gates that can be combined with the control circuit. Then there is no separate additive delay for the control circuit and the time T_0 is defined by Eq. (1).

5. SELECTION OF A CLOCK PERIOD T_0

The microoperation set of an average computer contains 120 to 150 different microoperations. Their typical distribution among different functional units is as follows: The processor accounts for 50 to 60 microoperations, the control unit implements 30 to 50 microoperations; I/O devices realize 30 to 40 microoperations; and the memory realizes 10 to 20 microoperations.

The time required to perform these microoperations may vary over a wide range. Thus a designer faces the problem of how to select the clock period time, T_0.

If T_0 is selected on the basis of the MID time of the slowest microoperation in the computer, all faster microoperations will leave the computer inactive at the end of T_0. This leads to a slower rate of information processing.

One solution to this problem is to execute slow microoperations during several clock periods. Since the delay MICD is practically the same for all microoperations, the time T_0 is then found using a typical MID time for a majority of the microoperations in the processor and the control unit. All microoperations with a greater MID are then executed during p clock periods, where $p = (\text{MICD} + \text{MID})/T_0$.

For the second case, described in Sec. 4.2, since EXD > MICD, $p = (\text{EXD} + \text{COD})/T_o = \text{MID}/T_o$.

The p value count is performed in the control unit to yield the time $p \cdot T_0$ needed for the long microoperation. The respective microcommand thus lasts continuously during p clock periods, T_0.

6. INSTRUCTION MICROPROGRAMS

Since a microoperation identifies an elementary processing operation, one machine instruction is usually assigned a sequence of several microoperations. This sequence is called the *instruction microprogram*.

There are two techniques for describing an instruction microprogram: (1) a symbolic technique using one of the existing hardware description languages, and (2) a directed graph in which each step in the microprogram is represented with one node. Arrows among nodes show sequencing among steps.

6.1 Microprogram Graphs

We will use the second technique for describing the microoperation approach to a computer's architecture since it is more illustrative.

Accordingly, a microprogram is a directed graph in which each node denotes one or several microoperations denoted by identifying numbers that are executed in parallel. Arrows show the sequencing of nodes, and are marked with any signals that activate transitions from one node to its successor.

Example 5. Figure I.8 shows the microprogram of an addition instruction. As shown in the figure, the instruction is executed in 13 clock periods. During the first clock period the computer executes microoperation 17 (the instruction address stored in the counter is transferred to memory, see the Microoperation Table and Fig. I.16). Microoperation 9 is executed during the second clock period (the instruction that was fetched from the memory is sent to the memory data register). During the following clock periods the instruction is stored in the instruction register, the numbers to be added are fetched from memory and stored in registers, and the addition itself takes place in node 10. During the eleventh clock period two microoperations, 7 and 14, are executed in parallel; 7 sends the result to the memory data register and 14 sends the address of the result to the memory address register, etc.

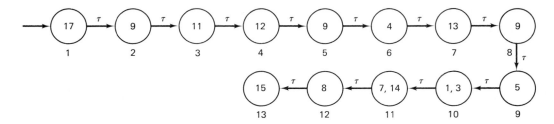

FIGURE I.8 Microprogram of addition instruction

Transition from one node to another is performed under the synchronization pulse τ that marks each arrow of the microprogram. ■

6.2 Conditional and Unconditional Transitions

Many instruction microprograms are characterized by multiple microoperation sequences in which the selection of each sequence depends on the results of conditional tests made during the microprogram's execution. Such tests may be the verification of the truth or falsehood of $A > B$, $A < B$, $A \geq B$, $A \leq B$, $A = B$, or $A \neq B$, where A and B are data words; or they may be the test of one or several flip-flops.

In a microprogram graph, multiple microoperation sequences usually originate from a node that contains one or several conditional test microoperations. This node will have several outgoing arrows connecting it with its successors. Selection of a successor node is determined by the value of the signal, listed by the arrow, that causes the transition. All transitions in the microprogram graph are either unconditional or conditional.

Transition $a \xrightarrow{x} b$ from node a to node b is called *unconditional* if node a has only one succeeding node b (Fig. I.9). Each unconditional transition is activated by the unconditional signal, a signal that may assume only one value. For instance, all the transitions in Fig. I.8 are unconditional and activated by the unconditional synchronization pulse τ which signifies the end of the microoperation executed in the present clock period and activates the transition to the next microoperation(s) executed during the next clock period.

Transition $a \xrightarrow{y} b$, $a \xrightarrow{\bar{y}} c$ is called a *simple conditional transition* if node a has two succeeding nodes b and c where b is selected when $y = 1$ and c is selected when $\bar{y} = 1$ where y and $\bar{y}$ are the true and false values of signal y that marks this transition. Thus, a simple conditional transition is activated by a single conditional signal y that may assume two values: a true value ($y = 1$) and a false value ($\bar{y} = 1$) (Fig. I.10).

Example 6. Consider the execution of a simple conditional transition in which the origin node executes an unsigned conditional test $A > B$ (Fig.

Historic Progress in Architectures for Computers and Systems 13

FIGURE I.9 Unconditional transition

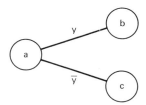

FIGURE I.10 Simple conditional transition

I.11(a)). If $A > B$ is true, microoperation (MO) 15 causes the program counter, PC, to generate the next address as PC + 1 → PC. If $A > B$ is false, MO 16 causes the jump address stored in the instruction register, *IR*, to be transferred to the PC (Fig. I.11(a)). To test if $A > B$, the origin node has to perform $A - B$ in the adder: Various adders may have different implementations of the basic operations they perform, such as $A + B$, $A - B$, $A \oplus B$, $A \cdot B$, etc. One of the possible realizations of $A + B$ and $A - B$ is shown in Fig. I.11(b) in which both operations require concurrent activation of MO 1 (that performs carry propagation) and MO 3 (that per-

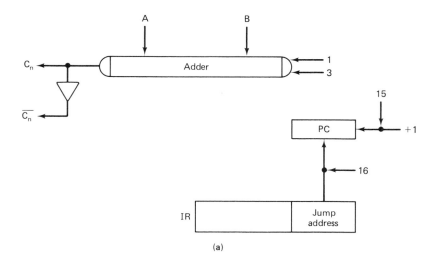

FIGURE I.11(a) Implementation of the test "If $A > B$" in a computer

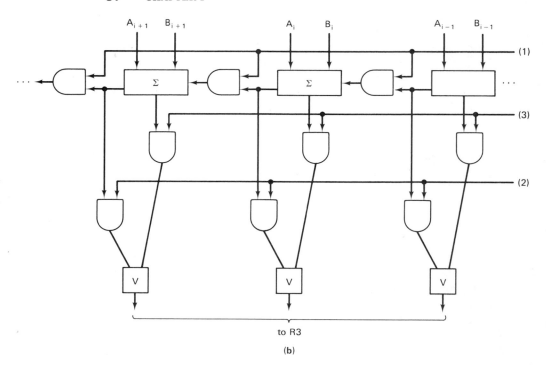

FIGURE I.11(b) Implementation of Mo 1 and Mo 3

forms mod 2 addition). Note that to do $A - B$, this adder has to receive the B operand in a complemented form (1's or 2's).

If $A > B$ is true, i.e., $A - B$ is a positive number, the adder generates an overflow ($C_n = 1$). If $A > B$ is false, the adder does not generate an overflow ($\overline{C}_n = 1$). Thus if $C_n = 1$, the node executing MOs 1 and 3 is succeeded by the node performing MO 15 (increment PC). If $\overline{C}_n = 1$, it is succeeded by the node performing MO 16 (Fig. I.12). ■

If microprogram node a has more than two successors, that is, is followed by nodes $b, c, \ldots, k$ in the microprogram graph (Fig. I.13), the transition $a \to b$, $a \to c$, $a \to d, \ldots, a \to k$ is called a *complex conditional transition*. For this case node a executes several conditional test microoperations so that each arrow of the transition is distinguished from others by a unique combination of conditional signals. If node a has three successors b, c, and d, then it must generate at least two values of conditional signals y and z so that transition from a to each successor is distinguished by a unique combination of y and z. For instance, $a \xrightarrow{yz} b$ is activated by yz; $a \xrightarrow{y\bar{z}} c$ is activated by $y\bar{z}$ and $a \xrightarrow{\bar{y}} d$ is activated by $\bar{y} = \bar{y}z \vee \bar{y}\bar{z}$. In general, if node

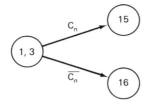

FIGURE I.12 Simple conditional transition during test "If A > B"

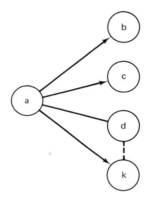

FIGURE I.13 Complex conditional transition

a has k successors in the microprogram graph, then it must generate at least $\log k$ conditional signals. This will allow assigning each arrow leaving node a with a unique combination of true and false values of the conditional signals generated in the node.

Example 7. Consider the execution of a complex conditional transition for the floating-point addition microprogram. This microprogram adds two floating point numbers, $A = AE \times AM$ and $B = BE \times BM$, specified by exponents AE and BE and mantissas AM and BM. Assume that exponents AE and BE are stored in registers $R4$ and $R5$ and mantissas AM and BM are stored in registers $R1$ and $R2$ (Fig. I.14).

Before adding mantissas the microprogram must execute a conditional test, $AE > BE$, aimed at finding out (a) whether mantissa AM or BM has to be aligned, i.e., shifted towards the least significant bit, and (b) how many shifts must be performed over AM or BM. To do this, the microprogram must execute the complex conditional transition shown in Fig. I.15. (Figure I.15 shows only the relevant portion of the floating-point addition microprogram.) The origin node, 10, transfers AE and BE to the adder (MOs 35, 36, Fig. I.14)

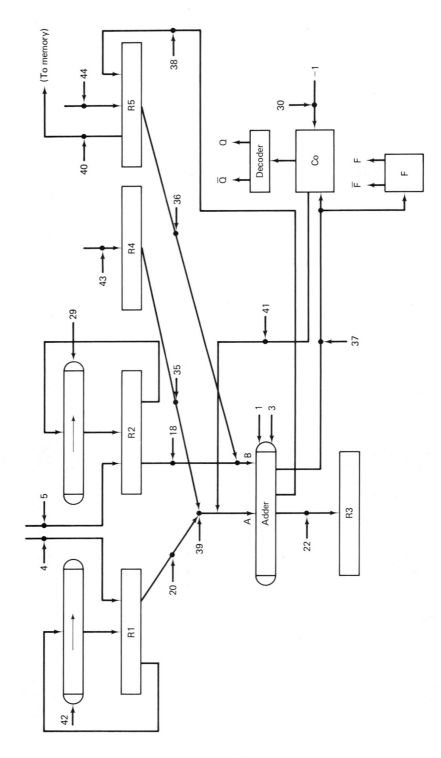

FIGURE I.14 Floating-point addition diagram

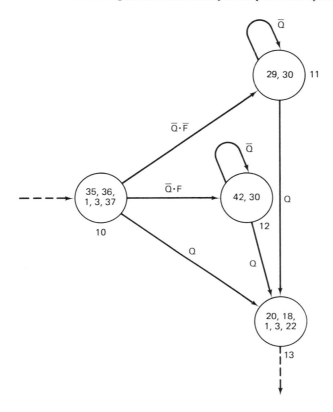

FIGURE I.15 Complex conditional transition for floating-point addition microprogram

and performs $AE - BE = \pm\Delta$(MOs 1, 3). Both the result Δ and its sign are sent to the counter Co and flip-flop F with MO 37, respectively.

There are three outcomes for this test:

1. $AE > BE$; $AE - BE = +\Delta$
2. $AE < BE$; $AE - BE = -\Delta$
3. $AE = BE$; $AE - BE = 0$

Case 1. If $AE > BE$, counter Co (Fig. I.14) will store Δ. Flip-flop F will store "+" encoded as 0; i.e., $\overline{F} = 1$. This will activate the combination of conditional signals $\overline{Q} \cdot \overline{F}$, where Q is the signal that decodes 0 in the counter. The signal $\overline{Q} \cdot \overline{F}$ will activate the transition to the next node, 11, that shifts the mantissa BM towards the LSB (MO 29, Fig. I.14) and decrements the counter (MO 30). This node is iterative and is performed Δ times until counter Co stores 0 and Q becomes 1. The next transition is then activated by signal Q.

Case 2. If $AE < BE$, counter Co will store Δ, Q will be 0 ($\overline{Q} = 1$), and flip-flop F will store 1. This will activate signal $\overline{Q} \cdot F$, enabling a transition to node 12 that shifts mantissa AM towards the LSB (MO 42) and decrements Co via M0 30. This node is also iterated Δ times until Co stores 0, Q becomes 1.

Case 3. If $AE = BE$, counter Co will store 0 and that will produce signal Q. This activates a transition to node 13 that performs mantissa addition. The mantissa AM in $R1$ is sent to the A terminal of the adder (MO 20); the mantissa BM in $R2$ is sent to B terminal of the adder (M0 18); the adder $AM + BM$ (MOs 1, 3), and it sends the result to $R3$ (M0 22). ∎

7. MICROOPERATION HARDWARE DIAGRAMS

A widely used technique for describing the hardware diagram of any functional unit of the computer (processor, control unit, memory, or I/O) is to specify all its registers, adders, counters, and their sizes and interconnections. However, it is very difficult to show what microoperations are executed in the architecture using this diagram, how two architectures may differ in their functions, each of which is identified with one microoperation set.

The drawback of a conventional hardware diagram is overcome if the computer is described with microoperation hardware diagrams, since in addition to listing all registers and logical circuits the microoperation hardware diagrams show which microoperations are executed during word transmissions between registers. Therefore, to obtain a conventional hardware diagrams microoperation analog, all the microoperations that are executed added to the conventional diagram. Their numbers are indicated on data paths between registers and between flip-flops and show which actions are executed during the execution of every machine instruction. The microoperation hardware diagram is supplemented by a microoperation table that describes each microoperation of the diagram. A single microoperation table is usually made for all the functional units of the computer; it contains the microoperation set that was detailed in the microoperation hardware diagrams.

The effectiveness of the microoperation approach is illustrated below for two different microprograms—conditional branch and floating-point multiplication. In spite of the fact that both are described by complex algorithms, their characterization via microprogram graph and microoperation hardware diagram allow us to receive a clear and consistent description of their clock-by-clock computation in the computer. Figures I.14 and I.16 show the microoperation hardware diagrams of a simple computer. We will use these later to explain some instruction microprograms. The Microoperation Table (Table I.1) describes the microoperations shown in these diagrams.

TABLE I.1
MICROOPERATION TABLE

Identification Number	Symbolic Notation	Description
1	$[\Sigma]$	Carry propagation in the adder.
2	$R1 \wedge R2 \rightarrow R3$	Logical multiplication of the words stored in the $R1$ and $R2$ registers with the result sent to the $R3$ register.
3	$R1 \oplus R2 \rightarrow R3$	Modulo 2 addition (Exclusive-OR microoperation).
4	$MODR \rightarrow R1$	Transfer of a data word from the memory output data register to the $R1$-register of the processor.
5	$MODR \rightarrow R2$	Transfer of a data word from the memory output data register to the $R2$-register of the processor.
6	$R1 \ominus R2$	Comparison by equality microoperation of the two data words stored in the $R1$ and the $R2$ processor registers, respectively.
7	$R3 \rightarrow MIDR$	Transfer of the adder result stored in the $R3$ processor register to the memory input data register.
8	$MIDR \rightarrow M$	Write to the main memory.
9	$M \rightarrow MODR$	Read from the main memory to the memory output data register.
10	Z	Start of the control unit.
11	$MODR \rightarrow R_{in}$	Transfer of the instruction from the memory output data register to the instruction register R_{in}.
12	$R_{in}(A1) \rightarrow MAR$	Transfer of the first address (stored in the first address field) to the memory address register.
13	$R_{in}(A2) \rightarrow MAR$	Transfer of the second address (stored in the second address field) to the memory address register.
14	$R_{in}(A3) \rightarrow MAR$	Transfer of the third address (stored in the third address field) to the memory address register.
15	$Co + 1 \rightarrow Co$	An increment by 1 of the program counter.
16	$R_{in}(A3) \rightarrow Co$	Transfer of the third address (stored in the third address field) to the program counter.

TABLE I.1 (Cont.)

Identification Number	Symbolic Notation	Description
17	$Co \to MAR$	Transfer of the address stored in the program counter to the memory address register.
18	$R2 \to \Sigma_B$	Transfer of the data word stored in the $R2$ register to the B terminal of the adder.
19	$R3 \to \Sigma_A$	Transfer of the result stored in the $R3$ register to the A terminal of the adder.
20	$R1 \to \Sigma_A$	Transfer of the data word stored in the $R1$ register to the A terminal of the adder.
21	$C_n \to \text{sign } R3$	Transfer of the adder overflow bit to the sign bit of the $R3$ register.
22	$\Sigma \to R3$	Transfer of the computational result from the adder output to the $R3$ register.
23	$R3 \overset{1}{\downarrow} \to R3$	One-bit shift of the $R3$ register to the LSB.
24	$R3 \overset{1}{\downarrow} \to R3$	One-bit shift of the $R3$ register to the MSB.
25	$\text{LSB } R3 \to \text{MSB } R2$	Transfer of the LSB $R3$ to the MSB $R2$.
26	$\text{MSB } R2 \to \text{LSB } R3$	Transfer of the MSB $R2$ to the LSB $R3$.
27	$\text{MSB } R2 \to \Sigma_A$	Transfer of MSB $R2$ to the LSB of the A terminal of the adder.
28	$R2 \overset{1}{\uparrow} \to R2$	One-bit shift to the MSB of the $R2$ register.
29	$R2 \overset{1}{\downarrow} \to R2$	One-bit shift to the LSB of the $R2$ register.
30	$Co_1 - 1 \to Co_1$	A decrement by 1 of the counter Co_1.
31	$0 \to Co_1$	Clearing of the counter Co_1.
32	$n \to Co_1$	Sending the quantity n to the counter Co_1.
33	$\text{Pr sign} \to T_3$	Transfer of the product sign identified by logic L_3 to the T_3 flip-flop.
34	$T_3 \to \text{sign } R_3$	Transfer of the sign of the result stored in the T_3 flip-flop to the $R3$ register.
35	$R4 \to \Sigma_A$	Transfer of the AE exponent stored in the $R4$ register to the A terminal of the processor adder.
36	$R5 \to \Sigma_B$	Transfer of the BE exponent stored in the $R5$ register to the B terminal of the processor adder.

37	$\Sigma \to Co_1$	Transfer of the adder difference obtained from the adder to the processor counter Co_1. Concurrently, the sign of this difference obtained from the $(n + 1)$-bit of the adder is transferred to the F flip flop.
38	$\Sigma \to R5$	Transfer of adder output which identifies the result of BE and AE exponent operation to the $R5$ register designated for storing the exponent of the result.
39	$B + 1 \to \Sigma$	Add 1 to the LSB of the A terminal of the adder.
40	$R5 \to MIDR$	Transfer of the exponent of the result to the memory input data register.
41	$Co_1 \to \Sigma_A$	Transfer of the counter Co_1 content to the A terminal of the processor adder.
42	$R1 \overset{1}{\downarrow} \to R1$	One-bit shift of the content of the $R1$ register to the LSB.
43	$MODR \to R4$	Transfer of the AE exponent of the data word to the $R4$ register.

8. THE USE OF THE MICROOPERATION APPROACH FOR THE DESCRIPTION OF COMPUTER ARCHITECTURE

Let us show that a combination of the microprogram graph and the microoperation hardware diagram is a very effective and simple way to describe arbitrarily complex instruction microprograms. These figures show how data words are moved during computation, which circuits are activated at each clock period, and the signals that activate transitions to the microoperations that are to be executed in the next clock period.

8.1 Conditional Branch Instructions

Each of these instructions fetches two operands, A and B; performs a conditional test between them; and depending on the results of the test, either steps to the next instruction or jumps to the instruction located at the jump address.

Consider the step-by-step execution of an instruction microprogram in the computer. To do so we will refer to the microoperation hardware diagram of Fig. I.16, the microprogram graph of Fig. I.17, and the Microoperation Table.

CLOCK period 1 (CLOCK 1). In accordance with the microprogram graph (Fig. I.17) the control unit activates MO 17 at this clock. Using the

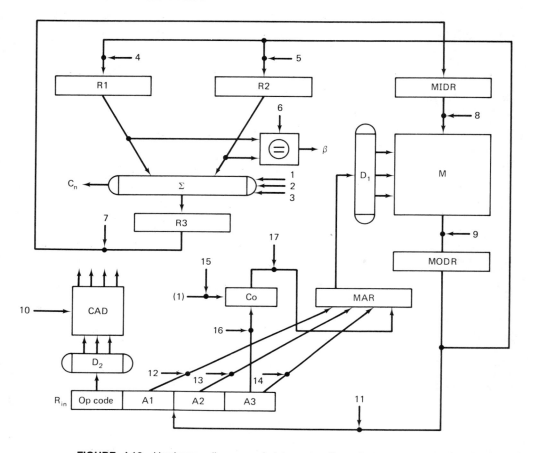

FIGURE I.16 Hardware diagram of interconnections between processor, memory, and control unit

Microoperation Table we find that this microoperation transfers the instruction address stored in the program counter Co to the memory address register, MAR; CO → MAR (Fig. I.16).

CLOCK 2 executes node 2 containing MO 9. This fetches the contents of the memory cell addressed to the memory output data register (M → MODR). The conditional branch instruction is now stored in the MODR.

CLOCK 3 executes node 3 containing MO 11. This transfers the instruction from the MODR to the instruction register, R_{in}.

CLOCK 4 activates MO 12, causing the address of operand A, stored in the first address field, $A1$, to be sent to the MAR ($R_{in}(A1)$ → MAR).

CLOCK 5 activates MO 9, thus performing a fetch of operand A from the memory to the MODR.

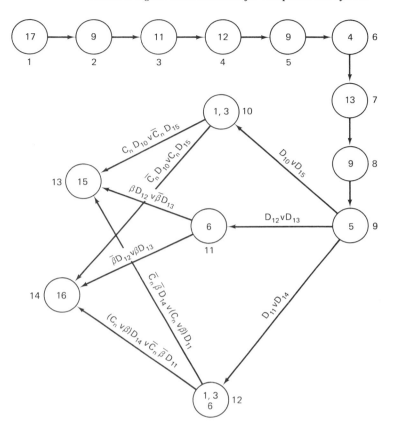

FIGURE I.17 Graph of conditional branch microprogram

CLOCK 6 activates MO 4, transferring the operand A to register $R1$.

CLOCK 7 activates MO 13, transferring the address of operand B, stored in a second address field, $A2$, to the MAR ($R_{in}(A2) \rightarrow$ MAR).

CLOCK 8 activates MO 9, a fetch of the operand B from the memory to the MODR.

CLOCK 9 activates MO 5, transferring the operand B to register $R2$.

CLOCK 10. At this clock the microprogram branches to one of three microoperation sequences depending on the conditional test being performed.

In a computer two types of conditional tests are distinguished—signed (arithmetic) and unsigned (logical). For a signed test, both operands A and B are signed numbers, so that each test involves comparisons of their signs and magnitudes, provided signs are the same. For an unsigned test only the magnitudes of A and B are compared. Thus A and B must be unsigned. If they are not, that is, they are signed and signs are different, a complemented number (with a negative sign) has to be transformed into magnitude form before an unsigned comparison may proceed. This reduces an unsigned conditional test into a signed test with equal signs.

Computers usually implement six unsigned test over words A and B: $A > B$, $A \leq B$, $A \geq B$, $A < B$, $A = B$, and $A \neq B$. Each test is usually implemented in a single conditional branch instruction. The six conditional branch instructions would thus be represented by six microprograms having common fetch portions (nodes 1 to 9). These are combined here to minimize the overall hardware outlay. It follows that node 9 must be followed at clock period 10 by either node 10, 11, or 12.

Nodes 10, 11, and 12 perform the following tests:
Node 10: $A > B$, $A \leq B$
Node 11: $A = B$, $A \neq B$
Node 12: $A \geq B$, $A < B$

The transitions from node 9 to node 10, 11, or 12 are activated by opcode D, that distinguishes among six conditional branch instructions. The pair of microoperations (1 and 3) cause $R2$ to be subtracted from $R1$ and MO 6 tests $R1$ and $R2$ for equality.

CLOCK 11. At this clock period, one of two alternatives is executed:

1. The true alternative, represented by node 13 (MO 15), in which the program counter is incremented by $+1$ ($Co + 1 \rightarrow Co$) to obtain the next instruction address.
2. The false alternative, represented by node 14 (MO 16), in which the jump address stored in the $A3$ field of the conditional jump instruction is sent to counter Co ($R_{in}(A3) \rightarrow Co$).

Let us consider how nodes 10, 11, and 12 select their true and false successors, nodes 13 and 14, respectively.

Transition from node 10 ($A > B$ and $A \leq B$). As was shown in Example 6, MOs 1 and 3 must be activated for testing $A > B$. (Since $A \leq B$ is a complement of $A > B$, it uses the same microoperations.)

Since true values of $A > B$ and $A \leq B$ are recognized by adder overflow C_n and no overflow $\overline{C_n}$, respectively, the true successor (node 13) of node 10 is determined by $C_n \cdot D_{10} \vee \overline{C_n} \cdot D_{15}$ where D_{10} and D_{15} are the opcode signals that distinguish $A > B$ and $A \leq B$, respectively. Similarly, $\overline{C_n} \cdot D_{10} \vee C_n \cdot D_{15}$ recognizes the false successor (node 14) of node 10 since the false values of $A > B$ and $A \leq B$ are represented by $\overline{C_n}$ and C_n, respectively.

Transitions from node 11 ($A = B$, $A \neq B$). To recognize $A = B$ and $A \neq B$, the two data words should be sent to the equality circuit (MO 6) which can either be dedicated (Fig. I.16) or obtained by ORing all the outputs of the mod 2 addition. Since true values of $A = B$ and $A \neq B$ are recognized by the equality signal β and the inequality signal $\overline{\beta}$, respectively, the true successor (node 13) of node 11 is determined by $\beta \cdot D_{12} \vee \overline{\beta} \cdot D_{13}$, where D_{12} is the opcode specifying $A = B$ and D_{13} specifies $A \neq B$. The false successor (node 14) likewise is activated by $\overline{\beta} \cdot D_{12} \vee \beta \cdot D_{13}$.

Transitions from node 12 ($A \geqslant B$, $A < B$). To find out if $A \geqslant B$ or $A < B$, words A and B must be sent concurrently to the adder (MOs 1 and 3) and to the comparing circuit (MO 6) since C_n alone is incapable of recognizing $A = B$. This is so because if $A = B$, $C_n = 0$ and if $A > B$, $C_n = 1$; however, if $A = B$, condition $A \geqslant B$ is true, and if $A > B$, condition $A \geqslant B$ is also true. The true value of condition $A \geqslant B$ is thus represented by both $C_n = 0$ and $C_n = 1$. To include the case of equality, words A and B are also sent to MO 6 where they generate equality signal β. If $A \geqslant B$, $C_n \vee \beta$; if $A < B$, $C_n \vee \beta = \overline{C_n} \cdot \overline{\beta}$. The true successor (node 13) of node 12 is therefore activated by $(C_n \vee \beta) \cdot D_{14} \vee \overline{C_n} \cdot \overline{\beta} \cdot D_{11}$, where D_{14} and D_{11} specify $A \geqslant B$ and $A < B$, respectively. The false successor (node 14) of node 12 is activated by $(\overline{C_n} \cdot \overline{\beta}) \cdot D_{14} \vee (C_n \vee \beta) \cdot D_{11}$.

8.2 The Floating-Point Multiplication Instruction

This instruction fetches two operands A and B, performs a floating point multiplication, $A \times B$, and sends the result to the memory.

The A and B operands are represented in floating point form as $A = AE \times AM$ and $B = BE \times BM$, where AE and BE are exponents and AM and BM are mantissas. Assume that when fetched to the processor registers, mantissas are stored in registers $R1$ and $R2$ and exponents are stored in registers $R4$ and $R5$ (Fig. I.14).

Consider the clock-by-clock execution of the instruction microprogram using the microprogram directed graph (Fig. I.18), and the microoperation hardware diagrams in Figs. I.16, I.19, and I.14. Figure I.16 shows the interconnections between the processor, memory, and control unit; Fig. I.19 shows the processor hardware involved in mantissa multiplication; and Fig. I.14 introduces the additional hardware needed for exponent manipulation. All microoperations shown in these three diagrams are described in the Microoperation Table. Since the floating-point multiplication microprogram contains complex conditional transitions in which one node may be followed by several successors, it is impossible to specify in advance which successor will be next. Thus, we will give a description of each node executed at each clock.

NODES 1 through 5 are similar to those for the conditional branch microprogram considered in Sec. 6.1. Nodes 1 through 3 fetch the floating-point multiplication instruction to the instruction register, R_{in}, and nodes 4 and 5 fetch the word A ($A = AE \times AM$) to the MODR (Fig. I.16).

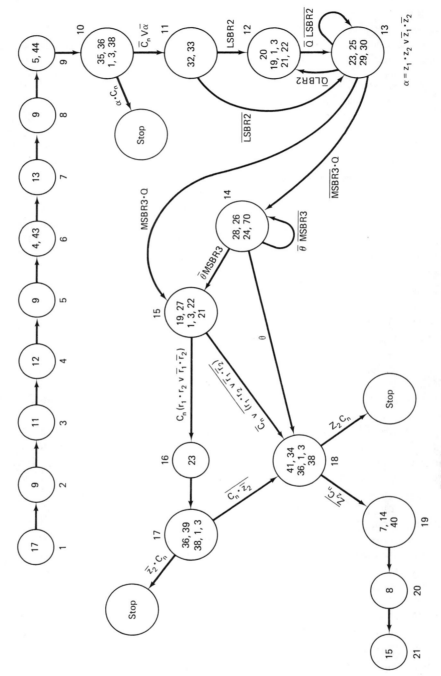

FIGURE I.18 Graph of floating-point multiplication microprogram

Historic Progress in Architectures for Computers and Systems 27

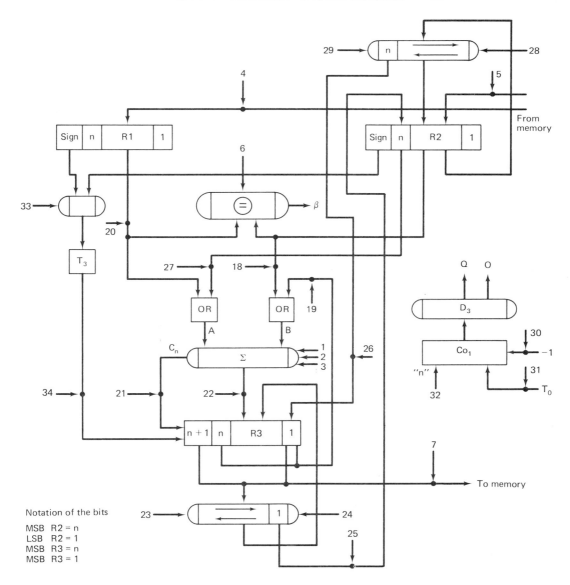

FIGURE I.19 Hardware diagram involved in mantissa multiplication

NODE 6 executes MOs 4 and 43; exponent *AE* is sent to register $R4$ by MO 43 (Fig. I.14), and mantissa *AM* is sent to register $R1$ by MO 4.

NODE 7 executes MO 13, which transfers the *B* operand address from the address field $A2$ of R_{in} to the MAR (Fig. I.16).

NODE 8 fetches the *B* operand from memory to the MODR via MO 9.

NODE 9 activates MOs 5 and 44, sending mantissa *BM* to $R2$ and exponent *BE* to $R4$ (Fig. I.14).

NODE 10 executes five concurrent microoperations that perform exponent addition $(AE + BE)$: MO 35 sends *AE* stored in $R4$ to the *A* terminal of the adder; MO 36 sends *BE* stored in $R5$ to the *B* terminal of the adder; the adder adds them (MOs 1 and 3); and the result is sent to $R5$ (MO 38).

Exponent addition is sensitive to exponent overflow since the result could be a $(p + 1)$-bit number that could not be stored in the p-bit exponent register $R5$. Thus, node 10 has two successors. If there is no exponent overflow the transition is to node 11. If there is exponent overflow, the transition is to node "Stop." This second transition occurs if conditional signal $\alpha \cdot C_n$ is activated, where $\alpha = z_1 \cdot z_2 \vee \bar{z}_1 \cdot \bar{z}_2$ shows that signs z_1 and z_2 of *AE* and *BE* are the same $(z_1 = z_2)$. If $z_1 \neq z_2$ then an adder overflow C_n signifies a complemented operation, $\overline{AE + BE}$.

If there is no exponent overflow, $\overline{\alpha \cdot C_n} = \bar{\alpha} \vee \bar{C}_n$ is true. This activates the transition to node 11.

NODE 11 originates mantissa multiplication, which is executed in nodes 11, 12, and 13. The microoperation hardware diagram of Fig. I.19 implements a multiplication algorithm that shifts both the partial product and the multiplier toward the LSB. Accordingly, MO 32 sends the size of mantissa, n, to the processor's counter Co_1 in node 11. The sign of the mantissa product is sent to flip-flop T_3 via MO 33, a sign microoperation that finds the sign of the mantissa product as a function of the signs of both mantissa operands.

The successor of node 11 depends on the value of the least significant bit of the multiplier stored in register $R2$, or briefly, LSB $R2$. If LSB $R2 = 1$, node 11 is followed by node 12, which performs an addition iteration.

NODE 12 executes six concurrent microoperations. It sends the multiplicand in register $R1$ to the *A* terminal of the adder with MO 20; the partial product in $R3$ is sent to the *B* terminal of the adder by MO 19; the adder performs binary addition with MOs 1 and 3; the result is sent back to $R3$ by MO 22; and possible mantissa overflow arising during the addition is sent to the sign bit of $R3$ via MO 21. Since the sign of the product of the mantissas is stored in T_3 during mantissa multiplication, the sign bit of $R3$ is used for temporary storage of the overflow C_n that may arise during the addition step. At the next clock period, during the shift operation, this overflow will be shifted right, releasing the sign bit of $R3$ for storing another possible overflow during the next iteration of addition.

NODE 13 executes one shift iteration using four MOs. The content of the product stored in $R3$ and $R2$ is shifted towards the LSB by MOs 23 and 29, respectively. Since $R3$ and $R2$ form a double-precision register to store a double-precision mantissa product during this shift, $R3$ is connected with $R2$ via MO 25. Thus, at each shift iteration the partial product will have one bit shifted in from $R3$ to $R2$ by MO 25, and the multiplier will have one bit shifted out from $R2$ by MO 29. Since node 13 is the end of one iteration of multiplication, the content of counter Co_1 is decremented by 1 via MO 30 ($Co_1 - 1 \rightarrow Co_1$) so that after node 13 is executed Co_1 stores the current number of iterations left.

Node 13 has four different possible successors, including itself (nodes 12, 13, 14, 15), depending on the value stored in the counter Co_1, the LSB of $R2$, and the MSB of $R3$. If Co_1 does not store 0 (signal Q is 0), multiplication iterations remain and must be executed. Therefore, if $\overline{Q} \cdot \text{LSB } R2$ is true, node 13 is followed by addition node 12, and if $\overline{Q} \cdot \overline{\text{LSB } R2}$ is true, node 13 is iterated again.

If Co_1 stores 0 (signal Q is 1), mantissa multiplication is over. When this happens node 13 is followed by either node 14 or 15, depending on the value of the most significant bit of the mantissa product stored in $R3$, otherwise called MSB $R3$.

If $Q \cdot \overline{\text{MSB } R3}$ is true, MSB $R3 = 0$ and mantissa multiplication must be followed by product normalization performed in node 14.

If $Q \cdot \text{MSB } R3$ is true, no normalization will be required since the most significant bit of the product in $R3$ is 1 (MSB $R3 = 1$). Node 13 will be followed by node 15 that rounds the mantissa product since it must be reduced to an n-bit fraction.

NODE 14 performs product normalization. To this end it shifts the least significant portion of the product in $R2$ towards the MSB with MO 28; it connects $R2$ and $R3$ during this shift with MO 26; and it shifts $R3$ towards the MSB with MO 24. This is one normalization iteration and the counter Co_1 is incremented by 1 with MO 70 to keep track of the total number of normalization iterations performed.

Node 14 has three successors depending on the count of the number of the normalization iterations already performed that is in counter Co_1 and the value of MSB $R3$. If Co_1 stores the number n recognized by signal θ where n is the size of the mantissa register, normalization ends, since the entire product stored in $R2$ was shifted into $R3$. Since no precision bits are left in $R2$, no rounding is necessary, and normalization node 14 will be succeeded by the exponent correction node 18.

If Co_1 does not store n (i.e., $\overline{\theta} = 1$), the end of normalization is signalled by MSB $R3$. Since some precision bits are left in R, normalization node 14 must be followed by rounding node 15.

Thus $\overline{\theta} \cdot \text{MSB} R3$ specifies transition from node 14 to node 15. If $\overline{\theta} \cdot \overline{\text{MSB} R3}$ is true, there is room for the next normalization iteration, and node 14 succeeds itself.

NODE 15 performs rounding; the product stored in register $R3$ is incremented by the MSB of $R2$ and the result is sent to $R3$. To do this, $R3$ sends its word to B terminal of the adder by MO 19; MSB $R2$ sends its value to the A terminal of the adder by MO 27; adder executes addition (MOs 1 and 3); and the result is sent to $R3$ by MO 22. Possible mantissa overflow during rounding is sent to the sign bit of register $R3$ by MO 21.

A floating-point operation should provide for automatic handling of mantissa overflow. Thus, the successor of the rounding node 15 is either node 18 provided no mantissa overflow occurs, or node 16, which handles mantissa overflow. Mantissa overflow during rounding occurs if there signal C_n is in the adder and the two mantissa signs are equal ($r_1 = r_2$); i.e., node 16 follows node 15 if $C_n \cdot (\overline{r_1}\overline{r_2} \lor r_1 r_2)$ is true.

NODE 16. Since this node is executed solely for mantissa overflow, the mantissa product stored in $R3$ is shifted towards the LSB by MO 23.

NODE 17. The exponent result is increased by +1 since node 16 reduced the actual mantissa value by one order of magnitude. To do this, the exponent result in $R5$ (Fig. I.14) is sent to the B terminal of the adder by MO 36, +1 is sent to the A terminal of the adder by MO 39, the adder executes MOs 1 and 3, and sends the result to $R5$ with 38. Node 17 may perform a transition to the node "Stop" if there is exponent overflow, recognized by $\overline{z_2} \cdot C_n$ where z_2 is the sign of the exponent result stored in $R5$. This condition would usually invoke a software error handling routine. Otherwise node 17 is followed by node 18.

NODE 18 corrects the exponent for errors caused by normalization of the mantissa. Since the mantissa product was increased during normalization by k orders of magnitude where k, the number stored in counter Co_1, shows the number of normalization iterations, the number k must be subtracted from the exponent result stored in $R5$. To this end, $R5$ sends its word to terminal B of the adder by MO 36, Co_1 sends a complement of k to the A terminal of the adder by MO 41, the adder uses MOs 1 and 3 to add them, and the result is sent to $R5$.

Since k is subtracted from the exponent result, CE, if CE is a negative number there can be an exponent underflow, recognized by $z_2 \cdot C_n$ and the corrected exponent will be a $(p + 1)$-bit negative number ($z_2 = 1$). In this case node 18 is followed by node "Stop" or "Error." Otherwise, if $\overline{z_2 \cdot C_n} = \overline{z_2} \lor \overline{C_n}$, it is followed by node 19.

NODE 19 sends the floating point result $CE \times CM$ to the memory. It sends exponent CE to the memory data register MIDR with MO 40; it sends mantissa CM to the MIDR with MO 7 (Fig. I.16); and the address of $CE \times CM$, stored in the $A3$ field of R_{in}, is sent to the MAR register with MO 14.

NODE 20 writes the result to memory via MO 8.

NODE 21 increments the program counter Co using MO 15 to obtain the next instruction address.

TRADITIONAL COMPUTER ARCHITECTURES

This section of the chapter concerns itself with traditional architectures—von Neumann computers and microprogrammed computers—and explains distinctions between them by using the microoperation approach developed in the first part of the chapter.

9. PECULIARITIES OF THE VON NEUMANN ARCHITECTURE

The first computers had an architecture subsequently named von Neumann since the basic principles of this architecture were described by von Neumann, and his colleagues in [1–5]. Although most old computers have been forgotten, the basic von Neumann architectural concepts are still alive in many modern computers and systems. Since these concepts are familiar to the majority of readers, we will focus here only on two major ones that distinguish the von Neumann from other types of architecture.

These concepts are:

1. Principle of addresses.
2. Fixed instruction microprograms.

9.1 Principle of Addresses

This principle of von Neumann architecture organized computations in a way such that each instruction fetched operands from memory and/or sent the computational result back to memory. As a result, it was necessary for each instruction to store data addresses; i.e., to have clearly defined address fields. The operation to be performed on the data words fetched from memory was written to its opcode field. Each instruction was accordingly partitioned into the two zones—opcode and address.

Depending on the number of addresses stored in the address field of an instruction, computers were divided into:

1. Three-address computers in which each instruction stored three addresses, two for operands and one for computational results.
2. Two-address computers in which each instruction stored only two addresses (either for two operands or for one operand and the result).
3. A one-address computer in which each instruction stored only one address, either of an operand or of the result.

The principle of addresses has evolved into various addressing schemes that allow each address field to store not only an effective memory address but also virtual and/or relative addresses. This tendency was caused by the necessity for the program to maintain its independence of memory locations as von Neumann computers began to work in a multiprogramming mode of operation. This mode of operation allowed program tasks involved with slow I/O operations to be interrupted to provide computer time to tasks involved in fast processor exchanges.

9.2 Fixed Instruction Microprograms

The partition of the instruction into two zones—opcode and address—led to the second major characteristic of von Neumann architectures: Each instruction activated a fixed microprogram, i.e., it could not change its sequence of microoperations. This means that a programmer could not alter any sequence of executing microoperations by writing a new code to the instruction field. Instruction microprograms could not be modified via software as was done later in microprogrammed computers.

Sections 8.1 and 8.2 considered two microprograms (conditional branch and floating-point multiplication) implemented in a typical von Neumann machine. Referring to Figs. I.17 and I.18, the only alterations in microoperation sequences that were allowed were those caused by changes in data words since such a change could lead to generate conditional transitions. The instruction fields were sacred and a programmer could not write any new code to the instruction that could have changed the microprogram.

9.3 Weaknesses of von Neumann Architectures

The two peculiarities of the von Neumann architecture considered above resulted in the following weaknesses:

Significant time losses caused by frequent memory access operations. Since each instruction had to fetch its operands from the memory and/or send the result back to the memory, an enormous amount of time was spent on memory access operations during which no computing was done. For instance, processing an instruction containing three address fields required four memory access operations—one for the instruction and three for the data words. Since as a rule, a memory access operation is much slower than the processor's operation, frequent accesses to memory sharply reduce the speed of computation.

Failure to compute algorithms under time restrictions. The assignment of fixed sequence of microoperations for each instruction microprogram significantly narrowed the area of applications for von Neumann computers. The execution of many programs can be accelerated if special programming techniques can be used. Typical special techniques include:

1. Reducing the number of memory access operations by storing temporary results in processor registers rather than sending them back to the memory.
2. Increasing the number of microoperations executed in parallel during one clock period.
3. Increasing the number of sequences of microoperations assigned to a program instruction.

While the use of these techniques is possible in microprogrammed computers, they can not be utilized in von Neumann machines since each instruction is assigned a fixed microoperation sequence. It follows that many algorithms having time restrictions are hampered by von Neumann machines since their computation cannot be enhanced by using factors 1 to 3.

9.4 Improving on von Neumann's Architecture

Let us consider how computer architects tried to overcome the weaknesses of von Neumann architectures indicated above.

9.4.1 Reduction in the Number of Addresses

The goal of reducing the time losses due to memory access operations has led to the appearance of von Neumann machines having a reduced number of addresses (one and two).

For a two-address instruction, address fields $A1$ and $A2$ may store either the addresses of two operands or those of one operand and the result. In either case the number of memory access operations is reduced by one. If the instruction fetches two operands the computational result is saved in a processor register for the next computation and thus requires no additional access to the memory. If, on the other hand, the instruction fetches only one operand, the second operand will be taken from a register where a previous computation stored it. The result of the computation is sent back to memory. Again, one memory access is saved. Thus, a two-address instruction can perform the same computation using only three memory accesses; one less than a three-address instruction uses.

One can similarly establish that a one-address instruction performs a computation using only two memory accesses, two less than a three-address instruction.

A reduction in the number of addresses assigned to one instruction may therefore lead to savings both in the number of memory accesses required and in the amount of memory needed to store the intermediate results of expressions that contain more than one processor operation executed sequentially (such as $(A^2 - B)/C + D$ or $(A - B)^2 > K$).

On the other hand, if an algorithm contains simple expressions containing not more than one arithmetic operation, and the result of this operation is not used in the next instruction (one instruction executes $A + B$, the next one $C \div D$, etc.), then computation of such expressions with a two-address or one-address instructions leads to an increase both in the number of accesses and in the amount of memory needed when compared to a three-address instruction.

For example, computation of $A + B$ performed with one-address instructions requires three such instructions, the first two fetch A and B, and the third computes $A + B$ and sends the result back to memory. It takes six memory accesses; three for instructions and three for the data words A, B, and $A + B$. The number of bits in memory such a computation requires is $3(p + n)$, where p is the size of the opcode and n is the size of one address. If $A + B$ is computed with a three-address instruction on the other hand, only four memory accesses are made (instruction, A, B, and $A + B$) and $p + 3n$ bits are needed. Three-address instructions may thus achieve a saving in both the number of memory accesses and memory size in comparison with one- and two-address instructions for executing of algorithms that contain simple expressions needing no more than one processor operation.

9.4.2 Increase in the Number of Processor Registers

Reduction in the number of addresses assigned to one instruction was accompanied by an increase in the number of registers in the processor designated for the storage of temporary results. It became conventional to provide processors with additional register memories called *general register sets* (GRS) containing several tens of registers specified by their addresses. GRS were filled during computation with intermediate results and the instructions fetched operands from the GRS and sent the results back to the GRS.

This was much faster than using the slower memories although it required that the instruction store the address of each word it needed to access in the GRS. Thus, before the operation could begin the operand(s) had to be fetched from the GRS to the registers connected with the adder.

The next step in reducing the time of memory access was to increase the number of processor registers connected with the input and output terminals of the adder. It appeared possible not to send each result to the GRS since it could be saved in a processor register, and the instruction that used this result had to connect the registers it used in the computation with the adder terminals via software.

The instructions thus had to store not only opcodes and addresses but also the so-called reconfiguration codes that specified what combination of registers needed by the instruction could be connected with each other and the adder terminals. However, this required a complete revision in the organization of the instruction microprograms adopted in traditional von Neumann architectures, in which only data words could alter the sequence of activated microoperations.

It became necessary to replace fixed microprograms with software-controlled microprograms in which some sequences of activated microoperations could be selected by the programmer when he or she wrote new reconfiguration codes to the instruction fields. This new organization of microprograms has led to the creation of a new class of computer architectures called *microprogrammed computers*.

10. THE ARCHITECTURE OF MICROPROGRAMMED COMPUTERS

A microprogrammed architecture can be viewed as the outcome of a process aimed at the improvement of von Neumann architecture by reducing the time involved in the nonproductive activity of accessing data words.

Historically, however, microprogrammed architectures have been considered as a separate class of computer architecture, because since their invention in 1951, their development paralleled that of von Neumann architectures [6, 7].

Let us now introduce some basic principles of a microprogrammed architecture.

10.1 Basic Principles of Microprogrammed Architecture

1. *Reconfiguration codes.* The instruction must store special reconfiguration codes that select various microoperation sequences in the instruction microprogram.
2. *Variable microprogram.* A programmer may change the instruction microprogram by selecting more task-oriented microoperation sequences.
3. *Separate control memory.* The instructions are stored in a separate memory called the control memory that is not used for keeping data words, with the exception of some addresses and constants.

These three principles follow each other since the variable microprogram property is implemented by the reconfiguration codes that must be stored in the instruction field. This leads to a significant increase in the instruction size, since, in addition to opcode and/or address, the instruction must store several reconfiguration codes. The width of the control memory should thus match the instruction size, which is as a rule much larger than the data word size.

If, on the other hand, the control memory is shared for the storage of both instructions and data arrays, each cell storing a data word will contain a number of unused bits. To increase the speed of computation, control memories must be very fast and thus very expensive. It therefore becomes prohibitively expensive to use the same memory for both instruction and data storage.

The main memory also stores the so-called *macroinstructions*. Since each instruction in the control memory performs a very simple computation, roughly equivalent to several parallel microoperations, the need arises for conventional microprograms specified by sequences of microoperations and be represented by separate instructions. Thus, in a microprogrammed architecture two types of instructions appear: *microinstructions*, each of which activates a set of parallel microoperations, and *macroinstructions*, each of which activates a microprogram represented by a sequence of microinstructions. Microinstructions are stored in the control memory; macroinstructions are stored in the main memory.

10.2 Reconfiguration Fields

As we saw above, a von Neumann architecture requires that two fields, the opcode and an address, be assigned to each instruction. In a microprogrammed computer there is the additional necessity of having a third field that stores the reconfiguration codes. This field is called the *reconfiguration field*. The reconfiguration field is partitioned into several zones, each assigned to one group of reconfiguration microoperations; i.e., those activated by reconfiguration codes. By writing a code into this zone, a programmer can select one or several microoperations from a group of microoperations.

Partitioning the microoperations into groups is, as a rule, performed on the following basis: Two microoperations belong to the same group if they have a common source or destination register or a common logical circuit.

Example 8. Figure I.20 contains the processor portion of a microprogrammed computer with 18 reconfiguration microoperations (their numbers differ from those given in the Microoperation Table). These microoperations may be assembled into four groups: Group GR_1, containing the microoperations that connect a common MODR register with registers $R1$–$R6$ (GR_1 contains MOs 1, 3, 5, 7, 9, and 11). Group GR_2 includes the microoperations that connect $R1$, $R2$ and $R3$ with the X terminal of the adder; i.e., GR_2 contains MOs 13, 14, and 15, all having a common destination logical circuit. Group GR_3 contains MOs 16, 17, and 18, which connect $R4$, $R5$, and $R6$ with the Y terminal of the adder (common destination logical circuit). Group GR_4 contains microoperations that connect the adder output (common source logical circuit) with different destination registers (MOs 2, 4, 6, 8, 10, 12, 19). ∎

Consider the selection of the size of one reconfiguration zone. Since one zone is assigned to one group, GR_i, its size depends on two factors: the number, n, of microoperations in the group, and the encoding technique used. There are two encoding techniques: *logarithmic* encoding and *linear* encoding.

Logarithmic encoding assigns each microoperation from the group with a p-bit code, where $p = \log_2 n$ bits; such an assignment leads to the minimal size of each reconfiguration zone.

Logarithmic encoding has the following serious drawbacks:

1. It allows execution of only one microoperation at a time from the group; i.e., it forbids parallel execution of several microoperations belonging to the same group.

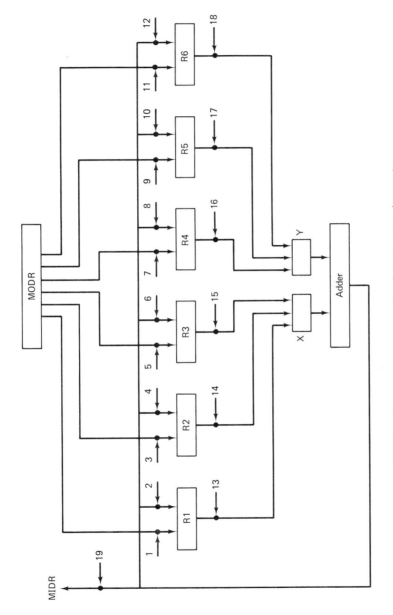

FIGURE I.20 Processor portion of a microprogrammed computer

2. It introduces decoding delays due to the logical circuits invoked to decode the various reconfiguration codes into microcommands. Thus, this technique leads to a delay in microcommand generation, MICD (see Sec. 3).

Linear encoding assigns each microoperation in a group to one bit. That is, for a group containing n microoperations, the size of reconfiguration zone is n bits. Linear encoding maximizes the size of the zone and leads to a very large microinstruction size. As a result, the width of the control memory also increases significantly.

However, linear encoding possesses two considerable advantages that sometimes outweigh its drawbacks:

1. All microoperations belonging to the same group can be executed in parallel; i.e., a programmer can provide for parallel execution of any combination of microoperations from the same group.
2. Linear encoding introduces no decoding delays since each microcommand is the output of one bit (flip-flop) assigned to the respective microoperation. Thus, no microcommand delays are generated with this technique.

Logarithmic versus linear encoding provides a classical time/cost trade-off. The width of the control memory is traded for fast execution speed.

Example 9. Figures I.21 and I.22 show the trade-offs between logarithmic and linear encoding. One group of eight microoperations encoded logarithmically yields a 3-bit reconfiguration zone in instruction register R_{in} (Fig. I.21). However, no pair of microoperations can be executed concurrently, since each 3-bit code written to the zone specifies a single microcommand, MIC, that activates a unique microoperation. For instance, a 100 written to this zone leads to the generation of MIC_5, which activates MO_5. This encoding introduces a microcommand delay (MICD) where MICD = t_d and t_d is the delay caused by one gate. The same group of microoperations encoded linearly leads to a 8-bit reconfiguration zone (Fig. I.22). However, here the MICD is 0, and any combination of microoperations may be executed in parallel. For instance, code 10010101 written to the zone causes the concurrent execution of MOs 1, 3, 5, and 8. ∎

The bit size of the reconfiguration field is generally the sum of the bit sizes of all the reconfiguration zones that can be specified in the computer; i.e., it is determined by the number of microoperation groups that can be selected by the programmer. Further, some groups must be encoded linearly since many of the microoperations they contain can and should be executed in parallel to enhance microoperation parallelism in the computer. For instance, the group of microoperations that connects the adder output to the

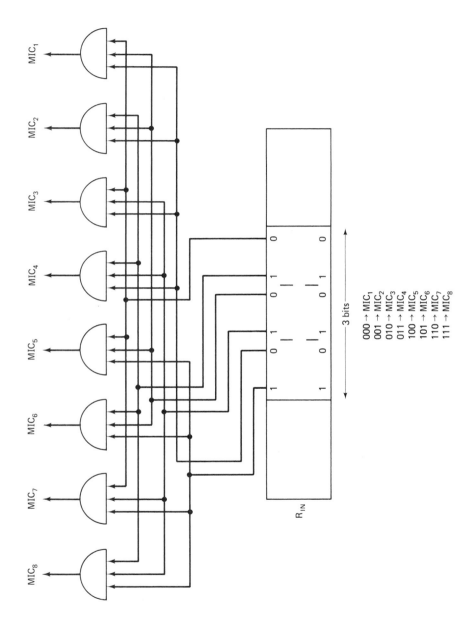

FIGURE I.21 Logarithmic encoding of microoperations

40 CHAPTER I

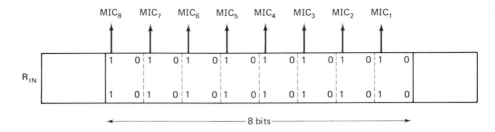

FIGURE I.22 Linear encoding of microoperations

destination registers is encoded linearly since it is important to send the temporary result concurrently to all the processor registers that may need it for future computation, and to the memory data register, MIDR. (Group GR_4 contains MOs 2, 4, 6, 8, 10, 12, and 19 and must be encoded linearly, Fig. I.20). As a result one obtains a very large reconfiguration field which significantly enlarges the overall instruction size since one instruction must contain other fields (opcode and/or address) in addition to the reconfiguration field.

10.3 Microinstruction Formats

Earlier we partitioned the instructions of a microprogrammed computer into macroinstructions that activate microprograms and microinstructions that activate sets of parallel microoperations. However, this still leads to an overly large microinstruction size and we can reduce it by further partitioning the microinstruction set into several classes, each of which includes microinstructions performing the same type of functions. For example, one may form a class of microinstructions that perform processor operations, or a class of microinstructions with conditional branches, or a class of microinstructions that perform addressing, etc. Such classes will have little or no overlap between the various groups of reconfiguration microoperations, and each of them will be represented by a unique reconfiguration field that is the sum of the reconfiguration zones it will use rather than the sum of all the reconfiguration zones required for a given architecture. This will lead to an overall reduction in the microinstruction bit size and, thus, the width of the control memory.

10.3.1 Microinstruction Classes

One may distinguish the following partitioning of the microinstruction set into microinstruction classes that fulfill this objective.

1. *Class of microinstructions that perform processor operations.* This microinstruction format that represents this class usually includes the opcode field and a reconfiguration field composed of zones that show reconfigurations between registers and the logical circuits in the processor.

2. *Class of microinstruction that perform conditional branches.* This instruction format includes three fields: opcode, reconfiguration, and address. The opcode field shows what type of conditional test must be performed; the reconfiguration field establishes the registers connected with the adder inputs; and the address field stores the address to a location in the control memory to be jumped to if the conditional test is true (or false).

3. *Class of microinstructions that perform addressing.* This microinstruction format also includes three fields: opcode, reconfiguration, and address. The reconfiguration field is composed of zones that establish connections between the register for data memory and processor registers. Other zones of the reconfiguration field establish connections between registers of the control unit that store base addresses, B, indices, I, and relative addresses, RA, with the memory address register that must receive the effective memory address, $E = B + I + RA$. The address field for this instruction format stores all the information (base addresses, indices, relative addresses, array dimensions, etc.) necessary to transfer a block of data words from the data memory either to processor registers or to a GRS.

We use a special *instruction-type* code belonging to the category of opcodes to distinguish among different instruction formats. This code allows a further reduction in the overall instruction size since it permits the assignment of the same microinstruction bits to new reconfiguration zones in each format. This, however, requires a multilevel decoding logic to decode the various reconfiguration zones since the instruction-type signals must be sent as inputs to the decoding gates that must in turn decode reconfiguration zones. As a result, the delay in microcommand generation, MICD, also increases.

Note that for many microprogrammed computers the microinstruction set may be partitioned into a larger number of classes in which each class is represented by a unique microinstruction format in order to provide an even greater reduction in the bit size of the microinstruction. Other computers, however, have no such partitioning at all. As a result their microinstruction sizes may be hundreds of bits.

10.3.2 Phased Execution of Microinstructions

In a microprogrammed computer one microinstruction does not activate a complex sequential microprogram as was the case with von Neumann architectures. To increase the amount of computation assigned to one microinstruction, it can be multiphased; however, i.e., the microcommands it generates may be activated during several clock periods. Thus, the microinstructions in microprogrammed computers may be one-phase, two-phase, three-phase, etc.

The duration of one phase equals that of the longest microoperation it executes. If all microoperations assigned to a phase last one clock period, then the phase takes one clock period. If the phase executes longer microoperations such as carry propagation or memory access, it may last several clock periods.

Termination of the phase may be made asynchronously by a completion signal that signifies the end of the process or by a count of the number of clock periods that the longest microoperation requires.

Example 10. Let us find the format of a processor microinstruction that may execute all microoperation groups described in Example 8. Since all microoperations were partitioned into four groups, GR_1 through GR_4, the reconfiguration field will include four zones, RZ_1 through RZ_4. We will find the size of each zone. Microoperations of GR_1 must be encoded linearly since a data word stored in MODR (Fig. I.20) can be sent concurrently to several processor registers. Since GR_1 contains six microoperations, the size of $RZ_1 = 6$ bits.

Groups GR_2 and GR_3 must be encoded logarithmically since both of these groups includes microoperations that cannot be executed concurrently. Since GR_2 and GR_3 contain three microoperations each, $RZ_2 = RZ_3 = \log_2 3 = 2$ bits. Group GR_4 should be encoded linearly since the adder should be capable of sending its result to several registers concurrently. Since GR_4 contains 7 microoperations, the size of $RZ_4 = 7$ bits.

Therefore the total size of the reconfiguration field, $RF = RZ_1 + RZ_2 + RZ_3 + RZ_4 = 6 + 2 + 2 + 7 = 17$ bits.

Let us find the size of the opfield. It must have two zones: the opcode zone, OP, that encodes the operation assigned to the instruction, and the instruction-type zone, IT, that distinguishes among various instruction formats. Suppose the adder is provided with 21 arithmetic and logical operations: $X + Y$, $X - Y$, $\underline{Y - X}$, $X + 1$, $X - 1$, $X \oplus Y$, $X \vee Y$, $X \wedge Y$, $X + 0$, $Y + 0$, $X + \overline{Y}$, $\overline{X} + Y$, $\overline{X \oplus Y}$, $\overline{X \vee Y}$, $\overline{X \wedge Y}$, $\overline{X} + 0$, $\overline{Y} + 0$, $\overline{X} \vee Y$, $\overline{Y} \vee X$, $\overline{X} \wedge Y$, and $X \wedge \overline{Y}$. Then the opcode zone, $OP = \log_2 21 = 5$ bits. If one assumes that a microprogrammed computer uses five different microinstruction formats, then the instruction-type size, $IT = \log_2 5 = 3$ bits. Thus the total size of opfield is 8 bits. The microinstruction format has 25 bits (Fig. I.23).

We will next find the number of phases in this microinstruction and duration of each phase. Microoperations from group GR_1 should be executed before those from other groups since registers $R1$ to $R6$ must be provided with operands before these operands are sent to the adder. Groups GR_2, GR_3, and GR_4 can be executed concurrently since together they establish one data

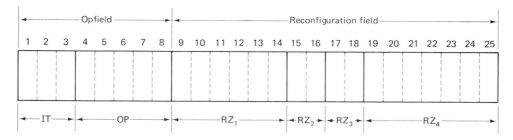

FIGURE I.23 An instruction format for a microprogrammed computer

broadcast between the source and destination registers. Thus, the microinstruction should have two phases. During the first phase GR_1 is executed and during the second phase GR_2, GR_3, and GR_4 are executed.

Let us establish the duration of each phase. The first phase is executed in one clock period, T_0, since all its microoperations activate a simple register-to-register transfer. The longest microoperation in the second phase is the carry propagation in the adder. Assume that it needs three clock periods to execute in the circuitry of our machine. Thus, the second phase will require time $3T_0$.

Let us consider the encoding of this microinstruction. Suppose that it must send the word A from the MODR to $R1$ and $R5$ (Fig. I.20); perform mod 2 addition $R1 \oplus R6$ where register $R6$ already stores word B; and send the result $A \oplus B$ to the MIDR and to $R2$. The following codes are used for the microinstruction zones: $RZ_1 = 100010$; the one in bit 1 is assigned to MO 1 and the one in bit 5 is assigned to MO 9. $RZ_2 = 01$; this code is assigned to MO 13. $RZ_3 = 11$; this code encodes MO 18. $RZ_4 = 0100001$; the one in bit 2 activates MO 4 and the one in bit 7 activates MO 19 (Fig. I.24). ∎

10.4 Block Diagram of a Microprogrammed Architecture

Let us now consider a simplified block diagram of a microprogrammed computer (Fig. I.25). Here the microinstructions are stored in the control memory; data words and macroinstructions are stored in the main memory.

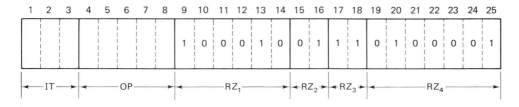

FIGURE I.24 Encoding example for the instruction $R1 \oplus R6 \to R2$, MIDR

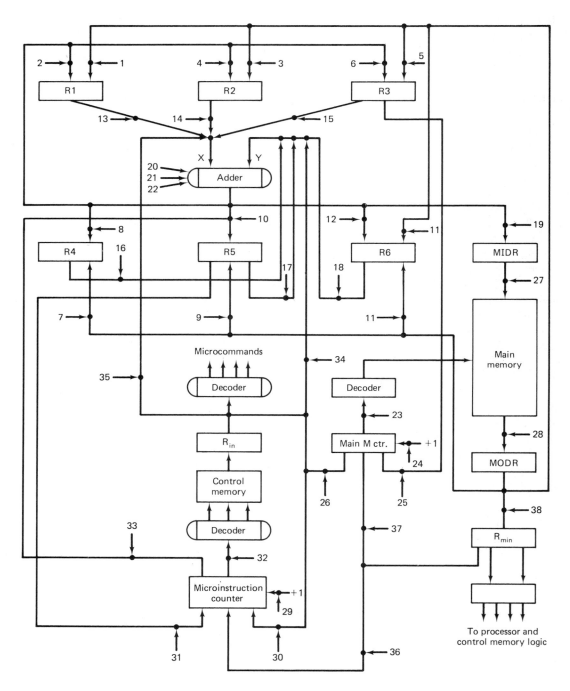

FIGURE I.25 A simplified block diagram of a microprogrammed computer

Microprograms specified as sequences of microinstructions are stored sequentially in the control memory and a current microinstruction address is stored in the microinstruction counter. Microinstructions are fetched to the microinstruction register, R_{in}, which is provided with a decoder that generates microcommands that activate the microoperation circuits denoted by their numbers. The next microinstruction address can be obtained in two ways:

1. Using the step microoperation (MO 29) that increments it by +1.
2. Using jump microoperations (MOs 30 or 31).

MO 30 is activated by the conditional branch microinstruction and transfers to the microinstruction counter a jump address stored in R_{in}, MO 31 sends to the microinstruction counter a modified jump address obtained in register $R5$ as a result of computation. If the current microinstruction address must be saved, $R5$ saves it with MO 33 and sends it back with MO 31. MO 26 transfers a data address stored in an addressing microinstruction to the main memory counter.

The entire program represented by a sequence of macroinstructions is stored in the main memory. When fetched from the main memory, each macroinstruction (MO 28) is sent first to memory data register MODR and then to the macroinstruction register, R_{min} (MO 38). The current macroinstruction address is stored in the main memory counter. The next address is obtained via MO 24 if it is a step address or via MO 37 if it is a jump address.

Fetches of data words written to consecutive cells of the main memory are also performed by a +1 increment of the main memory counter (MO 24). In case a data address is computed, it is transferred via MO 25 from $R3$ to the main memory counter. Various addends involved in the computation of an effective data address (base address, indices, relative addresses) are transferred from R_{in}, storing an addressing instruction, to the X and Y terminals of the adder with MOs 35 and 34. A computed address is sent to $R3$ by MO 6. MOs 20, 21, and 22 perform carry propagation, mod 2 addition, and logical multiplication in the adder respectively. The other remaining processor microoperations were discussed in Example 8.

More complex microprogrammed architectures provide for concurrency of processor activity and addressing computations. These architectures have a separate control processor that performs address modifications. As a result we obtain faster execution of microprograms.

10.5 Two-Level Control Memory

One of the ways of improving performance in a microprogrammed architecture is to select a very fast control memory. Its high cost was and still remains one of the major components of the overall cost of a microprogrammed architecture. Memory cost reduction therefore continues to be an important objective for this architectural design. Section 10.3.1 discussed one of the techniques for reducing memory cost. This consisted of partitioning the entire microinstruction set into several classes, each represented by its own microinstruction format.

Another technique is to divide each microinstruction of the microinstruction set into two instructions, $I1$ and $I2$, stored in the levels of control memory, $M1$ and $M2$, respectively (Fig. I.26(a)). Instruction $I1$ includes the opfield and the reconfiguration field. Instruction $I2$ has two address fields, $A1$ and $A2$. Field $A1$ always indicates the location of the respective instruction $I1$ in memory level $M1$; field $A2$ may be either a jump address to a new location in $M2$, or the address of a separate data memory.

Sometimes the $I2$ instruction has its $A2$ field filled with a constant that is used for computation if there is no need to store either a jump address for conditional branches or a data address for accessing a data word.

The division of each microinstruction into instructions $I1$ and $I2$ leads to a division of the entire control memory into the two levels, $M1$ and $M2$; instructions $I1$ are stored in the $M1$ level, and instructions $I2$ are stored in the $M2$ level. Such a division into two levels may lead to a real savings in the number of bits only if the size of instruction $I1$ exceeds a certain critical bit size. A technique for finding this critical bit size is discussed in 10.5.4.

One execution cycle for one macroinstruction is shown in Fig. I.26(b). Here macroinstructions are specified with storage addresses, S. The $I2$ instructions are specified with m addresses, while the $I1$ instructions are specified with n addressees. Therefore to fetch the $I1$ instruction requires that the $I2$ instruction store an n address; to fetch the $I2$ instruction requires that a macroinstruction store an m address. Thus a complete fetch time includes two phases aimed at fetching $I2$ and $I1$ respectively.

In general a microprogrammed computer has two elementary cycles of execution:

1. A cycle of macroinstruction which is the time required to execute a microprogram.
2. A microinstruction cycle which is the time required to execute one microinstruction.

10.5.1 One Microinstruction Cycle

In a microprogrammed computer with two-level memory, one microinstruction cycle, IC, includes two times to execute $I1$ and $I2$ instructions, respectively. The time IC is thus divided into three phases, $F1$, $F2$, and $F3$. $F1$ and $F2$ are fetch phases, and $F3$ is a computational phase (Fig. I.27). It is possible to perform a further partitioning of $F3$ into smaller phases, as was considered in Sec. 10.3.2 for processor microinstructions. For simplicity,

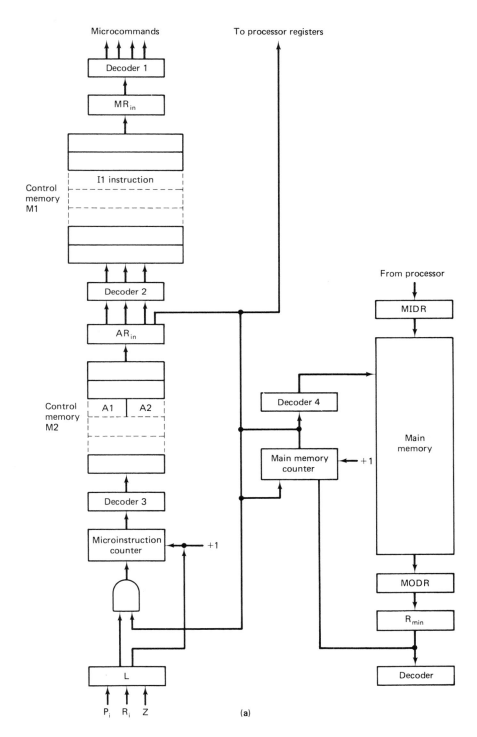

FIGURE I.26 Two-level control memory

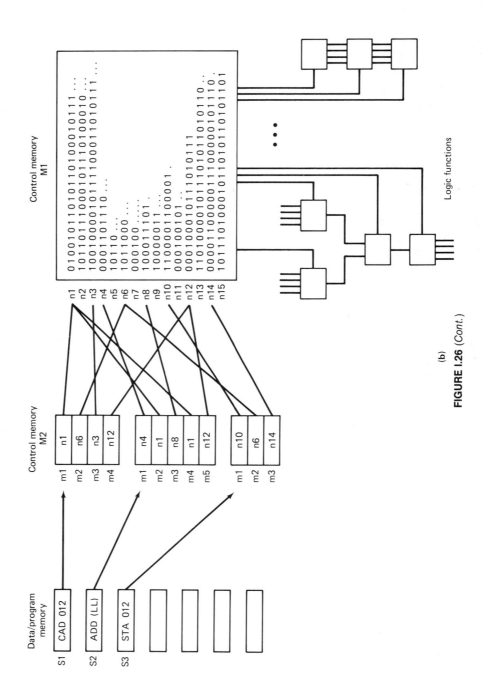

FIGURE I.26 (*Cont.*)

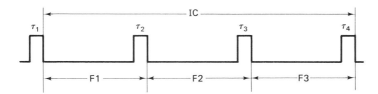

FIGURE I.27 Instruction cycle for a three-phase instruction partitioned into the address, the instruction, and the microinstruction

however, it will be assumed that $F3$ is a single phase not partitionable into smaller processes.

The first fetch phase, $F1$ (Fig. I.27), begins when the current address of instruction $I2$ is stored in the microinstruction counter (Fig. I.26) following the end of pulse τ_1. This address is then decoded in decoder 3 and sent to memory $M2$ to fetch instruction $I2$ to the address instruction register, AR_{in}. Thus the end of phase $F1$ occurs when instruction $I2$ is written to AR_{in}, during the pulse τ_2 (Fig. I.27).

A second fetch phase, $F2$, begins after the end of τ_2. During this phase the $A1$ field of instruction $I2$ is decoded in decoder 2 and then sent to memory $M1$ to fetch the instruction, $I1$. As for the $A2$ field, if it stores a new jump address, it can be written to the microinstruction counter only during the next phase, $F3$, since only this phase may execute an arithmetic operation and generate conditional signals.

If $A2$ stores a data word address, it is sent via decoder 4 to the main memory to fetch a new data word, which is written to memory data register MODR during pulse τ_3. Instruction $I1$ is concurrently written to the microinstruction register MR_{in}. A third phase, $F3$, generates microcommands in decoder 1 and executes the microoperations that are activated by these microcommands. Also during phase $F3$, if instruction $I1$ provides for a conditional jump to a new jump address stored in the $A2$ field of the AR_{in}, then conditional signals generated in the processor activate the writing of this address to the microinstruction counter. Phase $F3$ concludes the execution of one microinstruction cycle.

10.5.2 Overlapped Execution of Phases

In a microprogrammed architecture with a two-level memory, each microinstruction takes three phases, $F1$, $F2$, and $F3$, to execute. Of these three, only $F3$ performs useful computation since $F1$ fetches the $I2$ instruction and $F2$ fetches the $I1$ instruction.

If one assumes that all phases, $F1$, $F2$, and $F3$, take equal time, then the sequential execution of the phases assigned to each microinstruction when phase $F1$ of the next microinstruction occurs only after the completion of phase $F3$ of the preceding microinstruction, triples the overall time for the microprogram computation. This means that a sequence of 10 microinstructions will be executed in 30 phases of which only 10 phases perform computations over data words and 20 phases are spent for memory access.

To reduce the proportion of time spent for memory access, microprogrammed computers with two-level memories may employ overlapped execution of their phases. In any given time interval the computer is executing three consecutive microinstructions in three different phases ($F1$, $F2$, or $F3$) of their instruction cycles. This means that for every triplet of microinstructions the second microinstruction begins executing its instruction cycle delayed one phase behind the first one, and the third microinstruction begins its instruction cycle one phase behind the second and two phases behind the first microinstruction. For instance, if a third microinstruction is in phase $F1$, then the second microinstruction is in phase $F2$, and the first is in phase $F3$.

Example 11. Given a microprogram sequence of four consecutive microinstructions I_1, I_2, I_3, and I_4, consider its overlapped execution during the time intervals T_1 through T_6 (Fig. I.28).

Interval T_1: Microinstruction I_1 executes phase $F1$, in which its address, stored in the microinstruction counter, is sent to memory $M2$ to fetch its $I2$ instruction and store it in register AR_{in} (Fig. I.26).

Interval T_2: Instruction I_1 enters phase $F2$, in which the $A1$ field of the $I2$ instruction stored in AR_{in} is sent to memory $M1$ to fetch its $I1$ instruction to register MR_{in}. The next instruction, I_2, is concurrently in phase $F1$, in which the incremented content of the microinstruction counter is sent to memory $M2$ to fetch its $I2$ instruction to AR_{in}.

Interval T_3: Microinstruction I_1 enters phase $F3$, in which instruction $I1$ is decoded into a set of microcommands that are sent to the execution circuits, to perform the corresponding microoperations; microinstruction I_2 is in phase

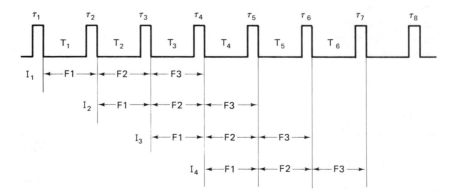

FIGURE I.28 Phase overlap in the execution of a three-phase instruction

$F2$ fetching instruction $I1$ to MR_{in}; and microinstruction I_3 is in phase $F1$, where it fetches its $I2$ instruction to AR_{in}.

Interval T_4: Microinstruction I_2 is in phase $F3$, microinstruction I_3 is in phase $F2$, and microinstruction I_4 is in phase $F1$. Therefore, beginning with time T_3 and at each time interval thereafter, one computational result—the response to one microinstruction—is generated because one microinstruction out of the three currently being executed is necessarily in phase $F3$. ∎

The phase overlap allows the computer to perform all the computations required by a microinstruction during the shorter time span of one phase, and thus excludes the time of memory access from the time of program execution.

10.5.3 Problems of Conditional Branches

Phase overlap is effective only for microinstruction sequences that do not contain conditional branch microinstructions. If any microinstruction in the sequence performs a conditional branch, then the phase overlap may be disrupted, since phase $F1$ of a possible successor microinstruction may follow the $F3$ phase of the conditional branch microinstruction. The example below demonstrates such a disruption in phase overlap.

Example 12. Let the microinstruction sequence in Fig. I.29 have I_1, a conditional branch microinstruction, testing the condition $A > B$ with two possible successors: I_{2T} if the condition is true, and I_{2F} if the condition is false. Suppose that the address of I_{2T} is obtained by a $+1$ increment of the microinstruction counter while the address of I_{2F} is a jump address stored in microinstruction I_1. Assume that words A and B used by microinstruction I_1 are already in the processor registers and their values are such that $A > B$ is false; i.e., microinstruction I_1 will be succeeded by I_{2F}.

Let us show that for this case there will be a disruption in the phase overlap; specifically that the $F1$ phase of I_{2F} will follow the $F3$ phase of I_1

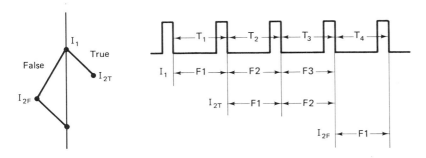

FIGURE I.29 Disruption in phase overlap during conditional branches

(Fig. I.29). During T_1, I_1 will execute its $F1$ phase accessing its address instruction from control memory $M2$. At the next clock period, T_2, I_1 will execute its phase $F2$ and its $I1$ instruction will be fetched to the MR_{in}; I_{2T} will execute its $F1$, since the next instruction address obtained by a $+1$ increment fetches the I_2 instruction to AR_{in}. At clock period T_3, I_1 executes conditional test $A > B$ and produces the false conditional signal ($\overline{C}_n$) since $A > B$ is false. (Generation of the C_n and $\overline{C}_n$ signals for recognizing $A > B$ is considered in Sec. 8.1.) This signal prohibits execution of $F3$ for instruction I_{2T} and writes the new jump address to the microinstruction counter.

Therefore, at clock period T_4, phase $F1$ of a new microinstruction, I_{2F}, begins, and it succeeds phase $F3$ of I_1. Accordingly, overlapping the phase of I_1 and I_{2T} did not create an execution speed-up since the microinstruction I_{2F} that follows I_1 in the program must wait until I_1 ends its execution to begin its own execution. ∎

10.5.4 One-Phase Microinstruction Cycle

As was shown above for a microinstruction sequence containing no conditional branches, phase overlap leads to a time of computation that is the sum of the $F3$ phases of all the microinstructions in the sequence. The time required for memory access is thus excluded from the total computation time.

However, since the phase overlap may be disrupted each time the microinstruction sequence contains a conditional branch microinstruction, many microprogrammed architectures with a two-level memory prefer not to use phase overlap, especially if the microprograms they are designed for are saturated with conditional branches.

These architectures may shorten the instruction cycle, IC, by reducing the number of phases to just one. This is achieved by excluding registers AR_{in} and MR_{in} from the architecture. A one-phase instruction cycle, IC^1, will consequently be shortened by the length of pulses τ_2 and τ_3 which were formerly used to write words into AR_{in} and MR_{in} in the three-phase instruction cycle, IC^3. We now have $IC^1 = IC^3 - 2\tau$ (Fig. I.27).

For this case, the microinstruction counter keeps the same microinstruction address throughout the entire microinstruction cycle, causing consecutive accesses of $I2$ instructions and $I1$ instruction, and execution of the $I1$ instructions in the processor.

10.5.5 Memory Economy Accomplished with the Two-Level Control Memory

This section will examine the conditions under which the total size of the control memory can be reduced if we use a two-level control memory instead of a one-level memory. We will show that we can achieve a real savings in the total number of control memory bits only if the size of the $I1$ instruction exceeds a certain critical size.

This means that there exists a critical bit size for the $I1$ instruction, $BS^*(I1)$, such that memory economy is obtained only if a selected size of microinstruction $BS(I1) > BS^*(I1)$. If $BS(I1) \leq BS^*(I1)$ the total size of the two-level control memory is no less than that of a single level memory.

On the other hand, since a two-level memory may require a longer time for program execution—due to frequent disruptions in the phase overlap—there is no need for using it if the size of the instruction $I1$ is below the critical size, $BS^*(I1)$.

The analysis below is dedicated to finding the critical size $BS^*(I1)$ and the total memory economy originating from the use of a two-level control memory instead of a single level memory.

Consider a microprogram having K microinstructions, its data array contains A words. To write this microprogram requires D different microinstructions; i.e., its microinstruction set contains D microinstructions.

As was shown in Sec. 10.2, each microinstruction, I, includes three fields: the opfield, OF, with bit size $BS(OF)$; the reconfiguration field, RF, with bit size $BS(RF)$; and the address field, AD, with bit size $BS(AD)$. The bit size of I is $BS(I) = BS(OF) + BS(RF) + BS(AD)$.

As was shown in Sec. 10.3.1 the microinstruction set was partitioned into several formats with identical bit sizes to reduce $BS(I)$. Accordingly, $BS(I)$ for processor microinstructions was reduced to the sum of the opfield and the reconfiguration field: $BS(I) = BS(OF) + BS(RF)$.

Since for an architecture with a two-level memory, instruction $I1$ includes only an opfield and reconfiguration field (inasmuch as all addresses are stored in the $I2$ instruction) $BS(I) = BS(I1)$.

The total bit size, $BS(M)$, of a single level memory M required to store a microprogram with K microinstructions is

$$BS(M) = BS(I) \cdot K = BS(I1) \cdot K \tag{3}$$

If the same microprogram is stored in two levels of memory, $M1$ and $M2$, then the $M1$ memory stores D instructions $I1$ since the microprogram contains D different microinstructions I; memory $M2$ stores K instructions $I2$ since the microprogram is composed of K microinstructions. Therefore, the bit size of the memory $M1$ is

$$BS(M1) = BS(I1) \cdot D \tag{4}$$

and bit size of the memory $M2$ is $BS(M2) = BS(I2) \cdot K$.

The bit size $BS(I2)$ should include two address fields, $A1$ and $A2$, where $A1$ stores an address in memory $M1$ and $A2$ stores either an address in memory $M2$ for conditional branches or an address in data memory for data fetches.

Since memory $M1$ stores D instructions $I1$, the bit size of field $A1$ is $BS(A1) = \log_2 D$. Since memory $M2$ has K words and the data array for the program has A words, the bit size of field $A2$ is $BS(A2) = \max[\log_2 K, \log_2 A]$.

It follows that $BS(I2) = \log_2 D + \max[\log_2 K, \log_2 A]$ and

$$BS(M2) = (\log_2 D + \max[\log_2 K, \log_2 A]) \cdot K \tag{5}$$

One achieves a total savings by using a two-level memory, $M1$ and $M2$, rather than a single level memory, M, if

$$BS(M1) + BS(M2) < BS(M) \tag{6}$$

Accordingly, no memory economy is achieved if

$$BS(M1) + BS(M2) = BS(M) \tag{7}$$

The critical size of instruction $BS^*(I1)$ that achieves this equality in memory sizes can be determined if Eq. (3), (4), and (5) are substituted into Eq. (7), leading to $BS^*(I1) \cdot D + K(\log_2 D + \max[\log_2 K, \log_2 A]) = BS^*(I1) \cdot K$ that is equivalent to $K \cdot (\log_2 D + \max[\log_2 K, \log_2 A]) = BS^*(I1)(K - D)$. Thus

$$BS^*(I1) = \frac{K \cdot (\log_2 D + \max[\log_2 K, \log_2 A])}{K - D} \tag{8}$$

The overall memory economy, ΔM, achieved by a two-level control memory is the difference between $BS(M)$ and $BS(M1) + BS(M2)$

$$\Delta M = BS(M) - BS(M1) - BS(M2) \tag{9}$$

By substituting into Eq. (9) the values of $BS(M)$, $BS(M1)$ and $BS(M2)$ given by Eq. (3), (4), and (5), and by assuming that $BS(I1)$ is the size of selected microinstruction we come up with

$$\Delta M = BS(I1) \cdot K - BS(I1) \cdot D - K \cdot [\log_2 D + \max(\log_2 K, \log_2 A)]$$
$$= BS(I1) \cdot (K - D) - K \cdot [\log_2 D + \max(\log_2 K, \log_2 A)] \tag{10}$$

In Eq. (10), the subtrahend $K[\log_2 D + \max(\log_2 K, \log_2 A)]$ is $BS^*(I1) \cdot (K - D)$ by of Eq. (8).

Therefore the total memory economy from using a two-level memory is

$$\Delta M = BS(I1)(K - D) - BS^*(I1)(K - D)$$
$$= [BS(I1) - BS^*(I1)](K - D) \tag{11}$$

As follows from Eq. (11), the memory economy in bits ΔM, depends on two factors

1. The difference between the size of the instruction, $BS(I1)$, and the critical one, $BS^*(I1)$, given by Eq. (8), and
2. The difference between the total size of the microprogram and the size of its microinstruction set.

The greater these differences are, the greater is the memory economy achieved by the use of a two-level control memory instead of a single level memory.

Example 13. Given a microprogram containing 2048 microinstructions and 4096 data words and described by a microinstruction set containing 256 microinstructions; let us find the critical size of an instruction $BS*(I1)$. This can be determined via Eq. (8) by letting $K = 2048$, $D = 256$, and $A = 4096$

$$BS*(I1) = \frac{2048 \cdot (8 + \max(14, 12))}{2048 - 256} = \frac{2048(8 + 14)}{1792} = 26 \text{ bits}$$

A two-level memory is economical only if the $I1$ instruction selected exceeds 26 bits.

Consider two options:
1. Each $I1$ instruction exceeds 26 bits.
2. Each $I1$ instruction is less than 26 bits.

Case 1. Assume that in a two-level memory, the width of level $M1$ is 32 bits; i.e., instruction $I1$ has 32 bits ($BS(I1) = 32$). We will find the memory savings gained by a two-level memory. Using Eq. (11) and substituting $BS(I1) = 32$, $BS*(I1) = 26$, $K = 2048$, and $D = 256$, one obtains $\Delta M = (32 - 26) \cdot 1792 = 6 \cdot 1792 = 10{,}752$ bits. This means that storing this program in a two-level control memory leads to a memory savings of 10,752 bits compared to the case in which the same microprogram is stored in a single level memory also having 32-bit cells.

Case 2. Assume that $BS(I1) = 16$ bits. Then $\Delta M = (16 - 26) \cdot 1792 = -10 \cdot 1792 = -17{,}920$ bits. This means that a single level memory requires 17,920 bits less than a two-level memory. So for this case a two level memory results in wasting memory bits rather than in saving them. ■

Note that this section dealt with a two-level control memory having two address fields in the $I2$ instruction. It is possible, however, to obtain further economy using a two-level memory if the address instruction, $I2$, has only one address field with a bit size that is the maximum of $BS(A1)$ and $BS(A2)$.† For this case a conditional branch microinstruction I requiring two addresses, $A1$ and $A2$, where $A1$ fetches the $I1$ instruction and $A2$ is the jump address, is represented by the two consecutive $I2$ instructions, $I2_1$ and $I2_2$, where $I2_1$ stores $A1$ and $I2_2$ stores $A2$. The $I2_1$ and $I2_2$ instructions have a special tag bit, $e = 1$, which marks them as representing a single microinstruction, I. In case an instruction I requires only a single address instruction $I2$, $e = 0$. The use of an $I2$ instruction with a single address field reduces the critical size of the $I1$ instruction; i.e., instead of $BS*(I1)$ given by Eq. (8) we obtain a smaller $BS*(I1)$ represented by

$$BS*(I1) = \frac{K \cdot (1 + \max[\log_2 D, \log_2 K, \log_2 A])}{K - D} \qquad (12)$$

†The microprogrammed architecture of the Burroughs Interpreter uses a *one-address field concept* for a second level of its control memory. It is called the *microprogram memory*. The first level is called the *nanomery*. That memory stores nanoinstructions that are equivalent to $I1$ instructions. A conceptual diagram of the two-level control memory utilized in the Burroughs Interpreter is shown in Fig. I.26(b). (Ed.)

where K is the total number of instructions in the microprogram, D is the number of microinstructions in the microinstruction set, and A is the number of data words in the microprogram data array.

Example 14. Using K, D, A of Example 13, find the critical size of the $I1$ instruction $BS*(I1)$ for the case that instruction $I2$ stores only one address.

$$BS*(I1) = \frac{2048 \cdot (1 + \max [8, 12, 13])}{2048 - 256} = \frac{2048 \cdot (1 + 13)}{1792} = 16 \text{ bits}$$

Thus, the use of $I2$ with one address saves memory when the $I1$ instruction size selected exceeds 16 bits. Also, as follows from Eq. (11), the larger the difference between the selected size of the $I1$ instruction, $BS(I1)$, and the critical bit size, the greater the memory economy achieved. For instance, if $BS(I1) = 32$ bits and $BS*(I1) = 16$ bits, $\Delta M = (32 - 16) \cdot (2048 - 256) = 16 \cdot 1792 = 28{,}672$ bits, whereas for the case of $I2$ with two addresses, selection of the same $BS(I1)$ produced $\Delta M = 10{,}752$ bits. ■

11. SUMMARY

This section was dedicated to considering the basic principles of the von Neumann and microprogrammed architectures. We saw that a microprogrammed architecture can be conceived of as an improvement on a von Neumann architecture since it reduces the time for memory access required by the von Neumann architecture.

Neither of these architectures, however, is capable of overcoming another problem they both face: *sequentiality of computation*. This means that each of the traditional architectures may execute only *one* operation at a time in the processor over *one* pair of data words.

This leads to the following restrictions:

1. In each type of computer (von Neumann and microprogrammed), program instructions are computed sequentially. If the program provides for some computational concurrency (several instruction sequences executed concurrently) neither of these architectures can implement instruction *parallelism*; i.e., execute several instructions at a time. The only parallelism they allow is *microoperation parallelism* in which each type of computer may execute several microoperations at a time.
2. Each instruction may handle only two operands at a time. If an array of operands needs to be handled, they can be computed sequentially using indexed addressing. This means that the addresses of the next pair of operands are computed in the control unit as sums of the addresses of the current pair of operands and an index increment.

These restrictions result in a slowed execution that is unacceptable for many algorithms. Further increases in computational throughput have been accomplished by assembling several computers into a parallel system. This subject will be considered in the next part of this chapter.

PARALLEL SYSTEMS

As we saw in the previous section both of the traditional architectures—von Neumann and microprogrammed—do not provide for parallel computation of instructions and data, which makes them inappropriate for many algorithms requiring high computational power.

As a result, near the end of the 1950s and during the first part of the 1960s, a great deal of architectural research was directed towards overcoming this sequentiality barrier [8–30] and constructing systems that could perform parallel computations. Such systems came to be called *parallel systems*.

By a parallel system we generally mean a system that implements either instruction parallelism, or data parallelism, or both types of parallelism in a single system.

Instruction parallelism is the capability of executing several instructions at a time.

Data parallelism is the capability of processing several data words at a time with a single instruction. This collection of words handled by one instruction is called a *data vector*.

Four types of parallel systems may currently be distinguished: multicomputer, multiprocessor, array, and pipeline. The appearance of each of these systems is dictated by the peculiarities of the complex parallel algorithms it was originally designed to compute.

12. MULTICOMPUTER SYSTEMS

A multicomputer system is a parallel system assembled from N computers in such a way that any two computers may communicate with each other via an interconnection bus connected with their I/O devices (Fig. I.30)[31–40]. The I/O device that performs this communication in each computer is, as a rule, a buffer memory (BM), that is connected to the local processor and the primary memory.

If computer A has to write to its primary memory, M_A, a block of data words stored in the primary memory of computer B, M_B, then the interconnection bus of the multicomputer system activates the following path between computers A and B (Fig. I.31): (1) The block requested in computer B is transferred from memory M_B to its buffer memory, BM_B; (2) from BM_B it is transferred to BM_A, the buffer memory of computer A; (3) from BM_A it

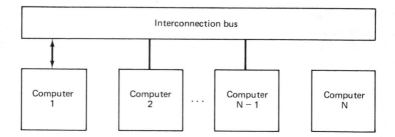

FIGURE I.30 A multicomputer system

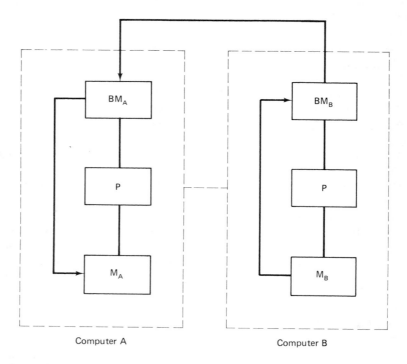

FIGURE I.31 Path between computers A and B activated in the interconnection bus

is sent to its destination, memory M_A. A block of data must therefore be written to two buffer memories, first to BM_B and then to BM_A, before going to its destination memory M_A. This slows down the information sent between computers significantly.

Another kind of bottleneck between computers is associated with the restricted bandwidth of the interconnection bus. If a system is assembled from a number of large computers then parallel word exchange among these computers leads to an interconnection bus of significant complexity. To reduce this complexity the bandwidth of the bus is often restricted. This may be done by replacing parallel word transmission by sequential byte transmission. For instance, in a 16-bit interconnection bus, a 64-bit word is transferred via 16-bit bytes, thus requiring four time intervals to transmit one word. Byte transfer therefore leads to another source of increasing the time of between computer communication.

It follows from the above that a multicomputer system assembled from N computers may compute N instruction sequences concurrently, where each instruction may handle a pair of operands at a time. A multicomputer system thus implements instruction parallelism but not data parallelism.† The relatively slow data transmission between computers makes a multicomputer system effective in computing those algorithms that have infrequent data exchanges between instruction sequences computed in different computers since each such exchange will require a significant amount of time to perform.

†Various issues pertaining to multicomputers are treated very briefly here. Refer to Chap. III, Sec. 8, which discusses detailed design techinques for multicomputer systems with dynamic architectures. (Ed.)

13. MULTIPROCESSOR SYSTEMS

A multiprocessor system is assembled from the same functional units (processors, memories, I/Os) that are used in a multicomputer system. However, the interconnection bus of the multiprocessor system allows activation of a direct communication path between any two functional units contained in the system (Fig. I.32) [32, 41–50].

A multiprocessor system can therefore provide direct data exchanges between any processor and any memory, or any two processors, or any two memories, etc. For instance, if processor P_2 needs a block of data words stored in the memory M_N, then the interconnection bus reconfigures into a direct path between P_2 and M_N, allowing such an exchange directly rather than by using two buffer memories as is done in multicomputer systems (Fig. I.33).

Such an organization of data exchanges will require the complexity of the interconnection bus to be even greater than for a multicomputer system. To make this complexity problem manageable, some systems have restrictions on direct communications among functional units. Their interconnection bus will, as a rule, support only direct processor–memory exchanges. Other exchanges, such as processor–processor or I/O–memory, may be prohibited. As with multicomputers, multiprocessors may reduce the bus complexity by restricting its bandwidth. This again introduces word transmission by bytes and leads to a slowing down of communications. In general, multiprocessor systems are effective for executing complex algorithms having a high interaction between the concurrent instruction sequences computed in different processors; i.e., one sequence needs blocks of data words computed by another sequence. As with multicomputers, multiprocessors imple-

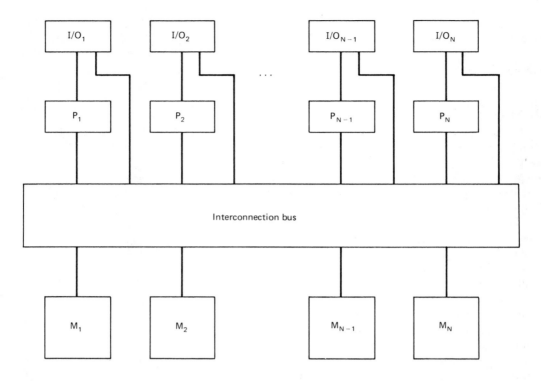

FIGURE I.32 A multiprocessor system

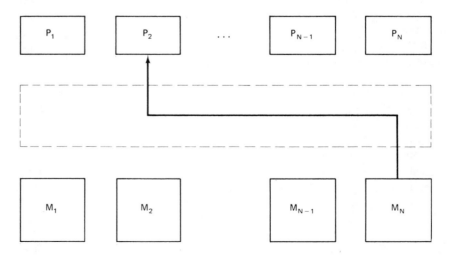

FIGURE I.33 Communication between P_2 processor and M_N memory in a multi-processor system

ment instruction parallelism but not data parallelism since a system containing N processors and N memories can compute N instructions at a time where each instruction can handle not more than two operands at a time.†

14. ARRAY SYSTEMS

An array system is generally understood to be a collection of N processors, $P_1, \ldots, P_N$, each handling the same instruction issued by a single control unit (Fig. I.34). Each processor, P_i, is equipped with a local memory, M_i, used by P_i for storing both its operands and the computational results it obtains.

Therefore, an array composed of N processors may concurrently execute N identical operations with one program instruction. Each instruction handles a data vector made of operands handled by $P_1, \ldots, P_N$, respectively. An array of N processors thus implements data parallelism but no instruction parallelism.

Since the number of data items handled by one instruction may change for different tasks or within the same task, the number of processors in the array should adapt to the current size of the data vector. This adaptation has been realized by different *reconfigurable array systems* [50–57].‡ The idea of such reconfiguration consists of the following: For a single array system, each processor, P, and its local memory, M, may reconfigure into several smaller size processors and memories, and vice versa. For instance, Fig. I.35 shows the hardware diagram of an array system containing three 64-bit processors. This system may handle three data arrays in parallel, stored in memories M_1, M_2, and M_3; i.e., each program instruction fetched from the control unit is sent concurrently to all processors of the array, causing execution of a data vector made up of three data items. The same system may

†See Chap. III, Sec. 9, for a discussion about designing multiprocessor systems with dynamic architectures. (Ed.)

‡Refer to Chap. II, Sec. 3, for a comprehensive insight into several representative types of parallel reconfigurable array systems. (Ed.)

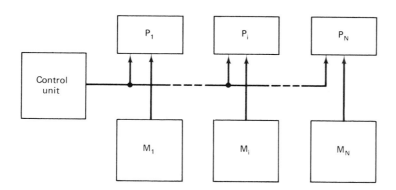

FIGURE I.34 An array system

62 CHAPTER I

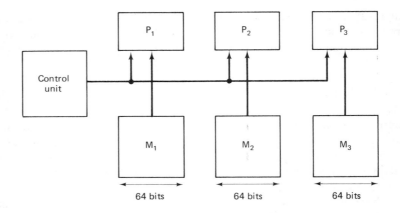

FIGURE I.35 An array system containing three 64-bit processors

†Here only major reconfigurations in arrays are presented. Chapter II, Secs. 1 and 3, gives a complete description of *all* reconfiguration techniques that are used in existing array systems. (Ed.)

reconfigure each 64-bit processor and its local memory into two 32-bit processors (Fig. I.36). Then the same system could perform identical parallel computations over six data arrays. Thus a reconfigurable array system may change the number of data items handled in parallel by changing its processor's sizes.†

The most widely known reconfigurable array system is ILLIAC IV in which each 64-bit processor may reconfigure into two 32-bit or eight 8-bit subprocessors. Thus, a system containing sixty four 64-bit processors can compute either sixty four 64-bit data words, or a hundred and twenty eight 32-bit data words, or five hundred and twelve 8-bit data words with a single instruction [56].

The major application of array systems is in the cost-effective computation of algorithms handling a large number of independent data streams such as the solution of identical partial differential equations for different geographical localities in the case of weather prediction, or matrix manipu-

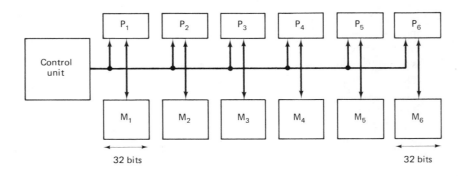

FIGURE I.36 An array system containing six 32-bit processors

lation and the solution of linear systems of equations in, for example, nuclear energy applications or air traffic control. Array systems have also been effective for data base management and for data processing with associative memories.‡

These systems have several essential drawbacks, however, The major one is that existing array systems may change only the number of parallel data streams; i.e., reconfigure to adjust to the size of a data vector. No array system has yet been produced which accomplishes reconfiguration to enhance instruction parallelism; i.e., increase the number of instruction streams computed as well as that of data words. Another drawback of array systems is the ineffective use of the available hardware resources caused by synchronous operation of all the processors working in the array. Since the number of parallel data streams may decrease, some processors may become idle. For array systems, however, it is fundamentally impossible to turn the idle processors to the computation of other programs since all of the processors in the array are controlled by a single instruction stream.†

‡The reader is encouraged to read Secs. 1 and 4 of Chapter II, which are dedicated to application of conventional and associative arrays. (Ed.)

†More information on limitations and benefits of existing and future array systems are in Chap. II, Sec. 6. (Ed.)

15. PIPELINE SYSTEMS

Since the organization of pipelined computations departs significantly from the traditional organization of computational process, and no separate chapter of this book is dedicated to pipeline systems let us look at the basic concepts of pipelined computing before we outline the essential principles of pipelined architectures.

15.1 Pipeline Computation

As we have shown, all architectures of computers and systems (von Neumann, microprogrammed, multicomputers, multiprocessors, and arrays) spend a significant portion of their time on memory access of both instructions and data. The reason for this is that execution of an operation in the processor always follows memory accesses aimed at fetching the instruction and its operands.

However, a reduction or even a total exclusion of the time of memory access from the time of computation can be obtained if the memory access for one instruction overlaps the execution of the operation of the preceding instruction. This is accomplished by introducing a pipeline organization of the computation.

The idea of such an organization consists of the following. Suppose one instruction computes the following expression: $a + b + c + d + e$, and word a is already in the processor (Fig. I.37). The execution of this expression can be partitioned into five computational phases; $F_0, F_1, \ldots, F_4$.

Phase F_0 is a preparatory phase in which the control unit sends the

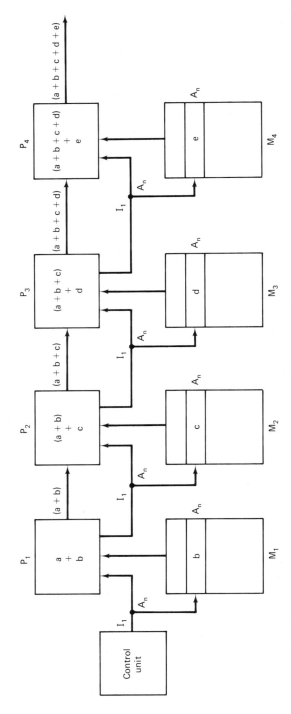

FIGURE I.37 Pipeline computation of $a + b + c + d + e$

instruction I_1 to the processor P_1. This instruction contains both the opcode and the data address, A_n, to fetch word from memory M_1 to processor P_1. F_0 ends with the writing of word b to a register of processor P_1.

Phase F_1 is a computational phase that performs the following actions concurrently:

1. Performs addition $a + b$ in P_1. At the end of the phase, the result of $a + b$ and the instruction I_1 are written to registers of the next processor, P_2.
2. Sends the same data address, A_n, stored in I_1, to memory M_2 and writes data word c, stored at this location, to a register of P_2.

Phase F_2 is also a computational phase that performs the same actions that characterized phase F_1.

1. Performs addition in processor P_2 of $(a + b)$ and c, and sends the result and the instruction I_1 to registers of the next processor, P_3.
2. Fetches operand d from memory M_3 for the next phase, F_3, to be performed in P_3.

Similar actions are performed during F_3 and F_4. At the end of phase F_4 the computational result $a + b + c + d + e$ is obtained.

Let us find the duration of each computational phase. Since each such phase performs two concurrent actions with times t_{op} and t_M, where t_{op} is the time of addition and t_M is the time of memory access from data memory, it requires the maximum of both times.

In order to exclude the time of operand access, t_M, from the time of computation one has to select a memory that has $t_M \leq t_{op}$. If $t_M > t_{op}$ then each addition in a processor will be delayed by the time $t_M - t_{op}$ since each processor must receive a second operand from its local memory before starting the operation. Thus if one selects $t_M \leq t_{op}$, it is possible to eliminate the time of operand access from the time of computation assigned to instruction I_1.

Note that the time of the preparatory phase F_0 can also be excluded from the overall time of execution when the processors $P_1, \ldots, P_k$ execute an instruction sequence made of instructions $I_1, I_2, \ldots, I_k$. When P_1 executes phase F_1 of instruction I_1, the control unit may fetch the instruction I_2. When instruction I_1 is in phase F_2 in processor P_2, P_1 executes phase F_1 and I_2 and the control unit fetches instruction I_3, etc. As a result of such an organization, the end processor of the pipeline will obtain the result of each instruction during a time of one computational phase, and the entire time of the memory accesses will be excluded from the computation.

In general, pipelining requires partitioning of the instruction microprogram into several phases and allows for the overlapped execution of consecutive phases assigned to consecutive instructions. If the microprogram is partitioned into K phases, $F_1, F_2, \ldots, F_K$, then the pipeline that executes it contains k stages, $S_1, S_2, \ldots, S_K$, so that S_i executes phase F_i of the microprogram. A pipeline stage is understood to be a separate resource unit assigned to the execution of one phase. It can be a processor or a dedicated resource unit (multiplier, divider, adder, etc.).

Figure I.38 shows a pipeline containing four stages that executes instructions I_1, I_2, I_3, and I_4, each of which is partitioned into four computational phases, F_1, F_2, F_3, and F_4. The current contents of the pipeline are phase F_4 of instruction I_1 in S_4, phase F_3 of I_2 in S_3, phase F_2 of I_3 in S_2, and phase F_1 of I_4 in S_1. A timing diagram of the computations performed by this pipeline is shown in Fig. I.39. The pipeline filling of Fig. I.38 corresponds here to time interval t_4, during which the pipeline produces the computational result of instruction I_1.

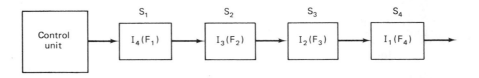

FIGURE I.38 Pipeline filled with instructions, I_1, in phase F_4, I_2 in phase F_3, I_3 in phase F_2, and I_4 in phase F_1

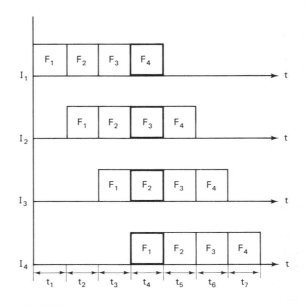

FIGURE I.39 Timing diagram of a pipeline computation

15.2 Organization of a Pipeline System

A pipeline system is a parallel system containing N pipelines. Since each pipeline computes one instruction sequence, a pipeline system can compute N instruction sequences concurrently and thus implement instruction parallelism. No data parallelism is implemented since each operation is executed over no more than two operands at a time.

The existing pipeline systems are divided into two categories: *instruction* and *arithmetic* [58, 59].

Instruction pipeline systems fragment the memory access process, aimed at fetching the instruction and its operands from the memory, into phases. For most instruction pipelines, the entire processor operation assigned to the instruction is treated as a single phase.

Arithmetic pipelines fragment the arithmetic operations assigned to an instruction into phases. These can be either complex expressions containing several arithmetic operations or single operations such as multiplication, division, and floating-point operations that are represented by iterative computational algorithms.

Example 15. Suppose that expression $[(A^2 - B)/C] \cdot D$ must be evaluated over four arrays of data words A, B, C, and D, each of which contains 100 words. The fastest way to compute this expression is to use the arithmetic dedicated pipeline shown in Fig. I.40. For this pipeline, each instruction realizes the sequence of arithmetic operations ($\times$, $-$, $\div$, $\times$) performed over words A_i, B_i, C_i, and D_i of the arrays A, B, C, and D, respectively. One hundred pipelined instructions are thus required to complete the entire computation. The entire expression $[(A^2 - B)/C] \cdot D$ is fragmented into four phases each of which is executed in a dedicated pipeline stage. The pipeline thus has four consecutive stages: multiplier, S_1; subtractor, S_2; divider, S_3; and multiplier, S_4. Since it takes four phases to fill this pipeline, one computational result will be obtained during each next phase, and the entire job over arrays A, B, C, D takes $4 + 100 = 104$ phases. This means that the pipeline loses no time for memory accesses (instructions and operands) since the accesses overlap with the processor's execution. ∎

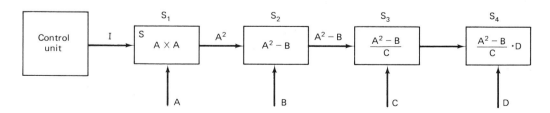

FIGURE I.40 Execution of $[(A^2 - B)/C \cdot D]$ in a pipeline with four stages

A pipeline system can be more effective for array computations than an array system having the same hardware complexity. This is illustrated by the example that follows.

Example 16. Let us find the time needed by an array system of roughly the same complexity to execute the expression in Example 15. Since the pipeline contained four stages and the control unit, the array system must contain four processors and the control unit in order to be compatible in complexity. Each processor executes one expression, $[(A_i^2 - B_i)/C_i] \cdot D_i$. Since the total number of such expressions required to complete the job is 100, each processor must compute $100 \div 4 = 25$ expressions. Computing each expression takes four arithmetic operations and four accesses of the operands A, B, C, and D. For simplicity assume that both an arithmetic operation and an operand access take the same time: one phase, F. Then it would take a time of $8 \cdot F$ to execute each expression. Thus, sequential execution of 25 expressions in each P_i takes a time of $200F$ versus the $104F$ required by the pipeline system. ∎

15.3 Problems of Pipeline Computation

The main problems faced in constructing an instruction pipeline are those of time overheads caused by conditional branches and of variations in the number of operand addresses and of the addressing procedures used in instructions [58, 59].

In existing instruction pipelines, reconfiguration is used mostly to offset the time overheads caused by the last mentioned factor. In the MU5 [60] instruction pipeline, for instance, instructions may bypass unneeded pipeline stages. This, however, creates dummy time intervals in order to resolve conflicts caused by the instruction encountering operands prepared for preceding instructions as a result of such bypassing.

The major problem with arithmetic pipelines is the time overhead introduced by the disparity between the pipeline(s) and the algorithm being executed. As a result, pipeline systems tend to become dedicated to certain types of computation and pipeline systems usually have limited applicability.

To broaden the range of their cost-effective application, arithmetic pipelines offer various software-controllable reconfigurations of the available hardware resources. The general idea is to reconfigure the resources, via software to reduce the dissimilarity between the pipeline and the sequences of operations assigned to various instructions.

Let us look at some existing systems and consider their use of reconfiguration. In the TI ASC [61] reconfiguration of the arithmetic pipeline means bypassing unneeded pipeline stages; i.e., the instruction propagates through a stage only if it implements the operation that is encountered in the instruction. Otherwise this stage is bypassed. This is similar to the way that the MU1-5 reconfigures its instruction pipeline. The weakness of this technique is in the time lost solving conflicts among instructions implementing different sequences of arithmetic operations.

In the CRAY-1 [62], 12 functional units are organized into dedicated pipelines. These are partitioned into four categories: address, scalar, vector, and floating-point. Although each pipeline cannot reconfigure, reconfiguration is used to chain several pipelines together to form pipelines with a larger number of stages. This allows the pipeline execution of instructions with long sequences of operations to be organized.

Reconfiguration is thus used in existing pipeline systems in a limited sense, as a means of bypassing or chaining pipeline stages [60–63].

Several authors have proposed theoretical pipeline systems with a deeper level of reconfiguration. Reddi and Feustel [64] describe a restructurable pipeline system that may reconfigure into different sequences of resource units. The architecture of such systems requires that programs to be executed be decomposed into separate program blocks and that the compiler determine which interconnections are to be established among the operational units assigned to the execution of each block. This pattern of interconnections is then encoded into a program instruction that activates the required interconnections among hardware units during execution of that program block.

Thomasian and Arizienis [65] propose a reconfigurable pipeline system consisting of an array of pipeline arithmetic processors that can employ different configurations between processors. To carry out pipeline computations, the tasks requiring the same configuration are put together. After executing all tasks appropriate for that configuration, the system reconfigures and starts executing a new block of tasks in the new configuration.

Since many programs have different sequences of operations following each other, a pipeline that executes such a program must reconfigure each time it switches from one sequence of operations to another. If Δt is the time required for each reconfiguration and N is the number of different operation sequences, then the pipeline loses time $\Delta t \cdot N$ reconfiguring its resources. If Δt or N is large, the speed advantage of the pipeline may be lost, removing the original reason for using it. To reduce N, programs are sometimes rearranged into tasks such that each task may be computed by a single configuration of the pipeline. But this requires special programming which may again restrict the class of programs that can cost-effectively be pipelined.

It is also quite difficult to compute via pipeline programs that require broad exchanges of temporary results between tasks. Each stage of pipeline must usually receive two operands to execute its operation: one from the preceding stage and the other from the local memory attached to the stage itself. Each local memory is therefore engaged in continuous fetches of operands required by that stage while the pipeline is working. It is thus impossible to load this memory with temporary results needed by the stage during computation, because access to operands stored in the local memory would then stop. Therefore, all local memories may be loaded with data words only before computation begins. Hence, if a pipeline stage needs a temporary result computed earlier, it must fetch this result from another memory designated for temporary results. Since a pipeline stage may require a temporary result computed by any other pipeline stage, it should be provided with fast information exchanges between stages that do not degrade pipeline performance. Such exchanges are poorly developed in existing systems, eliminating many programs from consideration for pipelining. This again narrows pipeline applicability. Limited applicability is therefore the most severe drawback of pipeline architectures.

THE IMPACT OF TECHNOLOGICAL PROGRESS ON THE ARCHITECTURE OF COMPUTERS AND SYSTEMS

This part of the chapter concerns itself with the analysis of the effects of LSI technology on computation and computer organization. The impact on computation in general is twofold: First, LSI technology has led to the cost-effective computerization of small processes, performed by microcomputers and microprocessors; i.e., it has shifted the lower bound of computerization downward. Second, by allowing the construction of complex modular computer systems that adapt their architectures to the problems being computed, LSI technology lights the way towards the computation of supercomplex parallel processes; i.e., it is shifting the upper bound of computerization upward.

†More information on these new architectures can be found in Chap. III. (Ed.)

In the area of computer organization, LSI technology has permitted new types of LSI modular architecture to be created that can adapt to the needs of an executing algorithm by reconfiguring module interconnections. Such architectures are characterized by the set of architectural states they can assume and software-controlled reconfiguration of the available system hardware resources from one architectural state to another. Each state is characterized by such things as the number and sizes of concurrently operating computers, the instruction set activated, and the type of architecture employed (array, pipeline, multiprocessor, or multicomputer).†

‡Different aspects of program preprocessing are treated in Chaps. III, IV and V of this book, where Chap. III and Chap. V discuss preprocessing aimed at resource assignment, and Chap. IV discusses program verification via preprocessing. (Ed.)

An essentially new opportunity is given the programmer: During the program's running time it is now possible to switch the system's architecture into states that match the computational specificities of the program. The consequences of such matching are additional performance gains obtained from the same resources [67]. To fully utilize the new adaptation capabilities provided by modular architectures, programs‡ have to be preprocessed in order to find the sequences of the best architectural states to be assumed in their execution. Thus, adaptation preprocessing emerges as a new stage in the overall organization of computing. A program written in a high-level language will need to be analyzed to find the architectures of hardware resources that can best execute it.**

To summarize the above one may say that LSI technology is causing a "quiet revolution" in computations and computer organization. Consider first the effect of this revolution on computation.

**Some techniques for such analysis are considered in Chap. III beginning on p. 245. (Ed.)

16. LOWERING THE LOWER BOUNDS OF COMPUTERIZATION

A decade ago, the cost of a simple computer (minicomputer) was in the thousands of dollars and this marked the lower bound of computerization. This separated the processes that allowed cost-effective computerization from those that did not. Since the fabrication cost for computer hardware has been reduced in each generation of computer components, this bound has moved steadily downward into the area of ever simpler processes and has made very simple applications amenable to cost-effective computerization.

The downward plunge of the lower bound of computerization was marked in the 1970s by advances in LSI technology that permitted the fabrication of simple computers (microprocessors and microcomputers) for less than $100. This shift has led to the computerization of an enourmous number of simple processes that were not appropriate subjects of automation even in the 1960s.

A sudden boom in the production and applications of microprocessors and microcomputers resulted. An immediate consequence of this boom was the appearance of an immense number of new computer companies involved in the manufacturing and application of such microdevices. Many computer firms that in the 1960s produced computer components now became important manufacturers of microcomputers and microsystems.

This flourishing microcomputer-oriented industry contributed to a rapid growth in the computer community and was further augmented by newcomers associated with the production, programming, and application of microdevices.

The 1970s saw the broad penetration of microdevices into diverse sectors of our modern consumer society. We can now name over a hundred areas where they are competing effectively.

17. RAISING THE UPPER BOUND OF COMPUTERIZATION

LSI technology, however, has caused not only a downward movement of the lower bound of computerization but is now also beginning to move the upper bound of computerization upward, leading to the automation of processes that were forbidden it before, due to their complexity.

17.1 Requirements for Supersystems†

†See a comprehensive treatment of this problem in Chaps. III and V. (Ed.)

There are a number of problems that lend themselves to the application of enormous computational power. Even in the beginning of the 1970s the need for a system with a throughput greater than scores of the then current large machines could provide was evident. Today, even this kind of throughput no longer satisfies the computational requirements of some of the problems that must be computed in real-time.† For instance, ballistic missile defense algorithms need a Supersystem capable of computing hundreds and even thousands of parallel program streams, each of which may have rates in the range of 5 to 100 million bits per second [68–72]. Even at these enormous speeds, the Supersystem must be absolutely reliable [68, 69].

†Chapter V provides a insight on how the entire system configuration—i.e., the number of separate computers and interconnections among them—can be obtained via analysis of program computational requirements. (Ed.)

Similar computational requirements are associated with some of the processes in energy and meteorology which require throughputs greater than any existing system can satisfy. The research on such Supersystems has proven conclusively that such extreme throughputs cannot be obtained through the mere merger of hundreds of large existing computers into a system [68].

The reason for this is that such a merger inevitably leads to enormous complexity in the computer system. However, as a system's complexity goes up its reliability goes down,‡ and the system thus becomes unsuitable for many complex algorithms requiring extreme reliability. Since many complex real-time algorithms are characterized by a steady growth in the number of information streams that need to be processed [68], the Supersystem must respond to this growth by increasing the number of computers it incorporates. Any modular expansion of an existing Supersystem by the addition of new components is limited by the delays it introduces and the erosion of the system's initial reliability.

‡See Chap. VI, Sec. 3, which discusses various reliability measures for computer systems. (Ed.)

It follows that for any Supersystem there exists a critical complexity of hardware resources such that any throughput increase due to the integration of new equipment into the system already constructed is offset by time and reliability losses incurred by it. This means that a Supersystem already built may become unsuitable for many algorithms requiring higher throughput.

17.2 New Architectures for Supersystems

To overcome this contradiction between a system's throughput and its reliability, Supersystems have to be equipped with new types of architectures that are capable of performing dynamic adaptations to the computational requirements of the algorithms being processed. This will allow the system to increase throughput using the same resources, without increasing its complexity.

A Supersystem must be able to respond to changes in the information streams in a complex algorithm by changing the number of computers it has;

i.e., the system must be capable of redistributing, via its software, the available hardware resources into a variable number of computers.†

For instance, suppose a system must compute two algorithms. The first algorithm processes 32-bit words and requires 100 instruction streams, and the second algorithm uses 64-bit words and 50 instruction streams. In order for the system to compute both algorithms one after another it must have fifty 64-bit computers and fifty 32-bit computers. This means that during the computation of either algorithm it has a redundant resource equivalent to fifty 32-bit computers, and this adds excessive complexity to the overall system. Should the system be capable of adapting to the computation requirements of the algorithms, each 64-bit computer could be partitioned into two 32-bit computers, and the entire resource of the system would then need only be equivalent to fifty 64-bit computers. While executing the first algorithm, the system establishes an architectural state characterized by a hundred 32-bit computers; while executing the second, it forms into fifty 64-bit computers. Therefore, the use of such an architecture allows us to achieve the required throughput while using fewer hardware resources [73–75].

An architecture for a Supersystem should also be able to perform instruction set adaptations. Each computer incorporated into the system could have hardware realizations, not of a single, but of tens of different instruction sets, each of which is oriented towards a separate class of dedicated applications. For instance, one set may be dedicated to handling trigonometric functions, another set to array processing, and a third could handle the fast Fourier transforms occurring in signal processing.‡ Each set could contain, for instance, 256 instructions partitioned into two categories: dedicated and general-purpose. The difference between the sets is in the dedicated instructions. A computer program would activate the dedicated instruction set that most closely matches its requirements. Such selective activation could be done by software by just writing a special code to a control unit [67]. Other forms of adaptation will be considered below.

The consequence of such a match between architecture and algorithm is additional execution speed achieved on the same resource. This means that a Supersystem will be able to augment its throughput without an accompanying increase in its complexity, and thus will be able to maintain the same level of reliability that existed in a less complex system.

†See Chap. III which discusses design techniques for such architectures. (Ed.)

‡Chapter II, Sec. 4, gives information on the requirements to a system configuration caused by signal processing. (Ed.)

17.3 Effect of LSI Technology on Supersystem Architectures

It follows that such adaptable architectures will lengthen the life cycles of Supersystems, which are distinguished by their extremely high costs and long fabrication times. Let us see how LSI technology permit the creation of adaptable architectures for Supersystems. LSI technology has introduced modularity as basic in the organization of a computer's architecture. Each LSI module from which a computer is assembled may be equipped with simple circuits for the software-controlled activation and deactivation of interconnections with other modules. For instance, processor modules may be switched among several main memory modules, which may reduce the time required for processor–memory communication. Or one processor may reconfigure into several smaller-size processors in an array-parallel system, increasing the size of the data vector processed by a single instruction. Finally, a modular architecture supports a complete dynamic redistribution of the available hardware by reconfiguring hardware resources into different numbers of variably sized computers.† This allows the computer to adjust to the changeable number of information streams encountered in complex algorithms. Thus, a computer architecture assembled from LSI modules may adapt much more closely to the needs of the algorithm than was ever achieved in traditional systems.

It is safe to predict that LSI technology will lead to a proliferation of adaptable architectures for Supersystems that can perform many new and cost-effective adaptations to algorithms that have never been attempted on traditional systems. Furthermore, since LSI technology significantly reduces the cost and enhances the reliability of computer components, it makes feasible the design of reliable modular computer systems‡ containing many more hardware resources than ever before. This should sharply increase the number of instruction and data streams that can be computed. Therefore, LSI technology has greatly influenced the emergence of the following techniques as powerful methods for increasing throughput:

1. Architectural adaptation.
2. Enhanced computational parallelism.

This encourages the creation of Supersystems with throughputs greatly exceeding that of systems from previous generations. Consequently, LSI technology is shifting the upper bound of computerization upwards.

†Various techniques for such redistribution are discussed in Chap. III, Secs. 5.6 and 6.6. (Ed.)

‡Design techniques for reliable modular computer systems are discussed in Chap. VI, Secs. 1, 2, and 4. (Ed.)

18. PROBLEMS IN DESIGNING A MODULAR COMPUTER

In addition to expanding the lower and upper bounds of computerization, LSI technology may lead to significant cost reductions in conventional or general-purpose computation performed by such popular computers as the IBM 360/370 and the larger Burroughs, DEC, and Univac machines. The reason is that it may allow for the replacement of the traditional hardware by low-cost LSI modular computers with a similar computational throughput.

Let us now consider some of the problems that arise in designing a powerful modular computer or a family of modular computers.

During the last few years the computer industry has manufactured a number of different microprocessors and microcomputers. Only recently, however, have announcements been made that commercial production of a sufficiently powerful LSI modular computer has begun [76, 77]. The question is why does the industry lag in the LSI modular implementation of an IBM 360/370 or the like? This surprising phenomenon needs a reasonable explanation due to the fact that this computer is popular throughout the world and is equipped with sophisticated software worth over 200 billion dollars [76].

If LSI modules are so cheap, then an LSI implementation of an IBM 360/370 would significantly lower its cost and enhance its reliability, and thus bring about a sharp drop in the costs of general-purpose computation. That this has not happened surely calls for an explanation, especially in the light of fashionable conversation which maintains that LSI technology allows one to obtain indefinitely low-cost hardware.

The problem is that, in spite of sound advantages, LSI technology unfortunately has several serious constraints that must be considered in computer design. The process of LSI module fabrication reminds us of that for producing a photograph: to begin the mass production of copies (reprints) one must first produce a module type (negative). The dollar cost of a module type is now in the hundreds of thousands of dollars range and takes a year to create [78–80]. Once a module type is produced, the cost of copy fabrication does not exceed $10, which has created the popular misconception of arbitrarily cheap hardware in the age of LSI technology.

What is ignored in this view is the cost of the module type, which must be spread among the copies. To obtain an LSI module for less than $10, the respective module type must be sold in the hundreds of thousands of copies. On the other hand, if a computer designer needs only a hundred copies of the same module type, each copy will cost no less than $1000, provided it took $100,000 to produce the module type.

These figures provide a conclusive answer to the question of why industry is still incapable of organizing the commercial production of popular mainframes from LSI modules. It is easy to believe that a microcomputer on a chip can easily be sold by the ten thousands, but who can realistically expect to sell the same number of IBM 370s on LSI modules? Far more credible is the assumption that at best a few thousand copies of this computer can be sold.

It then follows that the cost of each LSI module used in an IBM 370 will be less than several hundred or a thousand dollars, provided the entire computer is assembled from copies of the same module type. However, the IBM 370 is not a microcomputer. Thus, it is hardly conceivable that all its circuits can be mapped onto copies of the same type. It has been shown [80] that a sufficiently complex computer requires fabrication of tens of different module types and that each module type can be used only in a limited number of copies.

LSI technology, therefore, poses new restrictions on commercial production of complex computers: To create a pilot computer, we have to develop a set of module types that will be used in its assembly. However, the design of such a set requires enormous investments that can be recovered only through mass production. Since a powerful computer system can hardly be produced in volume, it is easy to see why the industry lags in manufacturing LSI modular computers capable of emulating existing general-purpose mainframes.

Let us now analyze several existing approaches that can reduce the cost of a set of module types.

19. BIT-SLICED MODULAR COMPUTERS

During the past few years the industry began manufacturing simple modular computers, often called "bit-sliced computers" [81–84].

The major characteristics of these computers are:

1. Their processor sizes grow in modular increments. If one processor module (slice) handles an h-bit word then the entire processor handles $k \cdot h$-bits where k is the number of modules used. The selection of h is dictated by the chip size of the logic family used and can currently range from 2 to 16. As for the value of k, it can now have a broad spectrum of ranges ($k=4$, 8, 10, etc.).

2. The majority of bit-sliced computers are microprogrammed. They include a control memory for storing instructions, a processor unit assembled from LSI modules in h-bit increments (slices), and a data memory for storing data words (Fig. I.41). To maintain parallel information exchange, the width of the data memory matches the size of the processor. As for the control memory, its width is determined by the microinstruction size. By changing the number of modules assembled into the processor and making corresponding changes to the width of the data memory, one obtains a family of bit-sliced computers; i.e., a set of computers with sizes in h-bit increments (h-, $2h$-, . . . ,$n \cdot h$-bits). To expand their applicability, the latest bit-slice modules do not implement instruction sets. Instead, they are microprogrammed with the use of sequencer, which is now included as a standard companion part to the bit-sliced modules [76, 84].

A bit-sliced organiztion leads to some benefits as well as some weaknesses.

19.1 Benefits

1. *Reduction in the Number of Module Types.* A microprogrammed architecture allows a reduction in the number of module types. A conventional von Neumann computer usually utilizes a powerful control unit, realized by random logic (flip-flops and gates). Unlike

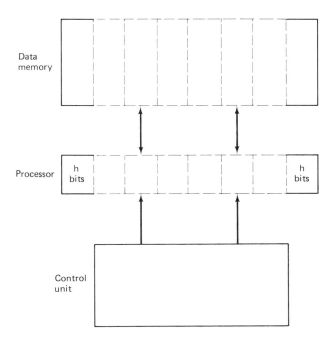

FIGURE I.41 A simplified architecture of a bit-sliced microprogrammed computer

a processor that is amenable to simple assembly from copies of the same module type, a control unit defies partitioning into identical portions. The assembly of a control unit, therefore, requires fabrication of many different module types.

In a simple microprogrammed computer, on the other hand, the control unit is reduced to a control memory and decoding logic. Since a control memory may be readily assembled from standard memory units and decoding logic can easily be included in the module used to assemble the processor, a microprogrammed architecture leads to a reduction in the number of module types used to assemble the modular computer.

2. *Adaptation to Computational Specificities.* A microprogrammed architecture allows a certain amount of adaptation to the computational specificities of an executing program because a programmer, by writing codes into instruction fields, may select microprograms that are more task-oriented than usual.

19.2 Weaknesses

1. *Requirement of Two Memories.* A bit-sliced microprogrammed computer requires two separate memories—for instructions and data

words whereas a von Neumann computer may use the same memory for storing both instructions and data. Since the processor size of a bit-sliced computer grows in h-bit increments, to preserve parallel data exchanges between the processor and the memory, it is necessary that width of the memory storing data and the width of the processor match. But to preserve software compatibility between all bit-sliced computers of the family, we have to maintain the same instruction size. It follows that the width of the control memory storing instructions should be permanent and independent of the processor's size. Since the processor size of a bit-sliced computer may not coincide with the width of the control memory, two separate memories are usually required, and this raises the cost of a bit-sliced computer.

2. *Simplified Instruction Set.* Because of current pin-count restrictions on an LSI module, a bit-sliced microprogrammed computer inevitably realizes a simplified instruction set, since only a limited number of pins in a module may be assigned to receiving instructions.† This number does not usually exceed half of the total pins on the module since the remaining pins have to be used, for example, for power, synchronization, and data exchanges. Since modern, commonly used modules contain from 40 to 64 pins (inasmuch as costs are disproportionately greater for larger pin counts) the instruction set of a bit-sliced microprogrammed computer is much smaller than that of conventional microprogrammed computers. Because they do not contain many of the instructions that are characteristic of conventional microprogrammed computers, it follows that the instruction sets of bit-sliced computers are simplified.

†Some bit-sliced families increase the size of microinstructions via spreading the amount of pins used to receive them among different module types. Thus, pin count restriction is overcome via fabrication of a new module type. (Ed.)

20. MODULAR COMPUTERS FROM OFF-THE-SHELF MODULES

The market now has a number of sets of different module types, otherwise called the off-the-shelf modules, that can be used for assembling microcomputers, multimicrosystems, and bit-sliced computers. Since the cost of these sets is very low, it is tempting to try to use them for assembling a complex modular computer. The attractiveness of this approach is that it would totally eliminate the stage of module-type fabrication from the manufacturing process, if it was successful.

It is interesting to analyze the results of the LSI modular implementation (on paper) of a powerful mainframe, a Univac 1108 [85]. The designers on this project decided to use the set of Motorola M10800 as a module-type set [86] since, according to their analysis, this unit provided the highest performance figures of all available sets. However, since off-the-shelf modules contain simplified circuits typical of a microprocessor, they could not be used to emulate all the parts of the Univac 1108. They were thus employed only in the ALU, in which each of the three processors was assembled from ten 4-bit LSI modules.

The remaining circuits of the Univac 1108 could not be mapped onto these LSI modules and required fabricating a large number of additional module types. All of these module types had a low coefficient of space utilization. One third of them had no more than 100 gates per chip and the remaining two thirds had at most 10 gates per chip. The reason for such low utilization of the chip area can be explained by the designers' attempt to reduce the number of additional module types they would be forced to produce. On the other hand, the small number of gates per chip qualified the additional modules (that the designers did have to produce) as computer elements of the third rather than the fourth generation.

This experience has shown the difficulties of using off-the-shelf modules for more than in a limited portion of complex computers employing traditional architectures.†

21. LSI TECHNOLOGY AND TRADITIONAL ARCHITECTURES

A major characteristic of the above two approaches towards designing a modular computer is that the designers attempted to map traditional architectures onto LSI modules. However, the current limits of LSI technology resulted in the first approach yielding to simplified modular computers and the second one requiring a large amount of additional logic to interconnect the off-the-shelf LSI modules used.

Let us analyze the problems faced by a computer designer when he or she attempts to map a traditional computer architecture onto LSI modules.‡ Since the problems of mapping a microprogrammed architecture were considered before, a typical von Neumann architecture will now be analyzed. Such an architecture assumes that control over all computer devices is performed by either a separate control unit (synchronous control) or by one central and several local control units (asynchronous control). LSI implementation of either synchronous or asynchronous computer architectures results in the following undesirable consequences:

1. The control unit has to generate all the microcommands for all other computer devices. This requires a large number of intermodule connections. Consequently, a large number of pins in each LSI module must be used for passing microcommands. Transferring microcommands in an encoded form, on the other hand, produces significant delays that slow the computer down considerably.

†See also Chap. II, Sec. 3.1.1.2 that describes another unsuccessful attempt of using off-the-shelf modules for implementation of a complex computer system. (Ed.)

‡The reader is invited also to look into Chap. II, Secs. 3, 4, and 6, that are dedicated to the problems of LSI implementation of the powerful parallel systems. (Ed.)

2. Whenever the control unit is mapped onto several LSI modules (because of pin-count and chip-size limitations) replication of these modules presents a problem, and usually increases the number of different module types. This is especially true for the asynchronous control organization.

3. Because of the pin-count limitation it is already very difficult to implement even a conventional execution device, such as a typical processor. This processor must perform at least 50 microoperations. Constructing this processor from similar hardware slices, each implemented as a single LSI module, requires the assignment of 50 pins in each just to receive microcommands. Thus the use of a 40-pin LSI module does not allow the mapping of one slice onto one LSI module. Likewise, the use of a more expensive 64-pin LSI module leaves only 14 pins for data word transfer, synchronization, and power supply. Therefore, the following options appear as a result of pin-count limitations in the processor module:

Its hardware has to be functionally simplified.

Microcommands must be encoded rather than transferred directly.

The bit size of each processor module must be reduced.

None of these alternatives is acceptable for constructing computers with significant throughput.

Existing technological constraints may be significantly relaxed, however, if new principles are applied to the design of a computer's architecture. One such approach, that eliminates the separate control unit from the computer's architecture is considered in Chap. III of this book. It provides that each LSI module be equipped with its own autonomous local control device which receives all program instructions and enables only those microoperations that are to be executed in that module [87–89].†

†See Chap. III, Secs. 5.2 and 5.3. (Ed.)

22. EVOLUTION OF COMPUTER DESIGN

As can be seen from the above, the opportunities and limitations of LSI technology have encouraged broad changes in traditional design techniques. To obtain a cost-effective design, a designer has to attempt to minimize not the component count, as was undertaken previously, but also the number of different module types used to implement his or her design. The hardware may become cheap only if the architecture is assembled from LSI modules specified by a minimal number of module types and has mostly pin-to-pin connections among modules; i.e., the number of external circuits connecting LSI modules is restricted.

Thus, when evaluating different architectures one has to consider:

1. The number of different module types used in assembling the system.
2. The percentage of circuits mapped onto LSI modules in the overall logical portion of the system.

The evolution of LSI technology indicates that pin-count restrictions per module and the high cost of developing new module types will continue to be the greatest limitations of LSI module utilization in the near future. Component count per module doubles each year; i.e., the complexity of circuits that may be mapped onto a single module grows exponentially, but the number of pins on a module grows at best linearly. Also, nothing indicates that the cost of developing a new module type will decrease significantly, although the cost of mass producing module copies continues to drop. So, LSI technology clearly encourages proliferation of new types of computer architectures assembled from small numbers of different module types which contain many copies of each type and have a restricted number of module interconnections.†

†One such architecture assembled from a universal module type is described in Chap. III, Secs. 5 and 6. (Ed.)

23. ADAPTABLE ARCHITECTURES—PAST AND PRESENT

As was indicated before, LSI technology leads to a proliferation of modular architectures that can establish reconfigurable interconnections among the modules. The consequences of this ability will be new forms of architectural adaptation to algorithms that were impractical in traditional systems. Therefore, a natural effect of architectural modularity is sure to be added momentum towards the further development of adaptable architectures.

Computer architectures may be divided into two categories: *static* and *adaptable*. Static architectures do not adapt via software to the programs being computed, and adaptable architectures do (Fig. I.42).

Adaptable architectures may at present be partitioned into three classes—*microprogrammable*, *reconfigurable*, and *dynamic*—depending on the level of reconfiguration performed.

23.1 Microprogrammable Architectures

The first adaptable architectures appeared in microprogrammable computers [6, 7] where it became possible to reconfigure interconnections between different devices, such as registers, adders, and counters, by software. Its net effect was the improved tuning of the microprograms to the algorithm it was executing. Peculiarities of microprogrammed architectures are considered in Sec. 10.

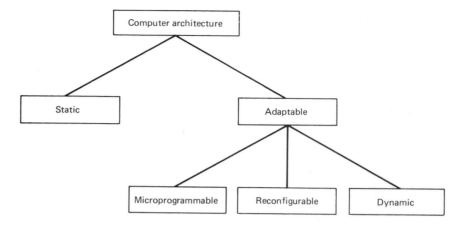

FIGURE I.42 Classification of modern computer architectures

23.2 Reconfigurable Architectures

Further performance improvements have been achieved by introducing reconfigurable interconnections between various functional units, such as processors, memories, and I/O devices. These units began to be called reconfigurable architectures. Although the first reconfigurable computer architectures were created in the early 1960s [15, 91], they were economically unfeasible. The reasons for this were that (1) additional switching circuits which were absent in dedicated interconnections had to be introduced to implement the reconfigurable interconnections, and (2) some hardware units (for instance, control circuits) had to be duplicated.

Thus, before the appearance of LSI modules with high throughput, pure economical considerations were against such architectures. The situation changed rapidly in the middle of the 1970s when the computer industry began to produce modules with high throughput. A relatively large module size allowed the introduction of additional circuits that implemented the reconfigurability of interconnections between modules. Nor did duplication of some circuits in each module lead to a significant cost overhead, provided the modules used were copies of one or a few prototypes.

Software control of interconnections may proceed in the following directions in reconfigurable architectures:

†See Chap. II, Sec. 3, describing several reconfigurable array systems that change the sizes of processors interconnected to an array. (Ed.)

1. It may be used to change the sizes of processors in array systems, giving the system the ability to compute a larger number of data items at the same time, thus enhancing data parallelism [50–57].†

2. Reconfiguration allows any processor to be connected with any memory unit or any other processor in a multiprocessing system. Thus all external communications between different processors by

way of slow I/O devices may be replaced by fast and direct processor–memory, processor–processor, or memory–memory exchanges [44–47].‡

3. In reconfigurable pipeline systems, reconfiguration allows changes to be made in the sequence of operational units (adders, multipliers, subtractors, dividers) connected into the pipeline in order to match the sequence of operations encountered in the program. The consequences of such a match are the minimization of the number of dummy time intervals caused by the disparity between program and pipeline structures [58–65].

4. Software control of interconnections between various computer nodes in a multicomputer system allows different topological configurations in the network to be established—such as star, closely connected graph, hierarchical pyramid, binary tree—depending on the structure of the algorithms being computed. [92–94, 107].

5. Since all faulty modules detected by diagnostic tests may be isolated via software and replaced by spares, software-controlled reconfiguration may lead to enhanced system reliability.†

‡See Chap. III, Sec. 6, describing the organization of such exchanges for dynamic architectures. (Ed.)

†For more information on fault-tolerant reconfiguration see Chap. VI, Secs. 2.1 and 4, where Sec. 2.1 describes various techniques of fault-tolerant design (redundancy, spare modules, and graceful degradation) and Sec. 4 covers the existing state-of-the art in diagnosis of modular systems. (Ed.)

‡This book contains a separate chapter (Chap. III) dedicated to the design and programming of dynamic architectures. (Ed.)

23.3 Dynamic Architectures

The advent of LSI modules with high throughput added momentum to the development of adaptable architectures. It became possible to reconfigure not only interunit but also intermodule connections. As a result, the available hardware resources could be redistributed among programs, leading to an increase in the number of programs computed by the same hardware [73–75, 95–100, 105, 106]. This was accomplished by switching the system into a number of independent computers that matched the number of program streams required. Such architectures were named "dynamic architectures" [66].‡ A modern modular architecture may accomplish all three classes of adaptation mentioned above: microprogrammable, reconfigurable, and dynamic. For instance, an architecture may reconfigure interconnections on a microlevel (registers, adders, conditional flip-flops) and thus perform a microprogrammable adaptation. It may also reconfigure on the level of separate functional units and perform a reconfigurable adaptation. And finally, it may reconfigure on the level of separate modules and perform a dynamic adaptation. One thus obtains microprogrammable, reconfigurable, and dynamic properties, all implemented in a single modular architecture [67].

24. MODERN PROGRESS IN ADAPTABLE ARCHITECTURES

This section will concern itself with a current progress in reconfigurable and dynamic architectures.

24.1 Reconfigurable Supernetworks

The greatest interest in developing LSI technology is currently focused on the further improvement of two types of reconfigurable architecture, array systems and multicomputer systems, each assembled from hundreds or thousands of microcomputers.

For reconfigurable array systems, LSI implementation of each processor via bit slicing makes cost-effective reconfiguration far more feasible than with previous technologies.† Consequently it is now possible to achieve a much better fragmentation of processor resources into different size processors. A single instruction may now operate on different word sizes and lead to a better utilization of the processor resource as a result [53, 57]. Second, a multicomputer system assembled from hundreds or thousands of microcomputers can be transformed into a complex Supernetwork capable of effectively performing various reconfigurations to conform to the structures of complex algorithms [93–96, 107]. Since the major problems of reconfigurable array architectures were considered in Sec. 14.3., and there is a separate chapter in the book dedicated to reconfigurable array systems (Chap. II), we outline only the essential problems of reconfigurable Supernetworks below.

During the past few years several multimicrosystems assembled from a few dozen microcomputers or microprocessors have been described [101, 102]. These are only the first steps towards the creation of reconfigurable Supernetworks containing hundreds or thousands of microcomputers. These networks will be able to switch into various topological configurations (array, star, binary tree, etc.) consonant with the computational structure of an algorithm. Let us show the effectiveness of several of the configurations that such networks may assume.

†See Chap. II dedicated to reconfigurable parallel array systems. (Ed.)

Example 17. Suppose a Supernetwork is faced with a meteorological task requiring that four hundred values of $A = (a_1, a_2, \ldots, a_{400})$ be incremented by b_0, i.e., $A + b_0$; and that four hundred different values of $C = (c_1, c_2, \ldots, c_{400})$ be given an increment d_0, i.e., $C + d_0$. Following this, all the results, $A + b_0$, are to be added to corresponding $C + d_0$ giving four hundred values, $A + b_0 + C + d_0$. If this computation is performed by a computer with a single processor it will require 1200 consecutive additions.

A Supernetwork, on the other hand, may solve this task significantly faster if it reconfigures its resources into two arrays. The first array, containing microcomputers MC-1 to MC-401, computes $A + b_0$ (Fig. I.43). The second array, assembled from microcomputers MC-402 to MC-802, computes $C + d_0$ and then computes $A + b_0 + C + d_0$, receiving computational results from the first array. The entire computation will take the time of two consecutive additions instead of 1200 additions as in the conventional computer. ∎

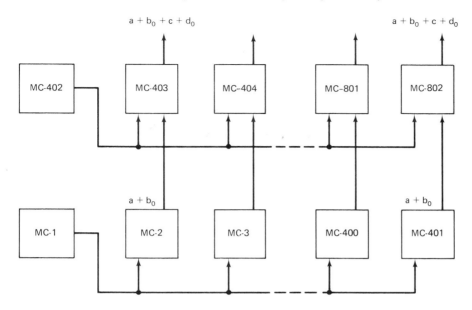

FIGURE I.43 A Supernetwork made up of 800 microcomputers

Example 18. Suppose each of 500 external devices receives simultaneously an address $A_i = B_0 + b_i$ ($i = 1, \ldots, 500$), where B_0 is the base address, and b_i is an incremental value that assumes individual meanings for each I/O device. Suppose the first I/O device requires four different increments during execution. Then b_i will assume the values, say, 02, 17, 31 and 49. Thus if base address $B_0 = 10000$ the following addresses are formed for I/O$_1$: $A_1 = 10002, 10017, 10031, 10049$. The increment b_2 assumes other values for the second I/O device, say, 08, 15, 23 and 41. So for the same base address $B_0 = 10000$, $A_2 = 10008, 10015, 10023, 10041$. And so forth for the remaining 498 I/Os.

To perform this task in minimal time, a Supernetwork has to assume a star configuration (Fig. I.44) in which microcomputer MC-1 computes base address B_0 and concurrently sends it to microcomputers MC-2 through MC-501, each of which computes individual addresses for their respective I/O devices. ■

24.2 Problems of Interconnections in Reconfigurable Supernetworks

As can be seen in the examples above, reconfigurable Supernetworks may speed up computation significantly when compared to conventional computers. The reason for this speed-up is an increase in either the number of data

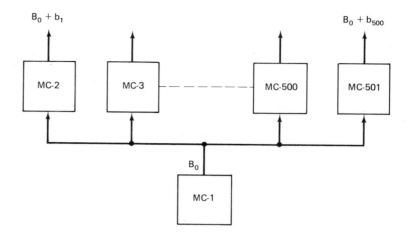

FIGURE I.44 Star configuration in a Supernetwork containing 500 microcomputers

words processed by a single instruction, or the number of instructions handled in parallel, or both.

The future proliferation of Supernetworks is ensured by the extremely low cost and high reliability of the microcomputers that will function as network nodes. In designing such networks, however, the most important problem requiring solution is that of cost-effective organization of the reconfigurable interconnections.

Since one microcomputer in a Supernetwork may need data words computed by any other microcomputer, every pair of microcomputers must be connected by a data bus. If we assume that each microcomputer handles h-bit words, then parallel transfer of an h-bit word from one microcomputer to another will take h lines. Each pair of microcomputers must also have at least two control lines connecting them in order that one member of the pair may send control signals to the other one. It follows that a Supernetwork containing a thousand 16-bit microcomputers will require each microcomputer to have at least $(16 + 2)\ 999 = 17{,}982$ interconnections, where each interconnection takes one pin. It is difficult to imagine an LSI module with this many pins.

To reduce the number of interconnections required for each microcomputer, dedicated connections between microcomputers have to be replaced by shared connections. For instance, suppose that microcomputer MC-1 has to send h-bit words to microcomputer MC-2, MC-3, or MC-4. To organize this transfer, MC-1 will require $3h$ dedicated connections, which will take $3h$ pins in MC-1 (Fig. I.45). The number of pins in MC-1 may be reduced to h, however, if these microcomputers are connected by a connecting element, MSE, containing a logical circuit that connects its input pins, A, with the output pins, B, C, or D (Fig. I.46). Such selectivity in

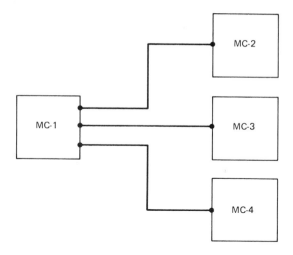

FIGURE I.45 Dedicated connections between microcomputers

connections is accomplished with the use of a special code written in the MSE; i.e., one of the values of the code activates an h-bit connection between pins A and B, or an A–B connection, while another code may activate an A–C connection, etc.

Since connecting elements may contain the same logical circuits, they may be mapped on copies of the same LSI module type. It follows that the cost of connecting elements in the network may be made very low; however, the pin-count restriction in an LSI module will limit the number of microcomputers connected to one MSE.

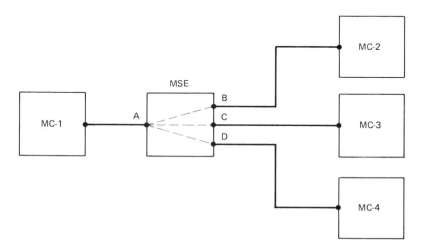

FIGURE I.46 Shared connections between microcomputers

To overcome this restriction, one can follow one of the two alternatives:

- Use networks with a one-level bus, otherwise called *one-stage networks*.
- Use networks with a multilevel bus, otherwise called *multistage networks* [95, 96].

24.3 One-Stage Networks

In this type of network, each microcomputer is connected with each of the other microcomputers through a dedicated connecting element, MSE, having h input and h output pins. Thus, a network containing n microcomputers, will have $n \cdot (n - 1)$ connecting elements. Figure I.47 shows a one-stage network containing four microcomputers and the 12 connecting elements used for establishing interconnections between any pair of microcomputers.

The advantage of this organization is that in order to communicate between any two microcomputers, only one connecting element has to be passed. This introduces a minimal delay into the signal's propagation; each h-bit transfer between microcomputers will be performed in minimal time. On the other hand, this Supernetwork will contain an enormous number of elements since the number of connecting elements approximates n^2, where n is the number of microcomputers. For instance, a Supernetwork incorporating 1000 microcomputers will take $1000 \times 999 = 999,000$ connecting elements, adding an inordinate amount of complexity to the hardware resources.

24.4 Multistage Networks

A pair of microcomputers could communicate through several consecutively connected MSEs in a multistage network. This reduces the overall number of MSEs in the network, but introduces longer delays into each h-bit transfer.

For instance, connecting eight microcomputers into a network requires the three-stage network shown in Fig. I.48. The network uses 12 connecting elements to establish the information paths between any pair of microcomputers. To transfer h-bits from one microcomputer to another, the word has to pass consecutively through a sequence of three MSEs. For example, to connect MC-1 with MC-8, the following path has to be formed: Input 1 of MSE_1 has to be connected with output 5; input 5 of MSE_6 has to be connected with output 7; and input 7 of MSE_{12} has to be connected with output 8. Had this been a one-stage network it would have taken $n \cdot (n - 1) = 8 \cdot 7 = 56$ connecting elements. Thus a reduction in the number of connecting elements was paid for by increased delay in each transfer.

The number of levels in the bus of a multistage network generally increases logarithmically with each increase in the number of microcomputers, i.e., a network of n microcomputers will have $\log_2 n$ levels and each data transfer between a pair of microcomputers will pass through $\log_2 n$

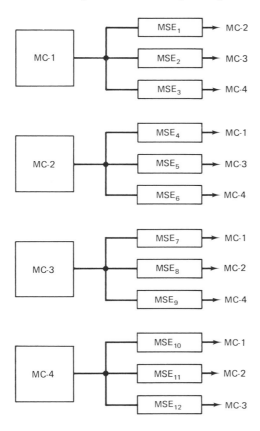

FIGURE I.47 A one-stage network containing four microcomputers

consecutive connecting elements. But this seriously reduces the speed of information processing in the network and is inadmissible in many applications.

24.5 Assessment of Supernetworks

One-stage and multistage Supernetworks are two examples of a classical tradeoff. A high rate of information processing in a one-stage network is traded for a less complex realization in the multistage network.

For those applications for which speed is a major concern† it is expedient to partition the entire network into several one-stage subnetworks in order to reduce the complexity of the resource. Each one-stage subnetwork may then function as a more complex node of the Supernetwork. It would then contain a much smaller number of such nodes, but each node could then be connected internally through a one-level bus. All microcomputers would be completely connected inside each subnetwork whereas data transmission from one sub-

†The reader who is interested in high-speed applications of complex parallel systems is referred to Chap. II, Secs. 1.6, 2.3, and 4. (Ed.)

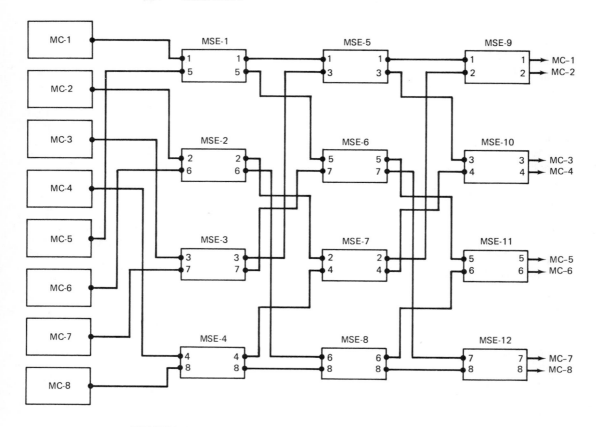

FIGURE I.48 A three-stage network containing eight microcomputers

network to another occurs h-bits at a time. Any one-stage subnetwork should be filled with tasks requiring high interaction, and algorithms with less interaction could run in separate subnetworks. Another peculiarity of Supernetworks is that data processing in the network can only proceed over word sizes not exceeding h bits because networks are assembled from h-bit microcomputers. This is a serious limitation, but it can be overcome if the network is equipped with a dynamic architecture that allows the merging of several nodes (microcomputers) into a larger size computer.

25. DYNAMIC ARCHITECTURES

As was indicated above, dynamic architectures are adaptable architectures that redistribute the available hardware resources into a variable number of computers with different sizes, using software. We will now consider a performance improvement possible with dynamic architectures that until now has been unexploited by existing reconfigurable systems.

Since programs handle different size words, the computer resource may have redundant equipment during computation (the most significant portions of the processor, memory, and I/O). None of the current reconfigurable architectures forms this redundant resource into additional independent computers that could be used to execute new programs. A dynamic architecture, however, permits software controlled reconfiguration of the available resources that form the system into new computers with different sizes; i.e., it creates an additional parallelism on the basis of existing equipment. For instance the system may exist as three 64-bit computers in one configuration; in another, it may switch into six 32-bit computers; and in a third, one 64-bit, one 48-bit, and five 16-bit computers may be formed from the available hardware.

Generally, dynamic architectures produce the following gains in performance.†

†Performance of dynamic architectures is described in Chap. III, Secs. 7–13. (Ed.)

25.1 Execution of Additional Program Streams on the Same Hardware Resources

By switching the hardware resources into the minimal size computers required, the architecture frees redundant resources and makes them into new independent computers. For instance, suppose the system functions as two 64-bit computers that execute two programs requiring not more than 48-bit words each (operands and results of computations). Therefore, there is a redundant resource equivalent to 32 bits of equipment (processor, memory, and I/O) in the system. With a dynamic architecture, however, the two 64-bit computers may reconfigure into two 48-bit computers, for processing the original programs, and an additional 32-bit computer, for a new program stream.

25.2 Speed-Up of Operations

Processor-dependent operations (addition, multiplication, subtraction, conditional branches on $>$, $<$, $\geq$, $\leq$, etc.) may be speeded up because they are executed by minimal size computers. The control unit of each such a computer may then be able to take advantage of this in its timing. Processor-independent operations (boolean, one-bit shifts, conditional branches on $=$, $\neq$, tests of flip-flops, etc.) may also be speeded up because they can be allotted the minimum operation period (for instance, the time for an 8-bit addition) in which to execute.

25.3 Speed-Up of Between-Computer Communications

If a computer needs data stored in the memory of another computer, then these two computers may merge into a single computer that has access to the desired memory units, i.e., a direct memory-processor path is established without the use of I/O channels or the need to construct a reconfigurable processor–memory bus.

25.4 Assignment of Hardware Resources†

To realize performance improvements for a set of concurrent programs, the system has to reconfigure into the sequence of the best architectural states, which take into account the computational specificities of each program. This implies that each program has to be preprocessed in order to find a sequence of minimal size computers that can execute it.‡ Next, the available hardware resources of the multicomputer system have to be distributed among user programs, each of which is executed by the sequence of minimal size computers found earlier. Hardware resource assignment then becomes a matter of finding a flow chart of architectural states that executes the given set of user programs with maximal concurrency [104].

The overall assignment of hardware resources (both processor and memory) is performed by the adaptation system. This system assigns the hardware resources to programs and constructs a flowchart of architectural states. A *monitor* system supervises the correct execution of the flow chart that was constructed by the adaptation system [67, 75].

26. EVOLUTION OF ADAPTABLE ARCHITECTURES

Research in adaptable architectures is heading in the direction of furnishing them with new adaptations that improve the performance of the available resources. The objective of these adaptations is to increase throughput of Supersystems without increasing their complexity, since it is already extremely high.

The computational specifics of some complex real-time algorithms provide computer designers with some ways to increase throughput through architectural adaptations. These algorithms sometimes have portions (tasks) requiring different types of computation: multicomputer, array, or pipeline. For instance, one task is characterized by a large number of parallel instruction streams with little interaction. This task could be computed by a multicomputer system with as many computers as instruction streams. Suppose that the next task of the algorithm employs a great deal of data parallelism; i.e, each program instruction has to handle a data vector of large dimension. This task requires an array system. The next task may require increased speed, which can be accomplished through pipelining.

To process this algorithm in minimal time, it must be computed by three separate systems employing different types of architecture: multicomputer, array, and pipeline. Such a resource would be exceptionally complex, with each of its systems engaged in computation of one task only, and idle during the execution of the remaining tasks.

---

†This resource assignment is adaptive inasmuch as it considers resource requirements of programs on memory and processor resources. Details of this resource assignment are in Chap. III, Secs. 7 to 13. In another chapter of this book (Chap. V), another resource assignment is described: the one that minimizes the number of communication links among different computer modules assigned to execute the same algorithm. (Ed.)

‡See also Chaps. IV and V dedicated to other types of preprocessing: (a) one that determines infinite loops in programs (Chap. IV), and (b) another one that minimizes the number of between computer data exchanges (Chap. V). (Ed.)

To reduce this complexity, the Supersystem should be provided with the ability to reconfigure its resources, via software, into different types of architecture: array, multicomputer, pipeline, or multiprocessor.

The performance of Supersystems, however, depends not only on the types of adaptations that are implemented but also on the quality of each adaptation; namely, on how precise the match between the dynamically created architecture and the algorithm is sustained. For pipeline architectures this means a minimization of dummy intervals created in the pipeline because of the disparity between the structure of the pipeline and of the algorithm. For arrays, the architecture must be capable of partitioning its resources into a variable number of concurrent arrays. Within each array it must be able to change the dimension of each data vector as well as the word size of each processor that computes one component of this vector.

For multiprocessors, optimization means reducing the time spent by any pair of resource units involved in communication; i.e., one must construct fast reconfigurable paths that transmit data words in parallel between any pair of resource units. For multicomputers, the architecture must provide fast parallel exchanges both between a pair of computers and between any pair of functional units from two different computers. Thus, the objectives for optimizing multiprocessors and multicomputers are similar and consist of creating fast reconfigurable busses that support parallel word exchanges between any pair of functional units, either from the same or from different computers.†

†See Chap. III, Secs. 6.3 and 6.4, that describes very fast communications between various functional units from the same or different computers. (Ed.)

CONCLUSIONS

This chapter analyzed the history of architectures for computers and systems. It has shown that the 1950s were the decade that perfected von Neumann and microprogrammed architectures, the 1960s created viable parallel computer systems, and the 1970s gave birth to the new modular architectures assembled from LSI modules. The complexity of the modular architectures may range from one or several LSI modules used to assemble a microcomputer to thousands of LSI modules which are to be included into a powerful parallel system (Supersystem).

Therefore, technological advances in the 1970s have given a new momentum to research and development of new architectures for powerful parallel systems and Supersystems. These architectures are capable of reconfiguring the available resources into variably sized computers (processors); different types of architectures (array, pipeline, multicomputer, or multiprocessor); processors with selectable instruction sets; etc. They allow for the creation of Supersystems with enormous throughput that yet maintain a high level of reliability.

New adaptation properties exhibited by these architectures will lengthen the life cycles of Supersystems since they allow the attainment of higher throughputs from the available resources. This is an extremely important goal in Supersystem design inasmuch as their cost is exceedingly high.

The appearance of Supersystems with adaptable architectures requires the creation of adaptation systems conceived as the software that performs preprocessing of algorithms and determines the best architectures for their execution. Research on adaptation systems may become a milestone in the computer sciences as happened in the past with operating systems, whose appearance in the 1960s was caused by creation of parallel systems.

REFERENCES

1. ARTHUR W. BURKS, HERMAN H. GOLDSTINE, and JOHN VON NEUMANN, "Preliminary Discussion of the Logical Design of an Electronic Computing Instrument" (Pt. I, vol. 1), reprint prepared for U.S. Army Ordnance Dept., 1946, in A. H. Taub (ed.), *Collected Works of John von Neumann,* **5**, pp. 34–79, Macmillan, New York, 1963.
2. HERMAN H. GOLDSTINE, and JOHN VON NEUMANN: "On the Principles of Large Scale Computing Machines," unpublished, 1946; in A. H. TAUB (ed.), *Collected Works of John von Neumann,* **5**, pp. 1–32, Macmillan, New York, 1963.
3. ———: "Planning and Coding Problems for an Electronic Computing Instrument" (Pt. II, vol. 1), reprint prepared for U.S. Army Ordnance Dept., 1947, in A. H. TAUB (ed.), *Collected Works of John von Neumann,* **5**, pp. 80–151, Macmillan, New York, 1963.
4. ———: "Planning and Coding Problems for an Electronic Computing Instrument" (Pt. II, vol. 2), reprint prepared for U.S. Army Ordnance Dept., 1948, in A. H. TAUB (ed.), *Collected Works of John von Neumann,* **5**, pp. 152–214, Macmillan, New York, 1963.
5. ———: "Planning and Coding Problems for an Electronic Computing Instrument" (Pt. II, vol. 3), Rept. prepared for U.S. Army Ordnance Dept., 1948, in A. H. TAUB (ed.), *Collected Works of John von Neumann,* **5**, pp. 215–235, Macmillan, New York, 1963.
6. M. V. WILKES, "The Best Way to Design An Automatic Calculating Machine," *Manchester University Computer Inaugural Conf.*, Ferranti Ltd., London, July 1951.
7. M. V. WILKES, and J. B. STRINGER, "Microprogramming and the Design of the Control Circuits in an Electronic Digital Computer," *Proc. Combridge Phil. Soc.*, Pt. 2, **49**, pp. 230–238, April 1953.
8. JAMES P. ANDERSON, "Program Structures for Parallel Processing," *Comm. ACM*, **8**, *12*, pp. 786–788, December 1965.
9. F. R. BALDWIN, W. B. GIBSON, and C. B. POLAND: "A Multiprocessing Approach to a Large Computer System," *IBM Sys. J.*, **1**, pp. 64–76, September 1962.
10. G. A. BLAAUW, "Multisystem Organization," *IBM Sys. J.*, **3**, *2*, pp. 181–195, 1964.
11. B. BUSSELL, and G. ESTRIN: "An Evaluation of the Effectiveness of Parallel Processing," *IEEE Pacific Computer Conf.*, pp. 201–220, 1963.
12. E. F. CODD, "Multiprogramming," in *Advances in Computers*, **3**, pp. 78–153, Academic Press, New York, 1962.
13. M. E. CONWAY, "A Multiprocessor System Design," *AFIPS Proc. FJCC*, **24**, pp. 139–146, 1963.
14. A. J. CRITCHLOW, "Generalized Multiprocessing and Multiprogramming Systems," *AFIPS Proc. FJCC*, **24**, pp. 107–126, 1963.

15. GERALD ESTRIN, "Organization of Computer Systems, the Fixed Plus Variable Structure Computer," *Proc. WJCC*, pp. 33–40, 1960.
16. S. GILL, "Parallel Programming," *Computer J.*, **1**, *1*, pp. 2–10, April 1958.
17. H. HELLERMAN, "On the Organization of a Multiprogramming-Multiprocessing System," *IBM Res. Rept. RC*-522, 52 pp., Yorktown Hts., N.Y., September 1961.
18. JOHN HOLLAND, "A Universal Computer Capable of Executing an Arbitrary Number of Subprograms Simultaneously," *Proc. EJCC*, pp. 108–113, 1959.
19. J. H. KATZ, "Simulation of a Multiprocessor Computing System," *AFIPS Proc. SJCC*, **28**, pp. 127–139, 1966.
20. M. LEHMAN, "A Survey of Problems and Preliminary Results Concerning Parallel Processing and Parallel Processors," *Proc. IEEE*, **54**, *12*, pp. 1889–1901, December 1966.
21. J. D. MCCULLOUGH, K. H. SPEIERMAN, and F. W. ZURCHER: "Design for a Multiple User Multiprocessing System," *AFIPS Proc. FJCC*, Pt. I, **27**, pp. 611–617, 1965.
22. R. J. MAHER, "Problems of Storage Allocation in a Multiprocessor Multiprogrammed System," *Comm. ACM*, **4**, *10*, pp. 421–422, October 1961.
23. J. C. MURTHA, "Highly Parallel Information Processing Systems," in *Advances in Computers*, vol. 7, pp. 2–116, Academic Press, New York, 1966.
24. LAWRENCE G. ROBERTS, "Multiple Computer Networks and Intercomputer Communication," *ACM Symp. on Operating System Principles, Gatlinburg, Tenn.*, Oct. 1–4, 1967.
25. DANIEL L. SLOTNICK, W. CARL BORCK, and ROBERT C. MCREYNOLDS: The SOLOMON Computer, *AFIPS Proc. FJCC*, **22**, pp. 97–107, 1962.
26. R. V. SMITH, and D. N. SENZIG: "Computer Organization for Array Processing," *IBM Res. Rept. RC* 1330, Yorktown Hts., N.Y., December, 1963.
27. J. S. SQUIRE, and S. M. POLAIS: "Programming and Design Considerations of a Highly Parallel Computer," *AFIPS Proc. SJCC*, **23**, pp. 395–400, 1963.
28. C. STRACHEY, "Time Sharing in Large Fast Computers," *Proc. ICIP, UNESCO*, pp. 336–341, June 1959.
29. S. H. UNGER, "A Computer Oriented Toward Spatial Problems," *Proc. IRE*, **46**, *10*, pp. 1744–1750, October 1958.
30. M. H. WEIK, "A Survey of Domestic Electronic Digital Computing Systems," Ballistic Research Laboratories, Aberdeen, Md., Rept. 971, December 1955.
31. JAMES P. ANDERSON, "Program Structures for Parallel Processing," *Comm. ACM.* **4**, *12*, pp. 786–788, December 1965.
32. C. GORDON BELL, *Computer Structures: Readings and Examples*, McGraw-Hill, New York, 1971.
33. R. E. PORTER, "The RW-400—A New Polymorphic Data System," *Datamation*, **6**, *1*, pp. 8–14, January/February, 1960.
34. GEORGE P. WEST, and RALPH J. KOERNER: "Communications within a Polymorphic Intellectronic System," *Proc. WJCC*, pp. 225–230, 1960.
35. S. ROTHMAN, "R/W 40 Data Processing System," *Intern. Conf. on Information Processing and Auto-math 1959*, Ramo-Wooldridge, Div. of Thompson Ramo Wooldridge, Inc., Los Angeles, Cal, June 1959.
36. A. BHUSHAN, R. H. STOTZ, and J. E. WARD: "Recommendations for an Intercomputer Communications Network for M.I.T.," Memorandum MAC-M-355, July 1967.
37. D. W. DAVIES, K. A. BARTLETT, R. A. SCANTLEBURY, and P. T. WILKINSON: "A Digital Communication Network for Computers Giving Rapid Response at Remote Terminals," *ACM Symp. on Operating System Principles, Gatlinburg, Tenn.* Oct. 1–4, 1967.
38. W. R. PLUGGE, and M. N. PERRY: "American Airlines' 'SABRE' Electronic Reservations System," *Proc. WJCC*, pp. 593–602, May, 1961.

39. J. E. ROBERTSON, "A New Class of Digital Division Methods," *IRE Trans.*, **EC-7**, *3*, pp. 218–222, September 1958.
40. R. J. SEGAL, and H. P. GUERBER: "Four Advanced Computers—Key to Air Force Digital Data Communication System," *AFIPS Proc. EJCC*, **20**, pp. 264–278, 1961.
41. R. G. DAVIS, and S. ZUCKER, "Structure of a Multiprocessor Using Microprogrammable Building Blocks," *Proc. of National Aerospace Electronics Conference*, IEEE Press, Dayton, Oh, pp. 186–200, 1971.
42. R. L. DAVIS, S. ZUCKER and C. M. CAMPBELL, "The Building Block Approach to Multiprocessing," in *AFIPS 1972 Spring Joint Computer Conf.*, AFIPS Press, pp. 685–703, 1972.
43. E. W. REIGEL, D. A. FISHER, V. FABER, "The Interpreter—A Microprogrammable Processor," *AFIPS Conference Proceedings*, AFIPS Press, **40**, pp. 705–723, 1972.
44. P. H. ENSLOW, JR., "Multiprocessor, Organiztion—A Survey," *ACM Computing Surveys*, **9**, *1*, pp. 103–129, March 1977.
45. ———, JR. (Ed.), *Multiprocessors and Parallel Processing*, John Wiley & Sons, New York, 1974.
46. R. J. SWAN, S. H. FULLER, and D. P. SIEWIOREK, "CM*—A Modular, Multi-Microprocessor," *AFIPS Conference Proceedings*, AFIPS Press, **46**, pp. 637–643.
47. R. J. SWAN, A. BECHTOLSHEIM, K. LAI, J. OUSTERHOUT, "The Implementation of the Cm* Multi-Microprocessor," *AFIPS Conference Proceedings*, AFIPS Press, **46**, pp. 645–655, 1977.
48. J. J. PARISER and H. E. MAURER, "Implementation of the NASA Modular Computer with LSI Functional Characters," *AFIPS Conference Proceedings*, **35**, AFIPS Press, pp. 231–245, 1969.
49. G. A. ANDERSON and E. D. JENSEN, "Computer Interconnection Structures, Taxonomy, Characteristics and Examples," *ACM Computing Surveys*, **7**, *4*, pp. 197–213, December 1975.
50. J. L. BAER, "Multiprocessing Systems," *IEEE Trans. on Computers*, **C-25**, pp. 1271–1277, December 1976.
51. D. SLOTNICK, W. BORCK and R. MCREYNOLDS, "The SOLOMON Computer," *AFIPS Proc. FJCC*, **22**, pp. 97–107, 1962.
52. J. GREGORY and R. MCREYNOLDS, "The SOLOMON Computer," *IEEE Trans. Electronic Computers*, **EC-12**, *6*, pp. 774–781, December 1963.
53. K. BATCHER, "STARAN Parallel Processor System Hardware" in 1974 NCC, *AFIPS Conf. Proc.*, **43**, pp. 405–410.
54. ———, "The Multidimensional Access Memory in STARAN," *IEEE Trans. on Computers*, **C-26**, pp. 174–177, February 1977.
55. S. REDDAWAY, "DAP—A Distributed Array Processor," *Proc. of the 1st Symposium on Computer Architecture*, pp. 61–72, 1973.
56. G. BARNES, R. BROWN, M. KATO, D. KUCK, D. SLOTNICK, R. STOKES, "The Illiac IV Computer," *IEEE Trans. on Computers*, **C-17**, pp. 746–757, August 1968.
57. Y. OKAGA, H. TAJIMA, R. MORI, "A Novel Multiprocessor Array," *Proc. 2nd Euromicro Symposium on Microprocessing and Microprogramming*," pp. 83–90, Venice, Italy, 1976.
58. C. V. RAMAMOORTHY and H. F. LI, "Pipeline Architecture," *ACM Computing Surveys*, **9**, *1*, pp. 61–102, March 1977.
59. M. J. IRWIN, "Reconfigurable Pipeline Systems," *Proceedings 1978 ACM Annual Conference*, **1**, pp. 86–92.
60. R. N. IBBETT and P. C. CAPON, "The Development of the MU5 Computer System," *Communications of the ACM*, **21**, *1*, pp. 13–24, January 1978.
61. W. J. WATSON, "The TI ASC—A Highly Modular and Flexible Super Computer Architecture," In *AFIPS 1972 Fall Joint Computer Conf.*, AFIPS Press, pp. 221–228, 1972.

62. R. M. RUSSELL, "The CRAY-1 Computer System," *Communications ACM*, **21**, pp. 63–72, January 1978.
63. D. W. ANDERSON, F. J. SPARACIO, and R. M. TOMASULO, "IBM System 360 Model 61, Machine Philosophy and Instruction Handling," *IBM Journal of Research and Development*, pp. 8–24, January 1967.
64. S. S. REDDI and E. A. FEUSTEL, "A Restructurable Computer System," *IEEE Trans. on Computers*, **C-27**, *1*, pp. 1–20, January 1978.
65. A. THOMASIAN and A. AVIZIENIS, "A Design Study of a Shared-Resource Computer System," *Proceedings of the Third International Symposium on Computer Architecture*, pp. 105–111, 1976.
66. S. I. KARTASHEV and S. P. KARTASHEV, "LSI Modular Computers, Systems and Networks," *Computer*, **11**, pp. 7–15, July 1978.
67. S. I. KARTASHEV, S. P. KARTASHEV, and C. V. RAMAMOORTHY, "Adaptation Properties for Dynamic Architectures," 1979 National Computer Conference, *AFIPS Conference Proceedings*, AFIPS Press, **48**, pp. 543–556, 1979.
68. C. R. VICK, "Research and Development in Computer Technology, How Do We Follow the Last Act?" Keynote Speech, *Proc. of Intel. Conf. on Parallel Processing*, pp. 1–5, 1978.
69. C. G. DAVIS and C. R. VICK, "The Software Development System: Status and Evolution," *Proc. Conf. on Computer Software and Applications (CompSac)*, pp. 326–331, 1978.
70. C. R. VICK, J. E. SCALF and W. C. MCDONALD, "Distributed Data Processing for Real-time Applications," *Proc. Sixth Texas Conf. on Computing Systems* pp. 174–191, 1977.
71. W. C. MCDONALD and J. M. WILLIAMS, "The Advanced Data Processing Testbed," Proc. Conf. on *Computer Software and Applications (CompSac)*, pp. 346–351, 1978.
72. H. FITZGIBBON, B. BUCKLES and J. SCALF, "Distributed Data Processing Design Evaluation Through Emulation," *Proc. Conf. on Computer Software and Applications (CompSac)*, pp. 364–369, 1978.
73. S. I. KARTASHEV and S. P. KARTASHEV, "A Multicomputer System with Software Reconfiguration of the Architecture," *Proc. Eighth Intl. Conf. on Computer Performance SIGMETRICS CMG VIII*, Washington, D.C., pp. 271–286, 1977.
74. ———, "Dynamic Architectures: Problems and Solutions," *Computer*, **11**, pp. 26–40, July 1978.
75. ———, "Multicomputer systems with Dynamic Architecture," *IEEE Trans. on Computers*, **C-28**, *10*, pp. 704–721, October 1979.
76. A. DURNIAK, "IBM Has A Message: The 4300," *Electronics*, **52**, *4*, pp. 85–86, 1979.
77. A DURNIAK, "VLSE Shakes the Foundations of Computer Architecture," *Electronics*, **52**, *11*, pp. 111–133, May 1979.
78. T. WELCH and S. S. PATIL, "An Approach to Using VLSE in Digital Systems," *Proc. 5th Annual Symposium on Computer Architecture*, pp. 139–143, 1978.
79. A. L. DAVIS, "A Data Flow Evaluation System Based on the Concept of Recursive Locality," *1979 National Computer AFIPS Conf. Proc*. **48**, AFIPS Press, pp. 1079–1088, 1979.
80. B. R. BORGERSON, "The Viability of Multimicroprocessor Systems," *Computer*, **9**, pp. 26–30, January 1976.
81. *AM 2901 AM 2909 Technical Data Manual*, Advanced Micro Devices Inc., Sunnyvale, Cal, 1975.
82. *Intel Series 3000 Microprogramming Manual*, Intel Corp., Santa Clara, Cal, 1975.
83. *990 Computer Family Systems Handbook*, Manual 945250-9701 Texas Instruments, Inc., Austin, Texas, 1975.
84. N. A. ALEXANDRIDIS, "Bit-sliced Microprocessor Architecture, *Computer*, **11**, pp. 56–79, June 1978.

85. B. R. BORGERSON, G. S. TJADEN and M. L. HANSON, "Mainframe Implementation with Off-the-shelf LSI Modules," *Computer*, **11**, pp. 42–48, July 1978.
86. *MC 10800 Product Description*, Motorola Semiconductors, Phoenix, Arizona.
87. S. I. KARTASHEV and S. P. KARTASHEV, "A Microprocessor with Modular Control as a Universal Building Block for Complex Computers," *Proc. 3rd Euromicro Symposium on Microprocessing and Microprogramming*, pp. 210–216, Amsterdam, The Netherlands, 1977.
88. ———, "Designing LSI Modular Computers and Systems," *Intl. J. of Mini- and Microcomputers*, **1**, *1*, pp. 14–24, 1978.
89. ———, "Selection of the Control Organization for a Multicomputer System with Dynamic Architecture," *Proc. 4th Euromicro Symposium on Microprocessing and Microprogramming*, Munich, Germany, pp. 346–357, 1978.
90. M. V WILKES, "The Best Way to Design an Automatic Calculating Machine," *Manchester Univ. Computer Inaugural Conf.*, Ferranti Ltd., London, England, July 1951.
91. G. ESTRIN, "Parallel Processing in a Restructurable Computer System," *IEEE Trans. on Electronic Computers*, **EC-12**, pp. 747–755, July 1963.
92. G. A. ANDERSON and E. D. JENSEN, "Computer Interconnection Structures, Taxonomy, Characteristics, and Examples," *ACM Computing Surveys*, **7**, *4*, pp. 197–213, December 1975.
93. Y. PAXER and M. BOZYIGIT, "Variable Topology Multicomputer," *Proc. 2nd Euromicro Symposium on Microprocessing and Microprogramming*, pp. 141–149, Venice, Italy, 1976.
94. M. C. PEASE, "The Indirect Binary *N*-cube Microprocessor Array," *IEEE Trans. on Computers*, **C-26**, *5*, pp. 458–473, May 1977.
95. H. J. SIEGEL, R. J. McMILLEN and P. T. MUELLER, JR., "A Survey of Interconnection Methods for Reconfigurable Parallel Processing Systems," *Proc. Conf. NCC AFIPS*, AFIPS Press, **48**, pp. 529–542, 1979.
96. H. J. SIEGEL, "Interconnection Networks for SIMD Machines," *Computer*, **12**, *6*, pp. 57–65, June 1979.
97. R. E. MILLER and J. COCKE, "Configurable Computers: A New Class of General Purpose Machines," In *Intl. Symposium on Theoretical Programming*, Lecture Notes in *Computer Science*, **5**, Springer-Verlag, Berlin, Germany, 1974.
98. S. I. KARTASHEV and S. P. KARTASHEV, "Designing LSI Metacomputer Systems with Dynamic Architecture," DCA Association, Lincoln, Nebraska, 1974.
99. S. S. REDDI and E. A. FEUSTEL, "An Approach to Restructurable Computer Systems," In *Parallel Processing* (lecture notes in *Computer Science*), **24**, pp. 319–337, Springer-Verlag, Berlin, Germany, 1975.
100. G. J. LIPOVSKI and A. TRIPATHI, "A Reconfigurable Varistructured Array Processor," *Proc. Intl. Conf. on Parallel Processing*, pp. 165–174, 1977.
101. R. MOREI et al, "Microcomputer Applications in Japan," *Computer*, **12**, pp. 64–74, May 1979.
102. W. L. SPETZ, "Microprocessor Networks," *Computer*, **10**, pp. 64–70, July 1977.
103. L. E. SHAR and E. S. DAVIDSON, "A Multiminiprocessor System Implemented Through Pipelining," *Computer*, **7**, pp. 42–51, February 1974.
104. S. I. KARTASHEV and S. P. KARTASHEV, "Software Problems for Dynamic Architecture: Adaptive Assignment of Hardware Resources," *Proc. Conf. on Computer Software and Applications (CompSac)*, pp. 775–780, 1978.
105. ———, "The Evolution in Dynamic Architectures," *Microprocessors and Microsystems*, July 1979.
106. S. P. KARTASHEV and S. I. KARTASHEV, "Adaptable Pipeline System with Dynamic Architecture," *Proc. Intl. Conf. on Parallel Processing*, 1979.
107. L. D. WITTIE, "Efficient Message Routing in Megamicrocomputer Networks," *3rd Annual Symposium on Computer Architecture*, 1976, 136–140, 1976.

Chapter II

Reconfigurable Parallel Array Systems

Hubert H. Love

PREVIEW

This chapter is intended for the working engineer or prospective engineer who wishes to gain an understanding of the nature and applicability of reconfigurable parallel array systems, or who merely wishes to obtain a few ideas concerning parallel hardware/software organizations that can be applied to more conventional system design tasks. The chapter is based on the author's own experience in system design and programming, and therefore pays special attention to associative processors, which represent one type of reconfigurable processor organization, and with which the author is most familiar.

The first section is an introduction to reconfigurable parallel array systems. The next section of the chapter gives a brief background on parallel array processors. It discusses some of the advantages to be gained from building and using these machines. Last, it gives a number of general criteria for determining the suitability of a given application to a parallel array system design.

The third section describes several representative types of parallel array designs giving, for each type: First, a simplified description of the organization; second, the author's opinion as to the nature of the applications best suited for the type; and third, a detailed description of an existing processor of that type. This section is intended to give the reader a feeling for the various tradeoffs between the complexity of the individual modules of the array, the number of modules, the complexity of the communication channels, the difficulty of programming the system, and the processing throughput.

The fourth section is a case study of an actual system design effort by the author and others. The starting point of the effort, and the end results that were desired and that were actually achieved are described. The progress of the effort is the substance of the discussion. The order in which the various phases of the design were undertaken is described, together with the justification for taking that order. Similarly, the tradeoffs between various approaches to system design are described, and justification is given for that which was finally adopted. It is hoped that this section will give the prospective system engineer an insight into the way in which system design tasks of this type are often conducted in industry, the restrictions under which an engineer must perform his work, and the tools and methodology which they have available to them. In addition to this, a number of specific design features applicable to associative processors are described which may be useful in future design efforts.

The fifth section of the chapter concerns the task of programming reconfigurable parallel array processors. The section first discusses the difference in the design, coding, and debugging phases of the programming task for parallel machines in general, as opposed to conventional machines. Then the particular problems and solutions associated with the programming of the Hughes ALAP, an associative processor with which the author has extensive experience, are described in more detail. The intention of the section is to give the system engineer enough of an understanding of programming tradeoffs so that he or she will not make serious errors in judging the applicability of the system from the programming standpoint. A lack of understanding of this sort has led to serious errors of judgment in the past.

The sixth section of the chapter discusses expected future trends in the design of reconfigurable parallel array systems. Limitations of current designs, and design aspects of greatest importance and greatest potential benefit as seen by the author are named.

The final section of the chapter contains references to a number of relevant books and papers, together with a brief guide to them.

RECONFIGURABLE PARALLEL ARRAY SYSTEMS

1. INTRODUCTION

When the computer systems engineer is given the task of designing an architecture for some unusual or unprecedented application, he or she may find that system designs using conventional processors are inadequate for handling the required data throughput rate, and perhaps can be expected to have an unacceptably high probability of hardware failure as well. Or it may be that he or she is seeking a solution which will permit taking advantage of the size, space, and cost savings that can be expected from the newly-emerging VLSI (Very Large Scale Integration) technology. In fact, it may be that the application has inherent parallelism of some sort, either in the input data format or in the processes for transforming the data.

In such cases, the engineer may be motivated to consider a system design which uses many similar logic modules to process the data simultaneously, and to redefine the input data organization and processing algorithms to suit the new architecture. This chapter discusses a general class of parallel computer system configurations, termed *reconfigurable parallel array systems*, that are being increasingly employed as solutions to difficult system design problems.

1.1 Classification of Parallel Architectures

Throughout this chapter, the term *data-sequential processor*, or simply *sequential processor* shall denote a computer processor which executes a single instruction at a time on a single operand or pair of operands. Such a processor may operate bit-serial or bit-parallel on each data item, depending on its design.†

Parallel processor systems, in contrast, have multiple processing units. These systems operate on many data items simultaneously. There are a number of ways in which parallel systems may be classified. One way that is widely used will be employed throughout this chapter. This classification is the following:

1. *Pipeline systems*. These systems have processors in which the instruction-processing logic is segmented in such a manner that several instructions can be processed simultaneously in an overlapped fashion. That is, each logic segment performs a different phase of the instruction execution process on a different instruction, with all segments operating simultaneously. The phases will typically be instruction fetch, instruction decode, operand fetch, execution, and result store. A well-designed pipeline processor can complete the

†The principle of sequentiality of computations for both traditional architectures, von Neumann and microprogrammed, is discussed in Chap. I, Part I, Sec. 2. (Ed.)

execution of an instruction almost every clock cycle, because of the efficiency of the pipeline process.

2. *Multicomputer systems*. These are systems which contain one or more computers. Generally, each computer contains processors, memory, and peripheral devices. The computers operate simultaneously, processing different tasks or different parts of the same task. System control may be centralized or distributed.

3. *Multiprocessor systems*. These are multicomputer systems in which the processors operate under integrated control. The processors share a common memory and common I/O devices. In general, many processors will operate on different parts of the same task most of the time.

4. *Parallel array systems*. These are multiprocessor systems in which the processors are organized into arrays by means of data communication channels which interconnect them. This chapter is concerned with an important subclassification of parallel array systems.

5. *Dynamically reconfigurable systems*. These system architectures are a more recent development. They can be reconfigured under software control into combinations of all the first four systems. The reconfigurability includes dynamically varying word lengths, as well as dynamically varying array structures and sizes.

1.2 Parallel Array Systems

The basic concept behind reconfigurable parallel array systems is that of the parallel array of processors. A *parallel array system* is a processing system containing a number of identical processing elements (called "PEs"). The system executes one instruction at a time under central control, like a conventional processor. However, unlike a conventional machine, the parallel array system processes a data vector with each instruction, with each element of the vector processed by a different PE. All PEs operate simultaneously. In many parallel array systems, the PEs are interconnected via communication channels in some regular fashion. This permits the simultaneous transfer of data among the PEs during instruction execution. All parallel array systems permit the simultaneous loading of all PEs with common data items.

Figure II.1 is the block diagram of a simple parallel array processor. Each PE is capable of processing single data elements, and contains a small local memory for data and working storage. A control unit is illustrated.† It has a local memory in which the system's program resides. It processes the

---

†Throughout this chapter, the terms "control unit," "controller," and "control console" are used as names for devices which exercise central control over some phase or phases of system operation. The various terms are those used by the system designers in every case.

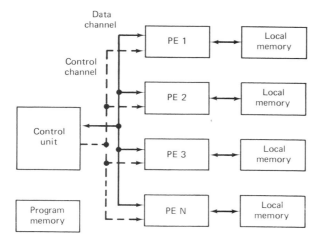

FIGURE II.1 A simple parallel array processor

instructions from this program sequentially, decoding them, and sending out the corresponding control signals to the PEs. The control channels for these signals are shown in the figure by dotted lines. This simple processor contains no intercell communication network. Instead, there is a common bus connecting all PEs to the control unit. The control unit can load common data items into all PEs simultaneously using this bus. The PEs can return status information over this bus to the control unit to assist it in making decisions (e.g., as to whether or not to branch in the program).

1.3 Applications of Parallel Array Systems

Simple parallel systems like the one just described are of limited applicability, since all PEs must execute every instruction, and are unable to communicate with one another except via the control unit. However, where they are applicable, they can give a great performance advantage over conventional architectures because of the simultaneous operation of the PEs. This advantage can be on the order of hundreds or even thousands if correspondingly large numbers of PEs can be effectively employed.

A simple yet useful application of the system shown in Fig. II.1 is that of parallel dictionary lookup. Figure II.2 illustrates the partitioning of the dictionary data among the PEs. Each PE contains a single dictionary entry. This consists of an alphanumeric identifier, which may occupy several contiguous words of memory, and the fixed-length (binary) numeric code that corresponds to the identifier. It will be noticed that the dictionary entries do not need to be sorted. If the array processor is given an identifier, it can retrieve the corresponding numeric code from the PE containing it, executing

104 CHAPTER II

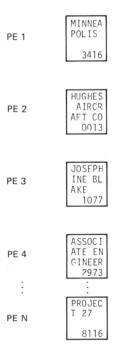

FIGURE II.2 Partitioning of data for dictionary lookup application

only a few parallel instructions. Similarly, if the processor is given a code, it can retrieve the corresponding alphanumeric identifier.

A parallel array processor with a dictionary stored in this fashion can be used as a peripheral device for a large conventional data processing computer hosting a query/response data management system. In such a system, the data base is stored in the host processor, and is organized in terms of the fixed length codes representing the items of interest. If there are many users on line, and if the query/response process is complicated enough to require a very high system throughput, the parallel dictionary-lookup capability furnished by the array processor can save great time in translating queries and responses for the system between the alphanumeric identifiers that represent the items of interest to the users and the codes that represent them internally. The dictionary entries shown in the figure might correspond to a data base dealing with employees, the projects to which they are assigned, the companies which employ them, and the cities in which the companies are located.

In translating from the alphanumeric identifiers to their fixed-length codes, the sequence of instructions executed by the parallel array processor is not unlike that which would be used for a single conventional processor, except that all instructions for the array are executed by all PEs. The given identifier, supplied a word at a time to all PEs via a common bus, is compared with all dictionary entries simultaneously. That PE (assumed to be unique) which contains the matching entry puts the corresponding numeric code in its output buffer. All other PEs put a string of binary zeroes in their output buffers. Finally, the logical OR of the buffers of all PEs is output from the array over the common bus through the control unit to the host processor (which is the conventional data processing system).

The improvement in processing time for this operation over that for a single conventional processor is about $n/2$, where n is the number of dictionary entries (assuming that the entries in the conventional processor are sorted).

1.4 Reconfigurability in Parallel Systems

Limitations in parallel systems such as the one just illustrated are apparent. Their applicability is limited because of the fact that all PEs must execute every instruction. For example, the simultaneous communication of data among the PEs, when it is implemented, will always be very special-purpose in operation, since data will always be communicated from every PE having a channel. The concept of *reconfigurability* is central to overcoming such limitations and in broadening the area of application of parallel systems.

A *reconfigurable parallel array system* is a parallel array processing system in which both the dimensionality of the data vector being processed and the length of the vector elements themselves can be varied from instruction to instruction under software control. The key to achieving reconfigurability is to permit the individual PEs to determine from their internal states whether or not to participate in each instruction execution.

Figure II.3 is a simplified block diagram of a reconfigurable parallel array processor. Sixteen processing elements are shown, interconnected by a two-dimensional communication network. A control unit for overall system control is illustrated. It decodes instructions for the system and feeds the resulting control information to the processor array. The control channels from the control unit to all of the PEs are shown as dotted lines. The data communication channels are shown as solid lines. The multiple input and output lines to the PEs are independent of the control unit with regard to data flow, although they are under its control. The independence of data flow is necessary so that high-speed parallel data can be handled.

The reconfigurability of the system shown in Fig. II.3 comes from the fact that the control of the operation of the individual PE is partly determined by the state of the individual PE itself, rather than being determined entirely by the global control signals from the control unit. That is, each PE has internal toggles which determine such things as whether or not the PE will execute the current instruction, and which instruction options will be executed if the PE does execute the instruction.

106 CHAPTER II

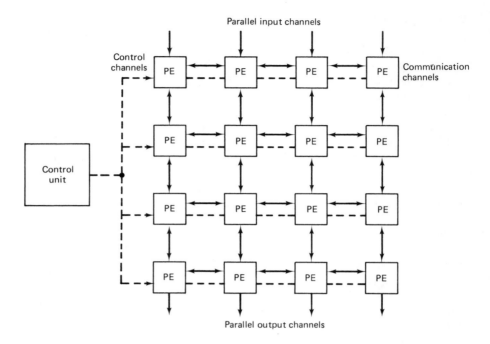

FIGURE II.3 Typical reconfigurable parallel array processor

1.5 An Example of Array Operation

When the algorithms for the application can be so structured that many data items of a similar nature can be processed simultaneously most of the time, the principal criterion for the applicability of parallel processors is met. The importance of reconfigurability in the system in meeting this criterion is illustrated in the following example.

Consider the solution by relaxation of the heat equation on a flat plate. This solution is accomplished as follows: First, the plate is represented by a grid of equally-spaced points. The solution of the equation at those points is then to be calculated. This is done by assigning initial temperatures (assumed in this example to be constants) to the points representing the boundary of the plane, and then repeatedly evaluating a function, called the *relaxation function* at all other points (i.e., the *internal points*) until the changes in temperature at all internal points between successive evaluations fall within a specified pair of limits.

A commonly used relaxation function is

$$x_{ij}^{k+1} = (x_{i-1,j}^k + x_{i+1,j}^k + x_{i,j-1}^k + x_{i,j+1}^k)/4 \qquad (1)$$

where k is the index over the iterations, x is the temperature, and i and j are the row and column indices, respectively, of the grid points. Using a reconfigurable array parallel processor, such as the one shown in Fig. II.3, this function can be evaluated at all internal points simultaneously for each iteration.

To accomplish this, each grid point is assigned to a processing element so that the rows and columns of PEs correspond to the rows and columns of the grid. This layout is illustrated in Fig. II.4, which shows part of the array of PEs with their assigned grid points. Since the system is reconfigurable, it is possible to select the PEs involved in each instruction. Thus the function can be prevented from being evaluated at the boundary points.

Only a handful of instructions is required for each iteration. The first of these transfers the temperature value at each grid point to the next lower PE in the array (excluding the PEs for the lower boundary). The communication channels are used in the transfer. The next instruction transfers the values at each PE to the PE to the left, at the same time adding these values to the values just transferred to the destination PEs (again excluding the appropriate boundary points). The next two instructions accomplish the same function for the upper and lefthand PEs, respectively, thus evaluating the sum on the righthand side of Eq. (1). Following this, the sums at all internal points are divided simultaneously by the common constant, 4, which is input via the common data bus from the controller. Last, the quotients at all internal PEs replace the original temperature values, thus completing the iteration. The function or functions for the boundary points are then evaluated (also in parallel) before proceeding to the next iteration.

1.6 Applications for Reconfigurable Parallel Array Systems

The advantages in processing throughput gained both from the parallel operation of systems like the one just described and from the reconfigurability of the systems is apparent from the example just described. For such systems,

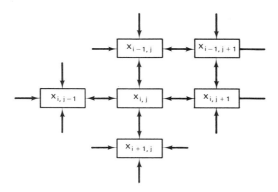

FIGURE II.4 Assignment of grid points to processing elements

there are a number of well-suited applications, all requiring throughput in a number of instances quite beyond the capability of conventional or pipeline systems employing single processing units. Among these applications are:

1. Advanced query-processing and question-answering systems, especially when the data base is very large, and is complex and dynamically varying, and when the query processing involves a large number of search operations (e.g., deductive and inductive inference).
2. Real-time text processing, again when the data base is very large, and when the processing is complex and fast response time is essential.
3. Applications dealing with large matrices: weather prediction, determination of neutron flux densities in nuclear reactors, etc. The process for the example previously described for Eq. (1) is similar to the last of these, although much less complicated.
4. Real-time radar data processing: track-while-scan, pulse interleaving.
5. Real-time data compression: very high data rates and complex compression algorithms.
6. Image processing in general.

1.7 References

Kuck [13] and Baer [23] are excellent surveys on parallel processors in general. Ramamoorthy [14] is a similarly valuable reference on pipeline processors. Enslow [15] is an excellent general reference on multiprocessing systems. In addition, [13] and [14] contain comprehensive bibliographies for those interested in further information on the respective topics. Dynamically reconfigurable architectures are described elsewhere in this book, and are discussed in Kartashev [26], [27], and [28].

2. BACKGROUND, GENERAL BENEFITS, AND APPLICABILITY OF RECONFIGURABLE PARALLEL ARRAY SYSTEMS

2.1 Background

Various reconfigurable parallel array processor designs have been in existence since the late 1950s,† and several actual systems were constructed during the 1960s.‡ The general acceptance of these systems, however, was hindered for many years by several factors:

---

†For example, the Holland machine, see Holland [6].
‡See, for example, Slotnick [7] and Barnes [8].

1. Array machines are unconventional and complicated, and were therefore both difficult and expensive to design and construct.
2. The expense of building parallel array systems made it necessary that strong justification exist for their development. For example, the intended application for such a system must be both highly important and beyond the capability of more conventional approaches.
3. There was insufficient study of applications for parallel array systems in conjunction with their design. As a result, most systems which have been designed and built have been too limited in their range of applications.

Starting in the late 1970s, however, a number of factors have caused greatly increased interest in parallel array systems. The most important of these are:

1. For reasons to be given shortly, parallel array systems are particularly amenable to VLSI fabrication. Thus the cost of these systems can be greatly reduced.
2. Several systems have been successfully applied to problems of importance,[†] thus demonstrating with actual systems the advantages inherent in highly parallel processing.
3. There is increased interest in those fields of application for which highly parallel techniques are most suitable. Among these are the aforementioned radar data processing[‡] and real-time image processing.[*]

2.2 Benefits in Building and Using Parallel Systems

When they are properly designed and employed, reconfigurable parallel array systems yield a number of benefits as compared with conventional systems. These benefits are realized both in the cost of design and fabrication, and in the performance of the resulting system. The next two subsections discuss the principal benefits of each of these two kinds.

2.2.1 Design and Fabrication

Array architectures are complicated, and usually have a higher gate count in proportion to their throughput than is the case for more conventional systems. However, there are certain characteristics of arrays which give

---

[†]See Boulis [16] and Berra [17].
[‡]See Love [18] and Hiner [19].
[*]See reference Vocar [20].

**The effect of LSI technology on design of complex parallel systems in general (not only arrays) is discussed in Chap. I, Secs. 18 to 23. (Ed.)

†Different techniques for software-controlled fault isolation with and without spare modules are discussed in Chap. VI, Sec. 2.1 (Ed.)

advantages in design and fabrication costs not found with other types of systems. With the advent of very large scale integration (VLSI), these characteristics, along with others to be named, work to the advantage of the system designer in a number of respects.**

First, most of the logic in a parallel array system, usually more than 90 percent, is the array itself. The array consists of repeated logical circuits (the PEs), together with a regular interconnection pattern (the communication channel network). Depending on their complexity, a number of these logical elements can be fabricated on a single LSI or VLSI wafer. This can mean a higher yield if discretionary wiring or other techniques can be used to disconnect faulty elements during the manufacturing process. Moreover, with proper circuitry, it may be possible to implement software-controlled fault isolation† and fault repair right on the wafer. Indeed, this has already been done with the ALAP, to be described.‡ Built-in fault isolation is of considerable benefit in some systems (e.g., those which must operate in space or in some other environment in which ordinary repair techniques are difficult or impossible). Moreover, such fault isolation will be of increasing benefit in all systems as growing system complexity increases the probability of run-time failure. The advantage of repeated circuit elements from the aspect of fault tolerance results from the fact that single failures on a wafer, or even multiple failures do not render the wafer worthless if the wafer contains spare elements that can be switched into the system under software control. Without this capability, complete spare wafers must be included in the system to give the same degree of fault tolerance. The switching reliability problems would therefore be increased, as well as size and cost for the entire system.

There are two additional cost benefits gained from the use of VLSI fabrication of parallel arrays. First, the number of different types of wafers is small for the complexity of the system, as compared with conventional systems. Furthermore, the number of wafer types does not increase as the parallel array system grows, since the system is made more powerful by simply increasing the size of the array.*

Second, the complexity of interconnections between wafers is low as compared with conventional systems of equivalent complexity. This is again the result of the array characteristics, and the fact that the number of pins on the wafer increases at a much lower rate than the number of gates as the number of PEs on the wafer increases.

Several additional advantages of array architectures from the standpoint of VLSI are discussed in Sec. 6.1. These have to do with the very high gate speeds expected from the new technologies, and the fact that the use of array architectures simplifies several of the problems encountered when circuitry is fabricated with these technologies.

‡See Sec. 3.2.3 and Finnila [9].

*This will be apparent from the description of array architectures in Sec. 3, as well as the general discussion of VLSI in Sec. 6.1.

2.2.2 Performance

The performance gains resulting from the use of parallel systems all derive from the simultaneous operation of the many processors. Ideally, the performance improvement over single-processor systems should be $n:1$, where n is the number of processors in the parallel system. For several reasons, however, this ideal improvement will not actually be realized except in very special instances, although an order of $n:1$ improvement should be the case.

First, there will generally be interprocessor communication during processing. This additional burden imposed on the processors will cause a reduction in processing efficiency. This reduction can be prohibitive, in fact, if the system is not properly designed.

Second, it is important to keep in mind that any capability given to one processor in the array will generally have to be given to all processors. When the array may contain hundreds or thousands of processors, the cost of any significant improvements can well be prohibitive. For this reason, the designer of the system will attempt to achieve the desired throughput from the parallelism of system operation to as great an extent as possible, rather than by attempting to maximize the performance of the processors. For example, the processors may be bit-serial in operation, rather than "broadside," or bit-parallel, in order to save arithmetic logic. If the designer has done a proper job, the order of $n:1$ improvement in system performance will be sufficient.

2.3 Determining the Suitability of the Application

There are a number of reasonably well-defined criteria which may be applied to a prospective application to determine its suitability for a reconfigurable parallel array system. These include not only the existence of elements of parallelism in the algorithms, but also the relative input and output loading on the processor, algorithm complexity, and even the environment in which the processor will operate. There are also a number of potential pitfalls to avoid in determining the type of processor best suited to the application. All of these points are covered in the remainder of this subsection.

2.3.1 Parallelism in the Algorithms

There are a number of ways in which parallelism in the application algorithms can be present. Some of these can be taken advantage of in all parallel system designs; others are applicable only to specialized designs.†

Many applications which are inherently parallel in nature have a charac-

---

†Section 3, which describes various types of parallel systems, discusses the applications best suited to each type.

teristic which shall be termed "block-oriented." Block-oriented applications are those which deal with a number of similar "objects" in the data base or in the outside world, generally a varying number, and on all of which the same general process is performed. In such applications, a "block" of memory or of cells in the processor will correspond to each such "object," and will contain the parameters and working space for that object—hence the name "block-oriented." The "objects" may be:

- Targets or threats being tracked by radar.
- Blocks of raw English text being searched or modified.
- Records in a file being searched or modified.
- Pixels or sets of pixels in an image being processed.

Depending on the particular architecture, separate PEs will be assigned to each block, a single PE will process several blocks, or a number of PEs will be assigned to each block.

2.3.2 Complexity of the Process

The existence of parallelism in the algorithm alone is not in itself sufficient justification for the use of parallel hardware. It is a requirement that the amount of processing on each item or each set of data be sufficient to justify the time required to load the data. If this is not the case, the complicated parallel hardware is not being effectively used, and the processing could perhaps be better performed by other means.

An example for which the use of parallel systems is well justified is the inversion of a matrix. The matrix need be loaded only once, and the number of operations performed to invert the matrix is very large indeed. Another example is dictionary lookup using an associative memory (i.e, the example of Figure III.3), when the same dictionary resides in the memory during a large number of lookups, and when the dictionary entries are of variable size.

An example of a poor application for parallel processing systems is a filtering task on real-time data in which only a few comparisons are made for each data item. Such a process may consist of only a "window" test—the arithmetic comparison of each data item against upper and lower limits. Such an operation, if it is time-critical, might best be performed by fast hardwired serial logic. If it is not time-critical, it should probably be performed by a serial processor of some sort, perhaps a single microprocessor.

2.3.3 Input Considerations

As previously mentioned, parallel array systems are most efficiently employed when the amount of internal processing is high in proportion to the quantity of input data. The actual input data rate which can be accommodated varies greatly with the system design. Some systems, having very powerful bit-parallel processors as their PEs,† can handle extremely high input rates when the data can be suitably partitioned among the multiple input channels. Other systems, such as many associative processors, which may have equally great internal processing capability, may not be able to handle such high input rates. This capability is of major importance in choosing a parallel system architecture.

2.3.4 Output Considerations

Parallel array systems are best used in applications requiring massive internal processing capability. However, it is extremely rare that output capability of such an order of magnitude is required if the system is properly designed. That is, the system can almost always be designed to perform all phases of the entire processing task, so that massive output rates are not needed.

This point is not an easy one to make, since it is not necessarily valid for applications requiring systems of more moderate capability. However, it is important in understanding the applications of very large parallel array systems. Consider the destination of the output of any extremely powerful processing system. In almost every application, it will be one of the following:

1. Output to a printer or display, to be absorbed by humans. Massive output streams do not fall into this classification. Selective output from a computer does. With its highly parallel operating capability, a properly designed reconfigurable parallel array processing system has an unusually great capability to select and sort output for direct use by humans.
2. Output to a mass storage device, for eventual selective retrieval and output as in 1 above. This is essentially the task of preprocessing the input to a large data management system. It is difficult to visualize a data-preprocessing task requiring a massive parallel processing system and which produces an output of the order of its input rate. If one such exists, it is a counterexample to the argument being made here.‡

---

†An example is ILLIAC IV. See Sec. 3.1.2.

‡Special preprocessing of a wideband stream of raw data for input to a computer is required in some real-time applications. Bulk filtering of radar data for input to a track-while-scan computer system is an example. However, it should be noted that the filtering process greatly reduces the quantity of data from input to output, as is indeed its purpose.

3. Output to another computer. If this other computer is necessary to the overall task, something is almost always lacking in the system engineering. The parallel array system should be designed to perform the entire task. High-speed data multiplexing is an exception. That is best performed by special hardwired logic.

4. Output to a wideband telemetry channel, for further computer processing at a remote location. This appears to be the only case in which the high-speed data multiplexing just mentioned could be required. However, it is again difficult to conceive of a task requiring such a high intermediate data transmission rate when the source of the data is itself a system with massive processing capability. This system should be designed to perform the entire task, and the need for expensive wide band telemetry thus made unnecessary.†

In short, the capability for great output data rates should not be a major consideration in the selection of a parallel array system design for the applications being considered. The capability to perform all phases of the task must be a consideration.

2.3.5 Other Parallel Attributes

For some system designs, it is possible to organize the processing so that some time-consuming operations can be performed for several different (and even unrelated) processes in parallel by a single hardware operation. This is true for systems, such as most associative processors, which are bit-serial in operation and which depend for their high processing throughput on a high degree of parallelism, rather than on being able to perform individual operations rapidly. Multiplication, division, and floating-point operations are examples of operations which are slow on bit-serial associative processors.† If it is possible to lay out the data and to construct the program so that a large number of multiplications, for example, can be performed with a single parallel hardware multiply operation, considerable execution time can be saved.‡ If many such combinations of operations can be performed in this fashion when the individual operations are time-consuming, a high degree of potential parallelism exists in the application, and, moreover, will be in addition to the parallelism gained from any block-oriented characteristics in the application.

The way in which the combination of operations can be made, and under what circumstances, will be made clear in the discussions on associative processors.

---

†In order to ascertain the validity of this statement, the reader is invited to examine a floating-point multiplication algorithm (Chap. I, Sec. 1.8) and to try to execute this algorithm in a bit-serial fashion. (Ed.)

‡If a sequence of arithmetic operations is to be performed over several data arrays, it can be computed by either an array or pipeline system. A comparison of such execution of the same arithmetic expression in these two systems (array and pipeline) is discussed in Chap. I, Sec. 15.2. (Ed.)

†If wideband telemetry is involved, the data will most likely be image related. It should be noted that there exists at least one technique (see Hilbert [21]) for compressing image data for which parallel array systems are quite well suited. Fourier and Handamard transform techniques are also well suited to some parallel array systems.

2.3.6 Communication Channel Considerations

In some block-oriented applications, the processing operations for each block are largely independent of those for the others. In such cases, the amount of interblock communications which must be performed is small, and the importance of the communication channel in the application is small. In fact, advantage of this characteristic is taken in the design of some parallel systems, such as PEPE† (to be described), by nearly dispensing with intermodule communication channels altogether. For many other applications, the capability to communicate large amounts of data is critical to system performance, and the communication scheme must be given special attention in the design. The problem which results from deficient communication capability is an unbalanced system, in which the individual PEs can operate independently with high throughput, but whose throughput is greatly reduced when they must communicate with one another. In this case, the designer has the difficult problem of designing the processing algorithm and the communication channel scheme together, and in great detail, before he or she can demonstrate that the system will have the required capability. Communication problems are the most difficult to handle, in short, when the designer is trying to determine the suitability of an application for implementation on parallel systems.

A more detailed discussion of communication channel schemes and their relation to system balance appears in the next section.

†See Evensen [22]. The book containing this reference also has several additional papers on the PEPE. Also see Sec. 3.1.3.

2.3.7 Avoiding Serial Operations

In determining the suitability of a task for implementation on a parallel array system, the designer must be aware of one important aspect of parallel processing. That is, it is not sufficient to demonstrate only that critical operations can be performed in parallel. The designer must also show that all sequences of operations can be performed in parallel fashion without the necessity of serial processing in between. If this is not possible, most of the advantage of the parallel processing capability will be lost. And it takes surprisingly few serial operations to seriously compromise the entire processing scheme. Associative processors are particularly subject to problems of this sort. In a typical case, after a parallel operation is performed, it may become necessary to reorganize the data so that the next operation can be performed in parallel. If the reorganization cannot itself be performed in parallel, the hardware and/or data organization must be redesigned. If this will not suffice, the parallel processing approach may have to be rejected.

2.3.8 Another Pitfall to Avoid

There is a well-known horror story in connection with an early effort to build a highly parallel system. The task to which the system was to be assigned was a file-management task and, because of the newness of the system concept, the designing organization attacked the task from two viewpoints. While one programming group constructed a program to perform the task on the parallel system, another group set out to program the same task on a conventional computer so that the performance improvement in the parallel system could be demonstrated. However, the latter programming group devised a clever file-organization-and-search scheme that so improved the conventional system's performance that it proved to be faster than the parallel system.† The reason for the failure of the parallel system to outperform the conventional system was that the parallel operation was a simple one, search-by-content at one level only, and that the parallel array was too small to permit the entire data base to be searched in one pass. If several parallel operations in sequence had been needed, the conventional system would have been inferior, and the use of the parallel system justified.

†The scheme employed was a hash-addressing, or "bucket" technique. Hash addressing permits fast content searches to be performed on conventional machines. The technique involves storing each item of data as close as possible to the memory location whose address is a known function of its data contents, interpreted as a number. The inverse of this address-encoding function, applied to the search criterion, enables the stored data having the desired contents to be retrieved with generally no more than one or two memory accesses. Knapp [29] shows an interesting application of hash addressing in radar data processing.

3. SURVEY OF PROCESSOR DESIGNS

In this section, a number of general reconfigurable parallel array systems will be described and analyzed. For each type, a general description of the system, its operation, and an application will be given. Then a description of a particular existing system of that type will be given, and further comparisons with other system approaches made.

The system types to be covered are:

1. Parallel network processors
2. Associative processors
3. Distributed-control parallel processors

3.1 Parallel Network Processors

3.1.1 General Description

The term "parallel network processor" shall refer to a reconfigurable parallel array processor in which:

1. The processing elements themselves are full fledged serial processors with sizeable local memories
2. The communication network connects each PE to several other PEs in a regular fashion
3. The PEs execute essentially the same program on different sets of data simultaneously, under central control

The concepts of the parallel network processor arises from the nature of the applications that are involved. These applications deal with "objects" in the real world that interact in the application in some orderly fashion. The communication network in the network processor interconnects the PEs in the same fashion, under software control, and each PE corresponds to one object. (Ideally only one—several are often accommodated by a single PE with some degradation in system throughput.)

Applications well suited to parallel network processors include:

1. The solution of sets of partial difference equations (by the relaxation technique, for example)
2. General matrix operations
3. Many image-processing applications—each PE corresponds to a pixel or set of pixels
4. Radar data processing—each PE corresponds to a target or threat being tracked.

The remainder of this subsection describes the general organization and operation of a parallel network processor and then describes an existing large scale network processor, the ILLIAC IV, and PEPE, which has many similar characteristics but is not strictly a network processor.

3.1.1.1 General organization and operation

The processor organization to be described here is essentially the same as that for the SOLOMON 1 computer,† which is the earliest of the well-known examples of a parallel network architecture.

The processor array. Figure II.5 shows the organization of the array of processing elements (PEs) for the system. Only 16 PEs are shown. The SOLOMON 1 design has 1024 PEs, organized into 32 rows and 32 columns. The array contains about 90 percent of the logic of the entire system. Each PE is a serial processor in every respect except for the absence of instruction-sequencing and decoding logic. The array hardware interconnects the PEs in a two-dimensional array configuration. In the original SOLOMON design, the PEs were bit-serial processors, and the communication channels were also bit-serial. In newer designs, because of the advances in LSI fabrication techniques, the PEs can be bit-parallel. In those cases, in order to maintain system balance, the communication channels are made bit-parallel also.

The operation of the communication channels during program execution is under software control. The control is largely central with respect to the general interconnection modes. The modes permit interelement communication by one or more of the following schemes simultaneously:

1. Each PE is connected to its two neighbors in the same row. The first

†See Slotnick [7].

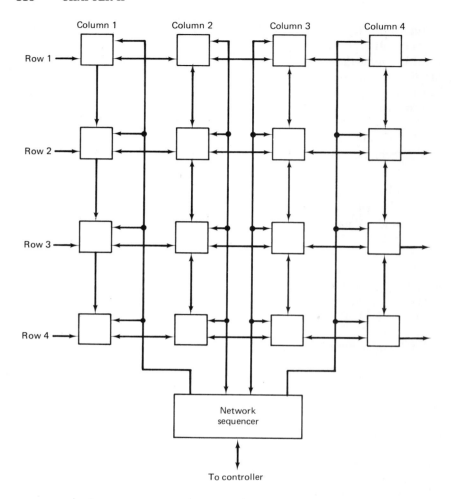

FIGURE II.5 Parallel network processor array organization, showing 16 processing elements

PE in each row is connected end-around to the last PE in the row, resulting in a "horizontal" cylindrical configuration.
2. Each PE is connected to its two neighbors in the same column, with the connection end-around as before, resulting in a "vertical" cylindrical configuration. Both this option and the first can be used together.
3. The PEs can all be connected in a single one-dimensional array, end-around if desired.

Other interconnection schemes, both hardware and software, can obviously be devised. For example, in some image-processing applications, each pixel could have hardware channels linking it to its eight nearest neighbors.

In addition to the array interconnections, all of the PEs are connected to common busses through which they can receive common data items simultaneously. There are common busses for each row and each column of the array. With this scheme, many common data items can be input to the array simultaneously, a different one for the PEs in each row or each column.

The *network sequencer* is shown in Fig. II.5. This device provides control signals to the I/O control unit during information transfers.

The processing element. Figure II.6 shows the organization of each processing element in the array. Each PE contains a set of working registers and a local random-access memory for data only. There is a logic module for performing arithmetic, boolean, and comparison operations. There is also instruction-execution logic, but there the similarity to a conventional serial processor ends. The PE receives its instruction control from an external source. It has no internal instruction sequencing or instruction decoding logic. In addition, the PE has several unique devices with functions as follows:

1. Routing Logic. This device controls the source or destination of data and control information passed between the PE and other PEs and

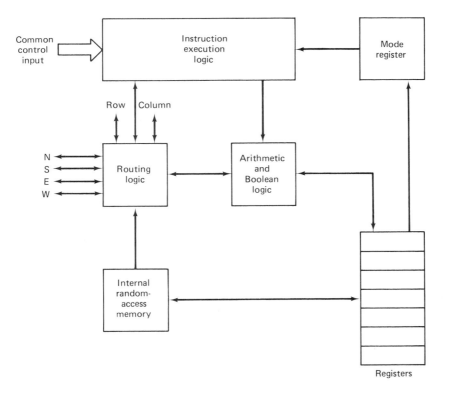

FIGURE II.6 The processing element

common busses. During the execution of an instruction, the routing logic state permits the PE to send or receive information to and from the common row or column bus.

2. **Mode Register.** This register (which is two bits long in the SOLOMON computer), is set from within the PE. The instruction control information received by the PE for all instructions contains a specification as to which mode or modes the PE must be set in order that the PE execute the instruction. (All modes may be specified, in which case the execution is mandatory.)

Through this device, then, each PE can determine in most cases whether or not it will execute an instruction. Each PE makes this determination as the result of the processing of its local data, and then sets its mode register accordingly.

The system organization. Figure II.7 shows the general organization of a parallel network processor. This system consists of the parallel network processor array just described, plus a number of modules for control of the array, operator interface, input, and output. These modules and their functions are as follows:

1. **The Control Unit.** This unit receives from the control memory each instruction to be executed by the system. It decodes the instruction and the addresses. For those instructions to be executed by the PE array, it sends the resulting control information and the row and column addresses to the PE array. These instructions will be executed by those PEs in the specified row(s) and column(s) whose internal mode register settings correspond to the mode settings specified in the instruction. Other instructions are sent to the system I/O control unit, to the operator terminal, or are executed directly, the results going to common system memory.

2. **The Control Memory.** This memory contains the system's programs, plus common data items and common working storage. The control unit has normal serial processing capability, which is used occasionally to calculate common data values and to determine the flow of the instruction execution sequence.

3. **The Parallel Buffer.** This buffer receives and sends data in parallel to and from all PEs in the array, or to and from all PEs in specified rows and columns. The buffer holds one data item for each row and one for each column. Data transfer between the parallel buffer and the remainder of the system is word-serial. If the data rate must be high, this latter transfer will still be serial for most applications, but high-speed multiplexors will be used to perform the conversion between serial and parallel forms, rather than word-by-word control by the control unit.

4. **The Input/Output Control Unit.** This device controls all system input

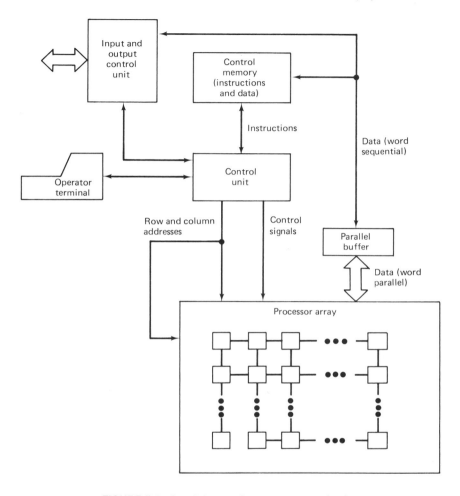

FIGURE II.7 Parallel network processor organization

and output for most applications. For some real-time applications, it is best to bypass such complex control units altogether for data being input directly to the processor array. Instead, the system should execute minimal control over such input to ensure maximum input rate and thus maximum throughput rate as well.

3.1.1.2 Fabricating a parallel network processor

Most of the remarks that can be made concerning the fabrication of reconfigurable parallel array processors in general apply to parallel network processors. Parallel network processors are modular in organization; that is, most of the logic is contained in the array of processing elements. Depending on the complexity of the individual PEs and on the size of the wafer, a number of PEs can be fabricated on a single wafer. If the number is great, there can be a problem with too many external connections; however, this is because the wafer must have communication connections to the number of rows plus columns times two. This can have a considerable effect on cost, since the cost of mounting a set of complicated wafers on PC boards can be as high or higher than the cost of the wafers themselves. In addition, system reliability is affected by the number of external connections to the wafers.

Some parallel network processors obtain their processing power from relatively few PEs of great individual capability. If these PEs can be fabricated on a small number of wafers each, these reliability and cost factors improve in proportion to the overall system processing capability.

One more general remark concerning parallel network processors as a class is appropriate. In the last few years, it has been possible to obtain off-the-shelf microprocessors of considerable processing capability on a single chip at very low cost. In looking for reasonable solutions to unusually demanding system tasks, many engineers have been examining the possibility of using large numbers of these conventional microprocessors in some sort of array configuration as a solution. Except in very few instances (the author knows of one†) this solution will not work with very high efficiency.‡ In any case, the designer should perform a very careful hardware and software timing analysis of the communications to be performed between the individual PEs. This is where most of the pitfalls in the entire scheme can usually be found. The trouble lies in the fact that the individual processors are completely serial in all aspects of their operation, including input and output. Unlike the PEs in the parallel array system just described, these microprocessors cannot obtain their instruction operands directly from the neighboring PEs as they perform each instruction, but must perform separate input and output operations in sequence for the purpose. In addition, there can be a bind on system throughput due to high input and output rates between neighboring PEs. This is usually the result of the fact that the operations performed by the separate PEs are not sufficiently independent of one another. This can happen even though the PEs are custom-designed for use in a parallel network array.

†See Minel [19]. The PEs in that system are built from AMD 2903 microprocessor chips.

‡Another experiment of using off-the-shelf modules for LSI implementation of a Univac 1108 is described in Chap. 1, Sec. 20. (Ed.)

3.1.2 The ILLIAC IV

As long ago as 1962,† it was realized that a very large, fast parallel network processor, built of the most advanced logic then available, would bring into the realm of the achievable a number of important applications which, up to that time, had been beyond the capabilities of any digital processing system yet built or envisioned. The capabilities, and the applications to which they apply, include:

†References [9], [1], and [18] give more detail on the ALAP and its applications.

†As described by Slotnick [7].

1. Manipulation of very large matrices (e.g., linear programming)
2. Solutions of large sets of linear equations (statistical analysis)
3. Solutions of large sets of partial difference equations over large grids (short-term weather prediction)
4. Fast data correlation (as when phased-array antennas are used in fast-response radar systems such as those for urban defense against ICBMs)

ILLIAC IV is a massive system, and has been designed and built to achieve the orders-of-magnitude increase in processor throughput necessary for such applications.

The similarity of the ILLIAC IV to the SOLOMON computer is direct. The ILLIAC IV implements the four principal features of the SOLOMON concept. These are:

1. Single instruction stream, multiple data stream.
2. Multiple processors simultaneously processing the same instructions under central control.
3. Local enable/disable flip-flops to permit individual processors to select the instructions that they will execute.
4. Direct communication channels from each processor to its nearest neighbors. The channels operate simultaneously.

3.1.2.1 System organization

Figure II.8 is a simplified diagram showing the general organization of the ILLIAC IV. The central component of the system is a set of four arrays (called "quadrants") of 64 processing elements each. Each array has its own central control unit. This permits the four arrays to operate on independent processes if desired. The four arrays can be connected together, each to the three others, through high-throughput communication channels. In this configuration, the entire array network can operate on different parts of the same application in synchronized fashion under common control. In both modes, overall control of the system resides in a general-purpose computer, the Burroughs B6500, which permits operator interface with the system, controls array configuration, supervises input and output operations, and controls the loading of programs into the arrays. Input and output is through the B6500 or, for real-time data, directly through an input/output switch for a real-time data link. A large, very fast parallel-access disk system acts as backup memory for the entire system. The instruction-by-instruction control of array activity comes from the arrays' own control units, rather than from the B6500.

124 CHAPTER II

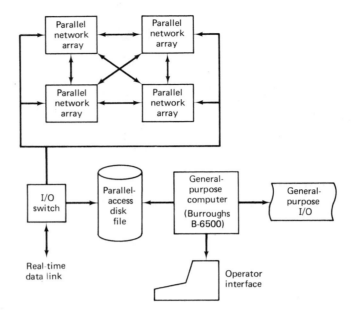

FIGURE II.8 General organization of ILIAC IV

3.1.2.2 Array organization

Figure II.9 shows the internal organization of one of the four 64-processor arrays. The figure shows the two common busses for the array. Both of these busses are 64 bits wide and connect the array's control unit to all of the 64 PEs. One of the busses, the "common data bus," carries common data items and their local PE memory addresses (which are always the same for all PEs) from the control unit to all participating PEs. The other common bus, the "control unit bus," conveys instructions and common operands from the PE memories to the control unit. The control unit performs the same functions for the array as the control unit in the SOLOMON; it decodes all of the instructions executed (simultaneously) by the PEs and sends the corresponding control signals to them. The control unit also senses the enable/disable mode bits of the PEs and thereby monitors the state of the array operation. (For example, it can sense if all of the PEs are enabled or if all are disabled.) The internal routing network for the PEs is not shown in the figure.

3.1.2.3 The array routing network

The internal communication network among the PEs in the ILLIAC IV array organizes the PEs into a combination of a two-dimensional array and a linear array. Figure II.10 shows the routing network. Each PE is seen connected to the preceding and following PE, and also to the PEs which are eight units away in both directions. The end-around connection between PE 63 and

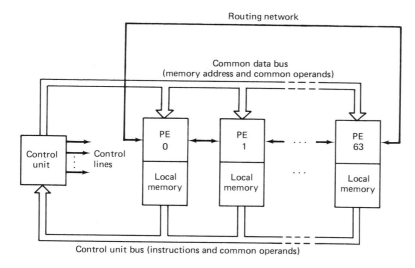

FIGURE II.9 ILIAC IV array organization, showing common busses

PE 0 can be broken under software control in order to connect several 64-element arrays in the system together end-to-end.

The communication channels in the network are 64 bits wide, to correspond to the 64-bit word length of the PEs.

3.1.2.4 Multiarray configurations

The reconfigurability of the communication among the PEs is not limited to those within a single array. The arrays may be linked to each other under software control in configurations of 128 and 256 PEs. These configurations are formed by connecting the individual arrays end-to-end. Figure II.11 illustrates the multiarray configurations and the single array configurations.

When several arrays are connected together into multiarray networks, all PEs in the combined array execute the same instructions simultaneously under central control. Bit-by-bit synchronization of the operation of the several control units, however, is only required for instructions in which data or control information must cross array boundaries.

3.1.2.5 The control unit

Figure II.12 shows the organization of the array control unit (CU). The various components and their functions are as follows:

1. Instruction Buffer (PLA): This 64-word buffer holds the current instruction and pending instruction, both of which are received from the PE memories via the control unit bus.

126 CHAPTER II

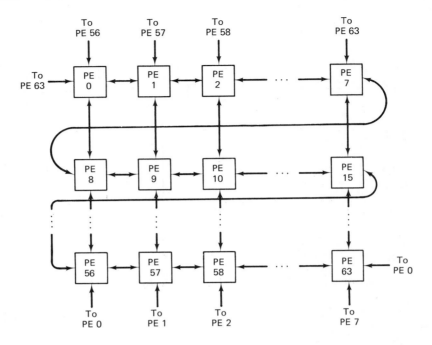

FIGURE II.10 Communication network for the ILIAC IV array

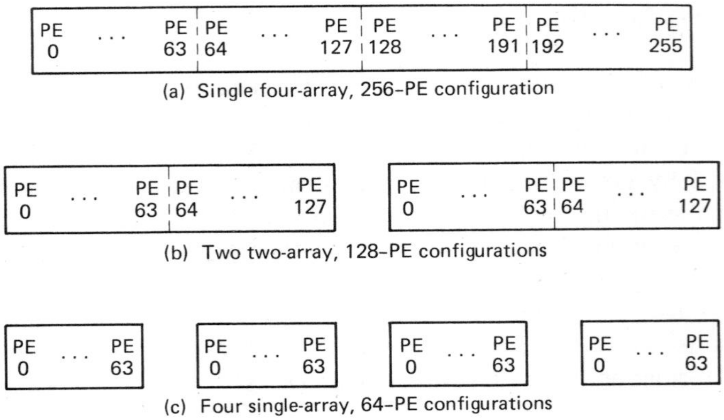

(a) Single four-array, 256-PE configuration

(b) Two two-array, 128-PE configurations

(c) Four single-array, 64-PE configurations

FIGURE II.11 ILIAC IV multiple-array configurations

2. **Local Data Buffer:** This 64-word buffer holds common operands received from the PEs via the control unit bus.
3. **Advanced Instruction Station (ADVAST):** The modules in this unit decode the instructions received from the instruction buffer. If the instructions are local ones (e.g., branch, loop) for execution by the

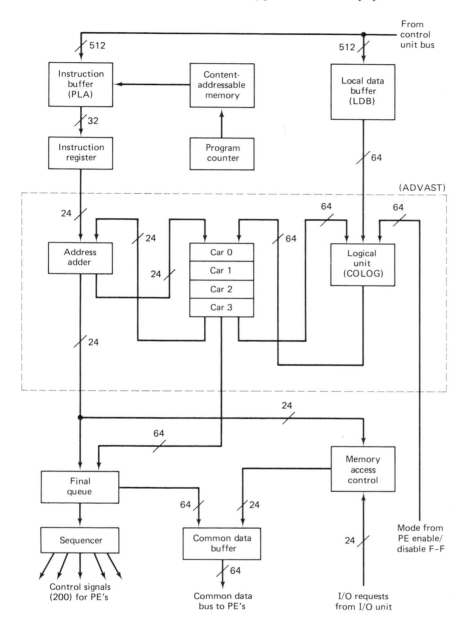

FIGURE II.12 Organization of the ILIAC IV control unit

CU, ADVAST executes them. If the instructions are for execution by the PEs, ADVAST puts address and control information for the instructions in a queue (FINQ in the diagram) for transmission to the PEs.

4. Accumulator Registers: The four accumulator registers, CAR 0 thru CAR 3, hold address information and active data for the instruction. These registers are central to communication within the CU.
5. Arithmetic Unit (CULOG): This unit performs the arithmetic and boolean operations for instructions executed directly by the CU.

3.1.2.6 Operation of the control unit

The great complexity of the CU is the result of eliminating as many delays as possible in program execution. The PE instruction queue permits overlap between the execution of local CU instructions and those executed by the PEs. The instruction buffer holds 128 instructions, enough for a medium-sized loop. This prevents delays in the execution of program inner loops due to the time required for fetching instructions from the slower and larger PE memories. Instructions are fetched in blocks to save delays due to initial transmission time. All of these time-saving strategies, however, must be known to the programmer if he or she is to make the most efficient use of the hardware. This results in high programming cost, and also in a proportionally great effort expended in writing optimizing assemblers and compilers for the system.

3.1.2.7 The processing element (PE)

The structure of the PE is simpler to understand than that of the control unit, since it has fewer functions to perform. These functions are not unlike those of a serial processor, minus the instruction sequencing and decoding logic, and with the addition of the routing logic. Figure II.13 shows the organization of the PEs. Inputs from the CU are shown, as well as the inputs and outputs for the communication network to the "nearest" four PEs in the array (designated N, S, E, and W in the figure). The components of the PE include:

1. Four 64-bit registers (A, B, R, S) for operands and results
2. An adder/multiplier unit (MSG, PAT, CPA)
3. A logic unit (LOG) for boolean functions
4. A barrel switch (BSW) for shift operations
5. An index register (RGX) for an adder (ADA) for memory address modification and control
6. A mode register (RGM) to hold the results of tests and the enable/disable state of the PE

The PEs can be partitioned into subprocessors of 8 bits or 32 bits when the full 64-bit precision of the PE is not needed. The number of PEs in the system is thereby effectively doubled or quadrupled. This makes more

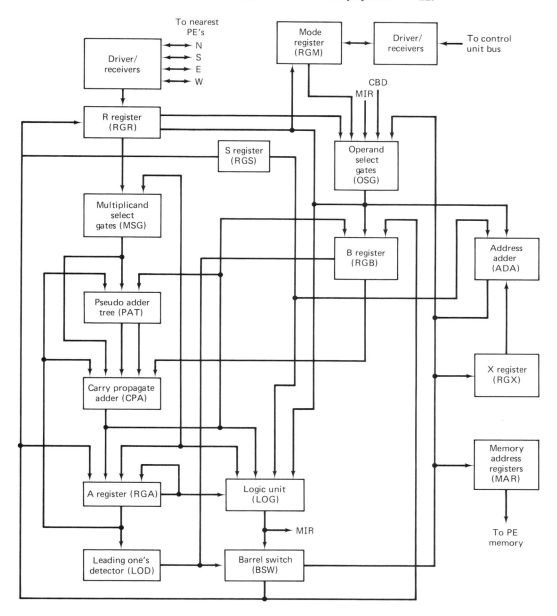

FIGURE II.13 Organization of the ILIAC IV processing element

efficient use of the hardware for some types of computation, even though the resulting bit-slice processors must share common data paths, indexing values, and mode flip-flops within each PE. (There are two mode flip-flops for the 32-bit partition, but not for the 8-bit partition.) This feature further increases programming difficulty, however, since the programmer seeking maximum throughput from the system will attempt to make use of the partitioning feature whenever he or she thinks it might help.

3.1.2.8 The PE memory

The individual PEs each have a local PE memory, which cannot be accessed by any other PEs. All of the PE memories together, however, constitute a single large memory that can be accessed by the control unit and the I/O unit. This combined memory contains the program for the array. Each PE memory has a capacity of 2048 sixty-four-bit words and an access time of 120 nsec.

3.1.2.9 Programming the ILLIAC IV

Since the construction of the ILLIAC IV, a great deal of effort has been spent in system and applications programming. Knapp [29] contains a thorough discussion on a particular application requiring the extremely high throughput of the ILLIAC IV, the processing of data from a phased-array radar system for urban defense. This appears to be an excellent application for the ILLIAC IV. The application required only one of the four quadrants of the system, and takes full advantage of the great computational capability of the ILLIAC IV. There is relatively little communication of data among the processors during computations. Thus the problems involving communication bandwidth which are often encountered with parallel network processors do not appear in this application.

3.1.3 PEPE, the Parallel Element Processing Ensemble

PEPE is an example of a successful parallel array system that has been built especially for use in applications having a particular characteristic, that of independence of process among the many processors. The applications for which PEPE is ideally suited are those in which either:

1. The input data can be factored into independent data streams to which the same algorithm can be applied simultaneously. The process for one data stream is independent of the results of the process for the other data streams. An example of such an application is the processing of image data from multiple sensors, when each sensor receives a separate and independent image.

2. The same input data stream can be processed independently and simultaneously against a number of internally stored data sets. Again, the processes for the data sets are independent of one another. An

example of such an application is track-while-scan radar, in which each processor is assigned to a particular radar target (for offense) or threat (for defense).

Note the use of the term "independent." PEPE has very little interprocessor communication capability, and is not well-suited for applications in which the processing elements must communicate simultaneously among themselves.

The same lack of interprocessor communication gives PEPE a great advantage over its competitors in the cost of its construction, since it is quite amenable to LSI fabrication and, in some of its variant architectures, can use nearly conventional microprocessors as its processing elements. The PEPE architecture, moreover, permits graceful degradation of its operation almost without special attention to the software design.

3.1.3.1 The PEPE system

The system to be described is the 288-element PEPE built for the ABMDA Research Center. The block diagram of the system is shown in Fig. II.14. The central component of the system is the *control console*. This control console provides control of instruction execution for the entire PEPE portion of the system. It interfaces with the array of processing elements (PEs), providing control of input, output, and instruction execution for the PEs. It also interfaces with the CDC 7600 data processing system for operator control and for sequential input and output to and from the real world. The system contains a maintenance control and diagnostic unit which also interfaces directly with the control console.

The control console contains a 32-Kword program memory from which all programs are executed. These programs consist of sequences of serial instructions to be executed within the control console itself, plus sequences of parallel instructions to be executed in the PE array. The serial instructions provide branch control of the program, based both on the results of its own sequential arithmetic and boolean processing and on the results of parallel PE activity. The parallel instructions are executed by some or all PEs simultaneously, depending on their internal states. The PEs are capable of performing simultaneous input and output operations together with their internal operations. The control console therefore has three control units which direct three simultaneous streams of control information to the PEs to support these three activities.

3.1.3.2 The PEPE processing elements

Figure II.15 shows the internal organization of a PEPE processing element. The four units shown in the figure will be described. The PE contains

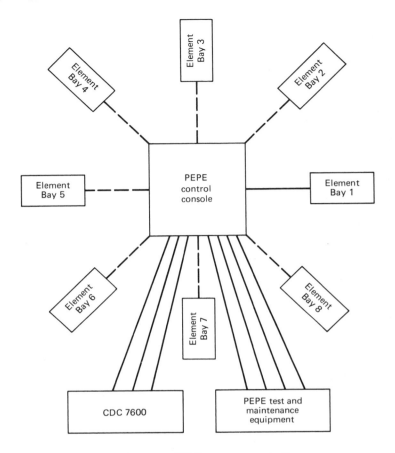

FIGURE II.14 PEPE system organization

no instruction execution control logic, and must receive all timing and control signals from the control console.

The arithmetic unit. The *arithmetic unit* (AU) contains conventional arithmetic and boolean instruction-execution logic, together with several operand registers. The operation of the unit is bit-parallel, and includes a floating-point instruction capability. The unit receives its control signals from the control console, and obtains its operands both from the control console via a data bus and from the PE's local memory. The data bus also provides output from the arithmetic unit to the control console.

The participation of the arithmetic unit in the execution of each instruction is determined by each PE according to the state of a local 1-bit "activity register." For each arithmetic instruction decoded by the control console, all of those PEs with a "1" in their activity register execute the

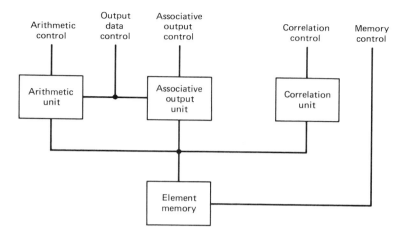

FIGURE II.15 PEPE processing element

instruction simultaneously; all of those with "0" do not. The activity register is supported by a stack, thereby enabling the PEs to retain and restore previous states of the register.

In addition, the arithmetic unit contains an 8-bit "tag" register which is used to perform associative matches on data received from the control console. The control console can execute special "select" instructions upon the PE's activity register, its stack, and the tag register to determine which PEs will take part in subsequent parallel arithmetic instructions.

The associative output unit. The *associative output unit* (AOU) contains conventional registers which support the execution of conventional integer arithmetic and logical instructions. In addition, the AOU implements associative output to the control console via the data bus which it shares with the arithmetic unit.

Like the arithmetic unit, the AOU contains an activity register, with its stack, and a tag register. These operate exactly as for the arithmetic unit. The AOU can receive its control signals from the control console at the same time as the arithmetic unit. Thus, the PE can perform associative output operations simultaneously with the arithmetic operations.

The correlation unit. The *correlation unit* contains an airthmetic register and 16 "correlation registers." It can execute integer arithmetic and logical instructions in the same way as the AOU. However, it does not perform output to the control console. The correlation unit also contains an activity register (but no stack) and a tag register which permit both the PE and the control console to determine its (CU's) participation in the execution of each instruction whose control signals are sent to it.

The correlation registers contain information during program execution which enable them to determine whether or not they will receive and store new data. There is no activity stack for the activity register, since the correlation unit needs to retain no history of previous activity.

The correlation unit can execute instructions sent to it by the control console simultaneously with the AU and AOU. This gives the PE array the ability to execute three instruction streams simultaneously; one for input, one for output, and one for internal data processing.

The element memory. The *element memory* consists of 1024 words of ECL (emitter-coupled logic) storage. It receives its address and mode information from the control console, rather than from any internal select logic. During the execution of a parallel instruction, the element memories in all participating PEs (as determined by the PE's correlation unit) receive the same data from the control console, through the correlation unit, for storage in the same corresponding memory and location.

Conflicts between memory accesses among the other three units in the PE are resolved by priority, with the correlation unit receiving the highest priority and the AOU receiving the lowest. Since the memory is much faster than the average instruction execution period, program execution time will increase by no more than five percent due to memory conflicts. The high priority given to input indicates the real-time nature of the applications for which the PEPE is intended.

3.1.3.3 Operation of PEPE

The principal application of PEPE is radar track-while-scan. In ballistic missile defense in particular, the processing task is extremely arduous. The PEPE implementation just described was designed to have correspondingly high processing throughput. The individual instruction-execution times are not unusually short by the standards of 1979, but average 200 to 300 nsec for each PE, with floating-point instructions requiring about 1.9 μsec. With 288 PEs in the system, and three instruction streams executing simultaneously, the maximum throughput is 3450 million instructions/sec. In practice, each PE operates on the data from a single object being tracked, or on more than one if the load exceeds 288 objects.

The reconfigurability of the PEPE is low compared to the other parallel systems described in this subsection; the selection of PEs is the main operation pertaining to reconfiguration. This, however, is suited to the application, since relatively little interaction exists in the real world between the objects whose activity is being monitored by the individual PEs.

From the standpoint of fault tolerance, the PEPE does very well compared to the SOLOMON and ILLIAC IV parallel network processors. There is no precedence or relative location for the PEs in the PEPE, so no rerouting of data need be performed in case of malfunctioning of a single PE. Another

PE automatically takes over the activity from the faulty one, and system throughput is degraded gradually on the average for each such failure. However, it should be noted, in case all of the PEs are already active when the failure occurs, all PEs will have to go into a multiobject tracking mode for the sake of the PE which must process two objects, even though the rest of the PEs are processing only one. This is due to the lock-step nature of the system operation.

It must also be noted that the PEPE pays for its advantages in fault tolerance and conceptual simplicity by having a smaller range of suitable applications than those parallel network systems with complicated interprocessor communication networks.

3.2 Associative Processors

3.2.1 General Description

The associative processor is one of the more unusual of the reconfigurable array systems, because its basic operating principle is in a sense the reverse of that for serial processors. The associative processor is based on a device known variously as "associative memory" or "content-addressable memory" which, in turn, is an outgrowth of a simpler device known as a "search memory."

3.2.1.1 The search memory and its operation

The search memory can be said to function in reverse fashion with respect to a random-access memory. That is, a RAM accepts the address of the desired memory location as input, and outputs the contents of that location in response. The search memory accepts the desired contents of the memory location or locations as input, and outputs flag settings indicating the memory location or locations having those contents. To prevent the operation from being trivial, the content search is masked; that is, the contents of some selected fields of the memory locations are specified, rather than the entire contents of the locations.

A content search can, of course, be implemented in any serial processor. The distinction in the case of the search memory is that the search is performed at all locations, or "cells" in the memory, simultaneously. There is nothing mysterious about this capability. It is simply the result of having compare logic at every cell in the memory. (The terms "cell" and "word" are sometimes used interchangeably in the discussion to follow, since the cell, which is a hardware device, contains a "word" of data in the usual sense.)

In most search memory implementations, the data for each cell resides in a shift register, and bit-serial compare logic is used at each cell. The desired speed advantage is gained if there are enough data items to be searched to require a large number of cells. Some search memories are fully bit-parallel; this implementation is used when the content search must be as fast as possible regardless of the cost of the logic. The same considerations govern the design of associative processors in general. This is discussed in greater detail in Sec. 3.2.1.2.

Figure II.16 illustrates the search-by-content operation as performed by a search memory. In the figure, the memory is shown as containing a number of cells, each of which contains seven characters (used instead of binary bits for clarity in the illustration). A corresponding *match flip-flop* is shown with each cell. This is a flag indicating the success/fail status of the cell as a result of a content search. The bottom of the figure shows two registers. One of these, called the "compare register" or "comparand register," contains the data item that is the object of the search. The other register, called the "mask register," determines by its contents the field or fields within the memory cells at which the content search will be made. In the figure, the first, fourth, sixth, and seventh character positions are masked out. The search will be conducted only at the remaining positions.

It is seen that only the first and second cells contain the same characters in the unmasked character positions as the compare register. As a result of the content search, the match flip-flops will be set only at those two cells as shown. (In fact, all of the match flip-flops are initially set, and the search operation resets the ones at nonmatching cells. However, this distinction is not important in this simplified illustration.)

Figure II.17 shows the basic logic function that is implemented at every

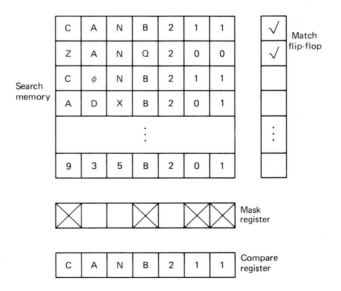

FIGURE II.16 The basic search memory operation: parallel search-by-content

cell in the search memory. Only two cells are shown, together with the compare register. This is the Exclusive-OR operation. At each bit position in every cell, the contents of that bit are Exclusive-ORed with the contents of the corresponding bit of the compare register. (The mask register is not involved in this illustration.) The results of that operation are Inclusive-ORed with the results of the Exclusive-OR at the remaining bit positions. The result of this combined operation is connected to the reset input of the cell's match flip-flop.

All match flip-flops are set true before the search operation begins. It is seen that if the result of the Exclusive-OR operation is true at any bit position, a mismatch has occurred at that position. The result of the combined operation will then be a true condition at the reset input to the match flip-flop, and the flip-flop will be reset. If no mismatch occurs at the cell, the match flip-flop will be left in its original state.

The figure shows logic at every bit position in memory. This simplifies the explanation of the operation, and also shows why such bit-parallel logic

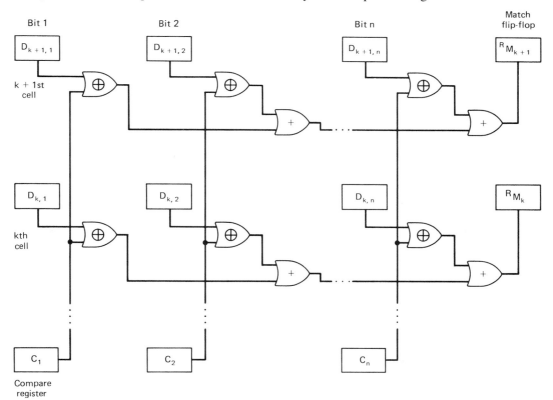

$$R_{M_k} = (D_{k1}\overline{C}_1 + \overline{D}_{k1}C_1) + (D_{k2}\overline{C}_2 + \overline{D}_{k2}C_2) + \cdots + (D_{kn}\overline{C}_n + \overline{D}_{kn}C_n)$$

FIGURE II.17 Basic search memory logic function

is too expensive for most search memory requirements. All of the associative processors to be described use bit-serial logic in their memory arrays.

3.2.1.2 The associative memory and its operation

The associative memory is an extension of the search memory and its operation is based on the same search-by-content function implemented in the search memory. The associative memory contains additional logic at each cell which permits it to perform word-parallel logical, arithmetic, input, and output operations, generally in bit-serial fashion. The logic needed for these operations is more straightforward, and will not be described in as great detail as for the content search operation.

The parallel-write operation. This operation is that of modifying selected bit positions at all cells in the associative memory, or at a selected subset of the cells. The bit positions to be modified are determined by the contents of the mask register. The contents to be inserted into those bit positions are the settings of the corresponding bit positions in the compare register. That is, the resulting setting at each selected bit position will be the same for all selected cells. The cells to be modified are selected by the settings of the match flip-flops.

It is the parallel-write operations and their extensions and variants that give the associative memory its capability to process data internally. This capability includes word-parallel arithmetic and logical operations as well as file searching and modification.

Arithmetic operations can be performed by using only the simple parallel-write and content search operations just described. Suppose, for example, that it is desired to add the contents of a specified field of the compare register to the contents of the same fields of all cells. This is done by performing repeated parallel-search and parallel-write operations at each bit position in turn, starting with the least significant bit. Each possible combination of states of the two operands, plus the states of the carry bits (stored in a spare bit position in each word) is processed in turn, with the corresponding states of the bit and the carry written into the set of cells having the particular combination of states.

This requires two search and two modify operations for the least significant bit position, and four each at every other bit position. The details of the algorithm are interesting to work out. However, the procedure is clumsy and, if it is desired to perform much arithmetic with the processor, processing efficiency is greatly improved by providing special arithmetic logic, including a full adder, at every cell. Note that only a single adder per cell is required for a bit-serial machine employing a shift register to contain the cell's data. Bit-parallel capability would require much more logic at each cell, with a correspondingly great increase in the cell's logical complexity and cost.

Serial-access logic. Nothing has been said so far about addressing individual cells within the associative memory, or about sequential processing of subsets of cells. Such processing must be performed, if only for sequential output of the results of the computation. It is in the nature of associative processors not to have conventional hardware addressing for accessing individual cells for input or output. Such capability would be contrary to the general philosophy of the concept, since the necessity to perform many individual cell accesses would itself indicate that the machine was poorly designed or poorly suited to the application.

More important, the presence of address-encoding and decoding logic on an associative memory wafer would make fault isolation and repair at the cell level impossible. The entire wafer would be rendered useless with the first fault, and an entire spare wafer would need to be switched in to continue operation. Such sequential processing as is necessary in an associative memory is usually limited to that of accessing a selected subset of cells in order (say, a set of cells whose contents have matched to a series of searches) so that their contents can be output a word at a time.

Sequential access capability can be implemented with a logical device known as a "ladder" or "propagating" network. The organization for such a network is shown in Fig. II.18. The figure shows several cells with their data registers (these are not directly involved in the ladder operation), their match flip-flops, their ladder logic modules, and a set of *ladder flip-flops*. The ladder flip-flops are set in sequence by each ladder operation, one flip-flop at a time, to indicate the next cell to be accessed.

In the first of a sequence of ladder operations, a "true" state is input to the first ladder module. It propagates from cell to cell at logic speed (i.e, unclocked) until it reaches the first cell at which a match flip-flop is set. At this cell, the ladder logic resets this signal so that it will not propagate further, resets the match flip-flop, and sets the ladder flip-flop. After the access to the cell's data register is completed for whatever purpose, the next ladder operation turns off that cell's ladder flip-flop and then sets the ladder flip-flop at the next cell having its match flip-flop set, and so on. For this and the remaining operations, a "false" state is input to the first ladder module. After the last operation, when there are no more match flip-flops set, the ladder signal is propagated out of the array and into the set input of an external latch, called the "catch flip-flop," where it indicates that the entire sequence of ladder operations is completed. The output labeled E from each ladder module goes to an error latch, and is turned on for an "impossible" set of input conditions to the module, namely $L_I Q_m Q_s$.

The basic logic equations for the ladder module are:

$$L = L_I \overline{Q}_M + L_I Q_L$$
$$R = L_I \overline{Q}_M Q_L$$
$$S = L_I Q_M \overline{Q}_L$$

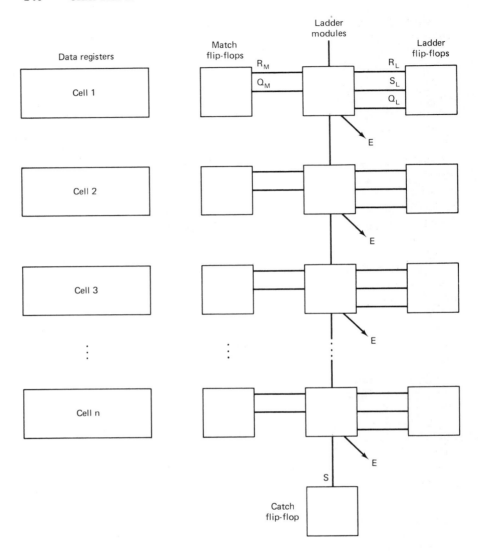

FIGURE II.18 Sequential access logic for a ladder operation

$$R = L_I Q_M \overline{Q_L}$$
$$E = L_I Q_M Q_L$$

Other operations. Other basic operations implemented in the typical associative memory are such auxiliary and support operations as setting or resetting all match flip-flops and ladder flip-flops, and the operation of outputting the contents of the (single) cell having its ladder flip-flop set.

Bit-serial requirement. The description of the associative memory cell makes clear a number of reasons why nearly all associative memories operate bit-serially rather than parallel-by-byte or by word, as sequential processors do. First, the limited data-handling capability of the cell makes it necessary for the memory to contain many cells if any real processing power is to be achieved. Therefore, in choosing the logical functions to be performed by the cell, the designer must be very careful of the required gate counts. Every capability considered for one cell must be included in every cell, or else the processor is not truly address-independent. (There can be exceptions to this rule, but only for very special purposes.) For example, where one would not hesitate to include a floating-point unit in a conventional design if it were needed, the inclusion of such complex logic in every associative cell would be nearly out of the question. It is possible, however, to implement some more primitive operations within the cell which could be used in combinations to perform floating-point arithmetic. For example, a left-shift-normalize-and-count function would be such an operation.

One solution to the problem of giving an associative memory great processing power without making the cells too complex is to have a separate array of complex logic modules, say, for performing complex or floating-point arithmetic, and to connect these modules to specified subsets of the cells, under software control, to perform those operations on the cell's data. This can be done in some designs, of course, but will be increasingly difficult as more and more cells are fabricated on individual LSI wafers. The wafers are pin-limited according to their size, and the number of cells contained in the wafer increases on the order of the area of the wafer, not its circumference.

In addition, fault tolerance is degraded when the wafers have a large number of pins. When all of the logic of the cell is contained on a single wafer, the number of pins on the wafer will be independent of the number of cells, since there is no need to connect any external circuitry to many cells simultaneously.

3.2.1.3 The associative processor and its operation

Figure II.19 shows the organization of a typical associative processor. An associative processor is a system that consists of an associative memory, together with a processor for controlling the operation of the associative memory and for interfacing with an operator. The controller is generally a sequential processor with a special interface for the associative memory and with some corresponding special instructions for implementing the control and monitor functions for the memory. The controller also has a conventional memory in which the controller's own program resides, and in which sequences of control settings for the associative memory may also reside. Some input and output operations are initiated by the controller directly to its own

I/O devices. Most input and output, unless it is very low rate, is directly to and from the associative memory, bypassing the controller altogether.

There have been applications of associative memories in which they were used essentially as high-speed peripheral devices for conventional systems for performing fast arithmetic.† Depending on the amount of special logic for controlling the associative memory and the percentage of the total system task performed by the associative memory, such systems can or cannot be termed associative processors. In the sense in which the term is used in this writing, the essential processing task should be conducted almost entirely within the associative memory for the system to be termed a parallel array processor.

3.2.1.4 Reconfigurable associative processors

It will be noticed that the associative processor just described has no reconfigurable characteristics and no intercommunication network except for the simple ladder network. For this reason, it is very limited in its applicability. For example, it is impossible in the system just described to transfer data from one cell to another except by means of the word-serial input and output operations using the ladder and parallel-write logic. For some file-management operations this is sufficient, and associative processors having no reconfigurability have been successfully applied to such tasks. These machines have very long data registers, generally 256 bits or greater, so that each cell can contain an entire record from the file being processed. In fact, by making the cell's arithmetic and logical unit complex enough (and not much complexity is required), such systems can perform complex data-retrieval operations based on the presence of logical combinations of states of the items within the record, searching all records in the file simultaneously. For example, the query directed to the processor by a user could be for the retrieval of all records in a file on military personnel for which the last name begins with any letter but K or L, whose age is between 25 and 28, who has not served in Alaska, and so on. Of course, the entire file must be present in the associative memory for many such queries, or else the previously mentioned problem of inefficient use of the system arises; that is, the problem of needing much input of data for each set of parallel operations performed.

In order that the associative processor have wider applicability, a communications network must be added so that data and also control states can be transferred between cells simultaneously (that is, between many pairs of cells or from one cell to many others). A number of techniques for accomplishing this through specialized hardware operating under software control have been devised. Two of these schemes will be discussed in this section in connection with the specific associative processors to be described. It will be

---

†Boulis [16] describes such an application.

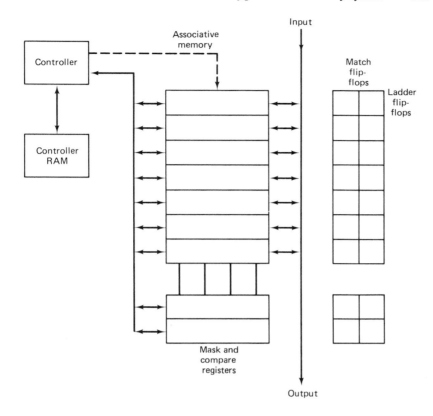

FIGURE II.19 Basic associative processor organization

seen that the nature of the communication network has a great effect in determining the suitability of the processor for the particular tasks. For example, one system is better suited for arithmetic processes, such as the fast Fourier transform, which require fast shuffling of data items. Others are better at complex retrieval operations in which the items being examined are of widely varying sizes (in bits). Still others have greater inherent fault tolerance and are better for space applications, and so on.

3.2.2 The STARAN

More effort has been devoted to the design, fabrication, and application of the Goodyear STARAN† than to any other associative processor. A number of full scale STARAN systems of various designs have been built, both for general-purpose application studies and for particular designated applica-

---

†TM, Goodyear Aerospace Corporation, Akron, Ohio.

tions. Some 155 different application functions have been studied for implementation on STARAN, and some 75 of these actually programmed and demonstrated. These deal with such general applications as surveillance systems, sensor signal processing, image processing, general scientific applications, communications processing, and data management.

This extensive effort has resulted in considerable refinement of the original STARAN concept, and particularly in the addition of a communication network, called the "flip network," for reordering the data among the cells in the STARAN memory.

3.2.2.1 The STARAN/635 system organization

The system to be described is the one installed at Rome Air Development Center (RADC), New York. This system consists of a STARAN system interfaced with a Honeywell 645 data-processing system.

Figure II.20 shows the system organization. The STARAN portion of the system contains the peripheral hardware (bulk storage, line printers, CRT terminal, etc.) needed for stand alone operation. The combined STARAN/635 system can run together under the control of the 635 multi-user time-shared operating system. Both the STARAN hardware and software are interfaced with the 635 in the integrated operating mode.

3.2.2.2 STARAN system organization

Figure II.21 gives a more detailed breakdown of the STARAN system organization. The widths of the data busses in bits are shown in the figure, giving an indication of the high data-transfer rates between modules. The description of the modules and their functions are:

1. **Array Control Unit.** This unit exercises direct control over the four associative memory arrays. This includes decoding of the parallel instructions received from the control memory, the loading of internal comparand and access mode registers accessed by the memory arrays, and the setting of control inputs to the arrays. The sequencing of the parallel instructions is also under the control of this unit.
2. **Parallel I/O Flip Network.** This communication network handles parallel data transfers among the four STARAN associative memory arrays, and also the parallel input and output of data to the arrays. The flip network permits the data to be permuted during these transfer operations. The way in which the permutations are implemented is the same as for the internal flip networks within the memory arrays; these will be described in connection with the arrays.
3. **The Parallel I/O Control Unit.** This module controls the data transfers performed by the parallel I/O flip network. This includes some con-

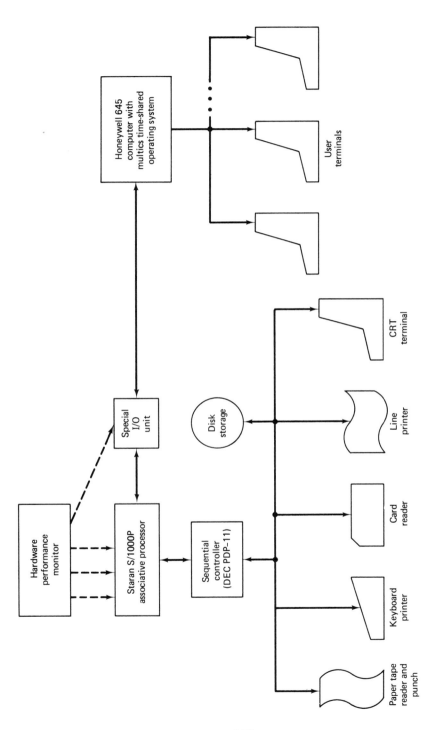

FIGURE II.20 STARAN/645 system organization

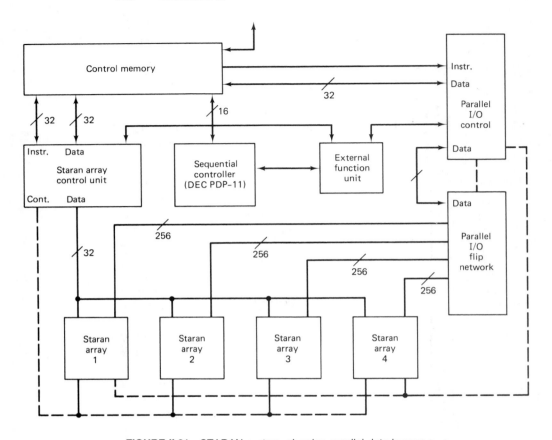

FIGURE II.21 STARAN system, showing parallel data busses

trol over the memory arrays themselves. The I/O control over some memory arrays is exerted simultaneously with the internal processing being performed on other memory arrays. Control instructions for these operations are obtained from the control memory; this requires that the control memory have a very fast access time.

4. **Sequential Controller.** This conventional sequential processor provides system interfacing with the system peripheral devices and with the operator.

5. **The STARAN Memory Arrays.** These four modules are the STARAN associative memory arrays. They each contain 256 associative memory cells of 256-bit data capacity each.

6. **Control Memory.** This fast random-access memory contains the parallel instructions for the array control unit and the parallel I/O control unit.

7. **External Control Unit.** There are three control units in the STARAN system. All three can operate simultaneously. The external control unit provides synchronization for their operation. Operation of this unit is initiated by commands from the control units.

3.2.2.3 The STARAN memory array

The associative memory of the STARAN is constructed from four memory arrays. Each array consistsof 256^2 bits of data storage, together with processing logic and intercell communication logic for the data.

The multidimensional access memory. In a sense, it is misleading to refer to the data bits in the array memory as being organized into 256 cells of 256 bits each. This is because the access to the data, for input, output, and computation can be made in a number of ways, some of which do not correspond to the square-array organization. In all cases, 256 bits are accessed simultaneously. If one thinks of the 256^2 bits in the array as being organized into 256 cells of 256 bits each, then there are two access modes which permit simultaneous access to the 256 bits constituting a single cell or the 256 bits constituting the kth bits of all 256 cells. For the 256-cell organization of the array memory, these two access modes are termed *word-slice* and *bit-slice*, respectively. Figures II.22(a) and (b) illustrate the memory of a STARAN array with these two access modes indicated by the shaded areas in the respective diagrams.

There are a number of additional memory-access modes. Three of these follow from the consideration of the array memory as being organized into 32 records of 256 eight-bit bytes each. These three modes permit access to:

1. Any set of 32 contiguous bytes in any single selected record
2. All eight bits of the kth byte in all records; $k = 1, \ldots, 256$
3. The jth bits of all bytes in any single selected record; $j = 1, \ldots, 8$

Figures II.22(c), (d), and (e), respectively, illustrate these three access modes. It is the existence of these modes and a number of other possible modes that gives rise to the name "multidimensional access memory" for the STARAN array memory.

Array organization. Figure II.23 shows the organization of the STARAN memory array module. The module contains a multidimensional array which communicates with three 256-bit registers in bit-parallel fashion through a permutation network called the *flip network*. The three registers are called the M, X, and Y registers.

The M register contains the write mask for the parallel-write operation to the memory. The M register can be loaded from the other modules of the array.

148 CHAPTER II

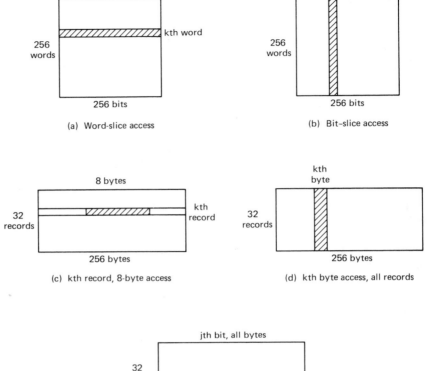

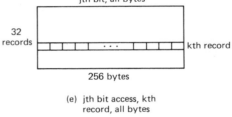

FIGURE II.22 Some multidimensional memory access modes

The X register with its logic implements the 16 boolean functions of two variables. One of the operands is the contents of the X register itself. The other is the 256 bits in the output of the flip network. The result of the operation goes to the X register.

The Y register operates like the X register. The X and Y registers can be operated together simultaneously, both implementing the same boolean function on their respective variable pairs.

The boolean operations on the X register can be masked by the contents of the Y register. That is, the operation on each bit of the X register and its corresponding bit from the flip network takes place if the corresponding bit of the Y register is set; the operation is not performed otherwise. The Y

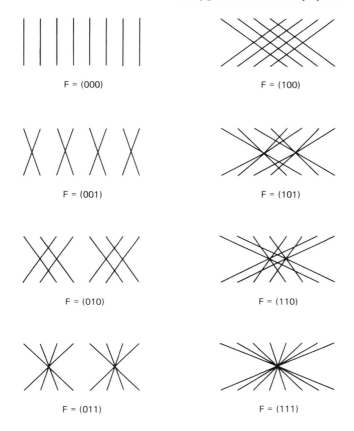

FIGURE II.23 STARAN array module

register logic may be simultaneously performing boolean operations on its contents while the masking operation is taking place. It is the former contents of the Y register that are used as the mask in this case.

Similarly, the contents of the X register may be used as a mask for boolean operations performed on the Y register contents.

Many parallel arithmetic and boolean operations can be performed using the logic and registers just described. For example pairwise-parallel arithmetic can be implemented by a simple programmed algorithm. The memory is considered to be organized into 256 words of 256 bits each for the operation. The operation adds the integers contained in one field of all selected words to the corresponding integers contained in a second field of the words, storing the sums in a corresponding third field of all of the words.

The algorithm is executed on each bit of the three fields in sequence, beginning with the least-significant bit. The bit-slice memory-access mode is used at each bit position. The algorithm requires four steps for each bit position. The first two steps consist of Exclusive-ORing the operands, one at a time, with the contents of the X register, which initially contains the carry bits from the previous bit position, and storing the results in both the Y and X registers, after each of the two operations. This puts the new sum and carry bits into the Y and X registers, respectively. The next two steps consist of first storing the sum bits back into the memory while selectively complementing the carry bits, and then Exclusive-ORing the contents of the X register into both the X and Y registers. This clears the X register and stores the new carry bits into the Y register in preparation for the next bit position.

Many other combinations of field-selective arithmetic and boolean operations, including floating-point, can be implemented by using similar techniques. The addition just described requires 700 nsec per bit, plus a small set-up time, with memory read and write cycles of 90 nsec and 100 nsec, respectively. At this rate, an upper limit processing rate of 40 MIPS (million instructions per second) could be achieved if all 1024 cells in all four arrays were to participate in every operation. It is true that the average activity rate in a typical application will seldom approach this, but the figure gives a meaningful indication of the power of the STARAN.

The flip network. The flip network is the communication system which gives the STARAN its software-reconfigurable array capability. With the flip network, the entire source item of 256 bits, which can be selected from the memory, the X, Y, or M registers, or from the outside, can be shifted and permuted in a number of ways, the results appearing at the output of the network.

The flip network is central to the programming and operation of the STARAN, and therefore deserves a somewhat more detailed description. There are two types of permutations performed by the flip network. These are called *flip permutations* and *shift permutations*, and will be described in order.

Flip permutations. The permutation function implemented by a flip network with 2^n input lines is specified by the value of an integer that is input to the flip network of n control lines. Let $F = (f_0 f_1 \ldots f_{n-1})$ be the n-bit binary vector that is input to the flip network control. Then the flip network will move the data that is input on the Ith data input line, where I has the binary representation $I = (i_0 i_1 \ldots i_{n-1})$, to the Jth output line, where the binary representation of J is the vector Exclusive-OR function $I + F$, defined as $(i_0 + f_0, i_1 + f_1, \ldots, i_{n-1} + f_{n-1})$. Figure II.24 shows eight diagrams illustrating the eight possible flip permutations which can be implemented for an 8-bit flip network. Here $n = 3$, so there are three control lines. The respective values of F for the eight flip permutations are shown beneath each of them.

As an example of the way in which the individual flip permutations are calculated from the definition just given, consider the permutation with the

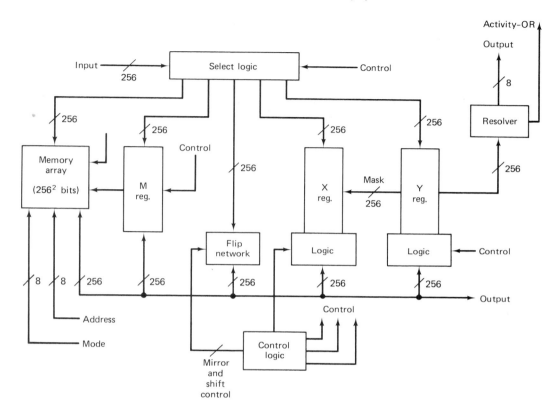

FIGURE II.24 All flip permutations for an 8-line flip network

control input value $F = 2 = (010)$. The data on the zeroth input line is transferred to the line whose number has the binary representation $(0 + 0, 0 + 1, 0 + 0) = (010) = 2$. Similarly, the data on the first input line is transferred to the third line, $(0 + 0, 0 + 1, 1 + 0) = 3$, and so on. The correspondence between the two values just evaluated for the function and the respective flip permutations in the diagram can be seen.

Rather than implementing all of these 256 flip permutations in one step, they can be implemented by means of a sequence of eight basic permutations performed in succession using eight levels of hardware logic. Each of these basic permutation networks implements one of the eight permutation functions whose numerical values have a binary representation containing exactly one zero; that is, the values of F are 1, 2, 4, . . . 128. It can be demonstrated that all 256 flip permutations of the data on the 256 input lines to the flip network can be performed by executing eight or fewer of these basic permutations in sequence. Thus the flip network logic can be controlled by inputting each bit from the control input as an enabling bit to the corresponding basic flip network, activating that network's function if the corresponding bit is 1, or not activating it if the bit is 0.

Figure II.25(a) shows the three basic flip networks which are connected in sequence to implement all eight flip permutations of eight input bits. These basic networks are called *two-way data selectors*. Figure II.25(b) shows how the same combined flip network can be constructed from three identical two-way data selectors, thus possibly saving design effort (depending on the size of the wafers). This is done by shuffling the eight data lines at the output of each data selector.

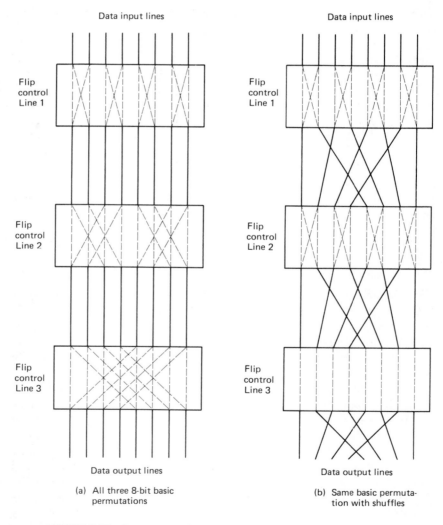

(a) All three 8-bit basic permutations

(b) Same basic permutation with shuffles

FIGURE II.25 General permutations performed with basic flip operations

One four-way data selector can be used in place of two two-way selectors. This will reduce the number of logic levels in the hardware by a factor of 2, and thus permit a faster clock time if the remainder of the logic permits. The flip networks in the STARAN memory arrays are constructed from four four-way data selectors.

Shift permutations. The flip network also implements a set of $(n^2 + n + 2)/2$ shift permutations, which are applied after each flip permutation. These consist of the identity permutation and $(n^2 + n)/2$ permutations which are shifts of 2^m (mod 2^p) places, where m and p are integers such that $0 \leq m \leq p \leq n$. One of these shifts divides the 2^n input data bits into subsets of 2^p bits each, and then shifts the items within each subset end-around 2^m places. Table II.1 shows the seven shift permutations performed in an 8-line flip network, with numbers, rather than diagrams, used to indicate the permutations. Notice by comparing the permutations with the flip permutations in Fig. II.24 that the shift permutations for 1 (mod 2), 2 (mod 4), and 4 (mod 8) are identical to the flip permutations for F = (001), (010), and (100), respectively. Figure II.26 shows a flip network for eight data lines which performs both the flip permutations and shift permutations. The names of the six control lines and the particular two-item data selectors which they control are shown. The right-hand part of Table II.1 shows the values assigned to those control lines to implement the shift-permutations shown in the center part of the table. A zero value input to a selector indicates that no permutation is performed (i.e., straight through); a one value indicates that the pairwise swap permutation is to be executed.

To perform combinations of flip permutations and shift permutations with the flip network, each control input is fed by the output of an Exclusive-OR gate. One input to each gate is the corresponding shift permutation control input; the other is the corresponding flip permutation control input. It is an interesting exercise to devise the control inputs to the 8-bit flip network to perform shifts in a negative direction. This requires two passes through the

TABLE II.1

SHIFT PERMUTATIONS FOR AN 8-LINE FLIP NETWORK

Permutation	Order of Data Lines	Control Signals 0A 1A 1B 2A 2B 2C
(Original Order)	0 1 2 3 4 5 6 7	
Identity	0 1 2 3 4 5 6 7	0 0 0 0 0 0
1 (mod 8)	7 0 1 2 3 4 5 6	1 1 0 1 0 0
2 (mod 8)	6 7 0 1 2 3 4 5	0 1 1 1 1 0
4 (mod 8)	4 5 6 7 0 1 2 3	0 0 0 1 1 1
1 (mod 4)	3 0 1 2 7 4 5 6	1 1 0 0 0 0
2 (mod 4)	2 3 0 1 6 7 4 5	0 1 1 0 0 0
1 (mod 2)	1 0 3 2 5 4 7 6	1 0 0 0 0 0

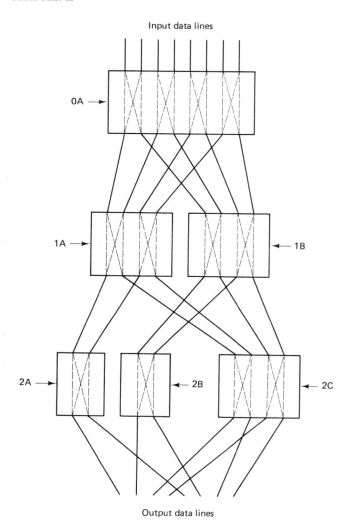

FIGURE II.26 Eight-line flip network for both flip and shift permutations

network. Reference [11] describes the flip network in STARAN in somewhat more detail. The reference illustrates a number of manipulations useful in various applications, such as the processing of digitized images and the solution of partial differential equations. These operations include the spreading of pixels over a larger number of cells, and also rotation and interpolation of the pixels. Interpolation of values when moving across mesh boundaries on a multi-mesh grid during the solution of differential equations is well suited for the flip network in processing data between different sets (or from and to the same set) of PEs.

3.2.3 The Associative Linear Array Processor (ALAP)

The Hughes Associative Linear Array Processor (ALAP) is another example of an associative processing system which has been built and programmed†: Like the STARAN, the ALAP implements the basic parallel-by-word, serial-by-bit operations common to most associative processors, and is effective in most applications for which the STARAN has been used. However, there are a number of basic differences in philosophy between the two machines, mostly originating with the difference in the communication techniques for interchanging the contents of the associative cells. These differences will be discussed following the description of the ALAP.

†References [9], [1], and [18] give more detail on the ALAP and its applications.

3.2.3.1 The ALAP system

Figure II.27 is a block diagram of the ALAP system. The principal component of the system is the ALAP memory, which is a set of cells, each having content-addressable memory (contained in a single shift register), arithmetic capability, and the capability to communicate with the others by means of a unique *chaining channel*. This chaining channel effectively organizes the cells into a linear array.

The ALAP memory interfaces with the rest of the system through an interface unit, the second major system component. The interface unit contains registers from which control and data information are fed directly to the ALAP memory, thereby causing it to execute instructions. The interface unit

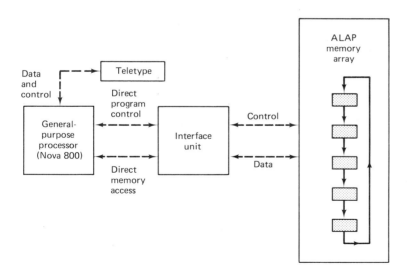

FIGURE II.27 The ALAP demonstrator system

stores data and control information for several instructions in advance, thereby enabling it to operate asynchronously with the system controller.

System control is exercised by a NOVA 800 computer, the third major system component. The NOVA interfaces with the user through a typewriter and a paper tape reader and punch. It interfaces with the ALAP memory via the interface unit. The NOVA contains, in its conventional memory, the ALAP application program.

The ALAP application program consists of:

1. A set of tables containing control and data information for the ALAP memory. These are the ALAP memory instructions.
2. A program of NOVA instructions which controls the general flow of system operation, initiates the transfer of tables of ALAP memory instructions to the interface unit as they are needed, and controls all system input/output operations.

The NOVA also performs some calculations incidental to ALAP memory operation, such as the occasional generation of data values. The overwhelming bulk of all system computation is performed within the ALAP memory, with very little communication of data or flag conditions occurring during program execution except for that necessary for input and output for the user. Such performance, in fact, is necessary for efficient use of the system, and is a measure of the applicability of the system to its tasks.

3.2.3.2 The ALAP memory array

Figure II.28 is a simplified view of the structure of an ALAP memory array, showing the data communication channels. There are four channels, all bit-serial, connected to each ALAP cell. Three of them are parallel busses. They are shown as a single bus in the figure. One of these is used to "broadcast" arithmetic operands to all cells, or to a selected subset of the cells simultaneously. The second parallel bus is used for input of a single data item

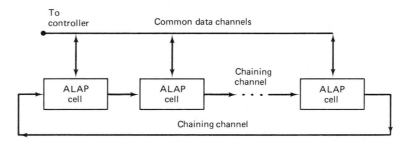

FIGURE II.28 The ALAP memory array

to some or all cells. The third is used for output of a single data item from a selected cell (or the logical OR of several items, if more than one cell is selected).

The fourth communication channel is called the *chaining channel*. It connects each cell to its neighbor (in one direction only), and thus effectively organizes the ALAP memory into a one-dimensional array. There is a "first" cell and a "last" cell in the array as determined by the chaining channel. These are usually connected together (under software control) during system operation. Both data items and the states of the cells' internal flags can be transferred via the chaining channel, the determination being made under software control. An individual cell, again under software control, can act as the originator or receiver of data on the chaining channel, or may simply act as a relay transferring the chaining channel input to the cell directly to its chaining channel output.

3.2.3.3 The ALAP cell

Figure II.29 shows a single ALAP cell with its major components and interconnections. Each ALAP cell contains a 64-bit shift register for its data, together with an arithmetic unit for bit-serial arithmetic and boolean processing of the data register contents. The second operand in arithmetic operations is external to the cell, and comes either from another cell via the chaining channel or from one of the common busses (the latter are not shown in the figure). The operations that are performed include addition, subtraction, step multiplication, division, square root, exact comparison, less-than and greater-than comparison, and Exclusive-OR. The result of an arith-

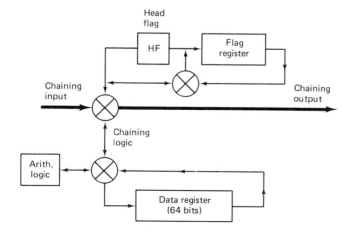

FIGURE II.29 General organization of the ALAP cell

metic operation may be stored in the cell's own data register or may be output on the chaining channel to the next cell in the array. All arithmetic operations are field-programmable, the operand fields being determined by the (bit-serial) input to two global control lines during the operation.

3.2.3.4 ALAP operation

The major ALAP instruction is called the *word-cycle* instruction. During the execution of this instruction, the data registers of a selected subset of the cells are shifted a specified number of bits, while arithmetic and data-transfer operations take place. The selection of the particular arithmetic operation to be performed during the word-cycle operation is globally determined. The selection of the cells at which the operation is performed is local; that is, it is determined by the settings of flags within each cell. The operation of the chaining channel during a word-cycle operation is determined by a combination of global and local states. The local states are the settings of the bits in a 6-bit shift register, called the *flag register*, in each cell. These flag bits are set and modified under program control by a set of *flag-shift* instructions. These instructions can also transfer flag register settings from cell to cell along the chaining channel. A *head flag* connected to the flag register permits the reordering and logical manipulation of the flag states within the cell, and also the transfer of these states between cells. Table II.2 shows the functions of the various flags in the flag register during word-cycle operations.

During word-cycle operations, input and output of data-register contents via the common busses may take place simultaneously with the arithmetic

TABLE II.2

FUNCTIONS OF FLAGS IN FLAG REGISTER DURING WORD-CYCLE OPERATION

Flag	Function
2	Together with Flag 3 and global state, determines chaining channel option during word cycle.
3	Together with Flag 2 and global state, determines chaining channel option during word cycle.
4	Determines participation of cell in globally specified arithmetic/boolean operation during word cycle.
5	Overflow flag for arithmetic operations. Also used in specifying type of operand in multiply and divide operations.
6	Specifies cell that is to receive input from a common bus during word cycle.
7	Specifies cell that is to output to a common bus during word cycle.

operations and the chaining channel data transfers. The cells which thus receive or output data are selected by their internal flag-register states.

Program execution in general consists of sequences of flag-shift operations which set up the local states of the cells, each such sequence followed by a word-cycle operation during which actual data transfer and computation takes place.

It is clear that in the ALAP, almost all processing is organized around the chaining channel, which takes part in nearly every operation. This is because the second operand in every arithmetic and boolean operation is always external to the cell, being accessed either through the chaining channel input to the cell or from the common operand bus (which is one of the three common busses in the array). The chaining channel is only one bit "wide," so data layout is of great importance in avoiding conflicts during parallel operations, and in minimizing the need to reshuffle the operands among the cells between sets of parallel operations. Some tricks in programming will be shown in Sec. 5.

3.2.3.5 Fault-tolerance in the ALAP

A major feature of the ALAP memory design is the fault-tolerant capability built into every cell. This capability is implemented by means of two special fault-isolation instructions.

Faulty cells can be identified by diagnostic programs, and flags set by the programs at all malfunctioning cells. Then, by executing one of the fault-isolation instructions, which set a number of toggles built into the cells, the software can switch all faulty cells to a no-op state, with the chaining channel set to route all data past them.

The ALAP array is equipped with an alternate chaining channel, and switching is provided at the chaining input to each cell for switching between the normal and alternate channels. Thus, if a cell's chaining channel malfunctions, the diagnostic program, by executing the second of the two fault-isolation instructions, can switch the chaining channel input to the cell following the faulty cell from the normal chaining input to the alternate one. The alternate channel comes from the cell prior to the malfunctioning cell, thus bypassing the bad cell altogether.

Once the fault-isolation instructions have been executed, the faulty cells cannot be reconnected to the array, or the chaining inputs switched back to the normal chaining channel except by turning off the power to the array and then reapplying power to reset it to its initial state.

Figure II.30 shows four ALAP cells with the two chaining channel inputs to each cell. A faulty chaining output from cell number $n + 1$ is illustrated. The chaining input to the next cell (cell $n + 2$) is shown switched to the alternate chaining channel. The chaining inputs to the remaining cells are shown switched to the normal chaining channels.

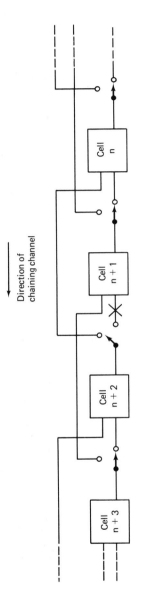

FIGURE II.30 Alternate chaining channels for fault isolation

3.2.4 Comparison of STARAN and ALAP

Although the STARAN and the ALAP both share the same basic associative memory capabilities and concept, the difference in philosophy between the two designs is considerable. Most of these differences are reflected in the intercell communication techniques. The STARAN was designed with the capability for fast parallel arithmetic in mind, and the flip network reflects this intention. It permits very rapid interchanging of operands among the STARAN cells, and it is particularly helpful in the important fast Fourier transform (FFT) operation, which requires substantial reordering of data. The ALAP cannot shuffle operands in this fashion because of the linear organization of its chaining channel. However, the chaining channel permits all of the cells in the ALAP memory to be effectively organized into a single long shift register with multiple ports for inserting new data and processing the memory contents in parallel. This is particularly helpful in data retrieval and text-processing operations, since the shift-register type of operation permits more flexible processing when the sizes of the individual operand (which are records in a file, or words and sentences in a block of text) are unrelated to the length of the cell's data register than is the case with STARAN.

STARAN has been successfully employed (see Boulis [16]) as a peripheral device on a fast, conventional system for performing a number of complicated data-reduction and analysis operations. The source of the data was the LANDSAT Earth-resources satellite, and the project was concerned with crop forecasting on a global scale. The capability of STARAN to output many bits of data at each clock cycle is important in this application. ALAP cannot output data at this rate, and is thus less applicable than STARAN in applications requiring shared processing with another computer.

ALAP was designed with the intention of reducing costs for large associative memories. The ALAP memory wafer, as a result of this, has few external connections (24 pins, to be exact), with this number being independent of the number of cells on the wafer. It is also the reason for the built-in fault-tolerant capability at the cell level. The STARAN architecture, because of the flip network design, cannot be made fault tolerant at the cell level. The number of pins on the STARAN wafers is relatively large compared to the ALAP, and is a function of the number of cells. The ALAP is thus more amenable to LSI implementation and to applications in which the data base is very large or in which fault tolerance is of great importance.

3.3 Distributed-Control Arrays: The Holland Machine

All of the reconfigurable parallel systems so far described have centralized control of program execution, with the program itself residing externally to the array of processors. (It is true that the ILLIAC IV has four control units; however, it is effectively constructed as four separate computer systems, with

162 CHAPTER II

a common operator interface and bulk storage; these four systems can be connected together for some applications to execute a single program stream.) There are machine designs in which program control is distributed throughout an array of parallel processors, and in which an arbitrary number of program streams can be in execution simultaneously. The best known of these is the Holland machine.

The Holland machine was originally proposed in 1958 as a conceptual device for the development of automata theory, rather than as the basis for a practical computer design. However, it is appealing to the designer because of its ability to execute multiple program streams, and for several other reasons which will be explained. As a result, a number of efforts have been made to modify the concept to produce a more practical design.

3.3.1 Organization of the Machine

Figure II.31 shows the organizational concept of the Holland machine. The machine consists of a rectangular array of processing elements (PEs), interconnected by two networks of communication channels. One of these networks is for communication of processing control. The other is for the communication of data. The indices of the PEs are shown in the figure.

Figure II.32 shows the internal organization of a PE. The PE is multi-purpose. It can execute an instruction or it can store an operand for an instruction in another PE. It can act as an accumulator for the result of arithmetic, boolean, or address calculations. And it can act as a node in a path

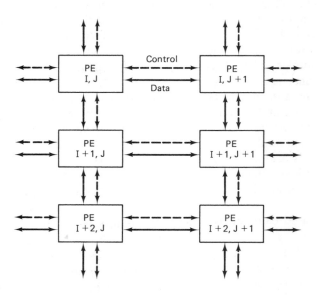

FIGURE II.31 Portion of Holland machine array

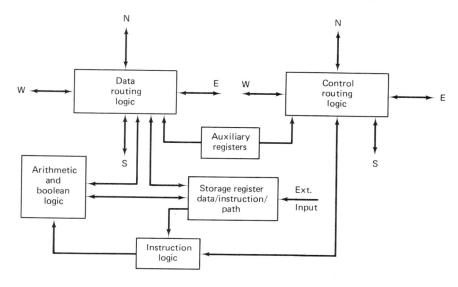

FIGURE II.32 A Holland machine processing element

over which data is routed between cells. This routing function can generally be performed by the PE simultaneously with one of the other three functions.

Storage for the PE consists of a single register. Depending on the function assigned to the PE, the contents of this register will be interpreted by the PE as an instruction, an operand for another instruction, an accumulator, or a data-routing node. As indicated in Fig. II.32, the PE contains its own instruction-decoding and instruction-execution logic, its own arithmetic and boolean logic, and the routing logic for both communication networks. It also contains a number of auxiliary registers, all of which are only one or two bits in length, and are used to flag the status of the PE and to contain some of the control data for the program in which the PE participates. The four external connections to each of the two communication networks connect to the four nearest-neighbor PEs in the array, denoted by N, E, S, and W in the figure.

3.3.2 Storage of Programs and Operands

In the Holland machine, each program residing in the array resides in a set of (generally) contiguous PEs, that is, in PEs that are directly linked to one another in string fashion by the communication paths. Figure II.33 illustrates a portion of an array of PEs with two program strings. Each PE in the string except the first has a predecessor PE. The relative "direction" of this predecessor (that is, N, S, E, or W) is stored in two bit positions of the PE's storage register. Each PE in the string except the last also has a successor PE, with its direction also stored in two bits of the register. Each PE except the first in

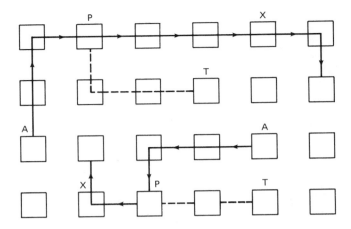

FIGURE II.33 Two program strings residing in the Holland array

the program string contains a single instruction, whose operation code is stored in three bits of the register. The successor PE stores the next instruction to be executed; the predecessor PE stores the previous instruction (with the exception of the last and first PEs in the string, respectively). In Fig. II.33, the lines of successors for the two programs are shown by solid lines interconnecting the PEs. The arrows point in the direction of program flow.

In Fig. II.33, one of the PEs in each program string is labeled with an X, indicating the *X status*. This is the active PE in the program, that is, the PE containing the instruction currently being executed. When the execution is finished, the active PEs will transfer program control to their respective successor PEs, putting them in an X status. The X status of a PE is denoted by the setting of a 1-bit internal register.

The PEs which act as accumulators for the programs are labeled with an A in the figure. They are the first PEs in their respective program strings. Operands for the programs can be stored in any PE not used as an accumulator, or to hold an instruction, or as a special node in a path, as will be explained. In the figure, two PEs, labeled with the letter T, each hold a single operand for their respective programs. These operands are contained in the main storage registers of the PEs, occupying all bit positions except the least significant one.

3.3.3 Path-Building

During program execution, the operand for an instruction being executed is directed to the program's accumulator by a process known as *path-building*. This is one of the more interesting concepts of the Holland machine, and deserves a more detailed description.

To access the operand, the active PE opens a set of gates along its line of predecessors to the nearest *P-module*. This is a PE with a bit set in its register denoting P status, which means that it is the origin of the path to the operand for the instruction. The active module then sends to the P-module the specification for the first segment of the path to the operand. This specification consists of two bits indicating the direction of the segment (N, S, E, or W), and a number of bits giving the number of PEs contained in the segment. (A path segment consists of a set of adjoining PEs all in the same row or column.)

If there is more than one segment, the PE following the last PE of the preceding segment contains the specification of the next segment in its register. The PE following the final segment contains the operand. The contents of one of the 1-bit auxiliary registers indicates whether or not the corresponding PE is the first one in a segment. If the PE lies at the end of a segment and that register is not set, the PE contains the operand.

All PEs have four 1-bit registers for indicating membership in one or more path segments. Any PE which is not at the termination of a path segment can belong to as many as four path segments, all in different directions. Each 2-bit register corresponds to a particular direction.

The PEs in Fig. II.33 which are labeled with the letter P are P-modules for their respective programs. The paths which originate at those P-modules are denoted by the broken lines connecting the PEs in the paths. Each P-module can be the origin of only one path. Each path segment is constructed by the process of setting the appropriate 1-bit path-direction registers in each PE belonging to the segment, meanwhile decrementing the count for the length of the segment. When the end of the segment is reached, either the next segment is constructed or else the end of the path has been reached. All of this operation takes place during a phase of the instruction execution, and takes place simultaneously for all programs currently being executed. After the operand is located, it is gated back along the path to the P-module, and then back along the line of predecessors to the accumulator PE, where it is used as directed by the instruction.

3.3.4 Storage and Auxiliary Register Functions

Figure II.34 shows the function of the PE storage register when the PE holds an instruction or the specification for the beginning of a path segment. The field descriptions are clear from the preceding discussion.

The eight auxiliary registers and their functions are the following:

1. The 1-bit E register. This register is set if the PE is active, and is reset otherwise.
2. The 1-bit A-register. This register is set if the PE is an A-module (i.e., an accumulator), and is reset otherwise.

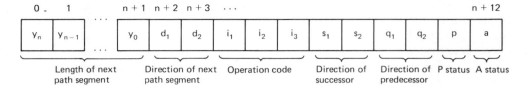

FIGURE II.34 The storage register and its control functions

3. The 1-bit D-register. This register is set if the PE is the initial PE in a path segment, and is reset otherwise.
4. The 2-bit (D_1, D_2)-register. If the D-register is set, the (D_1, D_2) register indicates the direction of the path segment. If the A-register is set, the (D_1, D_2) register indicates the direction of the successor PE.
5. The four 1-bit segment registers. Each of these registers corresponds to one of the four directions within the array. A segment register is set if the PE belongs to a path segment in the corresponding direction, and is reset otherwise.

3.3.5 Other Features and Conventions

There are a number of lesser features and conventions that will not be discussed in detail. The features include the capability to delete and add segments to a path and the capability for loading the programs into the array of PEs. The many conventions built into the hardware give rules of precedence for such things as avoiding conflicts in path-building operations, and for determining the order of operations when several programs attempt to modify the same PE at the same time. These features and conventions are described in the original paper on the Holland machine, which is Holland [6].

3.3.6 Execution of Instructions

Instruction execution in the Holland machine takes place in three phases. Each phase is executed simultaneously for all programs that are active. In the first phase, PEs which are flagged to receive input from outside of the array receive a common item of data (which is the same item for all flagged PEs). In the second phase, all modules in the X state interpret their instructions and cause paths to be built to their respective operands, if the instruction uses an operand. In the third phase, the active modules cause their instructions to be executed, moving the operands to their accumulators to perform the execution if required.

There are only eight instructions in the Holland machine. These are very general in purpose. They will not all be described here, but they include an ADD, a STORE, a NOP, and a STOP instruction. There is also a TRANS-

FER ON MINUS instruction, which provides a means of making conditional branches in program execution. If a reserved bit in the operand PE at the termination of the nearest path (nearest to the PE containing the branch instruction) contains a zero, the instruction in the successor PE is the next one executed. If the bit contains a one, the contents of the operand PE's storage register is the next instruction, and control is transferred to that PE.

There is a SET REGISTERS command which permits one program to alter the auxiliary registers of a PE external to the program. For example, using this instruction, one program can activate a PE in another program, thus causing it to execute its instruction during the next instruction-execution phase. This second PE can, for example, be a subprogram for the original program.

3.3.7 Construction of a Holland Machine

The original Holland machine required about 1000 gates per processing element in its construction, using a storage register of 40 bits. With current LSI technology, the construction of many PEs on a single wafer presents no problem except possibly with regard to the number of pins on the wafer. This number of pins increases with the number of PEs, although not as fast. There is one feature, however, which makes the Holland machine more attractive than most other multidimensional processing arrays. This is the fact that the machine has potential for built-in fault tolerance at the individual cell level.

3.3.8 Applying and Programming the Holland Machine

There are some very appealing aspects to the application and programming of the Holland machine. For one thing, the multiple programs which can execute simultaneously on the machine may be concerned with completely unrelated applications, or they may be subprograms of the same program operating together in some fashion. It is the latter cases which are interesting from the programming standpoint.

Consider, for example, an application in which the behavior of a physical system is being modeled; say, the system is an irregular surface with heat being applied at the boundary.† The usual practice is to approximate the continuous surface by a network of nodes spaced in as regular a fashion as possible. The partial difference equations for the heat function are then applied at all nodes simultaneously, thus approximating the change of temperature over the surface with respect to time.

With the Holland machine, the subprograms for the various heat functions would be laid out over the array of PEs in relative positions roughly corresponding to the positions of the nodes on the surface, one subprogram

---

†The same application is described in Sec. 1.5.

for each node. This would make it possible to program the interrelationships between the nodes by means of interactions between the corresponding subprograms.

Figure II.35 illustrates two subprograms that interact with each other (not necessarily for the application just discussed), as they might be laid out over a portion of the Holland array. Notice that both subprograms reside in closed loops of PEs. These loops will repetitively execute their respective functions, that is, they are program loops as well. It is seen that each subprogram has a path connecting it to a PE in the program string in the other program, thus enabling each of the two subprograms to modify the contents of a PE in the other. For the subprogram on the left, the PE being modified is the accumulator, which might contain a count over the number of iterations. The modification might be to reset the initial value of the count. For the subprogram on the right, the PE being modified might contain a NOP instruction which the program on the left changes to a STOP instruction in order to terminate the subprogram's execution. Notice that the subprogram on the right also can modify the contents of the same PE, perhaps to reset it to a NOP the next time the entire program is executed.

The appeal of such programming techniques is offset, unfortunately, by enormous difficulties in implementing practical programs on the Holland machine. It is the author's understanding that many of the problems encountered in programming the machine have to do with conflicts between paths connecting programs to their respective operands. When the programs are complicated or there are many subprograms, these difficulties multiply out of all proportion.

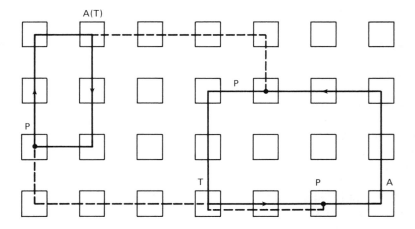

FIGURE II.35 Two interacting subprograms

3.4 References on Parallel Computer Architectures

References [6] thru [9] are standard references on their respective topics. Excellent references on STARAN are found in [5] and in several papers in the 1973 Proceedings of the Sagamore Computer Conference ([2], pp. 140–161.)

There are many interesting architectures which have not been discussed here. A very recent paper [10], describes a new architecture which is being built for advanced real-time image processing applications. Other architectures of interest are described by Hiner [3], 1976, pp. 140–144. The author [31] describes an associative processor architecture using a bulk memory for data base storage, and [31] an architecture for advanced question-answering systems. Lipovski [32] describes an advanced interconnection network and its use in dynamic restructuring of multi-microprocessor architectures. See also Reddaway [24] and Okaga [25] for other approaches to multiprocessor system design.

4. DESIGNING A RECONFIGURABLE PARALLEL ARRAY PROCESSING SYSTEM: A CASE STUDY

This section is a reconstruction of an actual system design effort by the author and others, an effort that involved reconfigurable parallel hardware. The section first describes the starting point for the effort. This includes the description of the application, some applicable algorithms, some assumptions that were necessary, the author's previous experience on related projects, and some of the tools and techniques which were available and applicable.

Next, the discussion follows the actual design effort to its termination, showing the order in which the various design tasks were undertaken and the justification for that order. It also describes the design tradeoffs which were made and the analyses which justified them. Since the effort was primarily a study, no hardware was built. However, the preliminary design needed to be justified in terms of cost and risk, and the means for achieving initial estimates for those factors are shown.

The application is a very demanding one, as will be seen, and certain features of the algorithms seemed to point to the need for parallel processing techniques. These two factors alone constituted sufficient justification for further analysis of the problem, and led to the author's involvement in the project effort.

4.1 A Note on the General Approach

The discourse in this section follows the order in which the actual design effort was conducted, rather than the order in which the design itself would be best explained. One will notice in particular that most factors in the analysis and design, such as hardware speed, overall system organization, software organization, and input processing are not discussed in full before proceeding to other such factors. The project effort, in fact, skipped from one subject to another with only a partial analysis, returning to each subject later for more thorough consideration. The reason for this deserves special explanation.

At the beginning of the project effort, no assumption was made that the system task would even be possible with the technology presumed to be available, much less that a reconfigurable parallel array processor would prove to be the best applicable system. This being the case, the author wished to obtain some assurance of feasibility at an early stage in the project, in order that his time and project funds not be wasted. In particular, the author did not wish to undertake a lengthy point-by-point analysis and design, only to discover near the end of such an effort that some obvious reason existed showing that the task would be impossible for any system in the foreseeable future.

Therefore, throughout the project effort, whenever some fact would become apparent which indicated a serious risk with regard to feasibility, the effort was diverted to the corresponding analysis until the question was resolved.

4.2 The Application

The application for which the system is intended is variously referred to as *radar pulse-deinterleaving*, or *signal-sorting*. This is the task of sorting out a stream of intermixed pulses, received from a number of radar and telemetry transmitters, into its constituent pulse streams. The objective is to identify or classify the sources of the individual pulse streams. These may be weapons under test, for example, or weapons actually being employed. A signal-sorting system is therefore a passive monitoring system, which can be employed in a number of different ways. Depending on its design and capability, it can be used in tactical situations to determine the types and numbers of weapons being deployed by an enemy, or it can be used to monitor the testing of weapons or other devices by a potential adversary.

The block diagram of a monitoring system using a signal sorter is shown in Fig. II.36. These systems employ wideband receivers to collect all digital signals over a frequency range wide enough to include all signals of interest. Analog devices and hardwired digital logic are employed to separate the legitimate pulses in the input from noise spikes, continuous analog signals, and other inputs of no interest. Following this, the distinguishing characteristics of each pulse are digitally encoded; that is, converted to binary numbers. This task is also performed by hardwired devices. The outputs of these devices are then input to the signal-sorting subsystem. The output from the

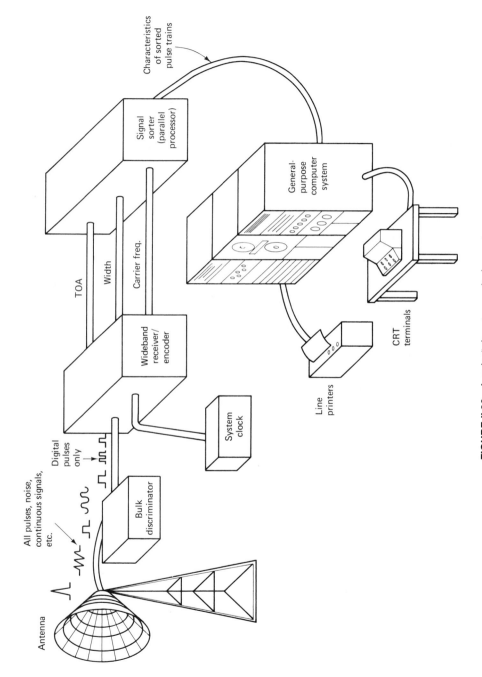

FIGURE II.36 A radar/telemetry monitoring system

signal sorter consists of the identifying characteristics of the individual pulse streams which were detected. This information is then analyzed by another system to identify the emitting device if it is known, to classify the device if it is not known, to extract information from the pulse stream if it is telemetry, or for some other related purpose. The requirements for the system to be designed by the author were among the most difficult requirements for systems of this type. For example, the input pulse rate could be as high as 1 MHz and the pulse stream was to contain data from up to 1000 separate emitters. In this system, three pulse characteristics were to be extracted from the pulses received by the system. These are:

1. The pulse time-of-arrival, or TOA. The system employs a high-resolution clock. At the instant that each pulse is received, the state of the clock is output to the signal sorter.
2. The width of the pulse. At the instant the pulse terminates, the state of the clock at the TOA is subtracted from the current state, and the difference is output to the signal sorter.
3. The pulse carrier frequency. The pulses of interest consist of short bursts of a high-frequency carrier. The wideband receiver determines the carrier frequency of each pulse, digitally encodes it, and outputs the resulting binary number together with the encoded pulse TOA and width.

Figure II.37 shows a digital pulse, with its carrier signal and its envelope. Figure II.38 shows a sequence of pulses from several emitters as they are received by the monitoring system.

The signal sorter receives the three pulse parameters for each pulse in real time, that is, at essentially the same time that the pulses are received at the antenna, and at the same rate. The rate is very high indeed for the scenario chosen for the project.

It is the task of the signal sorter to separate its input into separate pulse streams, keep track of the pulse streams for as long as they persist, and output the characteristics of all pulse streams once they have been obtained. These

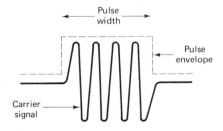

FIGURE II.37 A digital radar pulse, showing its carrier signal and envelope

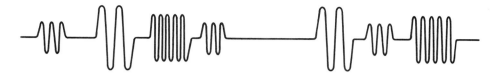

FIGURE II.38 A train of digital radar pulses from several emitters

characteristics include the pulse width and carrier frequency, the pulse-repetition frequency (PRF) of the pulse stream, the time at which the pulse stream began, and the number of pulses received.

It will be recognized by those familiar with signal processing that the foregoing description of the task is a simplified one. The emitters which have been described are the "nice" ones, in which the pulse-train parameters are reasonably well-behaved or regular. Figure II.39 shows a train of pulses with "normally" varying TOAs; that is, varying within narrow limits. In practice, the parameter values may be varied or "jittered" over wide ranges in order to hamper recognition. The scheme for doing this, of course, is designed into the intended receivers, and may be quite elaborate. For example, the pulses may arrive in pairs or triplets, rather than in equally spaced time intervals, or the carrier frequency and pulse width may both vary with time in some unusual fashion.

Figure II.40 shows two pulse trains with varying TOAs. The signal-sorting algorithms designed to overcome such schemes are correspondingly elaborate, and it is not within the scope of this section to deal with such elaborations. The sorting task for the "nice" signals is quite demanding enough. It is, in fact, quite beyond the technology of just a very short time ago, and barely within the present (1979) technology, as will be seen.

4.3 The Processing Concept

At the beginning of the project, no complete algorithm had been devised for the signal-sorting task. All that existed was a general concept of the way in which the processing would be performed. This subsection describes that concept.

Time marks

FIGURE II.39 Pulse trains with "normally" varying TOAs

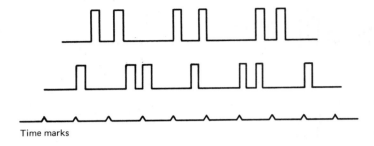

Time marks

FIGURE II.40 Pulse trains with deliberately jittered TOAs

There are two related tasks involved in the signal-sorting operation. These are:

1. Acquiring the pulse trains. This is the task of examining all incoming pulses which do not belong to any established pulse train and sorting them into new pulse trains if possible.
2. Maintaining the pulse trains. Once a sequence of pulses has been identified as belonging to a single pulse train, the signal-sorter must maintain that pulse train by assigning every subsequent matching pulse to it.

The second of these tasks seemed initially to offer no special problems. For each pulse, the pulse width and carrier frequency would be compared against upper and lower limits for those parameters stored in memory with each established pulse train. These limits would be determined at pulse-train acquisition time, and would probably not change appreciably thereafter. The pulse TOA would be compared against predicted upper and lower limits, which would then be updated. Successful comparisons for all three pulse parameters would constitute a match to the pulse train. The details could be worked out as a part of the project. The important thing was that it appeared initially that most of the pulse-train maintenance operations could be performed simultaneously for all established pulse trains using parallel hardware, and that no operations would involve the sequential processing of many pulse trains.

The process for pulse-train acquisition would consist of treating each pulse which did not match an existing pulse train as a "candidate" for a new pulse train, or as a component of a candidate. The characteristics of these candidates would be stored in memory in the same fashion as for the pulse trains themselves, and would be updated every time a pulse seemed to match them. Some candidates would acquire enough matching pulses to be reclassified as pulse trains. Other candidates (probably most of them) would not. These would eventually be deleted.

Most of these latter operations appeared to be amenable to parallel processing as well. Moreover, the number of operations performed would probably be independent of the number of candidates. This would give parallel-processing techniques a decided advantage over serial techniques, since there could be presumed to be many candidates in memory during the initial phases of operation.

The development of a complete signal-sorting algorithm would be one of the first tasks in the project.

4.4 The Project Objectives

The application and the starting assumptions have been described in the previous subsection. This subsection lists the project objectives, to be met at the end of a year of effort.

The end product of the project effort was to be a report containing the following:

1. A description of the application, similar to the one given here, but in more detail.
2. A system description, including a block diagram, for the proposed system for the application.
3. All special or original logic defined to the first level. This was not needed for conventional components such as off-the-shelf microprocessors.
4. A detailed description of the processing algorithms, where they are special, or the functions, where the processes are conventional.
5. A complete data-flow description. This is a listing of the data that passes between each system interface, including input and output, and a description of the general data format and data rates.
6. A timing analysis, especially of the processes which are critical with respect to time. This should show that the various subsystems can perform their respective functions without holding up the operations of the other subsystems or of the entire process. This obviously applies to the minor subsystems holding up the major subsystems, and not the other way around.
7. A comparison of the proposed system and processing algorithms with alternative approaches, including other parallel approaches.
8. Power, size, and weight estimates for the system, whichever of these is important to the application.†
9. A preliminary development schedule, together with labor and cost estimates.

†Items 8 and 9 will not be discussed in the section. Item 8 is not relevant in the application, and item 9 is not primarily technical in nature.

10. A description of expected risks in schedule, cost, or technology. This item and the preceding one are needed by management for making a decision as to whether or not to proceed with the development of the proposed system, or with further study and analysis.

All of this sounds like a tall order and, depending on the staff and funds available for the project, certain of these requirements will be met to a greater or lesser level of detail. The important thing for the designer is that he be certain himself that the approach is deserving of further effort and, if so, that he be able to convince his sponsors to that effect.

For the signal-sorting project, a year was not a very long time for meeting the listed objectives. This meant that some decisions would need to be made with a bare minimum of analysis, and that some parts of the design, those which were less critical, would be worked out to the rough functional specification stage only.

4.5 The Initial Task: Developing the Algorithms

A very good general procedure to follow in designing special-purpose digital systems is to:

1. Understand the application
2. Develop the processing algorithms in detail
3. Design the system around the algorithms

It would be satisfying to one's sense of order if these three steps could be followed one at a time as stated, each step taken after all of the previous steps were completed, and without consideration of the ones to follow. In practice, however, the designer will be thinking of possible procedures to follow while he has only a partial understanding of the system requirements, and will especially be thinking of possible system organizations while he is developing algorithms. Moreover, the designer will be shifting attention backward as well as forward during most of the development effort, as he seeks better algorithms to work with some system concept he has devised, or as he studies the system requirements over again as questions arise during the later phases of the project. In the description of the complete signal-sorting algorithms which follows, it will be apparent that parallel hardware was in mind to some extent as the algorithm was being devised.

One other point should be made with regard to the algorithm to be described. The objective of the project was to develop a system capable of performing general signal-processing tasks of a demanding nature, so as to demonstrate that the design could indeed satisfy those general requirements. This system could then be tailored to the requirements of a particular application and environment, with the particular maximum number of emitters, their range of characteristics, and the overall pulse rate specified, for example. For this reason, a number of minor requirements which would normally be included in the final design, such as the ability to accommodate particular forms of parameter jitter, have been left out of the algorithm. To include such minor requirements as part of the algorithm would serve only to complicate the project effort, and would contribute nothing to the confidence of the designer and the sponsors in the value of the design. (The designer must make very certain that the requirements are indeed minor ones, however, or the design may prove to be incomplete or deficient!)

4.6 The Signal-Sorting Algorithms

The algorithms to be described will accommodate an arbitrary number of emitters. It is capable of a "cold start," requiring no prior knowledge of pulse train characteristics to operate. It can accommodate pulse train characteristics which vary over a predetermined range with respect to time. "Windows" can vary in some circumstances to accommodate missed pulses (they are widened by the algorithm when a pulse is missed, so that the pulse train may more easily be reacquired). As previously mentioned, the algorithm does not accommodate deliberate jittering of the pulse parameters.

There are two algorithms, one for pulse-train acquisition and one for pulse-train maintenance. The following subsections describe both of the algorithms in detail.

4.6.1 Pulse-Train Maintenance

Each established pulse train is represented by a file, in which the matching criteria for the pulse train are maintained. These criteria are:

1. The upper and lower limits for the three pulse parameters. These respective limits are denoted by TOA_h, TOA_l, W_h, W_l, F_h, and F_l, where TOA, W, and F represent pulse time-of-arrival, width, and carrier frequency, respectively, and where the subscripts h and l represent the high and low limits, respectively. The limits for W and F are established at the time the pulse train is acquired, and remain the same thereafter. New limits for TOA are calculated after each successful match of a pulse to the train, using a simple predictor algorithm.
2. A predicted value for the TOA of the next matching pulse. This is used in calculating the predicted upper and lower TOA limits just described.
3. NM, the number of missed pulses. It is expected that some pulses will not be picked up at the system antenna with enough amplitude

to be recognized as legitimate pulses. The algorithm must take such missed pulses into account.

4. *NS*, the number of pulses in the pulse train so far. This will be needed for analyzing the final pulse-train data.

5. W_s and F_s, the cumulative sums of W and F, respectively, for all pulses in the train. These are also needed for final analysis of the pulse trains.

6. *A*, the "active flag." The state of this flag indicates that the pulse train is either active (A is set) or else is inactive or lost (A is reset).

The current input pulse is considered to match to an established pulse train if:

1. $TOA_l \leq TOA_{in} \leq TOA_h$
2. $W_l \leq W_{in} \leq W_h$
3. $F_l \leq F_{in} \leq F_h$

where the subscript "in" denotes the current input pulse from the antenna.

An update to a pulse-train file will be made under two conditions. They are:

1. All three parameters of the input pulse match successfully to the stored upper and lower limits for the train, as just defined.
2. The TOA for the input pulse exceeds the stored upper limit for the train, the parametes W_{in} and F_{in} fall within their respective tolerance "windows," as just defined, and the value of *NM* for the train is less than the permitted maximum for the algorithm. The permitted maximum is the same for all pulse trains. The pulse train is considered to have terminated or to be lost if *NM* consecutive pulses are missed.

For Case 1, the update to the pulse-train file is as follows:

First, the predicted TOA for the next pulse, TOA'_{pr} and the predicted upper and lower limits, TOA_h and TOA_l, for TOA are updated by the predictor algorithm:

1. $TOA_{err} = |TOA_{pr} - TOA_{input}|$
2. $TOA'_{pr} = TOA_{input} + \Delta TOA$
3. $TOA_h = TOA'_{pr} + TOA_{err} + C$
4. $TOA_l = TOA'_{pr} - TOA_{err} - C$

It is seen that the effect of these operations is to vary the width of the TOA window according to the error in the prediction of the TOA for the current pulse, increasing the size of the window when the error increases, and decreasing the size when the error decreases. The constant, C, in the third and fourth operations, is small. Its purpose is to prevent the window size from being zero in the event that there is no error in the predicted TOA.

The remaining update operations consist of the following:

1. Resetting NM, the number of missed pulses, to zero.
2. Incrementing W_s and F_s, the cumulative sums for W and F for all pulses in the train, by the respective values of W and F from the matching pulse.
3. Incrementing NS, the number of pulses in the train, by one.

Note that NM may have initially been zero, and thus need not have been reset. However, in parallel algorithms, it saves time to omit the test for zero values.

For Case 2, the update procedure is the following:
First, TOA_h and TOA_l are updated by the operations:

1. $TOA' = TOA + TOA + tol$
2. $TOA' = TOA + TOA - tol$

where tol_{TOA} is an increment for the permitted tolerance in the predicted values. The effect of adding and subtracting this fixed value each time that the pulse train misses a pulse is to further increase the width of the TOA window, and thus increase the probability that the next pulse in the train will be acquired.

The remaining operation in the update process for the missed-pulse case is to increment NM, the number of missed pulses, and to reset the active flag A for pulse trains which have exceeded the maximum permitted number of missed pulses. This deactivates the pulse trains, since A is tested as a part of the pulse-train match operation.

4.6.2 Pulse-Train Acquisition

The pulse-train acquisition procedure is followed for each input pulse which has failed to match any established pulse train. Such a pulse is assumed to belong to some new pulse train not yet acquired. For these pulses, there exist "candidates," which are files analogous in function to the files for established pulse trains. There are two kinds of candidates, called "one-pulse" and "two-pulse" candidates. One-pulse candidates represent single pulses to which no subsequent pulse has matched as yet. Two-pulse candidates represent pairs of pulses that have matched, i.e, they are updated one-pulse candidates.

The startup procedure first attempts to match the parameters of the current input pulse against the parameters for all candidates. If a match to a two-pulse candidate succeeds, the candidate becomes an acquired pulse train. A file is established for the new train. The "successful" candidate is then deleted, together with all other candidates which contain either of the pulses of the successful candidate. (They can no longer be part of any pulse train other than the new one just acquired.)

A file is presumed to exist for each candidate. The process of matching the input pulse parameters against the candidates is slightly different from that for the match against pulse trains. That is, the candidate files contain only the lower limits for the three pulse parameters. This is done to limit the size of the candidate files, as there will generally be a great proliferation of them during the initial period of processor operation. Instead of storing both limits for each parameter in the file, the tolerances which create the parameter "windows" are added to the input pulse parameter values. The test of the modified parameter values against the stored values then takes place.

After the limits comparisons have been performed, any matching two-pulse candidate is updated to a pulse train as previously described. Following the matching operations for one-pulse and two-pulse candidates, when no matching two-pulse candidates have been found, all matching one-pulse candidates are updated to two-pulse candidates. In addition to this, a new one-pulse candidate is created for the current input pulse. This is because it is not at all certain that any of the two-pulse candidates which now contain the current pulse will be acquired as a new pulse train even though there may be many such candidates. For each new pulse processed by the startup algorithm, a new candidate file will be created unless the pulse matches to a two-pulse candidate, thus forming a new pulse train for the candidate.

There will be occasional pulses which will never be a part of any pulse train. These may be the result of noise in the system or environment, for example. A test can be made from time to time on all one-pulse candidates having a very low TOA compared to the current time. All such candidates can be assumed to be extraneous, and therefore can be deactivated.

4.6.3 Example of Pulse-Train Startup

The operation of the startup algorithm can be illustrated by an example showing the way in which a "cold start" is performed. This is the case when system operation is just beginning and there are, as yet, no candidates or acquired pulse trains.

Figure II.41 shows sequences of pulses from four pulse trains displayed on a common time line. For convenience in the discussion, the pulses are numbered in order of arrival (at the receiver). Table II.3 is a list of the 30 pulses from Figure II.41, with respective values of their three parameters. Table II.4 shows the average values of the three pulse parameters for each of the four pulse trains. In the table, "PRI" is the acronym for

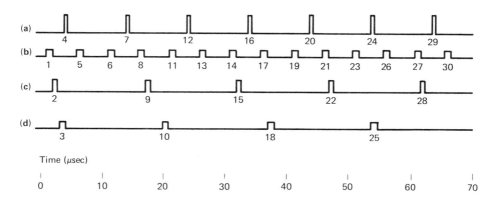

FIGURE II.41 Example of four concurrent pulse trains, showing order of arrival of pulses

TABLE II.3
PULSE PARAMETER VALUES

Pulse	TOA	W	F	Train
1	1.0	0.6	795	b
2	2.0	0.35	1010	c
3	3.0	0.6	790	d
4	3.7	0.15	600	a
5	6.0	0.65	795	b
6	11.0	0.65	798	b
7	13.8	0.1	610	a
8	16.0	0.6	805	b
9	17.0	0.3	1007	c
10	20.0	0.65	800	d
11	21.0	0.7	803	b
12	23.8	0.15	605	a
13	26.0	0.65	805	b
14	31.0	0.7	800	b
15	32.0	0.35	1007	c
16	33.7	0.15	607	a
17	36.0	0.65	798	b
18	37.0	0.7	797	d
19	41.0	0.6	800	b
20	43.7	0.1	605	a
21	46.0	0.6	802	b
22	47.0	0.35	1000	c
23	51.0	0.6	800	b
24	53.8	0.15	605	a
25	54.0	0.65	807	d
26	56.0	0.65	795	b
27	61.0	0.7	798	b
28	62.0	0.3	1005	c
29	63.8	0.15	603	a
30	66.0	0.7	798	b

TABLE II.4

PULSE TRAIN CHARACTERISTICS

Pulse Train	Average PRI (μsec)	Average W (μsec)	Average F (MHz)
a	10.0	0.125	605
b	2.5	0.65	800
c	15.0	0.325	1005
d	17.0	0.65	800

"pulse-repetition interval," which is the interval between two successive pulses in a pulse train.

Table II.5 gives the tolerances permitted for each of the three parameters by the startup algorithm. The parameter "windows," within which the parameter values must lie in order for a successful match to occur, are twice the absolute values of the corresponding tolerances. The startup algorithm uses the same tolerances for all candidates. For pulse width and carrier frequency, this algorithm uses the same tolerances for established pulse trains as well. A more elaborate algorithm would use tighter tolerances for established pulse trains, perhaps making them functions of the parameter values themselves. The predictor algorithm for pulse TOA for established pulse trains does permit tighter tolerance than can be used for TOA for candidates.

Table II.6 shows the assignments of the first 12 pulses to candidates and pulse trains, as they would be made by the startup and maintenance algorithms. In this table, each step corresponds to the processing of the corresponding pulse, and shows the situation after the processing of the pulse is completed. For example, after Step 7, during which pulse 7 is processed, there are five candidates and one established pulse train. The deletion of candidates at Steps 6 and 12, when new pulse trains are established, will be noticed. The enormous proliferation of candidates is apparent in this example. The pulse widths and carrier frequencies for pulse trains *b* and *d* were deliberately made the same to illustrate the way in which "false" candidates can multiply when there are many pulse trains in the input having similar characteristics. The processing requirements for the worst possible case, with 1000 pulse trains being received at an overall rate of a million pulses per second, can be appreciated.

TABLE II.5

TOLERANCES AND WINDOWS FOR PULSE PARAMETERS

Parameter	Tolerance	Windows
TOA	±0.5 μsec	0.1 μsec
Width	±0.05 μsec	0.1 μsec
Carrier frequency	±10 MHz	20 MHz

TABLE II.6

ASSIGNMENT OF FIRST 12 PULSES

Candidate File	Contents (Pulse Numbers)											
	Step 1	Step 2	Step 3	Step 4	Step 5	Step 6	Step 7	Step 8	Step 9	Step 10	Step 11	Step 12
1	1	1	1, 3	1, 3	1, 3		7	7	7	7	7	
2		2	2	2	2	2	2	2	2, 9	2, 9	2, 9	2, 9
3			1	1	1, 5		4	4	4	4	4	
4			3	3	3, 5				2	2	2	2
5				4	4	4	4, 7	4, 7	4, 7	4, 7	4, 7	
6					1				9	9	9	9
7					3	3	3	3	3	3, 10	3, 10	3, 10
8					5					3	3	3
9										10	10	10
Train File												
1						1, 5 6	1, 5 6	1, 5 6, 8	1, 5 6, 8	1, 5 6, 8	1, 5 6, 8 11	1, 5 6, 8 11
2												4, 7, 12

The justification for making three consecutive matching pulses the requirement for acquisition of a new pulse train, rather than two or four, should now be apparent. Two pulses would be unsatisfactory as the starting criterion because too many "false" pulse trains would be established. That is, there is too high a probability that two pulses would match by coincidence, rather than by virtue of belonging to the same actual pulse train. It is true that these false trains would be deleted later when no more matching pulses were found for them. However, in establishing a false match as a new pulse train, all candidates having the first of the two pulses would be deleted, thus preventing the true pulse train from being acquired until later.

If four matching pulses were adopted as the starting criterion, the number of candidates would be doubled (and the corresponding memory requirements also doubled), since there would now be three types of candidates rather than two. Moreover, the overall processing time-per-pulse would be increased because of the number of new operations required for processing the three-pulse candidates.

4.7 The Next Task: Determination of Suitability for Parallel Processing

Before any development of the system could be undertaken, it was necessary to determine that the algorithm satisfied the criteria for suitability for parallel processing of some sort. After this, then the examination of different approaches to implementation of the algorithm could begin. This subsection describes the analysis of the algorithm with respect to the parallel-processing criteria.

4.7.1 Satisfaction of Four Parallel-Processing Criteria

Six criteria discussed in Sections 2.3.1–2.3.8 for determining the suitability of an application for parallel-processing techniques were:

1. Inherent parallelism of a high order
2. Many operations required on each item of stored data, the same operations for all items
3. Large number of operations performed in proportion to the input rate
4. Relatively low output rate with respect to the input rate
5. Processes not overburdened with interprocessor communication difficulties
6. Absence of serial processes which would offset the gains in throughput achieved by the parallel processes

The application is easily seen to satisfy the first four of these criteria.

First, there is inherent parallelism in the application, and it is of a high order. The task is block oriented, in the sense previously defined. A block of memory will be assigned to each pulse train and to each candidate. Moreover, there will be many pulse files and many candidate files in memory simultaneously.

Second, the operations which search and modify these files are complex, and are clearly identical for each type of file. Some of them are identical for both types as well. These are, for example:

1. The matching of the parameters for the current input pulse against all pulse candidates and pulse trains. These operations can be done for all candidates and trains combined, if the file structures are suitably designed.
2. The deletion of all candidates corresponding to newly acquired pulse trains (done in several steps).
3. The deactivation of all pulse trains which have not received a new pulse within the allowed number of steps.
4. The update of all pulse trains which have missed a pulse.

This complexity is quite enough to justify parallel-processing techniques. Third, for each pulse, there will be a large number of pulse-train files and candidate files to be searched, and there are many sequences of searches and file modifications which must be performed on each file during its "lifetime."

The present application is not one in which only a few parallel applications will be performed for each item of data residing in the memory, or for each item of input.

The fourth criterion, that of low output rate with respect to the input rate, is also seen to be satisfied by this application. The 1-MHz pulse input rate is orders of magnitude greater than the 1000 pulse-train output from the system, even if processing only lasts for 1 sec.

4.7.2 Problems With Two Parallel-Processing Criteria

The last two criteria are not obviously satisfied for the application. These criteria are related, and the evaluation of both of them requires that the system design and operation be understood in considerable detail. Particularly important, for example, is the requirement that serial processes be avoided wherever possible throughout the performance of the task.

It was at this point in the project that comparison of alternative approaches to system design began. Three alternative approaches needed to be studied. The first of these was the use of a single, very fast sequential processor.† It would be unreasonable to propose a radically different approach to system design if a more conventional one would handle the task. The second and third approaches were the use of PEPE-like parallel systems and ALAP-like associative processing systems. The PEPE-type architecture had appeal because of the near-conventional structure of the processing elements, and because there appeared to be little need for interprocessor communication. The ALAP was appealing because of its high fault tolerance and because the analysis tools were familiar and available. The next three subsections give details of the initial analysis for these three approaches.

4.8 Preliminary Analysis of Three Candidate Systems

4.8.1 Fast Sequential Processors

In assessing the applicability of a sequential processor for the signal-sorting task, the obvious question to be answered was that of the processor's capability to keep up with the high pulse input rate. It is a safe bet with most sequential processor programs that 95 percent of the computer time is consumed executing five percent of the program, the so-called "inside loops." This simplifies the task of estimating computer applicability, since all that is generally required is the timing of the inside loops. In the present application, there are two such loops. One of these is executed for every input pulse; the other is executed for some input pulses. These are the comparison of the input pulse parameters with the upper and lower limits of the acquired pulse trains, and with the limits of the candidates, respectively. If the processor cannot handle these two operations, no further investigation is necessary.

†By "sequential" processor we mean a conventional processor which processes only a single item of data (or, perhaps, a pair of items) with each instruction. See formal definition of sequential computation in Chap. I, Part I, Sec. 2.

†Prerequisite information on pipeline computations is given in Chap. I, Sec. 15. (Ed.)

Let us assume a sequential processor uses a pipeline both for data and for instructions.† Such a machine, if optimally designed and programmed, can complete the execution of an instruction almost every clock time. We shall assume that to be the case for the present application. The tight loop for comparison of pulse parameters against the stored limits for existing pulse trains is then:

```
TAG1   LOAD          PARAM1, INDEXED
       COMPARE       LOWER LIM, INDEXED
       BRANCH LSS    TAG2
       COMPARE       UPPER LIM, INDEXED
       BRANCH GTR    TAG2
       LOAD          PARAM2, INDEXED
                .
                .
                .
       LOAD          PARAM3, INDEXED
                .
                .
                .
       DECREMENT     INDEX
       BRANCH POS    TAG1
TAG2   ETC.
```

This program is seen to consist of five instructions for each of the three pulse parameters, plus two for updating the index over the number of pulse trains and branching conditionally. At one clock time per instruction execution completion, the processor will consume 17,000 clock times for the maximum number of 1000 pulse trains, just to execute this loop. At the average pulse input rate of one pulse per microsecond, the requirements on the processor are orders of magnitude beyond the capability of any present or projected hardware technology. And it appears that no devices, such as multiple arithmetic units or multiple pipelines, will appreciably speed up the process as long as memory accesses are performed sequentially. If the capability for multiple memory access is added, we now have a parallel machine, and such a machine should be considered separately.

For some applications, when an existing sequential processor appears to be too slow by a factor of 10 or so, it is reasonable to examine a system employing a number of such processors, and to have the input data parceled out among the processors as it is received. However, the extreme requirements in this case suggest a highly parallel, reconfigurable system of some sort employing hundreds or thousands of specially designed processing elements. Such systems will be examined next.

4.8.2 Multiple-Processor Systems

With the advent of cheap microprocessors with surprisingly high throughput capability, it is reasonable, in fact almost mandatory, that systems employing many microprocessors be examined for high-throughput applications. It has generally been the case that when such systems are rejected, it is because of the inability of the conventional microprocessors to handle the interprocessor communication tasks. In the case of the signal-sorting application, however, an examination of the algorithms will show that very little communication of this type is required, provided that one microprocessor is assigned to each pulse train and candidate (or to several such).

Several problems did arise, however, when the use of large numbers of conventional microprocessors was considered in more detail. First, the microprocessors should operate in "lock-step" fashion for simplicity in programming and in synchronizing the system operation. However, this is not possible in many instances because some of the operations are data-dependent, and the data being operated upon is different in different microprocessors.

Second, the process of testing the parameters of an input pulse against the limits for a stored pulse train will still require 15 clock times. This is only part of the pulse-processing task, and already requires a clock rate of at least 15 MHz. Clearly, a clock frequency of several times that value, which is indicated as a requirement, is far beyond the capability of the present assortment of microprocessors.

Third, it is evident that whatever microprocessors are to be used in such a system, about 2000 of them will be required. Conventional microprocessors, one per chip, fail to take advantage of the capability of LSI technology, and will be subject to failures because of the large total number of external connections to the chips.†

All of this suggested that specially designed processors should be used in the system, and PEPE-like processing elements, many per LSI wafer, might prove to be the answer. The design of the PEPE processing element, after all, resulted from an examination of the requirements of a very similar application, radar track-while-scan, and embodied a solution to the first of the problems just described, along with many others.

The fifth of the six criteria for parallel processing appeared to be satisfied by the use of a PEPE-like system. To demonstrate the satisfaction of the sixth would still require considerable programming and system analysis, although there appeared to be no problems.

With regard to processing throughput, a preliminary analysis would need to be conducted before the PEPE could be considered for the application. This

†The problems of interconnections arising in computer networks assembled from microprocessors is discussed in Chap. I, Secs. 24.2 through 24.5.

analysis did not need to be extensive at this stage. All that was required was a ballpark figure for processing time-per-input-impulse, measured in clock times. If the resulting figure was within 10 or so of the requirement, the PEPE could still be a candidate, since it was presumed (as it should be in general) that devices and design tricks could be found to increase the throughput within that range. For example, multiple-PEPE systems could probably be used if nothing else would work.

The preliminary analysis assumed an architecture like that described in the section on the PEPE. The objective of the analysis would be to determine the throughput, in pulses per second, of the system and to discover any bottlenecks in the process or in the design if possible. The throughput could best be estimated by looking for the situation requiring the execution of the greatest number of instructions for a pulse and then estimating the corresponding execution time. Note that the situation must be one which occurs a reasonable number of times. If it is rare, it will have no effect on processor throughput, since the time lost for the pulse can be made up with the succeeding few pulses. That is, there can always be buffering at the data input, and it is therefore not necessary that every pulse be processed within one microsecond.

The case in which a new pulse stream is established, which requires the deletion of a number of candidates as well as setting up the file for the new pulse train, can be ignored as having a significant effect on throughput. The requirements state that there will be no more than a thousand of such pulses in an entire execution of the program, plus another thousand or so pulses representing "false starts."

The most time-consuming of the "normal" situations for an input pulse is that for which the input pulse does not match to an existing pulse, but does match a two-pulse candidate. This situation requires that the entire matching operation be performed for the input pulse parameters against the established pulse trains, and that the match then be attempted against the candidates. Many of the corresponding matching operations can be performed simultaneously for both candidates and pulse trains, however, so the times for the two processes are not strictly additive. In addition to the comparisons are the tasks of setting up a one-pulse candidate for the input pulse, and the operations required to upgrade the matching one-pulse candidates to two-pulse candidates. With the fact in mind that all of these operations are performed simultaneously for all candidates of each type, the number of instructions for the process were estimated. These are summarized in Table II.7.

A few remarks about Table II.7 are in order. First, for simplicity in the estimating process, no attempt was made to follow the coding practices for the PEPE. The estimation of the instructions was made as though the system consisted of a single sequential processor processing a single pulse train or candidate. The few special instructions, which perform such operations as setting or resetting the enable flags for the processing element, were not

TABLE II.7

TIMING ESTIMATES FOR PEPE SYSTEM

ITEM	Operation	Numbers and Types of Instructions
1.	Matching against pulse trains	6 compares, 6 conditional branches, 3 loads
2.	Matching against candidates	(simultaneous with 1.)
3.	Creating new 1-pulse candidate (8 items in candidate file)	10 loads, 10 stores, 4 adds, 3 subtracts
4.	Upgrading 1-pulse candidates to 2-pulse candidates	5 loads, 5 stores, 3 adds
5.	Other operations not known until detailed programming	55 instruction executions (as many as total of above) Total: 110 instruction executions, 330 clock times at 3 clock times per instruction

included. Second, the estimation of the number of instructions required for each of the five parts of the process was just that—an estimation. To accommodate operations which have been overlooked, a 100-percent overhead (item 5 in the table) was included in the total.

To estimate the number of clock cycles in the process, an average instruction-execution time of 3 clock cycles is assumed. This is the execution time for the add-from-memory instruction in a 16-bit processor developed at the author's company. The processor has considerable instruction-processing overlap, else the add-from-memory would require more like 9 clock cycles. At this rate, the 78 instructions estimated for processing each input pulse require a total of 330 clock cycles, or an equivalent clock rate of 330 MHz if the 1-MHz input pulse rate is to be accommodated by a single-array PEPE system. The extraordinary requirements of the application are indeed evident when it is realized that no existing conventional computer would be fast enough even if there were only one pulse train and one candidate to monitor.

The number of gates in the PEPE PE can be roughly estimated from the gate count of the 16-bit processor just described. That processor contains about 4200 gates, not counting memory. If the instruction-decoding logic is removed and the special PEPE logic (such as the content-addressed registers and their stacks) and a small memory are added, the gate count will increase to about 6400 gates per PE.

It is evident that no single-array PEPE will be fast enough for the application. However, the possibility of multiple-array PEPE systems using very fast logic still exists. The input data would be partitioned on some basis (perhaps on the basis of pulse width or carrier frequency) and each PEPE array would process only part of the data.

Consideration of the PEPE approach was suspended for the time being while the third candidate, the associative processor, was investigated.

4.8.3 Associative Processors

The author's experience with associative processors led him to consider such systems for the present application. In particular, one of the applications programmed for the ALAP was radar track-while-scan, and it showed the capability of the ALAP to handle block-oriented applications.

The analysis for determining the capability of a system to satisfy the fifth and sixth of the parallel-processing criteria is more difficult with an associative processor than with a PEPE type of system. That is, whereas it is possible in a PEPE-like system to process a single pulse train with a single PE independently of the other PEs, such is not the case with an associative processor such as the ALAP. With the ALAP, several cells will be required to contain the data for a single pulse train, one cell per data item, and communication between cells will be necessary in any process involving more than one data item. The ALAP chaining channel, however, is designed for such processes and had shown itself to be sufficiently flexible in the track-while-scan application.

With regard to the sixth criterion, which is concerned with the possibility of performance degradation from sequential processing, the author had not found such problems in his preceding applications work with the ALAP. However, there could always be a first time. Moreover, easy techniques for timing analysis, such as the one used for the single sequential processor, are not applicable to the ALAP (or the PEPE). This is because an ALAP program contains no loops that consume most of the program execution time. (The need for such loops would make it immediately apparent that the ALAP was unsuited for the application.) For the ALAP, the only technique for estimating program execution time requires that a rough design of the entire program be made. The next step in the project effort was the creation of such a preliminary program design for the ALAP. From this design, timing estimates were made.

The way in which ALAP programs are designed is somewhat unusual, and is described in more detail in the section on programming (Sec. 5). Briefly, the program design process consists of first making up drawings showing the initial layout of each type of block in ALAP memory, and then constructing a series of drawings showing the changes in the contents of the block after each stage in the operation of the program. The designer can tell during this process if the operations he or she wishes to use can be accomplished without conflicts on the chaining channel; the number of word cycle operations that will be required can then be calculated. For the signal-sorting application, there would be two principal types of blocks, one for established pulse trains and one for candidates. In addition, there might be a few cells (perhaps three or four) needed for ancillary purposes. Data items common to all blocks of a kind would generally be maintained in the controller memory, rather than the ALAP array, and input to the array through the auxiliary input channel when needed as operands.

The preliminary design for the signal-sorting program contained block layouts as just described for the processes for:

1. The comparison of the parameters of the input pulse against the stored limits for all pulse trains and candidates. (This is performed simultaneously for all pulse trains and candidates together.)
2. The updating of all pulse trains following the matching operation of (1).
3. The creation of new candidates for input pulses which have not matched as a result of (1).

Other processes in the program were considered to be similar to one of these three, or else would be executed so seldom as to have no significant impact on program execution time. The blocks for the ALAP for candidates were considered for the purpose of the analysis to contain the same number of ALAP cells as the number of words of memory required for the corresponding tables in the PEPE processing element.

The result of the initial analysis for the ALAP indicated that the most time-consuming sequence of operations commonly performed for a single input pulse was the same sequence as for the PEPE system. This was the case in which the input pulse did not match to a pulse train, but did match to one or more one-pulse candidates. The timings for this case are summarized in Table II.8. This table looks very much like the table for the PEPE timings. The design assumes 32-bit data registers, rather than the 64-bit data registers

TABLE II.8

TIMING ESTIMATES FOR ALAP

ITEM	Operation	Numbers and Types of Instructions
1.	Matching against pulse trains	6 word-cycles (arithmetic mode is LSS or GEQ)
2.	Matching against candidates	(simultaneous with 1.)
3.	Creating new 1-pulse candidate (8 cells in candidate block)	17 word-cycles (10 for input, 4 with ADD, 3 with SUB)
4.	Upgrading 1-pulse candidates to 2-pulse candidates	8 word-cycles (5 for input, 3 with ADD)
5.	Other word-cycle operations not known until detailed programming	31 word-cycles (total of above)
		Total: 62 word-cycle operations, 1984 clock times at 32 clock times per word-cycle

of the original ALAP hardware. This was done for simplicity in the analysis. The use of 32-bit registers permitted the assumption that a single number (for example, an upper limit) would be contained in each cell for one of the blocks. A 64-bit register would have had to contain several, thus making the count of the number of required word-cycle operations more difficult to estimate.

Notice also that the same assumptions as to the number of operations required were made for the ALAP as for the PEPE. For example, the matching operation against pulse trains assumes six comparisons in both instances. For the PEPE, this requires six conditional branches and six loads as well as the six compare operations. For the ALAP, only the six word-cycle operations, with the arithmetic mode set to GTR or LSS, are required, but they each require 32 clock times.

The next task in the project was the comparison of the PEPE and ALAP approaches to system organization, with some decision to be made as to which approach to pursue in greater detail. The following subsection discusses this decision.

4.9 Comparison of Results: Deciding on the System Approach

The preliminary analysis of the three preceding subsections, together with some additional considerations to be discussed here, was used to make the initial decision as to the way to proceed for the remainder of the project. That decision will be explained here.

4.9.1 Pipeline Processor Ruled Out

First, the system using the single fast sequential pipeline processor was pretty well ruled out as a candidate. The two highly parallel systems seem considerably closer to achieving the goal, although still too slow by an order of magnitude. The PEPE system, however, is seen to be about six times faster than an ALAP having the same clock rate.

4.9.2 Multiple Arrays of PEs as a Last Resort

In making the decision as to which system approach to take, it was desirable to consider ways in which the performance of both systems could be improved or the cost and complexity decreased. This was because both systems would be too slow when a single array of processing elements was used to perform all of the signal sorting. The use of multiple-array systems was to be considered only as a last resort, after consideration of all other approaches to increased system throughput.† That is, a system which doubles the throughput by simply doubling the number of gates is not a very appealing or cost-effective solution in any case. In the application being considered

---

†A general analysis of traditional and new sources that increase throughput in a complex parallel system is given in Chap. III, Secs. 2 and 3. (Ed.)

here, such a solution would require partitioning of the data to be fed to the individual arrays, and this approach introduces additional problems. The only reasonable partitioning technique is that of partitioning the range of one of the pulse parameters, say pulse width. Since the parameter values vary within the same pulse train, it is apparent that special consideration will need to be given to pulse trains having parameter values overlapping the range boundaries. This can be a messy situation indeed.

4.9.3 Improving PEPE Performance

It was difficult to think of many ways in which the throughput of the PEPE could be improved. The design analyzed in the preceding subsection made use of modern instruction-execution logic having considerable overlap. This is the reason why such operations as add-from-memory require only 3 clock cycles for both instruction fetch and execution. One way in which performance improvement might be achieved would be to install data pipelines in all PEs. Instruction pipelines, which are used in modern pipeline processors, would not be needed in the PEs since all instruction fetch and decode operations are performed in the system controller. Adding data-pipeline logic to a PE increases the gate count of the PE by a factor of about 3. Since the data pipeline enables the PE to complete the execution of an instruction essentially every clock time, system throughput would also be increased by a factor of 3 over the alternative PEPE array. The single instruction pipeline would be in the PEPE controller, and would have negligible effect on the system complexity.

The addition of data pipelines to the PEs does not result in a very favorable throughput-to-complexity improvement. However, a single-array PEPE system with data pipelines could process the 1-MHz input data with a clock rate of 110 MHz, as can be seen from Table II.7. This is a little closer to feasability, but will still require an extremely advanced gate technology. (Such fast gates will probably be required in any case, as will be seen.)

4.9.4 Improving ALAP Performance

Several improvements to the ALAP design which will increase performance or decrease complexity were considered. The simplest of these was to pack the data within the data registers. This is possible because of the number of bits needed for pulse width is 14 bits, and the number needed for pulse carrier frequency is 16 bits. This will save two 32-bit cells in each pulse train file and candidate file. A better idea, although it requires more careful analysis to be certain, is to increase the size of the data registers in the cells to 64 bits or perhaps even a larger size. This can cut the required number of cells in half. In addition, the number of gates in the larger cells will not be much larger than for the smaller ones (within limits, of course). This is because the data registers are dynamic, and consist of only one gate per bit position. The entire cell contains about 650 gates with a 32-bit register. Since the program's efficiency, and thus operating speed, is highly dependent on the data layout of the pulse train and candidate files, a careful analysis would be required before any final decision as to data register size. However, it seems likely that four 64-bit cells could be used for the files, rather than eight. This would make a much more favorable tradeoff in gate count with respect to the PEPE, since the four 64-bit ALAP cells needed for a pulse-train file would require 2856 cells, whereas using the PEPE PE for the same purpose requires 6400 gates. The performance would not improve by the change to 64-bit registers, but it need not decrease much either. It should be remembered that the ALAP arithmetic operations are field-programmable, and two 32-bit compare operations in separate halves of a data register require no more time than a single 64-bit comparison if they are correctly programmed.

There are a number of other tricks that would speed up the ALAP operation to some degree. Such tricks might be, for example, the inclusion of special-purpose registers in the ALAP cell, faster arithmetic logic, additional chaining channels, and so on. However, each of these tricks requires careful analysis to determine its value, and the decision on the system approach needed to be made soon so that a first-pass system design could be completed by the end of the first year of the project effort.

The rough comparison of the PEPE and ALAP approaches with respect to processing speed and gate count is summarized in Table II.9. The gate count involves the processor arrays only. By comparing the first and fifth lines of the table, one sees that the PEPE is twice as fast as the ALAP system having 34 percent more gates. Equivalently, by comparing the first and last lines, one sees that the PEPE attains the same processing throughput as the ALAP with only 37 percent of the number of gates. The comparison is made with the same clock rates for all systems, of course. On the basis of this performance comparison, then, the PEPE system is the first choice.

However, in determining the choice of system approaches, there are

TABLE II.9

COMPARISON OF PEPE AND ALAP PERFORMANCE

System	Number of Gates/ PE	Number of Gates/ Pulse Train File	Number of Clock Cycles/ Input Pulse Processed
PEPE (no pipeline)	6,400	6,400	330
PEPE (pipeline)	19,200	19,200	110
ALAP (32-bit cells)	650	5,200	1,984
ALAP (64-bit cells)	714	2,856	1,984
ALAP (2 arrays, 64-bit cells)	714	8,568	661
ALAP (6 arrays, 64-bit cells)	714	17,136	330

other factors which must be considered in addition to estimated performance. The understanding of these factors is important in understanding the way in which projects in industry are conducted. When a decision is to be made which may involve the expenditure of many millions of dollars to develop a new system, cost and risk are essential considerations. To make such a decision, then, the company's resources must be considered so that cost and risk may be minimized.

With regard to the signal-sorting project, consider the resources that are involved. First, an extensive ALAP development project was conducted under internal funding over a period of 3 years. As a part of this project, an ALAP demonstration system was designed, constructed, and checked out. This system included LSI wafers, containing numbers of ALAP cells each, which were designed and fabricated as part of the project effort. Programs to check out the wafers were written and executed on the system.

Along with the hardware development, an extensive software-development effort was conducted. This included the programming of a symbolic assembler for the ALAP and a simulator for exercising the object programs on another computer (the Xerox SIGMA 9) for checkout purposes. Using this assembler and simulator (the ALAP memory on the demonstrator system was too small), several applications programs were written and checked out.

All of this represented a considerable resource for the company in both experience and physical assets (such as the assembler and simulator). It also represented a reduced risk in any future parallel system development efforts if the ALAP approach was used, since advantage could be taken of the previous effort, including the services and advice of some of the previous participants.

In the case of the PEPE versus the ALAP decision, the cost and risk factors outweighed the better-than-2-to-1 estimated performance advantage of the PEPE over the ALAP. For that matter, if a fast conventional processing system could have been adequate for the task, the same considerations would have favored such a system over the ALAP.

Because of the considerations just explained, the decision was made in favor of the ALAP over the PEPE. The remainder of this section describes the system design effort and the resulting ALAP signal-sorting system.

4.10 Detailed Programming Analysis

Since the ALAP assembler and simulator were available to the project, and since risk minimization was an important objective of the project, it was decided to program the signal-sorting algorithm and to check it out with simulated pulse data. This would serve the dual purpose of verifying the applicability of the ALAP-type of array and the effectiveness of the algorithm. In particular, the program would serve as an excellent analysis tool. It would enable any deficiencies or bottlenecks in the ALAP cell and memory organization to be pinpointed. These could then be taken care of by suitable modifications to the memory design. Moreover, if the program showed the ALAP to be effective, it would constitute a convincing argument in favor of the continued development of the proposed system. Even if there had been no assembler or simulator, the coding and analysis of significant portions of the algorithm would have proven very useful for many of these purposes, and should have been done in any case. The capability for checkout was an added benefit.

The programming operation, along with the performance analysis, continued throughout the remainder of the program effort. During this same period, the system analysis was performed and the preliminary system design completed. The details of the program design and the programming techniques which were employed will be described to some extent in the following section (Sec. 5). The results of the programming operation will be discussed in this subsection.

When the design and coding for the program (hereafter referred to as ALAPSIG) was completed, the number of word-cycle operations for each phase of the program operation was counted, and the number of auxiliary operations (which were mostly set-up operations for the word-cycles) were counted. These latter operations amounted to an overhead on the word-cycle operations which was very close to 20 percent. Therefore, the 20-percent figure was used subsequently as an overhead on all word-cycle operations in estimating program performance as a function of word cycle count.

Table II.10 shows the principal phases of operations for ALAPSIG,

TABLE II.10

TIMINGS FOR PROCESSING ONE PULSE, ALAPSIG

Phase	Description of Operation	Number of Word-Cycles (32-bit data reg.)	Elapsed Time, Clock Cycles (incl. 20% overhead)
1	Match against all pulse trains (all pulses)	8	307
2	Update all matching pulse trains (all matching pulses)	14	538
3	Match against all candidates (all non-matching pulses)	6	230
3a	Create new pulse train (all matches to 2-sample candidates)	24	922
3b	Create new candidates (all nonmatches to 2-sample candidates)	13	499

together with the number of word-cycle operations (using the 32-bit ALAP cells required by the ALAP simulator) and the equivalent number of clock cycles (including the 20-percent overhead for set-ups). It will be noticed that different phases, or combinations of phases of the program will be executed for different input pulses, depending on the pulse parameter values. Table II.11 shows the processing time for pulses representing the three most common situations. In this table, under the heading "Pulse Type," the combination of phases from Table II.10 which are executed for the corresponding situation is shown. The elapsed time figures for the three cases include the 20-percent overhead.

From the three elapsed time figures in Table II.11, together with a few assumptions, the average processing time per pulse, expressed in clock cycles, was estimated. This figure was used as a guide in assessing the merits of proposed improvements to the system design.

Notice that the third case shown in Table II.11 occurs very rarely; in fact, only for about 2000 pulses for the entire program run (1000 pulses for pulse-train starts and about 1000 more for false starts). The overwhelming portion of the program execution time is taken up by the processing of pulses representing one or the other of the first two cases. These cases are for pulses that do match or do not match, respectively, to an existing pulse train. During the initial period of the program runs, the greatest number of pulses will be of the nonmatching case. Once most or all of the pulse trains have been established, almost all of the pulses will belong to the matching case. This second situation will be predominant throughout almost all of the program run. Nevertheless, if the system is to keep up with the input pulse rate, it must be designed to accommodate the worst of these two cases.

There are a number of possible approaches to improving the throughput of the original ALAP system, which is the one assumed thus far. These approaches are:

TABLE II.11

TIMINGS AND PROCESSING RATES FOR
MATCHING AND NONMATCHING PULSES, ALAPSIG

Pulse Type (Applicable Operation Phases)	Number of Word-Cycles (Aver.)	Elapsed Time (Aver.) in Clock Cycles
Matching pulses $(1 + 2a)$	22	845
Nonmatching pulses: no new pulse train created $(1 + 3 + 3b)$	27	1037
Nonmatching pulses: new pulse train created $(1 + 3 + 3a)$	38	1459

1. To employ faster gate technology
2. To improve the efficiency of the algorithms
3. To improve the efficiency of the program (i.e., fewer instructions in the processing sequence)
4. To improve the ALAP cell or array design
5. To improve the general system organization

The following subsections discuss these approaches in that order, since that is roughly the order in which they are considered.† It should be noted that it was often necessary to consider two or even three of the approaches simultaneously (such as the algorithm and the ALAP design, for example).

4.11 The Use of Faster Gates

The use of more advanced gate technology is always an appealing approach to a quick solution for improving the performance of an advanced processor design, since projections as to the performance of such technology are always easy to obtain, and the technology itself need not be available until the design is well under way. The engineers whose task it will be to make such technology practical, that is, to produce the LSI wafers at reasonable cost and with reasonable reliability, do not always take such a rosy view of such performance projections, however. They are acutely aware that much modern gate-fabrication technology is partly "black magic," making it difficult to predict results.‡

Likewise, the circuit designers who are trying to create circuit modules from the advanced technologies and put these modules together into functioning systems often take a dim view of optimistic projections made by system designers. They talk about such problems as "off-chip" delays. These are time delays caused when signals must pass from chip to chip. Off-chip delays constitute a primary problem when high-speed logic is used, and is a particularly difficult problem in MOS circuitry, which has little off-chip drive capability. If not properly compensated for during system design, the effect of off-chip delays is that the system, consisting of many chips operating together, must operate at a reduced clock rate as compared with that achievable for a single chip operating alone.

The author, at the time he was engaged in the project, had the choice of two technologies. One of these was CMOS Silicon-on-Sapphire (SOS) technology which could produce gates that operate at 50 MHz. This technology was predicted to be the standard CMOS technology by the time the fabrication of the signal-sorting system would begin. The other technology was D-ECL, a bipolar technology with gate speeds up to 1400 MHz (equivalent to a 700 psec delay). Chips containing complex circuity had been successfully fabricated from both technologies and were operating at the predicted speeds.

---

†This material is closely connected with the first part of Chap. III. Various methods are discussed here that increase throughput for associative reconfigurable arrays. In Chap. III, techniques that increase throughput of parallel systems in general are considered. It is interesting to note how a more general case of parallel systems makes modifications in existing methodology of increasing throughput for a parallel system (Ed.)

‡General problems in designing complex LSI-based systems are discussed in Chap. I, Secs. 21 and 22.

From this alone, it seemed to be advisable to propose the faster technology. However, the author preferred to propose the slower technology if at all possible and to attempt to achieve the desired system throughput by other means. The principal reason for this was that mentioned above, the problems with off-chip delays.

For the present system, in the event that it proved necessary to use the very fast D-ECL technology, the off-chip delay problem would be a very great one. At the extremely fast clock rates to be used with that technology, the lengths of even the on-chip connections must be carefully worked out. With connections made between different chips, the problems are much more severe. Particularly damaging are the cases in which "round-trip" signals pass between two or more chips. A round-trip signal is one which passes one way between two chips, causing a signal to be generated at the receiving chip which then passes the other way between the chips. Time and phase lags occur because of such signals, requiring that the clock rate for the system be decreased below that needed for single-chip systems.

If round-trip signals can be avoided, it is possible to compensate for off-chip delays by having each succeeding chip clocked by an appropriately delayed clock. The presence of any round-trip signals requires a synchronous clocking system with the clock period adjusted to account for all on-chip delays. Since it is nearly impossible, with conventional system designs, to avoid the continual occurrence of round-trip signals (when the entire processor cannot be fabricated on a single chip, that is), virtually all such systems are synchronously clocked machines.

The ALAP design has some considerable advantages over other systems with regard to the problem of off-chip delays. Inputs to the ALAP memory wafers are very orderly. All inputs except the input to the chaining channel originate at a single place (the controller) and from a set of identical shift registers. There are almost never any signals from the ALAP memory to the controller during instruction execution. For example, a 10,000-line radar track-while-scan program written for the ALAP has exactly one instruction in which a signal is sent from the ALAP memory array to the controller to assist it in determining a branch. Thus, if the chaining channel is not used during the execution of an instruction, the control and data lines from the controller to the ALAP memory wafers can be considered as a single delay line in effect (provided all these lines are exactly the same length, etc.), and the full clock rate of the on-chip logic can essentially be maintained for the system. The delay-line effect of the control lines simply is that the memory operation lags the controller operation by the length of the delay.

When the effects of the chaining channel are taken into account, and the fact that several wafers† will be needed for a large memory array, additional off-chip delay problems are introduced. These are the delays for propagation of the chaining channel data between each ALAP memory wafer and the following one. While these are not strictly round-trip delays, they do cause

†Throughout this chapter, the term "chip" is used for a circuit module cut from a large silicon "wafer" on which many such identical modules are fabricated. The wafer may be 2 in. or more in diameter. The ALAP memory, however, is fabricated from full wafers, which each contain many ALAP cells, and which are not cut up into individual chips after manufacture. This accounts for the use of the term "wafer" in the present discussion.

problems in maintaining the maximal clock rate of the individual wafers. However, these problems are considerably less than the general problems encountered with conventional processor designs using the same gate technology.

A lesson of considerable value to be remembered from all of the preceding discussion is that it is virtually impossible, in actual practice, to consider logic design as independent of circuit design and of the gate technology to be employed. Logic design, in other words, is an engineering discipline, and not a mathematical discipline.

4.12 Improving Algorithm and Program Efficiency

The design, coding, and checkout of the signal-sorting program, ALAPSIG, was one of the major efforts of the entire project. At every stage of the programming operation, it was possible to consider improvements in the algorithm and program design. A number of refinements that resulted from such considerations were actually incorporated into the program. One of these was the restructuring of the file organizations for the pulse trains and candidates so that the comparison of the pulse parameters with the stored data could be performed simultaneously for both pulse trains and candidates. This saved a total of eight word-cycle operations in the processing of all pulses which did not match to established pulse trains.

In addition, it was possible to consider modifications to the ALAP memory design during the program effort, since the effect of such modifications on processing throughput could be measured. One of these, the two-way chaining channel, was incorporated into the ALAP array design.

4.13 Improving ALAP Array Performance

A number of modifications to the ALAP cell and array organization were considered and discarded in an attempt to bring the performance of the array to the required operating speed. The reason for the rejection in most instances was either:

1. That the improvement in performance was not sufficient to justify the increased complexity of the cell or memory design.
2. The improvement in performance was offset by a decrease in the fault-tolerant capability of the memory.

Most modifications to the chaining channel fell into the second of these cases. It is very difficult, in fact, to gain any substantial improvement in the intercell communication capability of the ALAP without nearly or totally destroying the ability of the array to tolerate single cell failures without losing a number of other cells in the process (such as an entire row if the array is two-dimensional). One modification to the chaining channel was finally adopted, although it is a simple one.

Modifications in the ALAP cell design that were considered and rejected nearly all fell into the first class above. For example, the use of auxiliary registers in the cell, to be operated upon concurrently with the main data register, offered very little improvement in proportion to the increase in complexity. The reason for this was that the additional intercell and intracell communication logic, together with all of the switching and control logic for the additional cell functions, was much greater than the logic for the new register itself. It was more efficient in most respects to simply employ additional cells for the same function.

If the problem had been that of slow arithmetic operation, increasing the cell logic would have made sense. For example, two or more arithmetic units instead of one could be employed in each cell, resulting in a proportional increase in the speed of multiplication and division. This is an expensive modification, but could be a necessary one in some applications.

One modification which would be self defeating is that of building external fast-logic modules for performing critical calculations, and passing data and results between this logic and the ALAP cells. The I/O problems would be horrendous, both from the hardware and software standpoints. It will be remembered that the ALAP concept assumes a relatively low output rate compared with the rate of computation, with all critical processing to be performed internally.

Sometimes the nature of the application is such that very specialized ALAP circuitry, with special registers and flip-flops, can be added to the cell logic with a significant effect on performance. This was the case with a text-processing application, described in reference [11]. Nothing in the present application warranted such special logic. The ALAP memory was used quite efficiently in most respects. It was simply too slow.

The programming effort for ALAPSIG did make it possible to identify a source of inefficiency in the ALAP memory organization, and to accurately predict the effect of a corresponding design modification. The inefficiency resulted from the fact that the chaining channel operates in one direction only.

To explain this, consider a file-processing application, such as data management. In processing a set of identical files, the program processes the entries in the files in the order dictated by the direction of the chaining channel. As it does so, the status of each file (such as the success or failure of the file to meet some combination of criteria) will be maintained in some flag position (generally the head flag) at the cell currently being processed. When the processing of all files is finished, the status of the files will be tagged at the last cells of the files. Now, in general, it may be necessary to store the final status of the file-processing operation back at the head of each file, so that the the status itself may be used as a criterion in the next file-processing sequence. However, the status cannot be directly transferred

202 **CHAPTER II**

from the cell at the end of the file to the cell at the beginning. What must be done, is to shift all files down the chaining channel so that the header cells will be aligned with the corresponding status flags. This operation can be visualized by remembering that the chaining channel can be organized so as to effectively turn the entire ALAP memory into one big circular shift register.

Figure II.42(a) shows a set of seven contiguous ALAP cells, the first four of which contain the four entries of a file. The cell containing the last file entry is tagged in its head flag. Figure II.42(b) shows the same seven cells after the contents have been shifted 32×4-bit positions (requiring 32×4 clock cycles), thus realigning the tags with the first entry of the files. If the chaining channel could operate in the reverse direction, the states of the head flags could have been shifted back to the first words of their respective files in only 4 clock cycles.

The modification to the ALAP chaining channel logic to permit bidirec-

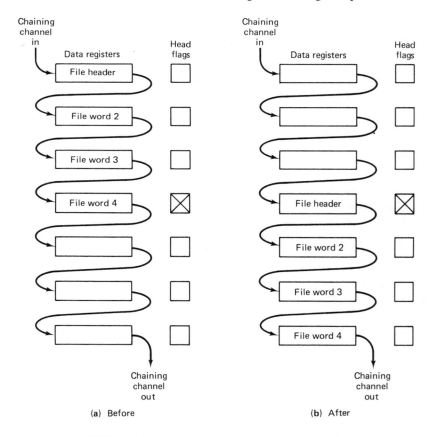

FIGURE II.42 Shifting of ALAP data register contents

tional chaining operations is a simple one, and requires only one additional external control signal (which remains in a constant state throughout each word-cycle operation). The modification is shown in Fig. II.43. It consists of the addition of a single toggle at the chaining channel input to each cell. This toggle permits the chaining input at the cell to be either the chaining output from the previous cell in the array or the chaining output from the following cell. The new toggle is placed at the cell's chaining input, rather than its output, following the general rule for maximizing fault tolerance in the circuitry. With the addition of the toggles and the single external control line, all cells in the array will perform their chaining in the forward direction during each word-cycle operation, or in the reverse direction, depending on the state of the control line.

The addition of the two-way chaining channel eliminates five word-cycle operations from Phase 2 of the signal-sorting task (see Table II.10), and eight word-cycles from Phase 3a. The resulting execution times for the three most common pulse situations, expressed both in word-cycles and in clock times, are shown in Table II.12, (part of which is an update of Table II.11). The figures for elapsed time look much better in this table than in the former one. The processing time for matching pulses has decreased by 23 percent, and the worst-case processing time, which is that for pulses which do not match either to an established pulse train or any two-sample candidate, has improved by 15 percent.

The last two columns of Table II.12 show the elapsed times for the CMOS and D-ECL technologies, respectively. The figures are derived from the gate speeds given for the two types of gate, and the fact that the ALAP memory has eight levels of logic (at most, eight gates will switch consecutively during each clock cycle). It can be seen that the CMOS logic is still orders of magnitude too slow for the application. However, the D-ECL logic now is within range. There is still a considerable risk associated with the use of the very fast D-ECL logic, so the optimum performance prediction from which the figures in the table were derived contains a high risk factor.

In order that a signal-sorting system employing the ALAP can be proposed with any confidence at all, the system design will need to be flexible enough to compensate (through expansion) for a large difference between the

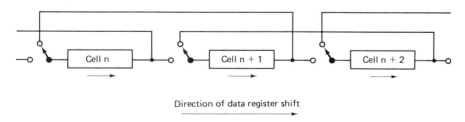

FIGURE II.43 Modifications to ALAP cell for bidirectional chaining channel

TABLE II.12

REVISED ALAP PULSE-PROCESSING TIMES

Pulse Type	Number of Word-Cycles	Average Elapsed Time in Clock Cycles	Elapsed Time μsec, 6-MHz Clock Rate	Elapsed Time μsec, 150-MHz Clock Rate
Matching pulses (Phases 1 + 2a)	17	653	104	3.7
Nonmatching pulses: no new pulse train created (phases 1 + 3 + 3b)	23	883	141	4.9
Nonmatching pulses: new pulse train created (phases 1 + 3 + 3a)	34	1306	209	7.3

optimum predicted performance of the ALAP array and the actual performance when the array is built and tested. In any case, the safer 50-MHz CMOS technology that was originally selected had to be abandoned at this stage in the project.

4.14 Improving System Performance

The ALAP memory array is the central component of the system in most respects. Almost all of the significant computation and data manipulation is performed in the array, and the array contains most of the system's logic. The ALAP array becomes the focus for all of the requirements for the remainder of the system. The two principal requirements are:

1. To permit system expansion to accommodate additional computational loading, or to compensate for an unexpectedly low rate of operation in the original system. The system design, in other words, should permit the use of multiple ALAP arrays to attain additional throughput.
2. To reduce to a minimum or eliminate all delays or idle time in the operation of the ALAP array. This means keeping the array supplied with data as it is needed, and having no pauses in array operation due to system overhead, for example.

In all of the discussion so far, the system has been assumed to meet the second of these requirements. This is because the original ALAP system design showed that this could essentially be accomplished. The original system design could not accommodate the first of these requirements, however, and suffered from a number of deficiencies and inefficiencies as well. The task of redesigning the remainder of the system was the last task in the project effort.

This subsection will first discuss the means for meeting the two stated requirements, in that order. Then it will show how the original system was redesigned and reorganized to both simplify it and to make it more efficient in operation and construction.

4.14.1 Adding Multiple Memory Arrays

There are only two general ways, to the author's knowledge, in which the throughput of an ALAP system can be increased by the addition of multiple ALAP arrays. (Also, both of these apply equally well to PEPE.) One of these is the technique of employing multiple arrays and then partitioning the input data among them on some basis. This has already been discussed. This technique was to be left until last, to be used to bring system throughput up to the requirements only after all other reasonable hardware and software improvements had been included in the design. This has already been briefly discussed.

The other technique, which the author employed in another ALAP system design, is to use the set of ALAP arrays, together with a set of controllers, in a pseudo-pipeline configuration. This technique works when both the input data and the working storage for the application can be divided among the arrays so that there is no interdependence among the contents of different arrays. It is also required that the process to be performed on each successive item of input data (or set of data) can be divided into successive, independent phases of approximately equal duration. This permits different controllers to process different phases of the task on different data simultaneously.

In the pipeline configuration, each of the multiple controllers contains the program for a different phase of the total procedure. As the first item of input data arrives, the first available memory array (call it Array 1) is assigned to it, and the controller for the first phase (call it the Phase 1 controller) is likewise assigned to it. The next input item is assigned to Array 2, by which time (if the system is correctly designed) the Phase 1 controller is finished with the previous data and can process the new data. At the same time, the Phase 2 controller is assigned to Array 1.

This process continues until all of the arrays are assigned to successive data items and all of the controllers are executing their respective phases of operation on the corresponding arrays. By the time the next data item arrives, the process for the data in Array 1 has been completed and the results have been output. Array 1 and the Phase 1 controller are now ready to process the new item. The similarity to the operation of a pipeline in a sequential processor is evident.

Figure II.44 shows a pipeline-array system employing four controllers and four memory arrays. Figure II.44(a) shows four data items, numbered in order of arrival. The switching for the assignment of the controllers and the switching of the input are both performed at the input to the arrays. Figure II.44(b) shows the assignment after the data in Array 1 has been completely processed, the array reassigned to the next item of data, and the controllers reassigned accordingly.

The pipeline approach works well for parallel systems when there is no relationship with respect to the procedure between the successive data items in the input stream. A sequence of independent images is a good example of such data. Moreover, many image-processing algorithms can be divided into phases which can be assigned to multiple controllers operating simultaneously on different images.

The author could find no way in the limited time available to make use of the pipeline technique in the signal-sorting application. This approach would have been preferable to that of partitioning the data among the arrays according to parameter range, since the previously mentioned problem of overlapping pulse trains would not have existed. However, due to time constraints, the decision was made to use the nonpipeline design for multiple-array systems.

4.14.2 Redesigning the Original ALAP System

The ALAP Demonstrator System, which was designed and built as a part of the original ALAP project, was intended to verify the basic ALAP concept, using conventional system components wherever possible. The present project had a different objective, that of developing a system design for a particular application. The justification existed for a complete redesign of the general system organization. The modifications to the ALAP memory and ALAP cell organization have already been described. The other two principal components of the original system were the special interface unit (or IFU) and the controller (which was a NOVA minicomputer). The ways in which functions of these two components were accommodated in the new system are described in the following two subsections.

4.14.2.1 Reassigning the IFU functions

Remember that the ALAP Demonstrator System was designed with the same objective in mind as this one. That objective was to keep the ALAP memory operating at full speed if possible, and free from wait periods. Since the decision had been made to use a conventional mincomputer as a controller, the burden of eliminating idle time in the ALAP memory operation fell entirely on the third component, the interface unit. To accomplish this, the IFU needed to be very elaborate indeed. Essentially, it had to be designed so that the ALAP memory could run asynchronously with the controller.

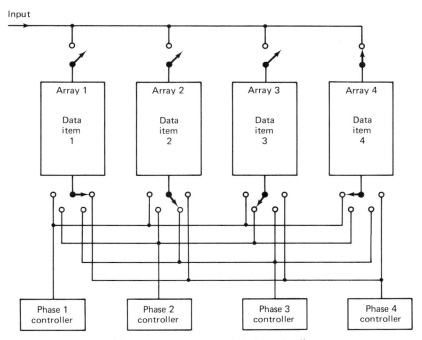

(a) Configuration before arrival of new data item

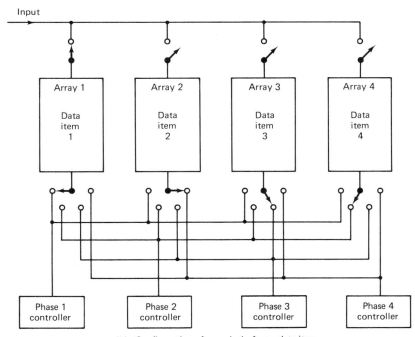

(b) Configuration after arrival of new data item

FIGURE II.44 Operation of a pipeline array system

Since the ALAP program resided in the controller memory, the asynchronous capability was achieved by having the IFU store several consecutive instructions at all times (two word-cycle instructions and eight flag-shift instructions). The instructions were loaded into the IFU from the controller memory well in advance of the execution. In addition, there were two sets of registers in the IFU for assignment to the ALAP data and control lines. Thus, at the end of an ALAP instruction execution, the set of registers containing the next instruction could be switched immediately to the ALAP memory and the next instruction executed without delay. There are almost never any conditional branches in an ALAP program and no fast loops, so the look-ahead scheme worked quite well.

However, the IFU logic was very complex, in fact, as complex as a normal central processing unit. In the present project, the author was free to design the controller from scratch and, as it turned out, the IFU as such could be eliminated altogether. Rather than having the ALAP instructions stored in the controller memory and fed to a set of registers for input to the ALAP memory, a specially designed separate memory was used for storing all ALAP instructions. This memory supplies the control and data information directly to the ALAP memory array as directed by the controller.

Two instruction memory designs were considered. The next two subsections describe them and show the reasons for the choice which was made between them.

The conventional ALAP instruction memory. This is a conventional addressable memory having the same word length (12 bits) as the number of control and data lines for the ALAP array. An ALAP word-cycle instruction, it will be remembered, is a table in which each entry contains the successive bit values to be fed to one of the ALAP array's control or data lines during each clock period of the instruction execution. In the instruction memory, the instruction tables are stored orthogonally, so that each word of the memory contains the input states for all ALAP control and data lines for a single clock period of the instruction. Figure II.45 shows the structure of the word-cycle instruction table as it appears in the instruction memory.

During the execution of an instruction, the controller and its associated clock logic cause the memory address to be incremented at each clock cycle of the instruction, thus putting the corresponding set of control states and data bits on the memory data bus. This bus is connected directly to the ALAP memory inputs. Needless to say, the ALAP instruction memory must be capable of operating at the same clock speed as the ALAP memory. The controller, as will be seen, can operate at a much lower speed.

The degenerate ALAP instruction memory. Some of the characteristics of the ALAP memory organization are quite applicable to the ALAP instruction memory. First, of couse, is its fault-to-tolerant capability at the cell level. A conventional memory is not fault tolerant except at the chip level,

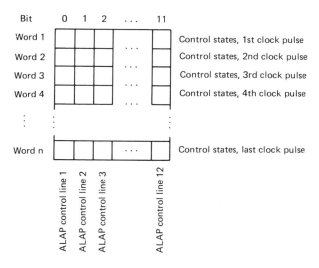

FIGURE II.45 ALAP instruction table organization in conventional instruction memory

and then only with much external switching (which itself will introduce reliability problems). Second, the ALAP cell uses a shift register to hold its data. Since the states for an ALAP control line are input serially at the same clock rate as the data register shift operation, using a shift register of the same organization for the instruction memory word is ideal.

As a result of these considerations, a "degenerate" form of the ALAP memory cell was designed as the basic instruction memory word, and a simplified chaining logic was used to provide extended shift capability and fault tolerance. The degenerate cell contains only the data register, data input logic, the basic chaining logic, and the fault-isolation logic.

Figure II.46 shows the organization of the degenerate ALAP instruction memory. Seven cells of the memory are shown, together with the chaining channel connections between them. The chaining channel need operate in only one direction. There are multiple common output channels, rather than only one, from each cell. Switching at each cell permits the cell to select one of the common output channels under program control.

There is also a chaining output to the next cell. Each cell, under program control, can output its contents either to the chaining channel for input to the next cell or to one of the common output channels, also selectable under program control, or to both the chaining output and the selected common output. Each of the common output channels is connected to one of the ALAP array's data or control input lines. In the figure, four common output channels lines are shown. Cells n_1 and n_2 are shown switched to two of the common output lines. The remaining cells are not so connected. In the complete

210 **CHAPTER II**

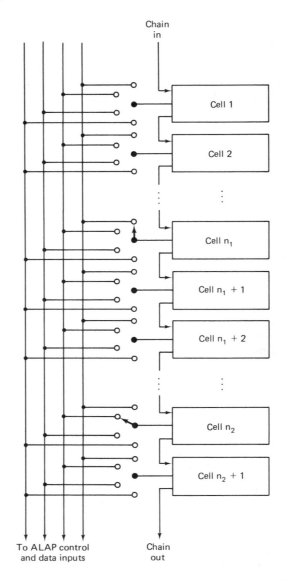

FIGURE II.46 Degenerate ALAP instruction memory

instruction memory, there are twelve common output channels, one for each ALAP control or data input.

Figure II.47 shows the way in which an ALAP instruction table is laid out within the degenerate ALAP instruction memory. All of the settings for each ALAP input line are stored in a set of contiguous cells in the instruction memory. In the figure, the contents of an instruction table for three ALAP

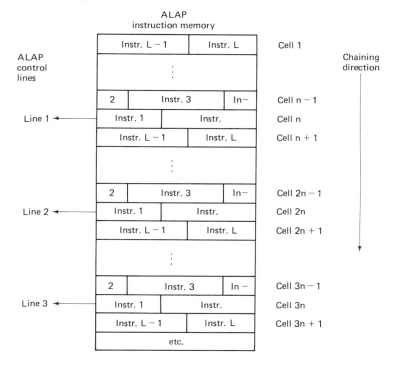

FIGURE II.47 ALAP instruction table organization in degenerate ALAP instruction memory

input lines are illustrated. Cells n, $2n$, and $3n$ have selected ALAP lines 1, 2, and 3, respectively. When the program execution begins, all of the cells will shift their contents together along the chaining channel. Cells n, $2n$, and $3n$ will also shift their contents to the corresponding ALAP input lines. Note that there is no need to have the data for each instruction contained within a single cell. The length of the cells' data registers need not correspond to the length of the ALAP data registers, for that matter. It is less expensive to have long data registers in any case, so the length of the registers can be chosen on the basis of cost to a great extent. Note also that this instruction memory has the same fault-tolerant capability as the ALAP memory itself, in contrast to the conventional instruction memory.

When one considers the way in which the degenerate ALAP instruction memory will be controlled, the picture is not quite so nice. For instance, when branches do occur in the ALAP program, it will be necessary to shift the instruction memory contents to align the new instruction with the first (leftmost) bits of their respective cells, and then switch the common output channel selections at all cells. The time consumed in this is not so important, since there are so few branches (but it is seen that the cells' data registers cannot be excessively long). However, it became apparent during the investigation that a control capability for the instruction memory almost as sophisticated as that for the ALAP memory itself would be needed. The same asynchronous operation would be needed for the instruction memory as well.

No doubt, if time had permitted, some reasonable way would have been found for simplifying the overall system organization and operation when degenerate ALAP instruction memories are used. However, in the final system design and construction, there would have been additional development costs for another LSI wafer, and for the additional complexity of the entire system. For this reason, the conventional ALAP instruction memory was chosen. The project effort then switched to the remaining system component, the controller, and to the overall system organization.

4.14.2.2 Redesigning the system controller

The task for the controller in the new ALAP system is much more demanding and complex than for the original ALAP Demonstrator System. First, the basic requirements for the controller are:

1. To determine the sequence of instructions to be executed and pass the required parameters and control signals to the proper system modules for the initialization of execution.
2. To control system mode and configuration, whatever that may entail. For example, there may be system readiness tests to be executed at prescribed intervals, with system configuration to be modified according to the results of the tests. There is also the task of initiating system operation (such as loading the ALAP instruction memory). These are all functions of the sort required for real world applications which are not important when the demonstration of a concept is the principal system objective.
3. The performance of all auxiliary arithmetic, logical, and data-manipulation operations. There is always some of this in a program. For example, there is the occasional calculation of program parameter values that are inputs to the ALAP array.
4. The communication with operators or other digital systems. This is a general data-processing system requirement, however, and will not be considered here.

The first decisions in the design of the controller were arrived at by considering the interfaces between the system components, and the data and control signals that pass over the interfaces. To simplify the immediate task, a single-array system was assumed.

As a basis from which the details of the system organization and operation could be worked out, a simplified preliminary system block diagram was worked out. The diagram is shown in Fig. II.48. This diagram would be modified as needed during the analysis. The principal interfaces for data and control were added during the analysis. These are shown in the figure, together with the types of information that pass across them.

The process for controlling and executing a word-cycle operation was the first thing considered, since it consumes 80 percent of the program execution time (as determined from ALAP programs previously written). The controller's memory must contain, in some form, sufficient information about the process to control the sequence of instruction executions. For this purpose, the sequential (i.e., controller) instructions cannot be stored in normal fashion. The parallel (i.e., ALAP) instructions cannot be stored in the controller memory, as they must be available immediately at the ALAP input when the array is ready for them. Instead, the controller memory need contain only the basic control parameters for the ALAP instructions. These are:

1. The address of the instruction table in the ALAP instruction memory
2. The clock rate for the instruction
3. The number of clock cycles for the instruction

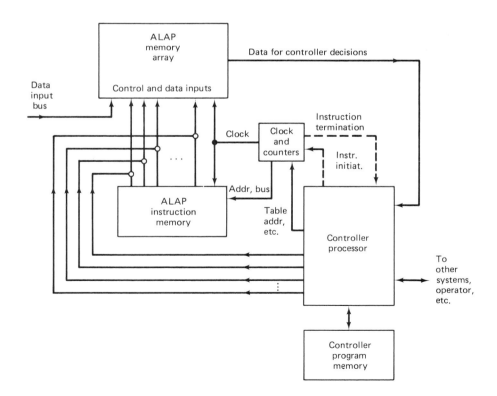

FIGURE II.48 Basic ALAP system components and interfaces

With this information, the controller should be able to initiate and control the sequence of operations for the entire system.

To initiate the execution of an ALAP instruction, the controller must put the control states for the first clock cycle of the instruction on the ALAP array input lines, give the ALAP clock the clock rate and clock count for the instruction, and then start the clock.

After each clock pulse, the ALAP instruction memory address must be incremented. Since this operation takes place at the ALAP clock rate, it cannot be performed by the controller, which operates from a different clock. The module containing the clock must therefore take on the additional function. Similarly, the clock pulse count must be decremented and tested at the clock rate. This function must also be assigned to the clock module. In Fig. II.48, the interfaces to the clock module show the instruction parameter and control inputs from the controller processor and the clock and address outputs to the ALAP instruction memory.

The controller, once it has initiated the execution of an ALAP instruction, has no further control over its execution, and is free to perform other tasks until the execution is completed. A control signal from the clock module informs the controller of the completion.

In the discussion thus far, all inputs to the ALAP memory have been treated as though they were data independent. This is not actually the case, of course, and provision must be made in the system to substitute other data, either calculated or input from outside the system, at the ALAP inputs in place of the stored data from the ALAP instruction memory. This capability is indicated in the figure by the lines from the controller processor which connect to the ALAP input lines. The logic (indicated in the figure by the small circles at the intersections) which selects the ALAP array inputs from either the instruction memory or the controller could be nothing more complicated than an OR gate at each input line, provided that the controller and the tables in the instruction memory were properly designed.

The controller processor must occasionally make decisions (i.e., for conditional branches) based on the results of the computation in the ALAP array. The figure shows a data input from the array to the controller for that purpose.

With this preliminary system configuration and operation procedure now worked out, a determination can be made as to the efficiency of the system, and modifications made accordingly.

The first consideration was the efficiency of the system in the control of the ALAP instruction execution, that is, the elimination or minimization of ALAP array idle time. The second consideration was the determination of the extent to which the controller could handle the other system operations without jeopardizing the efficient control of the ALAP array operations.

First, to minimize idle time between ALAP instructions, the controller must have everything ready for the initialization of each such instruction when the previous instruction is completed. Short delays would be permitted when a data-dependent conditional branch is made, or in the event that extensive peripheral calculations must be performed between ALAP instructions. This is because experience with previous ALAP programs, and with the signal-sorting program itself, showed that such situations are rare.

This did not seem like much of a requirement. However, it should be rememberd that the ALAP clock would be extremely fast, probably between 150 and 172 MHz. It was hoped that, for the sake of development costs, a fast, off-the-shelf microprocessor, or a set of such microprocessors could be used as the controller. A few of these microprocessors can run at 15 MHz.

Now a typical word-cycle instruction takes about 32 ALAP clock times (for processing a 32-bit field in the data registers). For the 15-MHz controller, this is the equivalent of only three controller clock times. This is not enough time to accomplish much. The prospect of designing a completely new controller to operate at a much faster clock rate was not an appealing thought.

A better solution was found by combining entire sequences of ALAP instructions into single tables within the ALAP instruction memory. This is possible because of the fact previously mentioned that long sequences of ALAP instructions are executed without conditional branches. More specifically, normal ALAP program execution consists of the execution of a sequence of flag-shift operations for setting up a word-cycle operation, followed by the word-cycle operation, followed by another sequence of flag-shift operations, and so on. The flag-shift operations comprise the 20-percent overhead mentioned previously. That is, they average a total of about seven clock cycles for each 32-bit word-cycle instruction.

It happens that two of the ALAP control lines specify the type of instruction being executed. Thus the combined table will work quite well from the standpoint of processing efficiency, with no pauses in ALAP operation occurring during the execution of the instruction sequence for each table.

To estimate the loading on the controller in the execution of the signal-sorting program, consider a typical subroutine from that program. This subroutine matches the input pulse parameters against the limits for all established pulse trains. The subroutine consists of seven word-chain operations plus the associated flag-shift operations, and contains no internal branches or loops. This comprises 269 ALAP clock times, or 26 controller clock times. This is sufficient time for the controller to execute about six instructions, which is enough for setting up for the next ALAP instruction sequence if some care is taken in the controller interface design. It appears that an additional controller processor will be needed for the auxiliary calculations, however.

Several relatively minor problems remained in the system because of the use of the combined ALAP instruction tables. One of these is the fact that the word-cycle operations in a single table will have different clock rates and counts. Rather than have these parameters stored in the clock module, it was decided to extend the word length of the ALAP instruction memory to permit them to be stored in the tables themselves for direct access by the clock module. This is inefficient use of the memory.

Another problem was the fact that the use of the combined ALAP instruction tables would make it impossible to conserve memory by combining the many similar sequences of flag-shift operations into subroutines. The same would be true for multiple appearances of identical word-cycle operations. All ALAP coding, in effect, would be straight-line coding as far as small subroutines would be concerned.

It seems certain that a little additional thought could have resolved these two problems if time had permitted. However, the big problems, that of making efficient use of conventional microprocessors for the control functions, appeared to have been solved. The resolution of these two minor questions (and, no doubt, a number of others which would arise), could be left until the next phase of the project. As for the design effort for the present phase, all that remained was the overall design for the multiple-array system. The next subsection describes that system.

4.15 The First-Pass ALAP System

The system design that resulted from the project was a first-pass design. The intention of the design was to demonstrate feasibility and to expose areas of risk. Many areas of the design, those which appeared to involve no serious questions of feasibility or risk, were left in only rough detail.

Figure II.49 is the block diagram for the first-pass ALAP signal-sorting system. Four of the five ALAP memory arrays are shown. The differences between this system and the preliminary system in Fig. II.48 are apparent. In addition to the multiple ALAP arrays, there is an extra controller processor for auxiliary computations. Both the auxiliary processor and the controller processor have their own local memories. It is the controller's memory that contains the signal-sorting program. The auxiliary processor operates in a slave mode to the controller processor. Its principal function is to generate single items of data, which are to be substituted at the ALAP array inputs in place of the data from the instruction memory.

Provision is included for loading the ALAP instruction memory from the controller processor memory. This permits the instruction memory to be volatile. No external interfaces to the system are shown except the data input.

4.16 Assessment and Explanation of Risks

There are three general types of risk in projects of the type proposed. These are technical, cost, and schedule risks. For the signal-sorting development effort which was proposed, the overriding risk was technical. The principal cost and scheduling risks were a result of the technical risks.

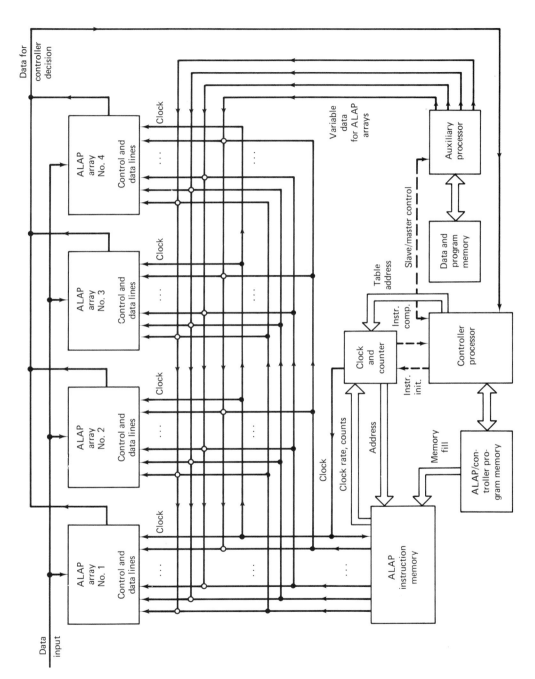

FIGURE II.49 Final first-pass ALAP system

The system concept and the ALAP concept itself were not high risk elements. The programming and checkout of the signal-sorting algorithm on the ALAP simulator helped to lower that element of risk. The fact that an ALAP processor had already been designed and built were big factors in lowering the risks having to do with concept or system. The one large risk was in the fact that very advanced technology was needed if the system was to operate in real-time at the required input pulse rate. This risk, it will be remembered, resulted from the fact that the recommended D-ECL technology had not yet (as of December, 1978) been employed successfully in full scale systems, although it was working at the chip level. Predictions as to the length of time for developing the signal-sorting system were thus not as reliable as they otherwise would have been.

The risk section of the final report explained the nature of the technical risk, and also discussed the reasons for assigning a much lower risk to other aspects of the proposed development effort.

5. PROGRAMMING RECONFIGURABLE PARALLEL ARRAY PROCESSORS

This section discusses the languages, tools, and techniques for the programming of reconfigurable parallel array processors. The discussion attempts to be as general as possible in most respects. While many of the ideas and suggestions in the section apply to all highly parallel systems, and some to unusual systems in general, there are many that will be of a much more restricted nature. There are great basic differences, for example, in the approach to programming PEPE and ALAP. A PEPE processing element is a conventional sequential processor in nearly all respects, and much of its internal programming can be designed and analyzed from that standpoint (as indeed it was in the preceding section). Programming ALAP or STARAN is a different matter altogether, and many of the techniques of analysis and design for sequential processors simply do not apply.

In any case, the discussion includes references as to the general or restricted nature of the points that are made. In other instances, there will be no such references. The bulk of the author's experience with the programming of parallel arrays has been with associative processors, particularly the ALAP, and some of the opinions expressed should be taken in that context. In addition, the major programming examples will be from ALAP programs.

5.1 Programming Languages

When it is possible to do so, an effective procedure for designing a computer system slanted toward a special class of application is:

1. Determine the data structures and processes which best apply to the class of applications.
2. Design a language which expresses these structures and processes concisely and thoroughly, preferably a high-order language.
3. Design the computer system to implement the language efficiently.

Using this processing, the result will be a system that is both efficient in the processing for the specified applications and easy to program. This latter is an important benefit if much programming is to be done. It has become well known that the major cost in most computer applications involving the design or purchase of a system has become the software cost rather than the hardware cost.

This chapter deals with a very special class of computer systems, the highly parallel reconfigurable array processors. In general, these machines are not presently competitors for conventional general-purpose systems. The need for highly parallel architectures almost always arises in connection with some application for which nothing else has the required throughput. The applications described in the preceding section are of this type. In such cases, the very desirable capability of the system to be efficiently programmable in a high-level language must generally be a secondary requirement to that of being powerful enough for the required task. Like it or not, then, the engineer must design the system directly from the requirements for the application, as expressed by the algorithms and data structures.

An approach such as this does not necessarily rule out high-level language programming for the system. It merely makes it a more difficult capability to achieve. The first programs for the new system will be written in some machine-level language. The experience of the programmers will then be used to determine the combinations of machine operations which encompass all major functions needed for the applications. These are then used as the basis for designing the high-level language.

For some systems, FORTRAN itself, or some equivalent language, can be used directly as the basis for the high-level language. To this basic language are added special statements and statement types for the functions peculiar to the machine and the application. It is less likely that this approach to language design will lead to the most efficient possible language from the programming standpoint. However, if it works reasonably well, it can more than justify itself by the greatly reduced cost and time for its development.

The ALAP was not designed with efficient high-level language programmability as a major requirement. Its objectives were low cost, efficient highly parallel operation over a wide range of applications, and built-in fault tolerance at the cell level.† The basic operations of the ALAP were developed at a very low functional level. It was known initially that the basic requirements for highly parallel operation were met by combinations of these basic operations. However, it was not known how the operations could best be

†General principles and state-of-the-art in the area of software fault-tolerance are discussed in Chap. VI, Sec. 2.2 (Ed.)

combined for efficient general programming. For this reason, no attempt was made to design a high-level language for the ALAP. Instead, the language was designed from the lowest-level standpoint possible, which was the equivalent of a microprogramming level. Essentially, the three basic ALAP array instructions were given symbolic names. The many instruction parameters were specified as the serial bit-by-bit inputs to the ALAP control lines. The language contains very few defaults, since it was not known what the most common parameter values would be.

The foregoing applies only to the language for specifying the ALAP array instructions. The controller is a sequential processor, and its internal programs, which specify the sequence of ALAP and controller instructions to be executed, are written in the controller's own assembly language. For ease in specifying the ALAP array instructions, almost the only tool given the programmer was the capability to combine common sequences of ALAP operations into subroutines, which can then be called out by their symbolic names. This capability has proven to be of great value in several ways. First of all, as soon as a few programs had been written a useful set of subroutines was developed and used in the succeeding programs. Second, an attempt has been made to define an all-encompassing set of subroutines from which entire ALAP programs can be written. The attempt has almost completely succeeded, and has given a considerable insight into the closely related problem of developing a high-level language for the ALAP. It has also led to the author's belief that good high-level language capability should not be expected from a computer design unless such capability was an original design objective.

5.2 Software Programming Tools

Software programming tools serve the same purpose for parallel reconfigurable array systems as for conventional systems. They enable the programmer to code programs in a symbolic language. They assist the programmer in editing source programs and in checking them out. When the object computer already exists, and has conventional peripheral hardware (e.g., card readers, high-speed printers, magnetic tape drives) and system software for programming support, the programming support software resides and executes on that machine. However, when the object machine is a reconfigurable array processor, the chances are that the object machine itself does not exist, or that it is being built at the same time that the support software is being written. Moreover, the new machine will most likely not have peripherals or system software for the support of programming activities. The support software should therefore be written for some conventional data-processing system. An additional advantage in doing this will be that the programmers and engineers will not be in each other's way during hardware checkout; they will be able to work independently to some extent.

The next subsections discuss the necessary programming support software, and explains the characteristics to be desired in the software.

5.2.1 Source-File Editors, File Managers, User Interfaces

All of the common support software, such as the source-file editor and the magnetic tape file manager, are available on the software support data-processing system as a part of its operating system. They are applicable to any programming task.

5.2.2 Symbolic Assembler

The symbolic assembler is a program that accepts a source progam module coded in a symbolic machine-level language and from it generates a relocatable binary object module. When the array processor is of a very unusual design, such as the associative processor, the assembler will be unusual in design and complicated in detail. It will probably be a very difficult and time-consuming task to program. Using a high-level language, such as Pascal, is certainly of assistance in simplifying the programming of the assembler; however, it will not make the task easy. The ALAP project was fortunate to have a software support computer available (a Xerox SIGMA 9) which had a very fine meta-assembler. Meta-assemblers are assemblers with the capability to assemble programs for computers other than the host computer, under the direction of a user-supplied set of assembly procedures. Even with this meta-assembler as an aid, however, it required a very skilled programmer, one who knew the meta-assembler's capabilities inside out, 5 months to program and check out the ALAP assembler.

5.2.3 Linkage Editor

The linkage editor is a program that combines a number of relocatable object modules generated by the assembler into a single object program ready for loading and execution on the object computer. The linkage editor is written in the software support computer's own assembly language or high-level language.

5.2.4 Functional Simulator

The only way in which a program can be checked out when the object machine is not available is by the use of a functional simulator resident on another computer. The simulator is a program that causes the host machine to behave as though it were the object machine. It accepts as input a binary load module put together by the linkage editor and simulates the instruction-by-instruction execution of the program. Whatever debugging aids are available to the programmer on the sofware support computer are those contained in the simulator. These will generally include the capability to set breakpoints, to display or print memory and register dumps at specified points within the object program and, ideally, to print or display a full trace of the object program execution. The trace contains the detailed results of the execution of each object machine instruction within a specified range of the object program. This includes the effective address or addresses for the instruction, if meaningful, and the contents of all registers and memory locations addressed and/or modified by the instruction. This may be the entire memory, if the object machine is an associative processor, or perhaps a single location from each processing element if the machine is a network processor.

The importance of the trace capability cannot be over-emphasized. For very unusual machines such as the array processors, such detailed debugging support is of immense assistance to the programmer in understanding the way in which the machine's hardware executes the instructions.

The design of the simulator itself is of considerable importance. If the rules of structured program design are followed, and if the simulator is designed with ease of modification as an objective, the simulator will become an effective design tool (and will also cost much more, incidentally, as a function of the increased flexibility). When design changes to the array processor are made or contemplated, the simulator can be modified accordingly and used to check the effectiveness of the changes. Unfortunately, the needed flexibility was not designed into the ALAP simulator, and the task of analyzing and justifying the proposed changes to the ALAP design has therefore been made much more difficult.

5.3 Programming Techniques for Array Processors

5.3.1 General

Programming for machines of unusual design is considerably more difficult than for conventional machines. The parallel processor to be programmed will most likely be new, and there will be little previous programming experience to be taken advantage of. Languages of advanced design which permit efficient use of the hardware will not generally be available. The need to gain the greatest possible efficiency in the operation of the machine will overrule most concerns for ease of programming, moreover, and the programmer will thus be faced with all of the complexities of the logic during program design and implementation.

With multiple-processor machines having no interprocessor communication network, the difficulty of programming the processing elements appears to be not much greater than for serial machines. (The author could very likely be wrong on this. The difficulties of programming often cannot be seen until the details are attacked.) The programming of the controller's part of the program, of course, will be more difficult, since the controller program contains the parallel elements of the entire program.

For multiprocessor machines having communcation channels between the processors, the task of writing efficient programs is much more difficult; in some cases extremely so. Parallel network processing arrays, for example, are particularly prone to difficulties having to do with keeping the communication rate between the processors great enough to match the processors' internal processing speeds.

5.3.2 Programming the ALAP

The ALAP is one of the most interesting and challenging machines to program. The parallel operation of the machine is interesting to observe in a trace printout of the program execution. The unidirectional chaining channel of the original ALAP presents a considerable challenge to efficient programming. The later design, with the bidirectional channel, is not such a challenge and is a better performer.

Since each ALAP cell contains only a single word of storage, the chaining channel is heavily involving the execution of the program, carrying operands and flag states to the cells where they are to be used. This will be shown in the examples.

The way in which a machine like the ALAP is programmed is not obvious at first glance, particularly the way in which the cells containing the operands for the parallel operations are located. The individual cells have no hardware address corresponding to word addresses on sequential machines. There are only two means whereby a particular set of operands may be identified and the cells containing them tagged. One is by the cells' contents, which can be checked by means of an exact-match word-cycle operation. The other is by the relative locations of the cells on the chaining channel with respect to a set of cells previously tagged. Flag-shift operations employing the chaining channel can locate cells whose relative locations are known. Once identified, a set of tagged cells can be set to perform parallel arithmetic or logical operations, to send or receive data to or from another set of cells, and so on.

An example from ALAPSIG, the signal-sorting program described in the previous section, will illustrate the way in which the ALAP performs parallel arithmetic operations using the chaining channel or the common operand channel. The example is a portion of a subroutine (all of it, in fact, except the entry and exit instructions) which updates the predicted upper limits, lower limits, and expected values of pulse TOA for all pulse trains with files in ALAP memory.

Figure II.50 shows the format of the pulse-train file for ALAPSIG. The chaining channel operates in the downward direction on the page. (The program was written for the ALAP simulator, which simulates the original

224 CHAPTER II

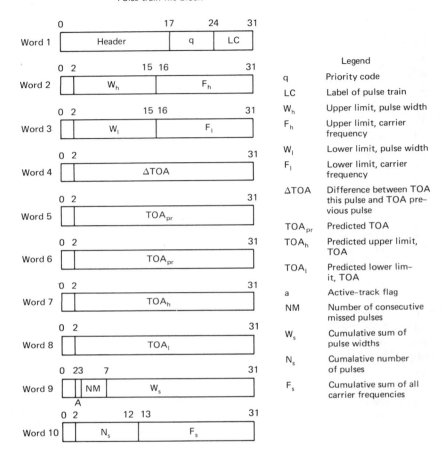

FIGURE II.50 Format of a pulse train file

ALAP one-way chaining channel.) The labels denoting the contents of the cells for the file are defined in the legend on the figure. The header words (the terms "word" and "cell" are used interchangeably in this discussion), which are the first words in the files, are the only words in the ALAP array which contain a "1" in the most-significant and next-most-significant bit positions. This makes it possible to locate the first cells of all files by a content search and to tag them. Once this is done, all second words, third words, and so on, in the files can be found by their relative locations with respect to the header words, as will be shown.

The reader should refer to the definition of the algorithm for updating matching pulse-train files (in Sec. 4.6.1, under Case 1) in connection with the following description of the operation of the update subroutine for pulse TOA.

The first operations in the update subroutine are the subtraction of the TOA value for the current pulse from the predicted value stored in the sixth words of the files, and the input of the new TOA value into the fifth words of the files. When the actual update subroutine in ALAPSIG is called, all pulse-train files to be updated (there will usually be only one) will have already been identified and will have their header flags set at the first words of the files. For the purpose of this illustration, however, it will be assumed that all pulse-train files are to be updated, so that the process for identifying all files can be described.

The first operation, that of identifying all pulse-train files, is performed as follows:

1. The file-header identifier, which is two 1's followed by 16 0's, is loaded into an external register for input as the operand for the word-cycle operation to follow.
2. All flag registers are recirculated by execution of a set of flag-shift operations totaling six clock cycles. During this process, the instructions set flag 4 at all cells, thus allowing the compare operation to be performed at all cells during the next word cycle.

Figure II.51 shows the cells for a single pulse-train file after these operations, with the states of the cells as just described. The auxiliary operand input register containing the comparand is also shown. The contents of cell 1 and cells 4 through 8 only are shown, since they are the only cells involved in the subroutine operations.

3. The word-cycle operation is performed with the exact-match operation specified. This will cause all flag 4's to be reset except at the file header words. These have all matched the comparand in the external register.
4. Another series of flag-shift operations is performed, again totaling six clock cycles (Most such series of flag-shift operations comprise six clock cycles, since the six-bit flag registers are thus shifted end-around back into their original orientations.) During this series of flag-shifts, the states of all flag 4's are transferred to their corresponding head flags, and also to their respective flag 7's. Flag 7 ordinarily denotes output from the cell to the common output bus. When no output is to be performed, it can be used as a "cheek pouch" for saving the tags for a desired set of cells. That is the purpose for setting flag 7's here.

All header words have now been tagged in their header words. Figure II.52 shows the states of the cells in the pulse-train file at the completion of these operations.

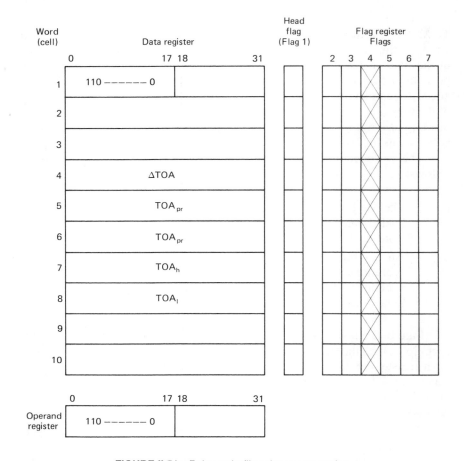

FIGURE II.51 Pulse-train file prior to processing

The first operations to be performed on the tagged pulse-train files (and the first in the actual update subroutine in ALAPSIG) are performed by means of a single word-chain operation. These are:

1. Subtracting the TOA value for the current pulse from the predicted values in the sixth words of all files. The absolute value of this difference will be TOA_{err}, the error in the predicted value.
2. Inputting the current TOA value, TOA_{in}, into the fifth words of the files, overlaying the former contents.

The process for setting up for this word-chain operation is as follows:

1. Perform a four-clock-cycle flag-shift operation to shift the states of all head flags four cells down the chaining channel. This leaves the

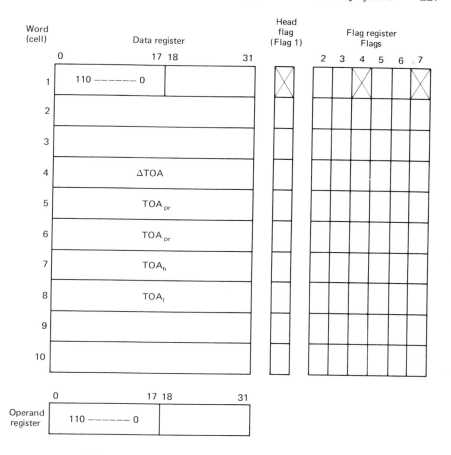

FIGURE II.52 Pulse-train file with header word tagged

head flags set at the fifth cells of the pulse-train files, and reset at all other cells.

2. Recirculate (i.e., shift for five clock cycles) all flag registers, while transferring the states of all head flags to their respective flag 6's. This tags the cells to input the contents of the external register connected to the common input channel during the next word-chain operation.

3. Load the TOA value for the current input pulse into the external operand register. This will be connected to the common input channel as well as to the common operand channel during the next word-chain operation.

Figure II.53 shows the state of the pulse-train file after these operations have been performed.

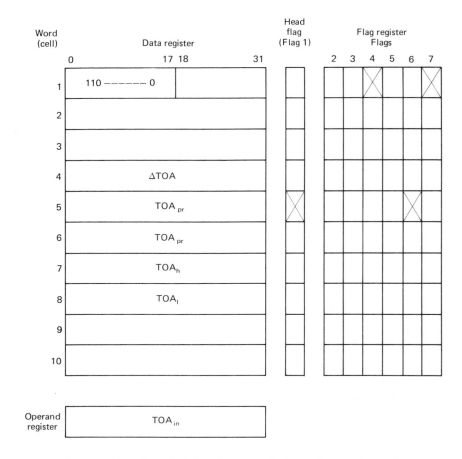

FIGURE II.53 Pulse-train file after set-up for input of new pulse TOA

The remaining set-up operations set the sixth cells of the pulse-train files to perform the subtraction of the input TOA value:

4. Perform a one-clock-time flag-shift operation to shift all head flag states one cell down the chaining channel (i.e., to the sixth cells).
5. Recirculate all flag registers while transferring the states of all head flags to their respective flag 4's. This sets the sixth cells of all pulse-train files to subtract the contents of the operand register from their contents.

Figure II.54 shows the state of the pulse-train file after this sequence of operations. The word-cycle operation can now be performed, inputting the TOA value into the fifth words of the files and subtracting that value from the

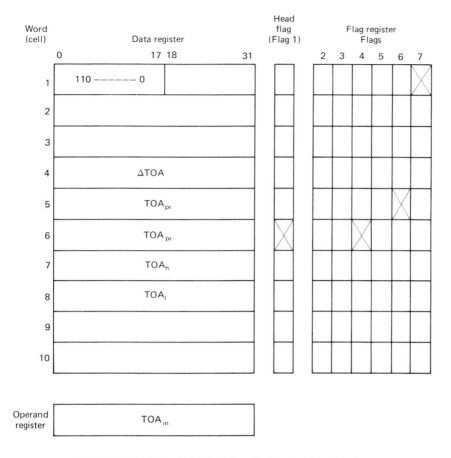

FIGURE II.54 Pulse-train file set up for input and subtraction

contents of the sixth words. Figure II.55 shows the state of the file after this operation. The entire procedure just described requires a total of 64 clock cycles for the two word-cycle operations, plus 6 clock cycles and 22 clock cycles, respectively, for the flag-shift operations, which set up the two word-cycle operations.

The set-up operations for the remaining word-cycle operations performed by the update subroutine will not be described. There are six such word-cycle operations. They are:

1. A combination less-than-compare/input operation. This resets flag 4 at all cells in which $TOA_{pr} - TOA_{in}$ is non-negative, and inputs the constant, C, into all word sevens.
2. A combination Exclusive-OR/input operation. This Exclusive-ORs

230 CHAPTER II

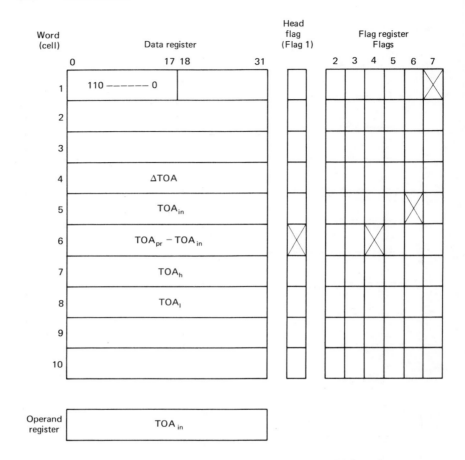

FIGURE II.55 Pulse-train file after input and subtraction

the remaining tagged values of $TOA_{pr} - TOA_{in}$, which are all negative, with a word of all 1's, thus obtaining the 1's complements of those values. At the same time, the constant, $-C$, is input to the eighth words of all files.

3. An ADD operation. This adds a 1 at the least-significant bit position of all tagged sixth words, thus obtaining the arithmetic complements of the negative $TOA_{pr} - TOA_{in}$ values. The sixth words of all pulse-train files now contain TOA_{err}, which is the absolute value of the difference between the predicted TOA and the actual TOA for the current input pulse.

4. A double-ADD operation. The delta-TOAs in the fourth words of the files are added to the TOA_{in} values in the fifth words, using the chaining channel to transmit the former operands. The sums, which

are the TOA'_{pr}, the new predicted TOA values, are retained in the fifth words. At the same time, the TOA_{err} values in the sixth words are added to the constants, C, in the seventh words. Notice that the head flag settings have been moved back in the reverse of the chaining channel direction to enable the fourth words to be set-up for this operation. This was actually done by setting all head flags to the flag 7 states, and then chaining the head flags down the chaining channel for three clock cycles. This is the reason for saving the head flags at the header words in flag 7.

5. A combination SUB/data-chaining operation, using the chaining channel. The TOA_{err} values in the sixth words are subtracted from the constants, $-C$, in the eighth words. At the same time, the new TOA'_{pr} values in the fifth words are transferred via the chaining channel to the sixth words, replacing the TOA_{err} values. Because of the way in which the ALAP cell logic operates, there is no conflict at the sixth words between the two chaining operations.

6. A combination ADD operation. This adds the new TOA'_{pr} values in the sixth words to the values in the seventh and eighth words, thus obtaining the new values, TOA'_h and TOA'_l.

This completes the update operations for TOA_{pr}, TOA_h and TOA_l as defined by the four functions listed in Sec. 4.6.1 under Case 1. Figures II.56 thru II.61 show the state of the pulse-train file after each of the six final word-chain operations just described. Flags 2 and 3 specify the internal state of the cell with respect to the chaining channel operations. Cells with both of these flags set will output their contents to the chaining channel, and will input the data received from the chaining channel. Cells with both flags reset will input the data from the chaining channel, and will also relay that data to the next cell. Other (global) parameters for the word-cycle instruction specify whether the operands for the arithmetic operations will come from the chaining channel inputs to the cells or from the common operand bus.

It will be noticed that the new values in cells 4 thru 8, resulting from these operations, will be the inputs for the same update operations the next time the update subroutine is executed (probably for the next input pulse). The evaluation of these four functions required three additions, three subtractions, and an absolute-value operation. All of this was accomplished by seven word-cycle operations. The computation of the absolute value required three of these operations, since it was not built into the cell logic as one of the basic arithmetic operations.

The example just given should help in understanding the way in which the ALAP programs are designed. The layout of the data structures and working storage is the first and most important task, once the general procedures for the task are understood. The objectives of most importance in doing this are:

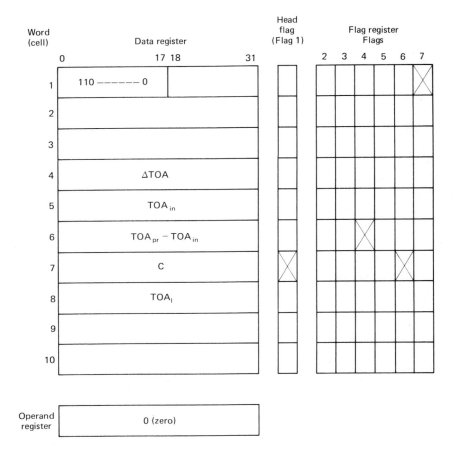

FIGURE II.56 Pulse-train file after less-than/input operation

1. To minimize the distances between pairs of operands involved in arithmetic or other operations. Long distances require increased time for the chaining operation, which is unclocked and operates at logic speed (about three gate delays per cell).
2. To eliminate or minimize the need to chain operands within the ALAP array in order to avoid conflicts on the chaining channel.

Second, "tight" coding is just as important in critical parts of an ALAP program as in programs for conventional machines. The example just illustrated is a case in point. The pulse-train file layout was reworked several times until a procedure employing a minimum number of word chains was devised. The update subroutine is executed for all pulses which match to a pulse-train file, and is therefore critical to program execution time.

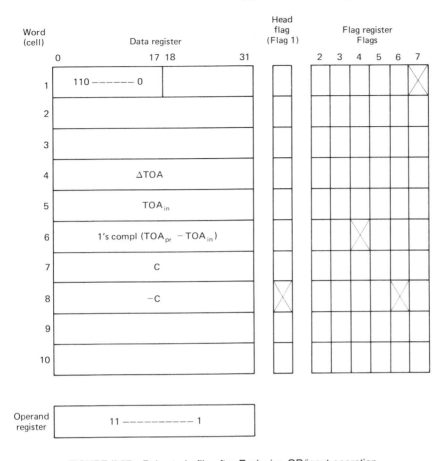

FIGURE II.57 Pulse-train file after Exclusive-OR/input operation

6. FUTURE TRENDS IN RECONFIGURABLE ARRAY DESIGNS

To assess the future of reconfigurable arrays as practical data-processing systems, one must examine:

1. The expected technology, its advantages, and the desirable characteristics of the circuitry that is to be implemented with it, especially with regard to VLSI.
2. The expected applications that will require very powerful array systems.
3. The limitations in the performance of present array processors with

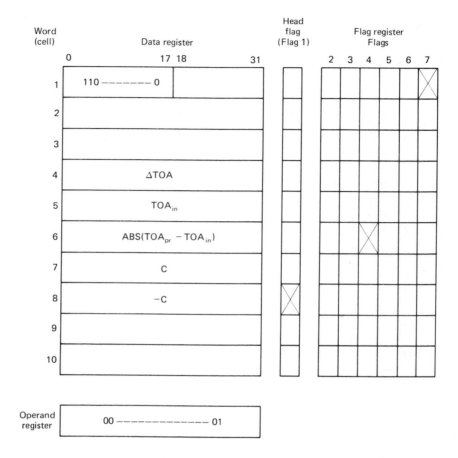

FIGURE II.58 Pulse-train file after ADD operation to yield TOA_{err} in sixth word

regard to throughput, fault tolerance, amenability to VLSI implementation, and suitability to the expected applications.

This section will take up these topics one by one, in the order stated.

6.1 The Emerging VLSI Technology

The two principal characteristics of some of the gate technologies that are presently being developed are extremely high speed (the order of 500 psec gate delays) and very large numbers of gates per wafer (several hundred thousand). This sounds like very good news for the system designer at first glance. However, when one looks at the problems being faced by the engineers trying to fabricate this circuitry and design systems using it, a quite

Reconfigurable Parallel Array Systems   235

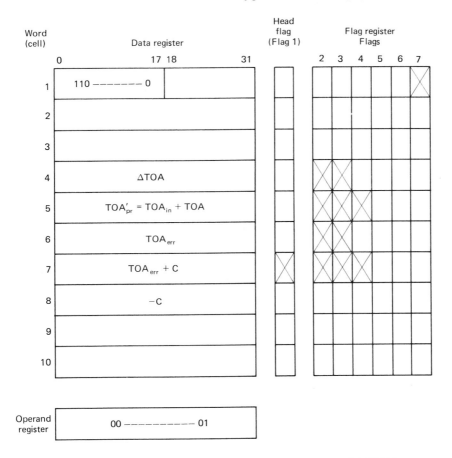

FIGURE II.59 Pulse-train file after double-ADD operation to yield TOA'_{pr}

different picture emerges. The big problem with off-chip delays has already been briefly discussed (Sec. 4.11). There are other problems almost as bad. Simply trying to pipe a clock from one wafer on a board to several others is a messy task at 250-MHz clock rates. For example, the wire lengths may be slightly different between the generating wafer and each of the others, leading to phase lags. The advantages of ALAP memory logic in minimizing the problems of off-chip delays have already been discussed. Since the gate-technology people are looking for suitable candidates for their products, this is an encouraging sign for the future of highly parallel array processors.

Large conferences are being held on the future of VLSI, with experts from all over the world in attendance. The big problem being discussed is that of how to take advantage of the capability to put so many gates on a single wafer. Conventional processors are not altogether suitable for these tech-

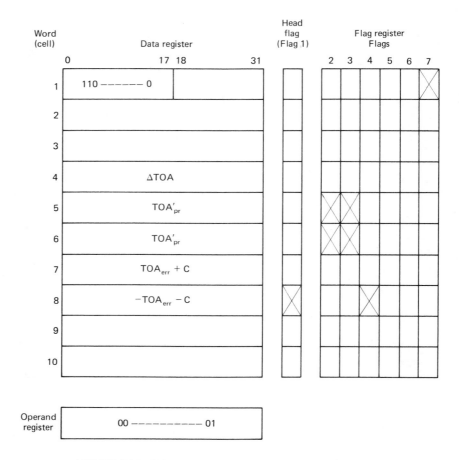

FIGURE II.60 Pulse-train file after SUB/data-chaining operation

nologies. The main reason, or one of the main reasons, is that it is difficult to factor the logic so that the interfaces between wafers in the systems are clean. This means relatively few pins (not easy when the wafer contains so many gates), similarity in the signal paths to the other wafers from the standpoint of phasing problems, loading, the generation of return signals, and so on. When VLSI fabrication is combined with the very fast gate technologies, the problems multiply.[†]

The ALAP memory design looks good to the VLSI designers when they find that the number of pins on an ALAP wafer is low, is independent of the number of cells on the wafer, and that almost all of the pins are for bit-serial inputs from identical external registers. The PEPE design looks good for many of the same reasons. With networks of processors, there are more problems, but still far fewer than for single fast serial processors.

On the negative side with regard to the use of array processors as a major

[†]Similar views on the proficiency of LSI based designs are discussed in Chap. I, Sections 18 to 22.

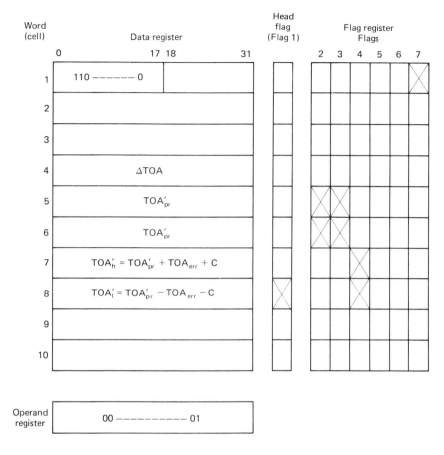

FIGURE II.61 Pulse-train file after combination ADD operation to produce TOA'_h and TOA'_l

candidate for VLSI implementation is the fact that these processors are not as general-purpose as conventional systems. It is evident from the standpoint of economics that wide applicability is a prime requirement for a VLSI wafer which will be very expensive to design. Array processors must therefore be designed with such versatility in mind, and an equal effort must be expended on applications analysis to discover new, highly parallel procedures for common applications. One promising aspect of the ALAP system organization (and also for the PEPE and STARAN) is that the controllers for the arrays can operate at much lower clock rates than the very fast rates for the arrays, since they have much less to do. This means that the serial controllers, with their relatively irregular logic, will not have the off-chip delay problems or the other problems associated with the extremely fast logic needed for the arrays.

Fault tolerance, both in the system architecture and within the wafers themselves, has historically been given far too little general attention. Future systems implemented on VLSI wafers will be extremely complicated and thus subject to failures of every type in relative proportion to that complexity (not absolute proportion, since the ratio of connections to logic is lower for VLSI).† VLSI fabrication will have great yield problems, for one thing. On-chip fault isolation and repair is a very desirable feature to have for this reason. It will increase yield, thus lower manufacturing costs. It will also extend the useful life of the wafers, since they will have the same fault-repair capability after they are installed and operated in a system.

In summary, circuit and system designs and technology are closely related. They cannot be thought of independently even now, and certainly not with very-high-speed VLSI technology. Regular, redundant logic, such as that found in parallel array sytems, makes them promising candidates for the expected technologies. The capability for software-controlled reconfigurability makes the redundant logic more fault tolerant, and therefore less expensive and more versatile with regard to the applications.‡

†Discussion on the types of faults in digital circuits is presented in Chap. VI, Secs. 1 and 2.1. Also Chap. VI Sec. 2.1. provides a comprehensive analysis on existing techniques for treating these faults.

‡Various techniques of software controlled reconfiguration aimed at enhancing fault-tolerance of digital circuits are discussed in Chap. VI, Secs 2.1.2. to 2.1.7.

6.2 Future Applications for Array Processors

The problem of versatility with regard to the applicability of highly parallel array processors has been mentioned. Consider the application areas for which the arrays are suitable or superior. These include the aforementioned block-oriented tasks, such as signal sorting. They also include real-time text-processing, although such systems will be very special-purpose designs. Advanced image-processing tasks, which are important as well as challenging, are high on the list. Very advanced file-retrieval applications, such as question-answering tasks employing deductive and inductive inference (also closely related to "knowledge-based systems") are potentially even more promising tasks, now that suitable algorithms and data structures are being developed. Any person familiar with one of these fields of application can name problems far beyond the capability of any existing computer. These problems are practical in nature, it should be said, not merely academic or theoretical. The great expense of system development and, unfortunately, the fact that only a few systems are needed for these advanced applications, has prevented these systems from being built in many instances. In other instances, development of very powerful array processors is underway, one particular system being the Goodyear Aerospace Massively Parallel Processor (MPP) System to be used in processing satellite imagery at high rates.‡

The author's pet application field is the aforementioned advanced question-answering systems. Such systems may open the door to such capabilities as context-sensitive language translation and interpretation, and thus

‡See Batcher [10].

to the simulation of human reasoning processes. To accomplish even a small step in such a direction is an objective of immense appeal. The context-sensitivity of the query-answering process seems to be a key factor. The interpretation of context must require, among many other things, a very large data base of possible contexts which can be searched very frequently and rapidly with complex search criteria. The potential applicability of associative processors to such a process is evident.

6.3 Limitations in the Performance of Present Designs

6.3.1 Throughput Limitations

It seems strange to discuss limitations on throughput for the array processors, when in fact their throughput surpasses in general that of any other system design (for suitable applications, of course). Nevertheless, they do have limitations of a particular kind for each design, and an understanding of these limitations is of value in predicting the trend of future developments.

The arrays having one-dimensional communication networks for transferring data among the processing elements are restrained by an "n-to-one" throughput barrier. The "n-to-one" reference is with respect to matrix operations, although it has general connotations. That is, in performing tasks in which square arrays of interrelated data items are involved, and in which all items can be processed simultaneously, the limitation means, in effect, that the processor can process the items at no better than an n-to-one rate as compared to a serial processor. ALAP and STARAN fall into this class. PEPE is poorer, since it lacks an interprocessor communication network. Two-dimensional arrays have a potential order of n^2-to-one throughput, although it is rarely achieved in practice. ILLIAC IV is a two-dimensional array that at best can only approach the n^2-to-one limit on the number of concurrent operations.

Conceptually, the removal of the n-to-one restriction by simply adding the additional dimension of communication capability seems like a reasonably simple task. In practice, it is anything but simple. It should be remembered, for one thing, that applications for which two-dimensional processor arrays are optimal are rare. The matrix applications will benefit from such arrays of processors; the other applications generally will not, or else will not make anywhere near full use of the two-dimensional communication capability. Second, the entire communication concept may have to change from that for the one-dimensional array when the dimensionality is increased. Consider, for example, the addition of additional chaining channels to the ALAP to achieve two-dimensional communication. Great changes would need to be made in the cell architecture to correspond to the new communication capability. For example, a single data register would hardly suffice when data was being received from several other cells simultaneously. If extra data registers are added, the selection logic for the communication operations would grow at an exponential rate, and so on. The result wold probably be a processor array having very little resemblance to the original ALAP, and would most likely lack a number of the ALAP's best features. The reader is urged to attempt such an extension of the ALAP, or of any other one-dimensional array processor with which he or she is familiar.

6.3.2 Limitations on Fault-Tolerant Capability

ALAP was conceived with built-in fault tolerance and repair capability as an objective. It achieves such capability at the level of the individual cell in part because the cells are not individually addressable by hardware, and partly because the interprocessor communication is one-dimensional. Extending, or attempting to extend, the dimensionality of the chaining channel would undoubtedly change the level at which fault tolerance could be implemented (unless, as implied, the entire ALAP concept were changed in the process). A 100-cell ALAP which experiences a failure in one cell becomes a 99-cell ALAP, regardless of which cell failed. For a square array, this cannot be the case. An individual cell is constrained in four directions. Attempting to route the communication channels around a defective cell will destroy the array structure. Attempting to substitute a single cell in place of a defective cell in general will require switching logic of impossible comlexity unless a spare cell is reserved for each cell in the array.[†] This is double-redundancy in the array logic, a level of redundancy not required in a one-dimensional array.

[†]The use of spares in modular systems is discussed in Chap. VI, Secs. 2.1.2. to 2.1.7.

The same problem exists for STARAN and for PEPE. (The PEPE was not designed with interprocessor communication in mind, and the question may be irrelevant for that machine.) A suggestion as to an approach to achieving better fault tolerance will be given later. It will require a complete departure in concept from existing multidimensional array designs. For present designs, the best that can be done in case of the failure of a single cell is to disable an entire row or column of cells. This solution may be acceptable in some instances because of necessity. It is not appealing in concept.

6.3.3 Limitations with Regard to VLSI Implementation

The dimensionality of the processor array is of crucial importance with regard to amenability to VLSI fabrication. The PEPE, STARAN, and ALAP are well-suited in this regard. The number of pins on a wafer is independent of the number of cells, or nearly so, for such zero-dimensional and one-dimensional arrays.[‡] For two-dimensional arrays in general, this is not the case. If many cells are fabricated on a wafer, the number of pins goes up as the square root of the number of cells if the array is two-dimensional. All of the cells on the edge of the wafer must have external connections. If they do

[‡]See Chap. III, Sec. 2.2. that discusses general questions of expanding the existing parallel system with new resource units and the effect of pin count restriction on this expansion (modular expansion).

not, array throughput will suffer because of bottlenecks in the interwafer communication. When the wafer contains, say, 300,000 gates, the problems with the higher number of pins can be imagined!

The only immediate solution to the problem for square arrays is to modify the algorithms to suit arrays of limited size. This is hardly an appealing answer to the problem. Another solution is to construct three-dimensional arrays by putting a set of two-dimensional wafers together in a stack like pancakes, with vertical communication channels fabricated directly on pads on the wafers. This may actually be done sometime in the future, when new technology can be devised, but does not seem like a solution which could be recommended for the near future.

6.4 Recommendations, Predictions, and Suggestions

For a solution to the problems just addressed, the author is tempted to look at the Holland machine concept. That machine has serious drawbacks as a practical system, as has been discussed. Nevertheless, it does not suffer from some of the limitations with regard to fault tolerance and pin count if it is used properly. Perhaps the user should not expect optimal efficiency in the utilization of the hardware. There can be cost-efficiency tradeoffs which can achieve equal results in throughput, while still retaining some of the advantages of the multi-dimensional array. It is these advantages which should be a prime objective of future efforts in parallel array design, in the author's estimation. Fault-tolerant capability is going to be of greater importance in these very complex systems, regardless of the environment in which they are used. Military applications do not represent the only such environment. Large-scale commercial applications need reliability as well, as any study of the effects of large system failures in such applications will reveal. The desirable effects of fault-repair capability on cost has already been noted.

Programming experience is important in assessing the value of a proposed system, and the virtues and limitations of present systems. This means detailed programming and checkout of application programs. It is often in the details of the program that the important virtues and limitations are discovered. Simulators and assemblers are invaluable as tools for this purpose, and for suggesting changes to system and component designs.

7. GENERAL REFERENCES

The books on parallel processor systems with which the author is familiar are collections of papers or chapters by various authors. Generally, these are related to a conference or symposium on the subject. These conference proceedings are as effective a way as any to keep up with the state of the art in parallel system design and applications. Reference [1] contains a number of papers on military applications for parallel processors in general. One of these (Knapp [29]) is a thorough exposition of the application of ILLIAC IV to radar data processing. There are also articles on algorithms; particularly of interest is the paper on parallel matrix operations (Katz [33]). All of the papers themselves contain short bibliographies on the subject of interest.

The author has relied on the proceedings of the International Conference(s) on Parallel Processing and its predecessors (references [2] and [3]) for up-to-date information on developments in the field. Also of great value are the proceedings of the annual (1976 and later) conferences on computer architecture held at Syracuse University. The author has not seen a more recent bibliography on associative processor publications than the one in reference [4], or one more thorough. The IEEE Transactions on Computers are hardware oriented in general, and contain papers of higher than average quality. There have been several issued devoted to parallel processors, one of which contains reference [9].

Reference [5] contains several excellent papers on ILLIAC IV, PEPE, and STARAN. One of them deals with ILLIAC IV's software development. For some reason, papers on applications software are hard to find. Most readers are evidently more interested in systems aspects of parallel processing, rather than software aspects. A better balance should be sought.

Most of the papers in references [6] thru [33] are discussed at the end of the topics which deal with them in this chapter. The reader is urged to contact the author or company if he or she desires particular detailed information on some aspect of an advanced computer organization or its application. It cannot be assured that such information is available in print, but the response is often very favorable.

7.1 Reference Books

1. HOBBS, L. C. et al, ed., *Parallel Processor Systems, Technologies and Applications.* New York, Spartan Books, 1970.
2. *Proceedings of the 1972 Sagamore Computer Conference on Parallel Processing.* Also 1973, 1974, 1975. Copies available from IEEE, New York.
3. *Proceedings of the 1976 International Conference of Parallel Processing.* Also 1977, 1978, 1979. Copies available from IEEE Computer Society, Long Beach, Cal.
4. CANNELL, M. H., et al, *Concepts and Applications of Computerized Associative Processing, Including an Associative Processing Bibliography.* U. S. Department of Defense Communications, Document No. AD879281, December 1970.
5. *Proceedings, 1972 Western Electronic Show and Convention.* Copies available from IEEE.

7.2 Reference Papers

6. HOLLAND, J. H., "A Universal Computer Capable of Executing an Arbitrary Number of Sub-Programs Simultaneously,"*1959 Proceedings of the Eastern Joint Computer Conference,* pp. 108–113.

7. SLOTNICK, DANIEL L. et al, "The Solomon Computer," *Proceedings, 1962 Fall Joint Computer Conference*, pp. 97–107.
8. BARNES, GEORGE H. et al, "The ILIAC IV Computer," *IEEE Trans. on Computers*, **C-17**, 8, pp 746–757, August 1968.
9. FINNILA, CHARLES A., and HUBERT H. LOVE, Jr., "The Associative Linear Array Processor, " *IEEE Trans. on Computers*, **C-26**, 2, pp. 112–125, February 1977.
10. BATCHER, K. E., "The Massively Parallel Processor (MPP) System," *Collection of Technical Papers*, AIAA Second Computers in Aerospace Conference pp 93–97, October 1979.
11. ———, "The Flip Network in STARAN, " *Proceedings of the 1976 International Conference on Parallel Processing*. IEEE Computer Society, Long Beach, Cal. 1976.
12. LOVE, HUBERT H., Jr., "A Modified ALAP Cell for Parallel Text Searching," *Proceedings of the 1977 International Conference on Parallel Processing*. IEEE Computer Society, Long Beach, Cal, 1976.
13. KUCK D., "Parallel Processor Architecture - A Survey," *Proceedings of the 1975 Sagamore Computer Conference on Parallel Processing*. IEEE Computer Society, Long Beach, Cal, 1975.
14. RAMAMOORTHY, C. V., and H. F. LI, "Pipeline Processor Architecture - A Survey," *Processings of the 1975 Sagamore Conference on Parallel Processing*. IEEE Computer Society, Long Beach, Cal, 1975.
15. ENSLOW, P. H., "Multiprocessor Architecture - A Survey," *Proceedings of the 1975 Sagamore Conference on Parallel Processing*. IEEE Computer Society, Long Beach, Cal, 1975.
16. BOULIS, ROGER L., and RUDOLF O. FAISS, "STARAN E Performance and LACIE Algorithms," *Proceedings of the 1977 International Conference on Parallel Processing*. IEEE Computer Society, Long Beach, Cal, 1977.
17. BERRA, P. B., and A. K. SINGHANIA, "Some Timing Figures for Inverting Large Matrices Using the STARAN Associative Processor, " *Proceedings of the 1975 Sagamore Computer Conference on Parallel Processing*. IEEE Computer Society, Long Beach, Cal, 1975.
18. LOVE, HUBERT H. Jr., "Radar Data Processing on the ALAP," *Proceedings of the 1976 International Conference on Parallel Processing*. IEEE Computer Society, Long Beach, Cal, 1976.
19. HINER, FRANK P. III, "Tracking Array Processor," *Proceedings of the 1978 International Conference on Parallel Processing*. IEEE Computer Society, Long Beach, Cal, 1978.
20. VOCAR, J. M., "Image Magnification," *Proceedings of the 1977 International Conference on Parallel Processing*. IEEE Computer Society, Long Beach, Cal, 1977.
21. HILBERT, EDWARD E., *Cluster Compression Algorithm: A Joint Clustering/Data Compression Concept, JPL Publication 77–43*. Jet Propulsion Laboratory, Pasadena, Cal, 1977.
22. EVENSEN, ALF J., and JAMES L. TROY, "Introduction to the Architecture of a 288-Element PEPE," *Proceedings of the 1973 Sagamore Computer Conference on Parallel Processing*. New York, IEEE, 1973.
23. BAER, J. L., "Multiprocessing Systems, " *IEEE Trans. on Computers*, **C-25**, pp. 1271–1277, December 1976.
24. REDDAWAY, S., "DAP-A Distributed Array Processor," *Proceedings of the first Symposium on Computer Architecture*, pp. 61–72, 1973.
25. QKAGA, Y., H. TAJIMA, and R. MORI, "A Novel Multiprocessor Array," *Proceedings of the second Euromicro Symposium on Microprocessing and Microprogramming*, pp. 83–90, Venice, Italy 1976.
26. KARTASHEV, S. I., and S. P. KARTASHEV, "Dynamic Architectures," Problems and solutions," *Computer*, **II**, pp. 26–40, July 1978.

27. ———. "Multicomputer System with Dynamic Architecture," *IEEE Trans. on Computers*, **C-28,** 10, pp. 704–721, October 1979.
28. KARTASHEV, S. I., S. P. KARTASHEV, and C. V. RAMAMOORTHY, "Adaptation Properties for Dynamic Architectures," 1979 National Computer Conference, *AFIPS Conference Proceedings*, AFIPS Press, **48,** pp. 543–556, 1979.
29. KNAPP, MORRIS A. et al, "Applications of ILLIAC IV to Urban Defense Radar Problem," *Parallel Processor Systems, Technologies and Applications*, L. C. HOBBS et al, Ed. New York, Spartan Books, 1970.
30. LOVE, HUBERT H., Jr., "An Efficient Associative Processor Using Bulk Storage," *Proceedings of the 1973 Sagamore Computer Conference on Parallel Processing*, New York, IEEE, 1973.
31. LOVE, HUBERT H., Jr., and D. A. SAVITT, "An Iterative-Cell Processor for the ASP Language," *Associative Information Techniques*, E. L. JACKS, Ed. New York, American Elsivier, 1971.
32. LIPOVSKI, G. JACK, and ANAND TRIPATHI, "A Reconfigurable Varistructure Array Processor," *Proceedings of the 1977 International Conference on Parallel Processing*. 1977, IEEE Computer Society, Long Beach, Cal, 1977.
33. KATZ, JESSE H., "Matrix Computations on an Associative Processor," *Parallel Processor Systems, Technologies and Applications*, L. C. HOBBS et al, Ed., New York, Spartan Books, 1970.

Chapter III

Designing and Programming Supersystems with Dynamic Architectures

Steven I. Kartashev

and

Svetlana P. Kartashev

PREVIEW

Today, there are many problems in science and technology whose solutions require enormous computational power. The systems that possess extreme computational power and are capable of solving such problems are now called *Supersystems*.

This chapter shows first that most of the traditional sources of increasing throughput rest on the horns of a dilemma—their use either limits the system's applicability or leads to excessive complexity of its hardware resources. The chapter analyzes the two new sources of augmenting throughput in a Supersystem that are a result advances of LSI technology:

1. Adaptation of hardware resources to instruction and data parallelism.
2. Reconfiguration of hardware resources into different types of architectures: array, pipeline, multicomputer, and multiprocessor.

Both these solutions may be accomplished by a Supersystem with dynamic architecture. The system will realize additional performance gains on the same resource by taking advantage of the following factors:

1. By reconfiguring the resource into minimal size computers, the system may maximize the number of programs computed by the same resource.
2. By switching the available resources into different types of architecture—array, pipeline, multicomputer, or multiprocessor—the system may speed up computations. This allows the available resources to be permanently involved in computations, even those that require dedicated subsystems such as array and pipelines.

The chapter next focuses on designing techniques for systems with dynamic architecture. To construct a system with dynamic architecture, one uses a new building block: the *dynamic computer group* (DC-group). Two types of design are featured:

1. DC-group with a minimal complexity memory–processor bus that interconnects processor and memory resources.
2. DC-group with a minimal delay memory–processor bus that introduces minimal delay in communication between any two modules of the resource: processor–memory, processor–processor, and memory–memory.

Each type of DC-group is studied using the following factors:

1. Basic architectural solutions.
2. Organization of internal and external communications between resource modules.
3. Organization of switches from one architectural state to another.

The third part of the chapter assesses the two dynamic architectures considered in the second part from the standpoint of major performance factors such as the time for architectural reconfiguration, delays introduced into signal propagation by reconfigurable busses, time of between computer communications, etc.

Finally, the fourth part of this chapter describes some programming techniques for systems with dynamic architecture. It introduces analysis techniques for a user program written in a high-level language that are aimed at finding the minimal sizes of computers which may execute this program. Next, the text considers techniques for assigning the DC-group hardware resource among several concurrent programs, thus allowing one to increase the executional parallelism obtained on the same hardware equipment. The presented methodology is shown to be simple and straightforward, requiring no complex computations. As a result, the resource assignment among user programs can be be performed quickly and efficiently.

The conclusions to this chapter summarize the usefulness of dynamic architectures for Supersystems and complex parallel systems since they lengthen the lifecycles of systems via increasing throughputs of the available resources.

ANALYSIS OF TRADITIONAL AND NEW TECHNIQUES FOR INCREASING THROUGHPUT IN SUPERSYSTEMS

1. INTRODUCTION

There are many problems in science and technology whose solutions require enormous computational power. For instance, partial differential equations in computational aerodynamics, real-time radar signal processing,[†] control of fast and complex nuclear or chemical reactions, accurate weather prediction, etc., necessitate throughputs several orders of magnitude higher than those of existing complex parallel systems.

The systems that possess extreme computational power and are capable of solving such problems are now called *Supersystems*. The problems for Supersystems may be nonreal-time or real-time.

A typical nonreal-time problem is exemplified by the simulation of fluid flow in the testing and validation of an aerodynamic design. Realistic aerodynamic flow requires an extremely high resolution of grid points appropriately spaced throughout the flow field. Further, it is necessary to simulate not only two- but a three-dimensional flow in order to have complete and accurate results. Although a powerful system such as ILLIAC-IV[‡] proved to be effective in simulating two-dimensional flows, a complex three-dimensional simulation requires a system with throughput nearly two orders of magnitude greater than that of ILLIAC-IV [1].

For complex real-time problems one can look at *ballistic missile defense algorithms*, which are characterized by thousands of information streams that have to be processed in real-time within very small time intervals [2–4]. This means that these algorithms require tht the Supersystem compute its responses during limited time intervals. Thus, there is a time restriction between the moment a particular data stream enters the Supersystem and the moment the Supersystem produces a response specified by this set. For instance, for a ballistic missile defense algorithm computing a response for a sensor (radar),

†Readers interested in real-time radar signal processing are invited to read the very informative Chap. II, Sec. 4, that describes a typical radar signal processing application and how a complex computer system for this application should be designed. (Ed.)

‡Chap. II, Sec. 3.1.2. gives a comprehensive description of the ILLIAC IV system. (Ed.)

there is a time constraint for computing a sequence of commands for the sensor that permits an improvement in the observation of the object entering its field of view [3].

Therefore, complex real-time problems impose more severe requirements on Supersystems. Indeed, in addition to the demands for handling an enormous amount of information, they require that this information be handled during specified and, as a rule, very small time intervals. Thus real-time problems like these require Supersystems with the highest throughputs attainable by current technology.

However, obtaining maximal throughputs is not the only goal pursued in designing such systems. Another objective of equal importance is to have highly reliable computations. This acquires special significance for many complex real-time problems. Indeed, ballistic missile defense algorithms necessitate a very high degree of computational reliability due to such factors as mission criticality, remote operation, and economics [3–5]. Similarly, an error in a computation of an algorithm that controls nuclear reactors, may lead to catastrophical consequences.

Designers of Supersystems are therefore confronted with two major requirements. On the one hand they are required to design systems of enormous complexity, while on the other the systems designed must provide a specified level of reliability.† These two requirements are contradictory because a system's reliability goes down as its complexity goes up [6].

†See Chap. V for a discussion of other requirements to complex parallel systems. (Ed.)

However, this is not the only contradiction encountered in the design of a Supersystem. Another contradiction, equally important, is associated with the inconsistency between long time periods for design and the requirement that the system perform the fastest computations that current computer technology can achieve. Since design periods usually span a 5 to 6 year interval, this amount of time may lead to the obsolescence of some of the basic architectural concepts as well as cause the inability of the system to take advantage of all the speed benefits that end-date state-of-the-art would allow.

Another well-known design problem is that at the beginning of the design stage the user cannot envisage all the computational subtleties of the problems to be solved by the future system, whereas near the end of this stage the computer designers are incapable of implementing into the system all the new requirements specified by the user that originate from a better acquaintance with the algorithms. Accordingly, failure to implement some crucial dedicated architectural solutions required by the algorithms may result in additional expensive software that increases the overall cost of system development, and also slows down computation. As a result, some complex systems may become obsolete even during their development stage.

The problems mentioned above are considerably intensified when designing a Supersystem for some complex real-time algorithms [2–4].

First, the complexity of a Supersystem will be several times greater than that of any existing complex system. This will lead to an increase in design time. As a result, the traditional gap between the computer technology used for the system implementation and technological opportunities available at the moment it is completed may become even wider.

Second, since some of the real-time problems are very complex, a computational algorithm that describes such a problem may be developed by means of a heuristic search which at each step gives a better approximation between algorithm and problem [3–5]. Therefore, it is difficult to grasp all the computational subtleties of such problems at once and implement them on the Supersystem's design. In addition, during system development there may be significant changes in the algorithms or even the appearance of entirely new algorithms which had not been considered by the system designers.

Thus the task of sustaining a continuous match between the architecture of the Supersystem specified in the beginning of the design period and the computational specificities of the algorithms in use at the time the machine is delivered may be far more difficult than was true for existing parallel systems designed for other applications [3].

The designers of Supersystems are then faced with the following major problems:

1. Obtaining maximal throughputs while the system maintains a specified level of reliability.
2. Maintaining the adequacy of the Supersystem for solving a class of complex algorithms during specified time intervals, while these algorithms change or even new algorithms are added to the class.

Since the Supersystem must possess an extremely high throughput, let us consider the sources of increasing throughput that it may rely upon. These sources may be divided into two categories:

1. Traditional sources used in existing complex parallel systems.
2. New sources provided by current technological advances.

The following two sections are dedicated to a comprehensive study of these sources and an assessment of their effectiveness in the Supersystem.

2. TRADITIONAL TECHNIQUES FOR AUGMENTING SYSTEM THROUGHPUT

One may increase the throughput of a complex system by drawing on the following traditional sources:

1. The use of high-speed components.

2. Modular expansion of the system with new equipment (adding new computers, processor, I/O units, etc.).
3. Equipping the system with a dedicated architecture that is very effective for a given class of applications.
4. Utilization of the maximal concurrency present in programs.
5. Application of special types of architectures: pipeline, array, and associative.
6. Optimization in data exchanges.

Consider now the effectiveness of these sources in increasing the throughput of a Supersystem.

2.1 The Use of High-Speed Components

The use of high-speed components means implementation of a system from the fastest components that are available at the moment the system construction begins.

This method for augmenting a system's throughput ceases to help the moment that system development is finished. Certain system units (for instance, I/O terminals) may be replaced with faster ones during the system's lifetime; however, massive on-line replacement of all system units with newer and faster counterparts would be highly unlikely.

2.2 Modular Expansion

Modular expansion means the integration of new resource units into an existing system. This source of throughput increase is presently acquiring an even greater importance as hardware resources become cheaper.

However, since each attachment of an additional hardware resource is accompanied by adding both new interconnections and new complexities to the system, the modular expansion of any existing system is limited by a combination of such factors as current technological restrictions, tradeoffs between delay and complexity of the interconnection logic, and tradeoffs between system reliability and complexity. Let us look at the impact of each of these factors on modular expansion.

†Problems associated with LSI implementation of complex systems are discussed in the last part of Chap. I, Part IV, and Chap. II, Secs. 3 and 6. (Ed.)

2.2.1 Technological Restrictions

Current technological constraints limit the number of new hardware components that can be attached to an existing system.

An LSI modular implementation of such units as powerful processors or control units is presently confronted with problems due to the limitation on the pin count in an LSI module.†

The following options appear in the processor module as a result of pin count limitations:

1. The design must be functionally simplified.
2. Instead of direct transfer of microcommands, they must be sent in coded form, which leads to additional delays for encoding and decoding.
3. The processor must be assembled from nonidentical LSI modules each of which implements a portion but not all of the totality of microoperations. This will lead to fabrication of new module types and a consequential increase in overall cost.

While alternatives (1) and (2) are unacceptable because they lower the system throughput, alternative (3) leads to a significant increase in the cost of a system.

A control unit, since it has to generate all microcommands for all the other devices it controls, requires a large number of intermodule connections. A large number of pins in each LSI module of the control unit must consequently be assigned only to generate microcommands. On the other hand, transfer of microcommands in an encoded form will produce significant delays that will slow processing down considerably.

LSI implementation of individual functional units therefore confronts computer designers with some difficulties associated with current technological restrictions. During modular expansion each LSI module has to be assigned additional pins to integrate all the resource units (new and old) into an existing system, and present LSI technological restrictions become considerably aggravated.

2.2.2 Increase in Delays or Complexities Introduced by Interconnection Logic

Expanding a system by adding resources may lead either to a significant increase in delays or to complexities introduced by the interconnection logic.

A Supersystem having any pair of its resource modules (processor element (PE), memory element (ME), I/O element (GE), etc.) communicating with each other through dedicated connections will require an enormous amount of interconnection.†

To reduce the number of interconnections required for each resource unit dedicated connections between every unit integrated into the system must be replaced by shared ones in which resource units share the same h-bit data bus, where h is the width of a transmitted word. But this requires that this data bus include connecting elements which selectively activate an information path between a pair of communicating resource units. Accordingly, to allow communications between any pair of resource units integrated into a Supersystem,

†Problems of interconnections in complex systems are considered in Chap. I, Sec. 24.2, p. 85. (Ed.)

an interconnection network including connecting elements must be provided. Two types of interconnection networks are possible:

1. One-stage networks in which any pair of resource units is connected via one dedicated connecting element. If a Supersystem has n resource units it will require about n^2 connecting elements to organize full one-stage connectedness among them. Each communication path will introduce the minimal communication delay, however, equal to that of one connecting element.

2. Multistage networks in which any pair of resource units is connected via a sequence of $\log_2 n$ connecting elements. The overall number of connecting elements and the complexity of the interconnection network will be reduced. However, each communication path between a pair of resource elements (PE–ME, PE–GE, ME–GE, etc.) will have a significant communication delay equal to that of the $\log_2 n$ consecutive connecting elements between them. For a Supersystem, n may be a very large number. For instance, for a Supersystem having $n = 1000$ resource units, a communication path between any pair of resource units will go through 10 connecting elements ($10 = \log_2 1000$) each of which will introduce at least one level of delay, t_d, and so each communication will be delayed at least by $10t_d$ which will lead to a significant reduction in the system's throughput.

Therefore, an organization with one-stage connections between different functional units of a Supersystem leads to enormous complexity of the interconnection busses, and on the other hand, the use of multistage connections leads to significant communication delays which reduce throughput of the Supersystem.

As a compromise, a *distributed* Supersystem appears to be the most plausible solution for many complex real-time algorithms, where speed of computation is a major design concern. In this case a distributed Supersystem is understood as a collection of several subsystems each of which uses one-stage interconnections between its resource units. Different subsystems may be integrated with each other via busses described in the literature: time-shared, cross-bar switch, multiport-memory [7].

Each subsystem maintains minimal communication times between any pair of resource units such as processor elements (PE), memory elements (ME), etc., inside it. This means that the following communications within each subsystem will be performed in minimal time: PE–PE, PE–ME, ME–ME, etc. Each subsystem must accordingly support maximal flexibility of interconnections between every pair of its resource modules.†

The modular expansion of each subsystem is limited by the exponential growth of the number of connecting elements included in it. If a subsystem contains n processor modules, PE, and n memory modules, ME, it will require n^2 connecting elements to achieve a one-stage connectedness between each pair of PE's and ME's.

Thus a distributed Supersystem will accomplish a reasonable tradeoff between the complexity (flexibility) of its resources and communication times between pairs of resource units.

†See Chap. V, Sec. 3, which contains a comprehensive treatment of partitioning a Supersystem into different subsystems. Such a criterion of partitioning is used as to minimize the communication overhead between different subsystems. (Ed.)

2.3 The Use of Dedicated Architectures

A *dedicated* architecture is an architecture that takes into account the computational specifities of some algorithm or class of algorithms. Such architectures are provided with special-purpose execution and control circuits. Special-purpose execution circuits are a set of dedicated microoperations† each of which speeds up execution of the algorithm. Special-purpose control circuits are hardware realizations of a collection of dedicated instructions, each of which implements complex sequences of operations encountered in the algorithm. Such instructions may execute complex expressions such as $(a + b)^2 - (a - b)^2 - k \div a$, or $(a + b - c) \times (a - b - c)$, etc. Each of these instructions is equivalent to several conventional instructions, so their use reduces the number of memory access operations required for fetching instructions and data.

†A comprehensive treatment of the microoperation concept is presented in Chap. I, Secs. 2–5. (Ed.)

Let us find the execution speed-up prompted by the use of dedicated instructions. An operation sequence composed of d consecutive operations is executed by a conventional microprogrammable computer as a subroutine composed of d instructions where each instruction executes one operation. (We exclude from consideration all instructions fetching data for the d operations because it is assumed that the number of such instructions is the same for a complex dedicated instruction as for a subroutine and thus depends solely on the way the storage of the data words required for the operation is organized.) Each instruction in such a subroutine takes a time t_M for instruction fetch and a time t_{op} for operation execution. It follows that if an operation sequence containing d operations is organized as one dedicated instruction, the speed-up, SIA_j, resulting from the use of this instruction is: $SIA_j = t_M (d - 1)$ [8].

To find the overall speed-up of execution caused by the use of one dedicated instruction one has to find the frequency F_j of its appearance in the program [12]. Then the speed-up is $SIA_j F_j$. If g dedicated instructions are used in the execution of an algorithm, each introducing speed-up SIA_j, then the overall speed-up resulting from their use, SPA, is $SIA_1 F_1 + SIA_2 F_2 + \cdots + SIA_g F_g$.

Example 1. Suppose that the time t_M of instruction fetch is 350 nsec and that three dedicated instructions are used in the program. The first instruction computes the expression $(k - b) \div (b + c)$ and is repeated 128 times in the program ($F_1 = 128$). The second instruction computes the expression $(a^2 - c^2) \times (a + b - c)$ and is repeated $F_2 = 175$ times. The third instruction executes $(a^2 + b^2) \times (c^2 - d^2) \div a + b$ and is repeated $F_3 = 44$ times. The speed-up caused by each instruction is:

$$SIA_1 F_1 = t_M(d_1 - 1)F_1 = 350 \cdot (3 - 1) \cdot 128 = 89.6 \ \mu\text{sec}$$
$$SIA_2 F_2 = t_M(d_2 - 1)F_2 = 350 \cdot (6 - 1) \cdot 175 = 306.25 \ \mu\text{sec}$$
$$SIA_3 F_3 = t_M(d_3 - 1)F_3 = 350 \cdot (9 - 1) \cdot 44 = 123.2 \ \mu\text{sec}$$

The overall speed-up for this program is:

$$SPA = SIA_1 F_1 + SIA_2 F_2 + SIA_3 F_3 = 89.6 + 306.25 + 123.2$$
$$= 519.05 \ \mu\text{sec} \quad \blacksquare$$

The major drawback of architectural dedication is that it lowers the system's applicability. If a new algorithm appears or the old algorithm is modified, the Supersystem may not be capable of completing it in the necessary time interval since all its dedicated instructions take into account the computational specifics of the original algorithms.

2.4 Utilization of the Concurrency Present in Algorithms

This is the technique of finding the maximal number of instruction and data streams present in a complex algorithm and assigning a separate computer (processor) to each such stream. Interest in this technique is caused by the advances in LSI technology that significantly reduce the cost and enhance the reliability of computer components. That makes feasible the design of reliable complex computer systems containing many more hardware resources than ever before. This leads to a sharp increase in the number of instruction and data streams that may be computed in parallel in a single system.

The major drawback of this technique is that it may lead to excessive complexity in the Supersystem and low utilization of the equipment in it. This is because the number of independent computers in a system is determined by the algorithm having the largest number of parallel information streams, while on the other hand, the bit size of each computer is determined by the maximal bit size of a computed result that can be encountered in all the algorithms processed by this computer. It follows that in computing other less complex algorithms requiring fewer instruction and data streams or smaller word sizes, a portion of the system resource becomes redundant.

If this excessive resource is part of a single computer that is processing a program requiring word sizes smaller than the computer is capable of, or is part of an array unused because the dimension of a data vector is smaller than the number of processors in the array, this resource cannot be freed for other computations and becomes idle.†

†Chapter II, Secs. 3 and 4, describe various organizations used in existing array systems to turn the unneeded processor elements into an idle state. (Ed.)

Example 2. Suppose a Supersystem contains one hundred 64-bit computers, eighty 48-bit computers, and fifty 32-bit computers, i.e., the system resource, $SR = [100 \times 64 \text{ bits}, 80 \times 48 \text{ bits}, 50 \times 32 \text{ bits}]$. Suppose also that this Supersystem has to compute three complex algorithms with the following resource requirements: The resource needs for Algorithm I, $RA_\text{I} = [150 \times 48 \text{ bits}, 25 \times 32 \text{ bits}]$, i.e., it needs one hundred fifty 48-bit computers and twenty-five 32-bit computers; the resource needs for Algorithm II, $RA_\text{II} = [200 \times 32 \text{ bits}]$; and for Algorithm III, $RA_\text{III} = [40 \times 64 \text{ bits}, 25 \times 32 \text{ bits}]$. The redundant resources, RR, of the Supersystem while executing Algorithms I, II, and III are shown in Table III.1.

TABLE III.1

SUPERSYSTEM WITH CONVENTIONAL ARCHITECTURE
SYSTEM RESOURCE = [100 × 64 bits, 80 × 48 bits, 50 × 32 bits]

Computational Modes	Resource Requirements of Algorithms			Redundant Resource (RR)	
	Algorithm I (RA_I)	Algorithm II (RA_{II})	Algorithm III (RA_{III})	Free Resources	Nonfree Resource
Computation of Algorithm I	150 × 48 bits 25 × 32 bits	——	——	30 × 64 bits 25 × 32 bits	70 × 16 bits
Computation of Algorithm II	——	200 × 32 bits	——	30 × 64 bits	70 × 32 bits 80 × 16 bits
Computation of Algorithm III	——	——	40 × 64 bits 25 × 32 bits	60 × 64 bits 80 × 48 bits 25 × 32 bits	——

Let us show how we can find RR. The Supersystem may execute Algorithm I in eighty 48-bit computers, seventy 64-bit computers and twenty-five 32-bit computers. That is, the system resource assigned to execute Algorithm I, $SRA_I = [80 \times 48 \text{ bits}, 70 \times 64 \text{ bits}, 25 \times 32 \text{ bits}]$. During execution of Algorithm I, the Supersystem has a free redundant resource, FRR_I, that can be used for other computation. $FRR_I = SR - SRA_I = [(80 - 80) \times 48 \text{ bits}, (100 - 70) \times 64 \text{ bits}, (50 - 25) \times 32 \text{ bits}] = [30 \times 64 \text{ bits}, 25 \times 32 \text{ bits}]$.

The Supersystem also has a nonfree redundant resource NRR_I that can not be used for other computations because some of the information streams of Algorithm I are processed in computers of nonminimal size. Since each 64-bit computer executes an information stream of 48-bit data words a $64 - 48 = 16$-bit portion is redundant. That is, for Algorithm I, NRR_I is found by taking the difference between the system resource assigned to Algorithm I, SRA_I, and the resource needs of Algorithm I, RA_I. Specifically $NRR_I = SRA_I - RA_I = [80 \times (48 - 48) \text{ bits}, 70 \times (64 - 48) \text{ bits}, 25 \times (32 - 32) \text{ bits}] = [70 \times 16 \text{ bits}]$.

For Algorithm II the following system resources are required: $SRA_{II} = [50 \times 32 \text{ bits}, 80 \times 48 \text{ bits}, 70 \times 64 \text{ bits}]$. Thus the redundant resources that can be freed for other uses are $FRR_{II} = [30 \times 64 \text{ bits}]$. The redundant resources that cannot be freed are $NRR_{II} = [80 \times 16 \text{ bits}, 70 \times 32 \text{ bits}]$. Similarly, for Algorithm III the system is found to have free redundant resources $FRR_{III} = [60 \times 64 \text{ bits}, 80 \times 48 \text{ bits}, 25 \times 32 \text{ bits}]$ and no nonfree redundant resources ($NRR_{III} = 0$). It follows that if the Supersystem employs a conventional architecture, even though it has redundant resources available, it can not execute any pair of these algorithms concurrently and is thus forced to execute them sequentially. ∎

2.5 Computation of an Algorithm in Subsystems with Different Types of Architectures

Complex algorithms may have portions (tasks) requiring different types of computation—multicomputer, multiprocessor, array and pipeline.†

To process such a complex algorithm in minimal time, requires a system with access to four separate subsystems employing the different types of architecture.

The major drawback of computing an algorithm in different types of subsystems is that it leads to excessive complexity of the resources. Since each of the subsystems is engaged in the computation of one task only, it may be idle during the execution of the remaining tasks.

Suppose a complex algorithm has 7 tasks. Tasks 1, 4, and 5 require array computations, tasks 2 and 6 may be pipelined, and tasks 3 and 7 require a multiprocessor. Resource requirements for these tasks as well as the sequencing of their execution are shown in Fig. III.1. Task 1 requires an array system with an overall complexity of 70,000 bits. Task 2 requires a pipeline

†Computation of one algorithm in systems with different architectures is also discussed in Chap. I, Sec. 26. (Ed.)

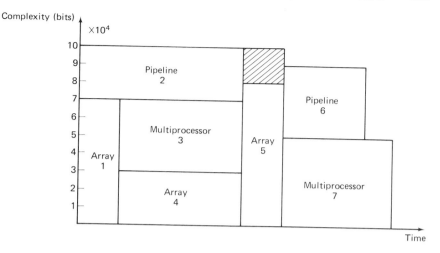

FIGURE III.1 Program computation in dedicated subsystems

system with a complexity of 30,000 bits. Task 3 requires a multiprocessor with a complexity of 40,000 bits, etc.

To minimize the total execution time, this algorithm must be executed in three dedicated subsystems—an array, a pipeline, and a multiprocessor. Figure III.2(a) shows the complexity and utilization of resources for the array subsystem. The overall complexity of 80,000 bits is determined by task 5 since its resource requirements are maximal. During the execution of task 1, requiring 70,000 bits of the array, 10,000 bits of the subsystem are idle. Similarly, 50,000 bits of the array subsystem are idle during execution of task 4. Figure III.2(b) shows the complexity and resource utilization for the pipeline subsystem, which must have 40,000 bits since that is the resource required in task 6. Figure III.2(c) shows that the complexity of the dedicated multiprocessor must be equal to 50,000 bits. The Supersystem incorporating these three dedicated subsystems will have an overall complexity of 80,000 + 40,000 + 50,000 = 170,000 bits. We can see from Fig. III.2 that a significant portion of the resource will be idle at times adding an excessive complexity to the overall system hardware.

2.6 Optimization in Data Exchanges

One of the ways to improve performance is to reduce the time of information exchanges on external I/O devices. This may be achieved if data transfers through I/O devices are replaced by direct communications between different functional units in a system.† In order to minimize the number of I/O exchanges needed, a system must thus be provided with a flexible processor–memory bus that supports direct processor–memory, processor–processor,

†Chapter I, Part III, Sec. 2, gives more discussion on the benefits of such exchanges. (Ed.)

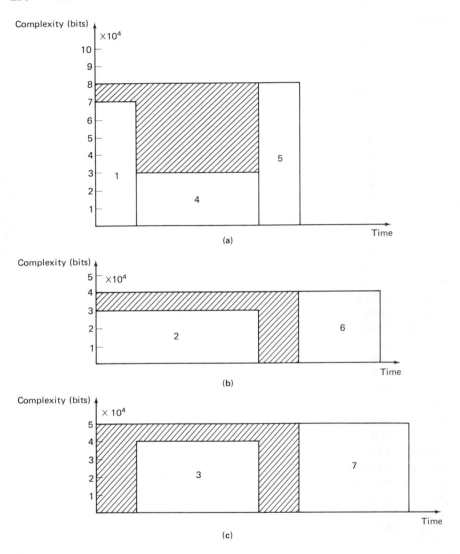

FIGURE III.2 (a) Engagement of the array subsystem; (b) engagement of the pipeline subsystem; and (c) engagement of the multiprocessor subsystem

and memory–memory exchanges in the system. To provide direct connections between all the processors, primary memories and I/O devices in a Supersystem would require a bus of enormous complexity. It is hardly conceivable that all the functional units of a Supersystem could be connected by such a bus. The complexity of the bus can be reduced if a hierarchy of interconnections is introduced: strong one-level connections between all functional units of each subsystem, and a strong connection between different subsystems using one of the traditional communication schemes: time-shared bus, cross-bar switch, or multiport-memory.

2.7 Section Summary

Most of the traditional sources of increasing system throughput considered above have distinct drawbacks; their use either limits the system's applicability or leads to excessive complexity of its hardware resources. One must therefore consider new ways to increase throughput that would be capable of augmenting a system's throughput without significantly increasing its complexity or restricting its applicability.

3. NEW SOURCES OF AUGMENTING SYSTEM THROUGHPUT

Advances of LSI technology in the 1970s made it possible to obtain LSI modules with high computational throughput. The use of such modules as the building blocks of a Supersystem provides new options for increasing its power. LSI technology has introduced modularity as basic in the organization of a computer's architecture. As was shown in Chap. I, each LSI module used for assembling the architecture may be equipped with simple circuits for the software-controlled activation and deactivation of interconnections with other modules. For instance, processor modules may be switched among several primary memory modules and in this way may reduce the time required for processor-memory communication. Or one processor may reconfigure into several smaller size processors in a parallel array system, augmenting the data vector size processed by a single instruction. Finally, a modular architecture can support a complete dynamic redistribution of available hardware by reconfigurng hardware resources into different numbers of variable size computers. This allows a computer system to adjust to the changeable number of information streams encountered in complex algorithms.

It follows that a computer architecture assembled from LSI modules may exhibit a much higher adaptation to the algorithm being executed than was ever achieved in traditional systems, and one may accordingly obtain additional throughput increases as a result of such adaptation.† The attractiveness of these increases is that they are not accompanied by a significant increase in the system's complexity or limits on the system's applicability.

These new sources of augmenting throughput in a Supersystem are:

1. Adaptation of hardware resources to instruction and data parallelism.
2. Reconfiguration of hardware resources into different types of architectures: array, pipeline, multicomputer, and multiprocessor.

†More on architectrural adaptation is in Chap. I, Sec. 23–26. (Ed.)

3.1 Adaptation of Hardware Resources to Instruction and Data Parallelism

During the past few years a number of publications [9-12] have described dynamic architectures for parallel computer systems. These architectures are capable of dynamically partitioning their hardware resources into a variable number of computers (processors). This allows a system accommodation to a current number of instruction and data streams it needs to handle. Thus, if the system has to process an increased number of information streams, it may partition its resources into a larger number of computers with smaller word sizes. For complex real-time algorithms a maximal number of information streams can be handled in real-time, although the precision of the computations may be reduced. In case a particular information stream with high priority requires a more precise computation, a larger size computer (processor) may be formed for handling it so the accuracy of computations will be enhanced.

The capability of dynamic architectures to perform dynamic partitioning of the resource allows the obtaining of a minimal discrepancy between the word size required by the program and that of the computer. This achieves a minimization of idle resources and allows formation of additional computers from the resources that are freed. As a result, a maximal number of programs may be computed concurrently by the available resources. It then follows that a dynamic architecture realizes a new source of throughput increase through dynamic adaptation to the current instruction and data parallelism present in a complex algorithm. This source was absent in traditional systems. Its advantage is that it is not accompanied by significant increases in system complexity, since it is accomplished through the redistribution of the available resources among programs.

Let us show that a Supersystem with dynamic architecture may handle a larger number of information streams in real-time than the Supersystem with conventional architecture having the same complexity of the resources.

Example 3. Let a Supersystem with dynamic architecture be provided with the same system resource, SR, as that of Supersystem from Example 2; that is, $SR = [100 \times 64 \text{ bits}, 80 \times 48 \text{ bits}, 50 \times 32 \text{ bits}]$. Suppose that this Supersystem has to compute the same Algorithms I, II, III as given in Example 2.

Let us show that if the Supersystem is equipped with dynamic architecture then it has enough resources to concurrently compute two algorithms, whereas the Supersystem with conventional architecture and the same resources could be used only for computing one algorithm at a time (Table III.1). A dynamic architecture may switch its resources into a set of architectural states distinguished from each other by the numbers and sizes of concurrently operating computers. Any computer is assembled from k computer elements, CE, each having word size h. Thus it handles $k \cdot h$-bit words where $k = 1, 2, \ldots, n$ [10–11]. If one assumes that $h = 16$ bits, then a 32-bit computer is assembled from two CEs, a 48-bit computer is assembled from three CEs and a 64-bit computer is assembed from four CEs. Since 100×64 bits = 100×4 CE = 400 CE; 80×48 bits = 80×3 CE = 240 CE; and 50×32 bits = 50×2 CE = 100 CE; the system resource, $SR = [100 \times 64$ bits, 80×48 bits, 50×32 bits] equals $SR = 400$ CE + 240 CE + 100 CE = 740 CE. Thus, the entire Supersystem resource is equivalent to 740 CEs.

Now we must find the number of CE's required for each algorithm. The resource needs for Algorithm I, $RA_I = [150 \times 48$ bits, 25×32 bits], is equivalent to 500 CE, because $RA_I = 150 \times 3$ CE + 25×2 CE = 450 CE + 50 CE = 500 CE. Algorithm II requires $RA_{II} = 400$ CE, because $RA_{II} = 200 \times 32$ bits = 200×2 CE = 400 CE. Algorithm III requires $RA_{III} = 210$ CE, since as $RA_{III} = [40 \times 64$ bits, 25×32 bits] = 40×4 CE + 25×2 CE = $(160 + 50)$CE = 210 CE.

The Supersystem with dynamic architecture has enough resources to compute concurrently the following combinations of algorithms (Table III.2):

1. Algorithms I and III, because $RA_I + RA_{III} < SR$ ($RA_I + RA_{III}$ = 500 CE + 210 CE = 710 CE < 740 CE) and

2. Algorithms II and III because $RA_{II} + RA_{III}$ = 400 CE + 210 CE = 610 CE < 740 CE.

Concurrent execution of Algorithms I and II is still impossible, since their resource needs are $RA_I + RA_{II}$ = 500 CE + 400 CE = 900 CE and the Supersystem has only 740 Ce, i.e., $RA_I + RA_{II} > SR$.

In executing Algorithms I and III, the Supersystem assumes the architectural state [150×48 bits, 50×32 bits, 40×64 bits]. The redundant resource equivalent $FRR = [30 \times 16$ bits] can be freed to other com-

TABLE III.2

SUPERSYSTEM WITH DYNAMIC ARCHITECTURE
SYSTEM RESOURCE = [100 × 64 bits, 80 × 48 bits, 50 × 32 bits]

Computational Mode	Resource Requirements of Algorithms			Redundant Resource (RR)
	Algorithm I (RA_I)	Algorithm II (RA_{II})	Algorithm III (RA_{III})	
Concurrent computation of Algorithms I and III	150 × 48 bits 25 × 32 bits	——	40 × 64 bits 25 × 32 bits	30 × 16 bits
Concurrent computation of Algorithms II and III	——	200 × 32 bits	40 × 64 bits 25 × 32 bits	130 × 16 bits

putations. In executing Algorithms II and III, the system assumes the architectural state [225 × 32 bits, 40 × 64 bits]. Its redundant resource equivalent to $FRR = [130 \times 16 \text{ bits}]$ can also be used for other computations. Therefore, the Supersystem with dynamic architecture allows an additional throughput increase using the same complexity of the resources. ∎

3.2 Reconfiguration of Hardware Resources into Different Types of Architecture

Let us consider one more source of increasing throughput in a Supersystem via reconfiguring available hardware resources into different types of architecture: multicomputer, multiprocessor, array, and pipeline, or *mixed* architecture which assumes coresidence of any combination of above architectures. As was shown earlier, many parallel real-time algorithms may be partitioned into portions (tasks) requiring different types of computations. To speed up execution, they are now executed by dedicated subsystems. This adds excessive complexity to the system and lowers its throughput, since each of the dedicated subsystems is engaged in execution of only one task with matching dedication.

To increase throughput of the Supersystem, it is necessary that all its resources be continuously involved in computations. This implies that the system must be capable of switching its resources into the different types of architecture. Or the system must allow coresidence of any combination of these architectures; that is, a portion of the resource functions as a multiprocessor, another portion behaves as an array or a pipeline, etc.

Therefore, to realize both sources of throughput increase, a Supersystem must be equipped with dynamic architecture and be able to switch its resources into different types of architecture.

Let us consider how such a reconfiguration of the architecture may speed up computations in Supersystem.

3.2.1 Multicomputer Computations

A Supersystem should assume a multicomputer mode of operation in executing those complex real-time algorithms that are characterized by a large number of independent instruction streams with little interaction so that each stream requires no data words computed by another. To perform this adaptation, the system should reconfigure into a state of the multicomputer architecture specified by a required number of concurrent computers. This will allow it to implement all the instruction parallelism present in algorithms because the system will follow all changes in instruction streams by switching into states characterized by matching number of computers [11].

As for the precision of computations, this will depend on the priority of a program requesting a high precision computation. If the priority of this

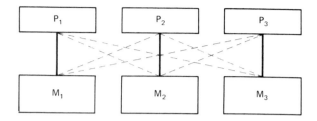

FIGURE III.3 Processor–memory exchanges

program is high, then the Supersystem should redistribute the resource assigned for othe programs computed by multicomputer architecture and form a computer of the size needed by the high priority program [12].

3.2.2 Multiprocessor Computations

The Supersystem should assume a multiprocessor architecture while executing complex algorithms with high interaction among instruction streams; that is, where each program may require data words computed by other programs. The use of a multiprocessor mode of operation allows a Supersystem to establish direct connections between different functional modules it incorporates: processor–memory, processor–processor, memory–memory, etc. This will reduce the time of communication among these modules.

Figure III.3 shows what processor–memory interconnections may be established in a system containing three processors, P_1, P_2, and P_3, and three memories, M_1, M_2, and M_3. In addition to local exchanges $P_i \leftrightarrow M_i$ ($i = 1, 2, 3$), the system must maintain cross exchanges $P_i \leftrightarrow M_j$, $i, j = 1, 2, 3$, $i \neq j$. The next figure (Fig. III.4) shows the type of processor–processor exchanges that need be organized in the same system: $P_i \leftrightarrow P_j$. Figure III.5 shows all possible memory–memory exchanges that the same system must support: $M_1 \leftrightarrow M_2$, $M_1 \leftrightarrow M_3$, and $M_2 \leftrightarrow M_3$.

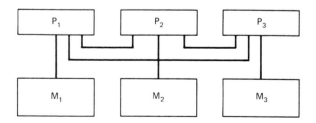

FIGURE III.4 Processor–processor exchanges

264 CHAPTER III

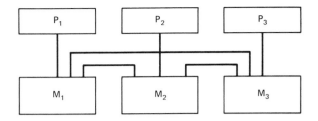

FIGURE III.5 Memory–memory exchanges

Example 4. Let program A computed in the P_1 processor need a data array obtained by program B which was computed before by the P_3 processor and stored in its local memory M_3 (Fig. III.6). To perform this interaction the P_1 processor should establish direct communication with the M_3 memory. This will allow the P_1 processor to compute in the mode when it fetches the first operand from its local M_1 memory and the second operand from the M_3 memory. Thus, this mode will eliminate time overheads associated with transfers of blocks of data words from the memory of one processor to that of another, which is extremely important for complex real-time algorithms requiring very fast computations. ∎

3.2.3 Array Computations

†The reader interested in the in-depth study of the array systems is referred to Chap. II which discusses the designing and programming of reconfigurable array systems. (Ed.)

In executing complex real-time algorithms requiring array computations,† a portion or the totality of the Supersystem should assume array architecture. In this case the system must be able to change the number of concurrent arrays, the dimension of a data vector computed in each array, and the word sizes of processors working in each array. For each array, its resource should be redistributed among its processors, so that a data vector computed in each array may contain different size data items. Therefore, the architecture of the Supersystem should be able to perform array computations with variable precision and variable dimensions of data vectors.

Let us consider how array computations might be organized in a Supersystem with dynamic architecture.

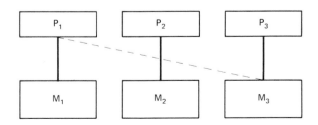

FIGURE III.6 P_1–M_3 exchange

First, a portion or the totality of the resource must be partitioned into several concurrent arrays. For each array one $16 \cdot k$-bit processor P must assume the function of processor-supervisor and broadcast instructions fetched from local memory to all other processors working in array. Since processor-supervisor P contains k processor elements, PE, its primary memory contains k memory elements ME's. For dynamic architectures forming different processor sizes in 16-bit increments, to maintain the universality of programs, a 16-bit instruction format must be maintained [10,11]. Since the word size of one memory element ME is 16 bits, one instruction is stored in one cell of an ME. Thus, the processor-supervisor must be capable of storing instructions in any of its k memory elements and broadcast them only to all PEs in the array. That is, the processor elements from other arrays or computers should not receive instructions executed in an array other than their own. Since for a given processor resource there exists different partitions into various arrays, the array architecture may be characterized by a set of architectural states, which differ from each other along one or several of the attributes that follow:

1. The number of concurrent arrays
2. Dimensions of data vectors computed in each array
3. Word sizes of processors working in each array

Example 5. Consider a set of array states assumed by dynamic architecture having $n = 4$ processor elements, PE_1, PE_2, PE_3, and PE_4. These states are shown in Table III.3. For every array, a processor-supervisor function may be assumed by any of its $16 \cdot k$-bit processors which then begins instruction fetches from one of k memory elements, ME, contained in its local primary memory. In Table III.3, current positions of processor supervisors are marked with (*). For state N_0, one array is formed. It contains four 16-bit processors each of which is equivalent to one PE; i.e., PE_1, PE_2, PE_3, or PE_4. This means that each instruction handles a four-dimensional data vector (a_1, a_2, a_3, a_4) made of 16-bit words. If PE_1 is processor supervisor (in Table III.3 it is marked with asterisk), PE_1 broadcasts instructions to PE_2, PE_3, PE_4; i.e., $PE_1 \rightarrow PE_2, PE_3, PE_4$.

For state N_6, one array is also formed. It handles a two-dimensional data vector (a_1, a_2) where a_1 is a 48-bit word processed by 48-bit processor assembled from PE_1, PE_2, and PE_3, and a_2 is a 16-bit word handled by the 16-bit processor PE_4. The current position of the processor-supervisor is the 48-bit processor (PE_1, PE_2, PE_3). It may fetch instructions either from ME_1, ME_2, or ME_3. For the N_7 state two concurrent arrays are formed. Each of them handles a two-dimensional vector (a_1, a_2) made of two 16-bit operands. Current positions of processor-supervisors are PE_1 for the first array and PE_3 for the second. For state N_{15}, the array formed $(PE_1 \rightarrow (PE_3, PE_4))$ uses only a portion of the total resource. This array is made of processor elements that are not adjacent. ■

TABLE III.3

State Code	Architectural Configuration	Symbolic Notation of the Architecture
N_0	PE$_1$ PE$_2$ PE$_3$ PE$_4$	PE$_1$ → PE$_2$, PE$_3$, PE$_4$
N_1		PE$_1$ → (PE$_2$, PE$_3$), PE$_4$
N_2		PE$_1$ → PE$_2$ (PE$_3$, PE$_4$)
N_3		PE$_1$ → (PE$_2$, PE$_3$, PE$_4$)
N_4		(PE$_1$, PE$_2$) → PE$_3$, PE$_4$
N_5		(PE$_1$, PE$_2$) → (PE$_3$, PE$_4$)
N_6		(PE$_1$, PE$_2$, PE$_3$) → PE$_4$
N_7		PE$_1$ → PE$_2$; PE$_3$ → PE$_4$
N_8		PE$_1$ → PE$_2$, PE$_3$
N_9		PE$_1$ → PE$_2$, PE$_4$
N_{10}		PE$_1$ → PE$_3$, PE$_4$
N_{11}		PE$_2$ → PE$_3$, PE$_4$
N_{12}		(PE$_1$, PE$_2$) → PE$_3$
N_{13}		PE$_1$ → (PE$_2$, PE$_3$)
N_{14}		(PE$_1$, PE$_2$) → PE$_4$
N_{15}		PE$_1$ → (PE$_3$, PE$_4$)
N_{16}		PE$_2$ → (PE$_3$, PE$_4$)
N_{17}		(PE$_2$, PE$_3$) → PE$_4$
N_{18}		PE$_1$ → PE$_2$
N_{19}		PE$_2$ → PE$_3$
N_{20}		PE$_3$ → PE$_4$
N_{21}		PE$_1$ → PE$_3$
N_{22}		PE$_1$ → PE$_4$
N_{23}		PE$_2$ → PE$_4$

Note that to speed up array computations, each array must contain $16 \cdot k$-bit processors that are not necessarily adjacent. Then one instruction may handle a data vector whose components (data words) are processed by arbitrary $16 \cdot k$-bit processors formed in the resource; i.e., there will be no limitation on where data items handled by any one array are stored.

Let us show that each $16 \cdot k$-bit processor of the array which contains k processor elements, PE, and handles $16 \cdot k$-bit words must be assembled only from adjacent PE's. To perform fast computations, a reconfigurable processor bus must be dedicated [13]. Otherwise, if it shares interconnections with the memory–processor bus [14], the bus introduces significant delays, since in propagating from one PE to the next-most-significant PE, carry signals must pass through $2 \log_2 n$ connecting elements where n is the number of processor elements in the resource.

However, if the processor bus is dedicated and includes not only adjacent PE's, the number of possible combinations of PE that may be contained in one $16 \cdot k$-bit processor grows exponentially. This leads to an exponential increase in the complexity of the bus. Therefore, if the processor bus is not dedicated, it becomes very slow; if it is dedicated but assembled not from only adjacent PE it becomes very complex.

On the other hand, as was shown in [13], should each $16 \cdot k$-bit processor be assembled from adjacent PE's, one may construct very simple and fast processor busses which perform very fast propagations of carries and overflows and take the minimal complexity of the hardware.

3.2.4 Pipeline Computations

Pipelining is an attractive choice for very fast computations, since it allows elimination of the time of memory accesses from the time of execution.† Thus the architectures of Supersystems must be capable of switching into pipeline states. Consider what properties must be implemented in pipeline states in order not to introduce time overheads into execution.

†Basic principles of pipeline computations are presented in Chap. I, Sec. 15.1. (Ed.)

3.2.4.1 Adaptation to parallel streams

Since a computed program may be decomposed into several pipelined streams, the architecture must be capable of assuming different pipeline states where each state is specified by the number of concurrent pipelines and the number of stages in each pipeline. A transition from one state to another should be performed via software, during the time of program computation. This allows the performing of dynamic redistribution of the resource by selecting pipelines which take into account the specificity of the computed programs.

Example 6. Let a system be equipped with S pipeline states, N_0, $N_1, \ldots, N_{S-1}$. Then for state N_0, the entire resource goes into formation of, say, 3 pipelines each containing 5 stages; i.e., $N_0 = (3 \times 5)$. For another state, N_1, $= (2 \times 4, 2 \times 4)$. For the third state N_2, it forms 3 pipelines with 6, 5, and 4 stages, respectively; i.e., $N_2 = (1 \times 6, 1 \times 5, 1 \times 4)$. ∎

3.2.4.2 Adaptation on operation sequences

As was shown in Sec. 2.3, a significant execution speed-up may be achieved if the computer architecture is equipped with dedicated instructions, each of which realizes a complex sequence of operations. When such an instruction is executed by the pipeline, each stage of the pipeline must be assigned to one operation in the sequence. It then follows that each stage must be capable of executing any operation. This requires that each pipeline stage function as an $h \cdot k$-bit computer capable of executing any operation provided by an instruction.

Example 7. A single pipeline must be capable of implementing different sequences of operations such as $(-, +, \times)$ in the arithmetic expressions, $(a - b) \times (c + d)$, or $(\times, -, \times, +)$ in $a \times b - c + e^2$, etc. ∎

3.2.4.3 Adaptation on the number of pipeline stages

One of the basic problems of modern pipelined systems is their inability to quickly vary the number of stages contained in a single pipeline. Indeed, existing techniques [15–20] provide that instructions bypass the unneeded stages. But this may involve conflicts when one instruction, as a result of such bypassing, jumps to the operands prepared for preceding instructions. On the other hand, if no such bypassing is implemented, the pipeline contains a permanent number of stages specified by the instruction containing the maximal sequence of operations. All other instructions which implement shorter sequences are slowed down because they have to pass through all stages.† To alleviate this problem, each pipeline stage must be capable of executing any operation encountered in the instruction and each instruction must propagate through the number of consecutive stages in the pipeline that matches the number of operations it realizes. Consequently two instructions that realize a and b operations, respectively, must activate a and b consecutive stages in the same pipeline.

†More information on problems of existing pipeline systems is in Chap. I, Sec. 15.3. (Ed.)

Example 8. In a pipeline of five stages the instruction that implements the formula $(a + b) \times c - d$ which contains three operations $(+, \times, -)$ must be executed only in the first three stages; i.e., the computational result should be produced by the third stage and be prevented from going to the fourth and fifth stages. ∎

3.2.4.4 Adaptation to operation time in each stage

Each stage of a pipeline may speed up execution if it changes the time of operation. Such a variation may be accomplished if each $h \cdot k$-bit computer, implemented as a stage, is equipped with modular control organization

[21] which can change the time of operation with a special control code p, where p is the number of small clock periods, t_0, that a processor-dependent operation requires to execute in a given size computer. Another code, k, may also reconfigure the processor into a minimal size. Therefore, if each pipeline instruction stores its own codes k and p, then each pipeline stge will be able to reconfigure into the minimal size computer and generate the minimal time interval for a processor-dependent operation. As for processor-independent operations (boolean, shifts, conditional branches on $=$, $\neq$, tests of flip-flops, etc.), the modular control organization may speed up their executions as well. The reason for this is that for any size computer the same clock period, t_0, the time of 8-bit addition, is used. That is, a processor-independent operation may be executed during one t_0 rather than during the longer time of a processor-dependent operation.

Therefore, if each stage is realized as an $h \cdot k$-bit computer, then the time of operation in this stage can be either one t_0 or $t_0 p$. For instance, in a 64-bit computer with no CLA circuits, $h = 8$, $p = 8$ and the time of operation can be either 1 t_0 or $8 t_0$. If CLA circuits are used, a $p < 8$ is selected. It follows then that if the time of an operation executed in the ith stage is changeable one obtains additional execution speed-up in the pipeline.

However, the problem that must be overcome is the problem of so-called pipeline races. Indeed, the variation of operation times in pipeline stages may lead to a situation where an instruction containing a shorter operation executed in the ith stage may race the preceding instruction which executes a longer operation in the $(i + 1)$th stage. Thus, the result of the ith stage will be routed to the $(i + 1)$th stage before the $(i + 1)$th stage finishes its operation. As was shown in [21], pipeline races can be easily solved by introducing asynchronous movement of instructions from one stage to another.

Example 9. Let us compare two types of pipeline: P_1 with a permanent operation time in a stage and P_2 with a changeable one. Assume that P_1 and P_2 have pipeline stages implemented as 64-bit computers so that P_1's operation time is that of a 64-bit addition whereas P_2's time depends on the sizes of data words and operation types. Let the time of 64-bit addition, $T(64) = 400$ nsec, and the time of 32-bit addition be $T(32) = 200$ nsec. Find the times, $T(P_1)$ and $T(P_2)$, required by both pipelines for the execution of the same operation sequence $((A \wedge B) + C - F) > K$ corresponding to a conditional branch instruction where A, B, C, F, and K are 32-bit operands. Since the first pipeline has the same time for each stage and the operation sequence ($\wedge$, $+$, $-$, $-$) takes four operations, $T(P_1) = 400 + 400 + 400 + 400 = 1600$ nsec. The second pipeline executes 32-bit addition in $t(32) = 200$ nsec and logical multiplication in $t_0 = 100$ nsec. Therefore it executes instruction $((A \wedge B) + C - F) > K$ during the time $T(P_2) = 100 + 200 + 200 + 200 = 700$ nsec. Thus, implementation of changeable operation times in pipelines is a source of additional speed-up. ∎

3.2.4.5 Adaptation on conditional branch

A general weakness of pipeline architectures is pipeline drains due to conditional branching. Indeed, a pipeline architecture may have dummy time intervals when no processing is performed if, as a result of a conditional test, the program switches to another instruction sequence that was not already being processed by the pipeline stages. The problem of the conditional branch may be solved if the architecture switches into architectural states containing several independent pipelines, of which one is selected as the main pipeline and the others are subsidiary. For this case "true" (incremental) and "false" (specified with jump address) program sequences may be computed, respectively, by two independent pipelines, a main and a subsidiary, where a subsidiary pipeline is switched into operation only during the respective conditional branch instruction; i.e., its instruction memory replicates the entire program. If the conditional branch is made to the instruction sequence computed in the subsidiary pipeline, then it transfers all computational results necessary for further computations to the main pipeline and stops computation.

Note, that if several subsidiary pipelines are available, the architecture may effectively organize multiport branching, minimizing the number of dummy time intervals.

Example 10. Let the conditional branch instruction I_1 requiring execution of the conditional test $(A - B)^2/K + C > W$ store jump address A_F. Since the decision on selection of either true or false program sequences is made only at the fifth stage of the basic pipeline, it begins execution of instructions I_{2T} and I_{3T}, which immediately succeed I_1 in the true sequence, before it completes the conditional test. However, before entering the basic pipeline, instruction I_1 transfers the jump address A_F to the subsidiary pipeline, which begins execution of the false instruction sequence I_{2F}, I_{3F}, etc. If the fifth stage of the basic pipeline transfers control to the false sequence executed in the subsidiary pipeline, then each ith stage of the subsidiary pipeline ($i = 1, \ldots, 4$) broadcasts its computational result to the $(i + 1)$th stage of the basic pipeline. Therefore, the basic pipeline continues computation as if it had contained the false sequence. ∎

3.2.5 Mixed Computations

A Supersystem may have concurrent coresidence of several types of architectures while executing a set of complex real-time algorithms requiring different computations. Or concurrent tasks within the same algorithm may need different architectures in order to be executed during minimal times. Thus the Supersystem should assume a mixed architecture mode when portions of its resources reconfigure into different types of architectures. This feature will allow the performing of concurrent multicomputer, multiprocessor, array, and pipeline computations using the same hardware resources instead of implementing separate dedicated subsystems. This means that the same resource will be continuously involved into different types of computations and a Supersystem will realize an additional throughput increase using the same complexity of the available resources.

In a Supersystem multicomputer and multiprocessor architectures merge so that desirable characteristics of each such architecture will be taken advantage of. Indeed, since dynamic architecture is provided with reconfigurable interconnections on the level of separate modules, any functional module of the resource (processor element PE, memory element ME, I/O element GE) may be directly connected with any other functional module without the use of I/O devices. Therefore, dynamic architecture possesses all attributes of reconfigurable multiprocessor architecture. On the other hand, dynamic architectures may form variable size computers from the available resources thus taking advantage of the multicomputer architecture to operate with $16 \cdot k$-bit data words in parallel. That is, for a $16 \cdot k$-bit computer assembled from k computer elements, CE, [13] shows how the reconfigurable bus may establish:

1. The instruction path, whereby an instruction stored in one CE may be transferred to all other CE's of the computer.
2. Parallel data path, whereby the processor and memory of $16 \cdot k$-bit computer may have parallel exchanges with $16 \cdot k$-bit words.

Accordingly, dynamic architecture is able to utilize positive attributes of both multicomputer and multiprocessor architectures.

Example 11. Table III.4 shows some of the architectural states that may be assumed by the dynamic architecture containing four CE's in the mixed mode of operation. Since multicomputer and multiprocessor architectures merge, $C_i(k)$ means one $16 \cdot k$-bit computer, A means one array. For instance, stage N_4 forms one array A and one 16-bit computer $C_4(1)$. The A-array includes a 32-bit processor, (PE_1, PE_2), and a 16-bit processor, PE_3; i.e., $A: [(PE_1, PE_2), PE_3]$. This array handles a two-dimensional data vector (a_1, a_2) where a_1 and a_2, respectively, are 32-bit and 16-bit data words. State N_{10} forms one A-array, $A: [PE_1, PE_2]$, and one 32-bit computer $C_3(2)$. The A-array contains two 16-bit processors, PE_1 and PE_2. ∎

3.3 Section Summary

The appearance of LSI modules with high throughput makes it feasible to organize a cost-effective reconfiguration of module interconnections. This allows obtaining new types of architectures, dynamic architectures. A Supersystem with dynamic architecture may realize additional performance gains on the same resource by taking advantage of the following factors:

TABLE III.4

State Code	Architectural Configuration	Symbolic Notation of the Architecture
N_0	A — A — A C	$A: [PE_1, PE_2, PE_3]; C_4\ (1)$
N_1	A — A C A	$A: [PE_1, PE_2, PE_4]; C_3\ (1)$
N_2	A C A — A	$A: [PE_1, PE_3, PE_4]; C_2\ (1)$
N_3	C A — A — A	$A: [PE_2, PE_3, PE_4]; C_1\ (1)$
N_4	A ——— A C	$A: [(PE_1, PE_2), PE_3]; C_4\ (1)$
N_5	A A ——— C	$A: [PE_1, (PE_2, PE_3)]; C_4\ (1)$
N_6	A ——— C A	$A: [(PE_1, PE_2), PE_4]; C_3\ (1)$
N_7	A C A ———	$A: [PE_1, (PE_3, PE_4)]; C_2\ (1)$
N_8	C A ——— A	$A: [(PE_2, PE_3), PE_4]; C_1\ (1)$
N_9	C A A ———	$A: [PE_2, (PE_3, PE_4)]; C_1\ (1)$
N_{10}	A — A C	$A: [PE_1, PE_2]; C_3\ (2)$
N_{11}	A C A	$A: [PE_1, PE_4]; C_2\ (2)$
N_{12}	C A — A	$A: [PE_3, PE_4]; C_1\ (2)$

1. By reconfiguring resource into minimal size computers, it may maximize the number of programs computed by the same resource.
2. By switching the resources into different types of architectures, it may speed up respective computations. This allows the available resources to be permanently involved in computations, even those that require dedicated subsystems (array and pipelines).

Therefore, the use of dynamic architectures in Supersystems leads to the realization of new sources of throughput increases heretofore unused in traditional parallel systems.

DESIGNING SYSTEMS WITH DYNAMIC ARCHITECTURES

4. INTRODUCTION

The evolution of computers and parallel systems has shown a tendency towards an increased use of between module interconnections that are reconfigurable via software. The reader is referred here to Chapters 1 and 2 which contain a comprehensive survey of reconfigurable parallel systems and performance gains accomplished via reconfiguration.

This section will focus on designing systems with dynamic architectures that reconfigure resources into a variable number of computers (processors) and thus perform resource adaptation on instruction and data parallelism.

To construct a system with dynamic architecture, one uses a new building block: *dynamic computer group* (DC-group). Two types of design are featured:

1. DC-group with the minimal complexity memory–processor bus that interconnects processor and memory resources (*DC-group with the Minimal Complexity*) [9–11].
2. DC-group with the minimal delay memory–processor bus that introduces minimal delay in communication between any two modules of the resource: processor–memory, processor–processor, memory–memory (*DC-group with Minimal Delay*) [13].

Each type of DC-group is studied using the following factors:

1. Basic architectural solutions
2. Organization of internal and external communications between resource modules
3. Organization of switches from one architectural state to another

5. DC-GROUP WITH MINIMAL COMPLEXITY

This DC-group has the minimal complexity of the resources since it uses only $(n - 1)$ connecting elements, MSE, in the memory–processor bus that interconnects n PE with n ME [22, 23, 9–11]. This number of MSE is minimal since it may separate as many as n independent computers. We start describing this DC-group by giving details on its resource diagram.

5.1 DC-Group Resources

The DC-group with minimal complexity contains n computer elements (CEs), (CEs), $n - 1$ connecting elements (MSEs), and one monitor (V) (Fig. III.7). One CE processes h-bit words in parallel. Assume $h = 16$. A pair of adjacent CEs is separated by one connecting unit which may assume the following modes: right transfer, left transfer, or no transfer. For a right transfer mode, a 16-bit byte is transferred from pins a^1 to a^2; for a left trasfer, the byte is transferred from pins a^2 to a^1; and for no transfer, there is no data path between them.

A DC-group may form computers of variable sizes, labeled $C_i(k)$. Each $C_i(k)$ computer ($k = 1, 2, \ldots, n$) integrates k computer elements and $k - 1$ connecting elements. The i shows the position of the computer's most significant CE. This computer handles $16 \cdot k$-bit words and has a primary

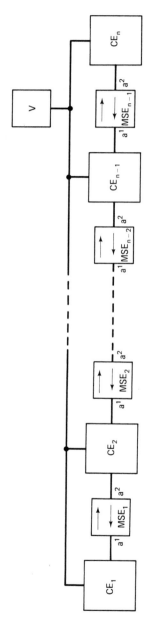

FIGURE III.7 DC-group with minimal complexity

memory $16 \cdot k$-bits wide. By changing the number of computer elements in $C_i(k)$, one obtains 16, 32, 48, ..., or $16 \cdot n$-bit computers.

A DC-group is a dynamic architecture as we have defined the term: a DC-group containing n CEs may assume 2^{n-1} states, $N_0, N_1, \ldots, N_{2^{n-1}}$. The difference between states is in the sizes, positions, and numbers of computers operating concurrently.

Example 12. A DC-group with $n = 5$ CEs forms the $2^{5-1} = 16$ different states shown in Table III.5. Column 1 lists architectural states; for every state, columns 2 and 3 show, respectively, the positions and bit sizes of the computers comprising it. For instance, for state N_1, the $C_1(4)$ computer (64 bits) and $C_5(1)$ computer (16 bits) function concurrently; that is, $N_1 = (1 \times 64, 1 \times 16)$. All connecting elements, MSE_1, MSE_2, MSE_3, integrated by $C_1(4)$ (Fig. III.8) are in a transfer mode ($\rightleftarrows$), which may be either left or right. The instruction fetched by one CE is transferred to all the others. MSE_4 is in the no-transfer mode ($\not\rightleftarrows$) and it separates the program and data computed by $C_1(4)$ from that computed by $C_5(1)$. For the $C_1(4)$ computer, the

TABLE III.5

States	Positions of Concurrent Computers	Bit Sizes of Computers
N_0	$C_1(5)$	1×80
N_1	$C_1(4), C_5(1)$	$1 \times 64, 1 \times 16$
N_2	$C_1(3), C_4(2)$	$1 \times 48, 1 \times 32$
N_3	$C_1(3), C_4(1), C_5(1)$	$1 \times 48, 2 \times 16$
N_4	$C_1(2), C_3(3)$	$1 \times 32, 1 \times 48$
N_5	$C_1(2), C_3(2), C_5(1)$	$2 \times 32, 1 \times 16$
N_6	$C_1(2), C_3(1), C_4(2)$	$1 \times 32, 1 \times 16, 1 \times 32$
N_7	$C_1(2), C_3(1), C_4(1), C_5(1)$	$1 \times 32, 3 \times 16$
N_8	$C_1(1), C_2(4)$	$1 \times 16, 1 \times 64$
N_9	$C_1(1), C_2(3), C_5(1)$	$1 \times 16, 1 \times 48, 1 \times 16$
N_{10}	$C_1(1), C_2(2), C_4(2)$	$1 \times 16, 2 \times 32$
N_{11}	$C_1(1), C_2(2), C_4(1), C_5(1)$	$1 \times 16, 1 \times 32, 2 \times 16$
N_{12}	$C_1(1), C_2(1), C_3(3)$	$2 \times 16, 1 \times 48$
N_{13}	$C_1(1), C_2(1), C_3(2), C_5(1)$	$2 \times 16, 1 \times 32, 1 \times 16$
N_{14}	$C_1(1), C_2(1), C_3(1), C_4(2)$	$3 \times 16, 1 \times 32$
N_{15}	$C_1(1), C_2(1), C_3(1), C_4(1), C_5(1)$	5×16

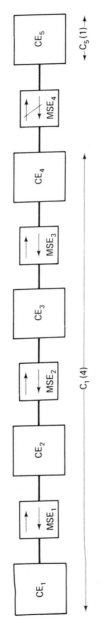

FIGURE III.8 Architectural state $N_1 = [C_1(4), C_5(1)]$

transfer mode of its connecting units depends on which CE stores a currently executed program segment. If the instruction is fetched from CE_1, all MSE's (MSE_1, MSE_2, MSE_3) assume the right transfer mode, if it is fetched from CE_2, MSE_1 must transfer left while MSE_2 and MSE_3 retain their right transfers, etc. ∎

Any DC-group transition from one state to another is performed by a special *architectural switch* instruction. A DC-group may perform any transitions, say, $N_0 \rightarrow N_7$ or $N_6 \rightarrow N_2$, i.e., no transitional constraints exist. Since for each transition, at least one computer in the present state must give up its resource to form another computer in the next state any architectural transition occurs at a moment when all computers that sacrifice their resources finish their tasks. This moment is identified by the V monitor.

Another function of the V monitor is to resolve conflicts between concurrent programs. Two programs may request transitions into two different states, causing conflicts in the system. To resolve them the V monitor checks the priority of a requesting program and allows or denies its call for the next transition. If the program request is allowed, the V monitor allows execution of the architectural switch instruction. It fetches special control codes to be written to the computer elements of all newly formed computers, and writes new transfer modes to the connecting elements which integrate or separate these computers. The codes activate a new pattern of interconnections specific to the next architectural state. If the required resource to establish the next state is free, then the architectural switch instruction lasts as long as 64-bit addition. Since a DC-group is assembled from computer elements, let us now describe a CE.

5.1.1.1 Computer element

Each CE includes a 16-bit processor element (PE), a 16-bit wide memory element (ME), and a 16-bit I/O element (GE), equipped with a small memory (M(GE)), having the same width as a PE (Fig. III.9). Generally the word size h of one CE depends on the current restrictions of LSI technology (chip size and pin count) and may be 4, 8, 12, 16 bits, etc. To increase h, each CE may have its processor and I/O elements assembled from several LSI modules.

5.1.1.2 DC-Group size

The DC-group hardware resource is characterized by an F parameter (expressed in bits) that shows the maximal size of the computer that can be formed in one DC-group. By fixing h, the number of bits processed by one CE, one gets $n = F/h$, where n is the number of computer elements and $n - 1$ is the number of connecting elements.

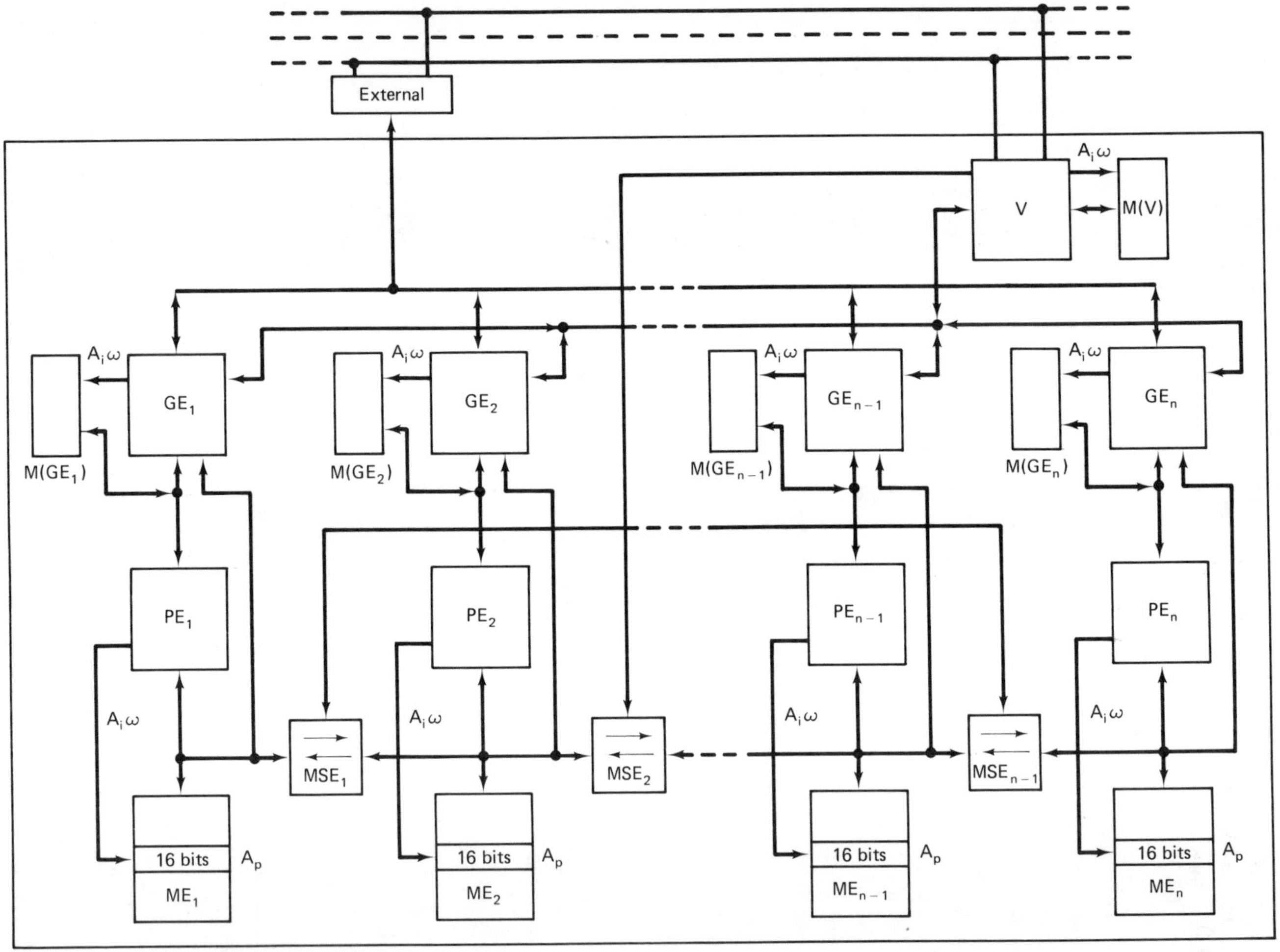

FIGURE III.9 Resource diagram of DC-group

Example 13. Let $F = 80$ bits and one CE process 8 bits ($h = 8$). Then the DC-group resource contains 10 computer elements ($10 = 80/8$) and nine connecting elements. Such a DC-group may assume $2^{10-1} = 512$ different states. ∎

5.1.1.3 Internal communication

For one computer element the following communication paths are present: a 16-bit byte fetched from ME that can be written either to its PE and/or GE (Fig. III.9). The address of this byte (A_i) and read and write signals (w) are generated by the respective PE. The PE exchanges 16-bit bytes with GE, and GE generates address and read/write signals for its local memory M(GE). A byte fetched from M(GE) can be written either to GE or to PE or to both.

Different CE's may communicate in two ways: through the connecting units and through the time-shared bus which connects all GE's of one DC-group. With the use of connecting elements, a 16-byte fetched from a given ME may be written not only to the PE and GE associated with it, but passing through connecting elements to its right in right transfer mode, it may be written to computer elements which are to the right of the given CE. Similarly, a byte may be broadcast to computer elements which are left of the given CE provided that the left connecting elements are in the left transfer mode. Further propagation of a 16-byte (left or right) is blocked by the first connecting element in no-transfer mode in its path.

With the use of a time-shared bus, a 16-bit byte retrieved by one GE may be received by any other GE, provided the V monitor allows this transfer. In case several GE's make concurrent communication requests, the V monitor resolves conflicts on the basis of priority codes that have been assigned to the programs being computed. To this end the V monitor first identifies which $C_i(k)$ computer a requesting GE belongs to. Thereafter, its requests for communication are allowed or denied depending on the priority of the program computed by this computer. The V monitor is also connected via a separate bus to every element GE. Using this bus, the most-significant GE of every computer transfers to the V monitor the control codes necessary for architectural transitions. In addition, the V monitor is connected with all connecting elements, MSE, in order to establish in them the transfer modes required by the next architectural state.

5.1.1.4 External communication

In a multicomputer system, all DC-groups are connected by an external bus which is organized using a conventional scheme (time-sharing or crossbar). The bus width is a multiple of 16. Each computer element is connected with this external bus through the External unit (Fig. III.9). External communication is controlled by the V monitor. The V monitor of the given DC-group is also connected with monitors belonging to other DC-groups via dedicated communication lines. In case of concurrent requests from several other DC-groups, the V monitor makes a selection based on the priority codes assigned to a requesting DC-group. These codes are stored in the M(V) memory and may be changed by the operating system.

5.2 Control Organization in DC-Group Computers

The control organization of dynamic architectures must take into account *common* LSI technological restrictions—high cost of a module type and restricted pin count—and *specific* restrictions that originate from the dynamic partitioning of the resources into a variable number of concurrent computers. Since the effect of LSI technological restrictions on control organization was described in Chap. 1, we will discuss only specific requirement here.

5.2.1 New Requirements for Control Organization

The requirements are:

1. Each formation of a new computer must be accompanied by establishment of its control unit via software. This control unit must generate variable operation times for processor and memory access operations.
2. The control organization of a DC-group computer must function independently of the number of control units it contains.

Here are more details on these requirements.

5.2.1.1 Variable operation times

A system with dynamic architecture may assume a set of different architectural states. Each state is distinguishable from others by at least one new computer. Every formation of a new computer must be accompanied by the establishment via software of its control unit. It must generate minimal operation times for its processes; i.e., they must be functions of either processor size for processor-dependent operations or memory speed for memory-access operations. Let us find the number of different time intervals that must be generated by the control organization.

A dynamic architecture, may form n different-size computers ranging from 16 to $16 \cdot n$ bits. For instance, for $n = 4$, computer sizes of 16, 32, 48, and 64 bits may be formed. Since the time of processor-dependent operations, t_{op}, depends on the computer size, the control organization must generate n different t_{op}'s. Also, the n memory elements in the primary memory may all have different speeds, so the control organization must be capable of generating n different fetch times. Therefore, to satisfy both, the control unit must be capable of producing $2n$ different time intervals. Of those, however, it need select only three (via software) when a new computer is formed: one for processor operation, another for data word fetch, and the third for instruction fetch.

5.2.1.2 Control independence

If each computer element, CE, functions as an independent computer, then the corresponding architectural state has n independent computers operating concurrently. For instance, if the resource contains four CEs, all of them may function as four independent 16-bit computers. Thus each CE must be provided with its own control unit. On the other hand, if all the CEs are integrated into the largest possible computer (for this example, the 64-bit computer), its control unit must contain all the control units of these CEs. Therefore, in any computer, its control organization must function independently of the number of control units it contains.

These two requirements for software-controlled control organization are met in the modular control organization considered below.

5.2.2 Modular Control Organization: Basic Concepts

Contrary to the synchronous and asynchronous control organizations which assume that control over all computer devices is performed by either a unique control unit (synchronous control) or by one central and several local control units (asynchronous control), the modular control organization eliminates the separate control unit in the architecture. Instead, every LSI module is provided with a local *modular control device* (MCD), and the functions of the control unit of any computer are performed by concurrent operations of all the MCDs contained in this computer [9–11, 24, 25].

Each program instruction is written concurrently to all modules in the computer. It is executed during one instruction cycle (IC),† which consists of d intervals: $IC = T_1 + T_2 + \cdots + T_d$. Since the control organization must generate variable time intervals which are functions of operand word size for processor dependent operations and memory speed for memory access operations, the MCD generates a variable time interval T in the following way: T contains b clock periods t_0; i.e., $T = bt_0$, where t_0 is the time of the longest operation (8-bit addition) in one LSI module. Variation of T is achieved by variation of b, which for different types of operations assumes the value of one of several control codes stored in each LSI module. The instruction cycle may contain the following types of intervals:

†Prerequisite information in instruction cycles and instruction intervals can be found in chap. I, Sec. 1.6. (Ed.)

1. Intervals, T_p, for processor-dependent operations: $T_p = pt_0$ where p is the number of clock periods a processor-dependent operation requires to execute in a given size computer.

2. Intervals, T_0, for processor-independent operations (boolean, equality, inequality, shift, etc.). These are of permanent duration and equal to one clock period t_0.
3. Intervals, T_I, for instruction fetch: $T_I = m_I t_0$ where m_I is the number of clock periods required to fetch an instruction from a given memory element.
4. Intervals, T_E, for data-word fetch: $T_E = m_E t_0$ where m_E specifies the speed of the slowest memory element contained in the primary memory of the given computer. This originates from the fact that a $16 \cdot k$-bit word is stored in a parallel cell made of k 16-bits bytes. Thus to fetch this word, one has to take the access time of the slowest memory element.

Variation of these intervals is accomplished with several control codes stored in each LSI module. For processor-dependent intervals, $T_p = bt_0$, b assumes the value of *processor code p*, which changes each time when a new computer size is formed.

Example 14. Suppose that dynamic architecture forms a 48-bit computer assembled from three CEs. Let us find the processor code p to be written to each CE. Assume that fast CLA circuits are used in the adder and the size of one group is 4 bits. Since each CLA circuit may be expressed as a disjunctive normal form, its delay is $2t_d$, where t_d is one gate delay. This yields a time, t_0, of one clock period (the time of an 8-bit addition) equal to $4t_d$. Since for 48-bit addition, carry signal must pass through $48 \div 4 = 12$ CLA circuits, they will delay signal propagation $2t_d \times 12 = 24t_d$. Since each $t_0 = 4t_d$, 48-bit addition takes $24t_d \div 4t_d = 6t_0$ and the processor code $p = 6$. ∎

For an interval $T_I = bt_0$ in which an instruction is fetched, b assumes the value of *fetch instruction code* m_I. For an interval $T_E = bt_0$ in which a $16 \cdot k$ bit word is fetched, b assumes the value of *fetch data code* m_E.

Example 15. Let a state N_1 have two computers: 48-bit, $C_1(3)$ and 16-bit, $C_4(1)$. For the 48-bit computer $C_1(3)$ including three CEs (CE$_1$, CE$_2$, CE$_3$) the primary memory contains three ME: ME$_1$, ME$_2$, and ME$_3$ (Fig. III.10). Let the speeds of these MEs be: $13t_0$ for ME$_1$, $12t_0$ for ME$_2$, and $3t_0$ for ME$_3$. Thus CE$_1$ stores instruction fetch code $m_I = 13$, CE$_2$ stores $m_I = 12$, and CE$_3$ stores $m_I = 3$. However, all these CEs store the same $m_E = \max(13, 12, 3) = 13$. If instruction is fetched from CE$_1$, $m_I = 12$ controls the instruction fetch intervals $T_1 = m_I t_0 = 12t_0$. If the instruction is fetched from CE$_2$, $m_I = 13$ and $T_1 = 13t_0$. If a 48-bit data word is fetched in parallel from 48-bit primary memory, assembled from ME$_1$ through ME$_3$, it is fetched during time $T_E = m_E t_0 = 13t_0$, since this is the speed of the slowest memory element of this primary memory, etc. ∎

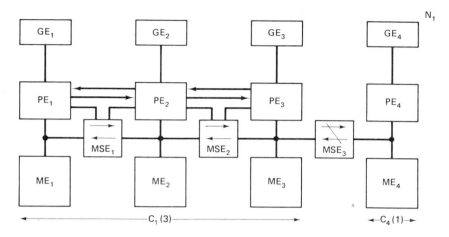

FIGURE III.10 DC-group in state $N_1 = [C_1(3), C_4(1)]$

For one instruction, sequencing of the intervals is defined by an *interval sequencer* which is included in the MCD of every LSI module. In this sequencer, each interval is identified with one state, and a sequence of states is selected by the instruction op code. For each interval, variation of its duration is achieved with the *operation sequencer*. This sequencer is controlled by one of the three codes (p, m_E, or m_I) considered above. This is achieved by executing a loop which lasts either p, m_I, or m_E clock periods. If the interval sequencer initiates the operation sequencer, the interval sequencer retains its state (interval) until the operation sequencer completes its operation. Therefore, the interval duration is a function of one of three codes stored in an LSI module of the $C_i(k)$ computer.

When a new computer is formed, it is sufficient to write new values of codes p and m_E into all its LSI modules for the control unit to start generation of new time intervals. The m_I code is not changed during architectural transitions, since it characterizes the speed of the local memory element associated with the computer element. This code is written only once, during the assembly stage, into all LSI modules of the respective CE.

Consider now how modular organization meets the second requirement mentioned above (independence of the number of control units). This requirement means that all modules of computer must execute the same sequences of instruction intervals, each interval has to have the same duration, and the switch from one interval to another has to be performed at the same moment. Since one program instruction is fetched concurrently to all LSI modules of the $C_i(k)$ computer, they all receive the same op code. Because the MCD is the same for all modules, in every module the same sequence of intervals is activated by this code. Processor dependent and data fetch intervals last the same time in all modules because they store the same processor and data fetch codes, p and m_E.

As for the instruction fetch interval, different CEs of computer $C_i(k)$ may store different m_I codes because dynamic architecture may use memory elements with different speeds. However, since the instruction is fetched from one ME, the instruction fetch interval is specified by the speed of this ME. Other CEs of the $C_i(k)$ computer receive a completion signal which specifies the end of this interval. This signal is sent through the path formed for the equality signal which exists only for CEs belonging to the given computer. Finally, all LSI modules of the computer perform concurrent transitions from one instruction interval to another because the instruction is received concurrently by all of them. Thus, they initiate a particular interval sequence at the same time.

Therefore, we have shown that the number of modules contained in a $C_i(k)$ computer affects neither the sequencing of intervals nor their durations. It then follows that the modular control organization provides independence of the control of a computer from the number of control units in it.

5.2.3 Hardware Realization of the Modular Control Device

The modular control device (MCD) is housed in each LSI module of the computer. It includes three registers (Y_1, Y_2, and Y_3), the 8-bit counter (Co), the interval sequencer CAD-I, the operation sequencer CAD-0, logical circuits (MIC, $L(r)$, L_1, $-L_3$), and flip-flop B_0 (Fig. III.11). In the MCD, register Y_1 (16 bits) stores a program instruction fetched through pins 3 to 18, the Y_2 register stores variable *control* codes which are changed during each reconfiguration, and Y_3 stores invariant control codes which are unaffected by reconfiguration and are written to each module only once during the assembly stages. MIC is the circuit for generating local microcommands and $L(r)$ produces completion signals which are used to signal the end of some instruction intervals that last longer than one clock period (instruction fetch I/O operation, etc.). These signals are always produced only in one CE and sent to all other CEs through pin 1 and logic L_2. All modules of the given computer have their 1 pins connected via connecting elements. L_1 enables the transfer of a byte of memory address made up of register R_A (8 bits) and counter Co, to pins 19 to 30. This is performed during a memory access interval. L_3 selects whether code m_I or m_E should control the work of the CAD-0 sequencer. B_0 is the flip-flop which recognizes the CE that fetches program instructions from its memory element.

Consider now how control codes are designated. The variable codes are as follows:

1. *Processor code p* sets the time for a processor dependent operation. (Let a 64-bit computer use 4-bit CLA circuits, the time for 64-bit addition be selected as 600 ns, and the time of one clock period t_0 to be equal to 150 ns. Then processor code $p = 600/150 = 4$; i.e., in

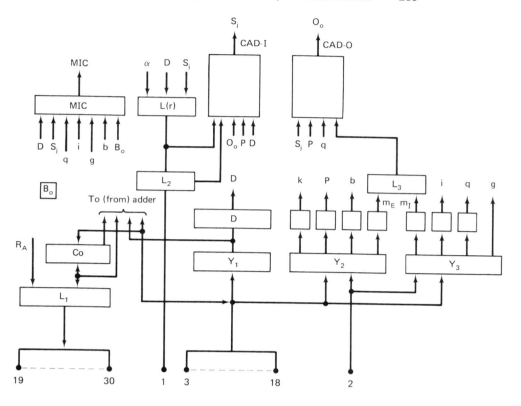

FIGURE III.11 Modular control device

the computer all processor dependent operations are allowed four t_0 to execute.)

2. *Data fetch code* m_E is the time for fetching a $16 \cdot k$-bit word from the primary memory of a $16 \cdot k$-bit computer. It specifies the speed of the slowest ME contained in this computer. For the 48-bit computer containing ME_1 (600 ns), ME_2 (300 ns), and ME_3 (900 ns) and having $t_0 = 150$ ns, $m_E = 900/150 = 6$.)

3. *Computer size code* k specifies the number of computer elements contained in computer. This code participates in forming a variable $16 \cdot k$-bit adder provided each PE is assembled from more than one module. (A 64-bit computer contains four CEs, therefore $k = 4$.)

4. *Significance code* b marks the most-significant, least-significant, and middle computer elements contained in the computer. (For 64-bit computer, containing computer elements CE_1, CE_2, CE_3, and CE_4, the most significant, CE_1, stores $b = 11$, the least significant, CE_4, stores $b = 01$, and the middle ones, CE_2 and CE_3, store $b = 00$.) This code establishes connections for a variable, $16 \cdot k$-bit adder.

Note: All variable codes except b are the same for all LSI modules of the computer.

The invariant codes are as follows:

1. *Function code q* determines the function performed by a given LSI module. (For all modules forming the processor, $q = 10$; for I/O, $q = 01$; for connecting elements, $q = 00$; and for the monitor, $q = 11$.)
2. *Postion code i* shows the position of the given CE among n other CEs. Its size is $\log_2 n$ bits where n is the overall number of CEs. (For a hardware resource containing $n = 6$ computer elements, all modules of CE_1 store $i = 1$, in CE_2 all have $i = 2, \ldots$, in CE_6, $i = 6$.)
3. *Fetch instruction code m_I*, specifies the speed of the ME contained in the given CE.
4. *Slice code g* is used only when a functional element is assembled from more than one LSI module. (For PE_1, assembled from two LSI modules, the most-significant module stores $g = 1$ and the least-significant stores $g = 0$.) When a 16-bit byte is received by a PE or GE, this code forces the more- and less-significant 8 bits to be processed by the more- and less-significant modules, respectively.

Thus, the modular control organization introduced allows one to use the same MCD contained in all modules to form the control unit of any computer. When a new computer is formed, one has to write only two new codes, p and m_E, and the control unit of this computer begins generating new time intervals specified by the new size of its processor and the new speed of primary memory. This organization completely precludes any necessity for hardware alterations in order to form new time intervals. Also, the use of an insignificant clock period, t_0 (time of addition over 8-bit words), allows one to form any time interval with a precision which does not exceed one t_0. Consequently, the modular control organization achieves a good match between the time intervals it generates and the minimally required times of operations† executed during these intervals.

†See Chap. I, Sec. 3, which desribes a technique for finding the minimally required time of operation. (Ed.)

5.2.4 Modular Control Organization and LSI Technology

In Sec. 5.2.2., we saw that the modular control organization satisfies two basic requirements of dynamic architecture—it is able to generate variable time intervals for computer processes and provide independence of the computer control organization. Here we will show that the modular control organization is amenable to LSI implementation since it takes into account major technological restrictions—the high cost of fabrication of a chip type and the pin count constraint.

Indeed, as was noted in [26], to construct a sufficiently complex computer system one has to produce tens of different LSI chip types. This leads to significant investments since the cost of one chip-type fabrication may be expressed by a five digit number. On the other hand, modular control organization allows obtaining a universal LSI module for implementation of all functional units of dynamic architecture. Indeed, since each logical module is equipped with the same MCD, its hardware consists of two portions: control and execution. Accordingly, if one equips each module with the same executional portion, containing the logical circuits characteristic of all functional units in the resource (processor, I/O, monitor, and connecting element), one obtains a unique building block, which we call a *universal module* (UM), that can be used for implementing all functional units. A functional orientation of a UM will then be distinguished by a special function code q considered above.

Obtaining a universal module for implementing all functional units in a resource sharply reduces the initial investments related to a system's fabrication. As was shown in [27], in spite of the fact that the modular control device occupies 30 to 35 percent of every LSI chip, the cost figures (numbers of LSI modules and chip types necessary to fabricate a system with dynamic architecture) for modular control organization are about one half that for the best synchronous alternative and 40 percent of the best asynchronous alternative. Also, since the entire architecture is assembled from copies of the same chip type with only pin-to-pin connections between modules, the assembly costs are also reduced.

Inasmuch as each UM executes only those microoperations that are activated inside, the only signals that are transferred between modules are a set of completion signals sent in a time-shared mode through a single pin, pin 1 (Fig. III.11). Since the modular control principle provides that each microoperation requires no pins for its implementation, the computational throughput of a UM no longer depends on its pin count but only on its component count. Because each year the component count per LSI module grows exponentially whereas its pin count grows at best linearly [28], the independence of module throughput from pin count sharply increases the gate-to-pin ratio in a module. This leads to increased utilization of the chip space.

Therefore, modular control organization meets two basic requirements of control organization in dynamic architecture and overcomes restrictions of LSI technology.

5.3 Description of a Universal Module

As was shown above, a natural consequence of the modular control organization is the creation of a universal module (UM), a unique building block for all functional units of the DC-group. Each UM contains two parts: control and execution. The control portion is the modular control device (MCD) (Fig. III.11).

5.3.1 Execution Portion

The execution portion is implemented as an 8-bit parallel microprocessor, (Fig. III.12). Since the complexity of execution portion does not affect major design concepts of dynamic architecture, Fig. III.12 gives the hardware diagram of a typical microprogrammed processor.† If the chip size allows mapping of additional hardware circuits, the execution portion of the UM may be extended with additional registers and/or microoperation circuits.

The functional orientation of the UM is distinguished with a two-bit function code q, that activates only the appropriate functionally oriented microoperations in each UM. For instance, if a UM performs the function of a processor element, it stores $q = 10$ which activates only processor microoperations, whereas for the V monitor operation, $q = 11$ activates only monitor microoperations, etc.

The execution portion of UM contains five 8-bit registers $(A, B, D, E,$ and $H)$ one 9-bit register, C, and one 8-bit adder (Ad). A UM may receive or retrieve an 8-bit byte which, having passed through the adder, may be written to any of its registers or to the counter, C_0 that belongs to the MCD. A universal module is synchronized by a single clock sequence τ_0 with clock period t_0. The t_0 time interval equals that of the longest operation (addition) over two 8-bit words. A shifted sequence τ with the same period t_0 is used for for clearing (Fig. III.12).

All registers are connected with the inputs of the 8-bit adder Ad through the logical circuits X and Y. The adder, Ad, result may be written into any of the registers or the 8-bit counter C_0. The $L(Ad)$ circuit performs a 1-bit shift either to the MSB or the LSB. Pins 33 to 40 and circuit L_5 receive or retrieve an 8-bit data from or to another UM.

Consider organization of $16 \cdot k$-bit processor. Since it is assembled from universal modules, one has to provide correct broadcast from one UM to another of the following processor signals: carry, overflow, and equality. As [11] shows, to organize a reconfigurable path for $16 \cdot k$-bit adder, one may use only four pins in each UM: two pins for carry-in and carry-out for the local adder slice and two pins for overflow-in, overflow-out when it is in transit through the given UM. Thus, in each UM, pins 41 and 42 are used to pass carry-in and carry-out for the local adder slice. Overflow-in and overflow-out pins in each UM are not shown in Fig. III.12 since the overall organization of the reconfigurable adder with all the pins and logical circuits will be discussed below.

The $L(\beta)$ circuit performs an 8-bit word comparison,‡ used in conditional branching $Z_1 = Z_2, Z_1 \neq Z_2$, where Z_1 and Z_2 are two $16 \cdot k$-bit words. Each UM of the $16 \cdot k$-bit processor compares two 8-bit bytes of Z_1 and Z_2, respectively. If these bytes are not equal, $L(\beta)$ generates the inequality signal $\overline{\beta}$ which is sent out to pin 43. For the UMs contained in the resource all pins

†Peculiarities of microprogrammed architectures are described in Chap. I, Sec. 10. (Ed.)

‡Processor signals generated during conditional branch tests are discussed in Chap. I, Sec. 8.1. (Ed.)

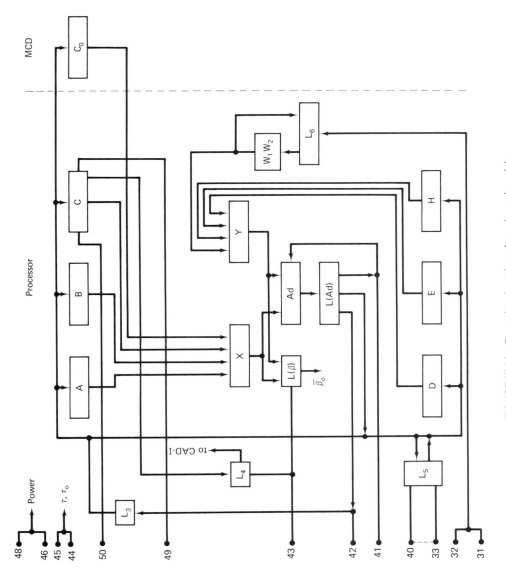

FIGURE III.12 Executional portion of a universal module

43 are connected via connecting elements. Accordingly, since connecting elements integrated into a $16 \cdot k$-bit computer assume a transfer mode, then pins 43 of all UMs forming a $16 \cdot k$-bit processor are connected together. Therefore, each UM of the computer receives the $\bar{\beta}$ signal if it is generated at least in one UM of the processor. This means that at least two 8-bit portions of the two $16 \cdot k$-bit words compared do not match. The $\bar{\beta}$ signal received through pin 43 is sent in via logic L_4 to the input of the local sequencer CAD-I of the MCD. If two $16 \cdot k$-bit words are equal, then no $\bar{\beta}$ is produced in every UM. This means that all pins 43 have no signal and CAD-I performs branching on a true $Z_1 = Z_2$.

Each UM is supplemented with several circuits which implement multiplication of two $16 \cdot k$-bit words. For multiplication, the multiplicand can be placed in either the A or B registers, whereas the multiplier can be placed only in the C register. A double precision product is placed in the H register (most significant $16 \cdot k$-bits) and in the C register (least significant $16 \cdot k$-bits). The hardware implements the algorithm for multiplication which provides for the shifting of both the multiplier and partial product to the least-significant bit (LSB). Since the multiplication algorithm performs† a shift of the partial product, if the LSB of the multiplier is 0, and the add and shift of the partial product is LSB = 1, then the LSB of the multiplier stored in register C of the least-significant UM must be transferred to all other UMs of the computer in order that the instruction sequencer perform either a shift instruction interval or an add and shift interval.

†Organization of a floating-point multiplication in a computer is discussed in Chap. I, Sec. 8.2. (Ed.)

To avoid the assignment of a special pin for this purpose, each UM is supplemented with the L_4 logic circuit which transfers the content of LSB stored in the ninth bit of C register located in the least-significant UM to pin 43, which during multiplication can receive no $\bar{\beta}$ signal. Because all k pins 43 are connected together, the signal received by each pin 43 is fed to its respective CAD-I.

The shift-out of the partial product slice in H register during multiplication is performing via the adder, and pin 41, which is connected with pin 42 of the next less-significant UM. Thereafter, through logic L_3, it is transferred to the next less-significant H register.

Since a double precision product is stored in the two $16 \cdot k$-bit registers H and C, the portion of the product stored in C register is shifted out from the UM through pin 49 and shifted into the next less-significant UM through pin 50. To form a double precision product, pin 41, in the least-significant UM of the computer that receives the LSB of its H register is connected with pin 49 in the most-significant UM of the computer that connects it with the MSB of C register. To maintain the dynamic nature of this connection it is performed via connecting elements.

Each UM is provided with two conditional flip-flops (W_1 and W_2) which may be connected with external devices through pins 31 and 32 and the L_6 logic.

5.3.2 Organization of Instruction Sequencing and Variable Time Intervals

As was indicated in Sec. 5.2.3 each program instruction fetched to UM is interpreted as a sequence of intervals in the internal sequencer, CAD-I, of MCD.†

The CAD-I sequencer is the asynchronous sequential machine having 32 states ($S_0, S_1, \ldots, S_{31}$). It contains five flip-flops, the next state logic, and the decoder. Each output S_j is maintained for the time CAD-I stays in the state S_j. This means that CAD-I executes interval T_j of the instruction cycle.

Variation in T_j intervals is performed with the operation sequencer CAD-0, which is a synchronous sequential machine having 16 states ($O_0, \ldots, O_{15}$). Its memory portion contains four flip-flops. Each state (except the initial O_1) lasts one clock period, t_0. From each state O_j, the CAD-0 may perform transition to the O_0 state followed by transition $O_0 \rightarrow O_1$ to the initial state O_1. Every $O_j \rightarrow O_0$ transition is activated by one of the three variable control codes (p, m_I, or m_E) stored in each UM. Therefore, CAD-0 passes through the loop $O_1 \rightarrow O_2 \rightarrow O_3 \rightarrow \ldots \rightarrow O_j \rightarrow O_0 \rightarrow O_1$ which lasts j clock periods.

For the instruction intervals executing a processor dependent operation the loop is terminated by the processor code p which activates $O_p \rightarrow O_0$ transition. For instruction fetch intervals, code m_I terminates the loop. For data fetch intervals, the loop is terminated by code m_E.

Because the CAD-I sequencer is implemented as an asynchronous sequential machine, it executes any transition from one state to another during a single clock pulse. Thus the instruction microprograms contain no dummy clock periods and can be implemented as complex iterative sequences of operations. This allows one to increase greatly the amount of computation assigned to a single program instruction. Consequently, the overall size of the program is shortened and a speed-up in its execution is achieved. The CAD-I design can be formalized if one uses systematic procedures for designing sequencers.

†Organization of instruction sequencing in conventional computers (von Neumann and microprogrammed) is discussed in Chap. I, Secs. 6 and 13. (Ed.)

Example 16. For the 48-bit computer each UM stores processor code $p = 6$ (see Example 14). Accordingly, the 48-bit addition is executed during $6t_0$. Code $p = 6$ is decoded as signal p_6 which activates the $O_6 \rightarrow O_0$ transition and CAD-0 passes through the loop $O_1 \rightarrow O_2 \rightarrow \cdots \rightarrow O_6 \rightarrow O_0 \rightarrow O_1$ which lasts six clock periods. If $p = 3$, signal p_3 activates the $O_3 \rightarrow O_0$ transition and CAD-0 passes through the $O_1 \rightarrow O_2 \rightarrow O_3 \rightarrow O_0 \rightarrow O_1$ transition of $3t_0$ duration. ■

The following interaction exists between the interval and operation sequencers, CAD-I and CAD-0. If the interval T_i of the interval sequencer, CAD-I has to last more than one clock period, then transition to the next

292 CHAPTER III

interval T_{i+1} is activated by signal O_0 generated by the operation sequencer CAD-0. This means that CAD-0 passes a loop controlled by one of the control codes p, m_E, or m_I, depending on the type of interval. When CAD-0 establishes O_0 state, CAD-I performs transition to the next instruction interval, whereas CAD-0 is switched to the initial state O_1 ($O_0 \rightarrow O_1$). Therefore, this organization allows that some instruction intervals last the time specified by the meanings of codes p, m_I, and m_E.

5.3.3 Generation of Microcommands

The instruction cycles have intervals of various durations. Thus the MCD has to generate microcommands, MIC, of varied duration. Apart from the time variations, some MICs are produced locally; i.e., not in all UMs of the computer. Indeed, each instruction is interpreted as the same sequence of CAD-I in all UMs of the computer. However, UMs performing I/O functions interpret this instruction as a sequence of I/O microoperations, whereas in processor UMs the same sequence is executed as a sequence of processor microoperations, etc.

In all, in as much as one uses the universal module concept, the same logical circuits (available in each UM) must be selectively activated and deactivated depending on the functional orientation of each UM. Such selectivity in microoperation execution is accomplished with the use of microcommand circuit, MIC, which outputs activate execution of microoperations in each UM.

The MIC is controlled by the following codes (signals)(Fig. III.11):

1. Instruction opcode D
2. The outputs S_j of CAD-I sequencer that perform time variation in MICs
3. Function code q that activates the functionally oriented MIC in each UM
4. Position code i that generates MICs in a single or several UMs specified by their positions
5. Slice code g that allows selective reception by two 8-bit UMs of a single 16-bit data byte, where more- and less-significant UMs receive more- and less-significant portions of this byte, respectively
6. Significance code b distinguishes the MICs executed by the most-, middle-, and least-significant UMs of the $16 \cdot k$-bit processor
7. Bit B_0 achieves selectivity in the generation of memory addresses: if $B_0 = 1$, the counter and base address register (R_A) have the effective memory address sent to pins 19 to 30; otherwise it is not sent

Example 17. If microcommand MIC_1 is activated during interval T_3 (S_3 output) of the instruction with opcode D_{25}, then logical circuit MIC contains the function $\text{MIC}_1 = S_3 \cdot D_{25}$. This means that MIC_1 is generated in all UMs of the computer during interval T_3 of the instruction D_{25}. If MIC_1 is generated in the T_1 and T_4, and T_7 intervals of D_{25}, then the MIC logic contains the following function: $\text{MIC}_1 = (S_1 \vee S_4 \vee S_7)D_{25}$. Local generation of MICs is accomplished with codes q, i, g, B_0, and b. If the same MIC_1 has to be generated only in processor modules, then its logical function specified above is gated with function code q_p; i.e., $\text{MIC}_1 = (S_1 \vee S_4 \vee S_7)D_{25}q_p$. This means that it is activated only in processor UMs during intervals T_1, T_4, and T_7 of instruction cycle D_{25}. ∎

5.4 Memory Management

In a multicomputer system with dynamic architecture the same program may be computed by a sequence of different size computers. Such computations may achieve a performance improvement due to the factors considered in Sec. 3.1.

However, this speed-up may be entirely lost if before each computation on a computer of different size it becomes necessary to change some codes or addresses stored in the instruction fields. This is why programs should satisfy the important property of program universality [10, 11] which requires the following:

1. Instructions should not store codes or constants which change meaning depending on the computer's word size. This is achieved with the modular control organization which provides that all control codes be stored in the MCD of all UMs.
2. None of the addresses contained in instructions should undergo any changes. This condition is satisfied if the same instruction array or data array contains, respectively, the same number of cells when the program is run on different size computers and if the array is specified by the same base address in both cases. Permanent sizes in either case may be maintained if one uses a new memory allocation technique (*PS exchanged*) considered below.
3. All DC-group computers should be equipped with a unique instruction format and unique instruction set.

5.4.1 PS Exchange

In order to maintain the universality of executed programs and to achieve a complete filling of the primary memory, DC-groups use a memory allocation technique known as *parallel-serial exchange* (PS exchange). Its basic features are as follows:

1. In any $C_i(k)$ computer one $16 \cdot k$-bit word is stored in a single parallel cell of k memory elements, all with the same address. (For

instance, for computer $C_1(4)$, a $16 \cdot k$-bit word is stored in all MEs under the same address A_n (Fig. III.13).) Consequently, a data array containing d $16 \cdot k$-bit words will take d parallel cells for its storage in different size computers. When a $16 \cdot k$-bit word is accessed, each of k computer elements generates the same address, resulting in a concurrent fetch of all k 16-bit bytes comprising the data word.

2. Since the minimal size computer is equivalent to one CE, the instruction size has been made coincident with that of one CE. The instruction sequences are stored in consecutive cells of one ME, and complex programs may be stored in several MEs. The instruction fetched from one ME would then be sent concurrently to all modules of the $C_i(k)$ computer. Therefore, the same program occupies the same number of cells in any size computer.

However, for dynamic architectures such an organization of instruction storage requires a considerable number of pins to be assigned in each LSI module. In order to provide separation of concurrent instruction streams computed by adjacent computers, each ME should be connected with each LSI module through a dedicated instruction bus (16 bits). Since for the $16 \cdot n$ bit computer, instructions may be stored in every ME, each LSI module should have dedicated connections with all n memory elements. It then follows that this requires $16 \cdot n$ dedicated pins in every LSI module, which becomes non-implementable even for small n due to current pin count restrictions on LSI modules.

To reduce the number of pins per module required for instruction reception, dedicated busses should be replaced by a shared-bus containing $(n-1)$ connecting elements, MSE (Fig. III.14). This reduces the required number of pins from $16 \cdot n$ to 16 pins for a module, used for both instruction and data byte reception. As was noted above, if one MSE is in no-transfer mode, it separates two concurrent instruction streams computed by two adjacent computers. For a $C_i(k)$ computer, all its MSEs assume either:

1. A transfer mode (left or right) during an instruction fetch, which connects the memory element that stores the instruction to all LSI modules of the $C_i(k)$ computer, or
2. A no-transfer mode during a $16 \cdot k$-bit word fetch, which connects each memory element with the PE and/or GE belonging to just the same computer element, effecting a parallel exchange of $16 \cdot k$-bit words.

Thus, the introduction of connecting elements into the memory–processor bus results in a sharp reduction of the number of pins required and preserves the permanency of array sizes for instructions and data. Consider now the issue of address generation both for instructions and data words.

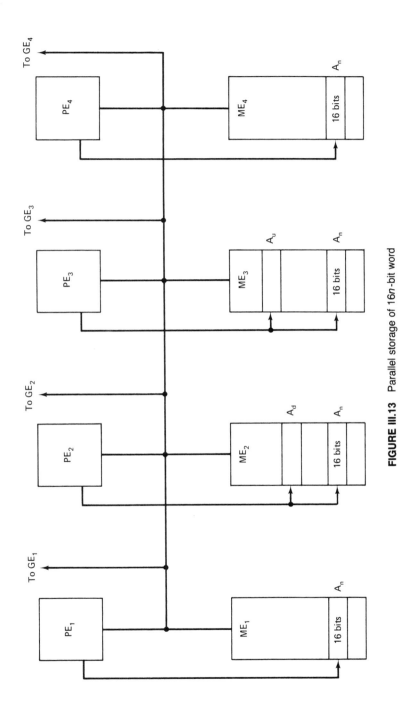

FIGURE III.13 Parallel storage of 16n-bit word

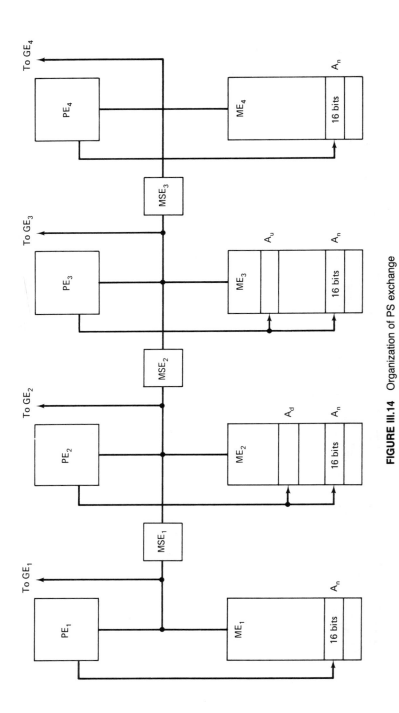

FIGURE III.14 Organization of PS exchange

5.4.2 Addressing

Let us determine which processor element should generate instruction addresses in a DC-group computer. First, assume that addresses are generated by the most significant PE. Since instructions may be stored in any memory element, the most significant PE should be connected through the dedicated address bus not only to the ME contained in the same CE but also to any other ME to the left of it storing less significant bytes of a $16 \cdot k$-bit word. Since for each PE there exists at least one DC-group computer in which this PE is the most significant, and the DC-group contains n PEs, the least-significant memory element, ME_n, should be connected through n dedicated address busses with n PEs, requiring $A \cdot n$ pins where A is the bit size of the address; ME_{n-1} should be connected with $n - 1$ PEs, requiring $A \cdot (n - 1)$ pins; . . . ; ME_1 should be connected with PE_1, requiring A pins. Thus, such a solution is not adequate since it requires a large number of pins per memory element. (For $n = 4$, $A = 32$, and ME_4 needs 128 pins.)

To reduce the number of pins required for an address bus, assume that for each ME an address is generated only in the PE contained in the same CE. This requires that during instruction fetches addresses be produced only in the CE which stores the current program segment, whereas during $16 \cdot k$-bit word fetches, the same address be generated concurrently in all k computer elements, each fetching a byte stored in its own ME. This allows a sharp reduction of pins required for each memory element because any ME is now connected through the address bus only with its respective PE and requires A pins for address transfer.

Consider now generation of an effective storage address E with an addressing procedure slightly modified from the IBM 360/370 [29, 30]; i.e., E is formed as a concatenation of three components, B, X, and D, where the base B (6 bits) and the index X (6 bits) are stored in two 8-bit registers, R_A, respectively, of the most- and least-significant modules of the same PE; displacement D (12 bits) is stored in the address counter made from the two C_0 counters belonging to the two LSI modules of a PE. The effective 24-bit address E is obtained as an output of logic circuits L_1 of the two modules and is sent through pins 19 to 30 to the respective ME (Fig. II.11).

Every DC-group computer uses five instruction formats. These formats are made compatible with basic instruction formats used by the IBM 360/370 in order to take advantage of the extensive software developed for these machines. Also, a 16-bit instruction is stored in one cell of one ME, a 32-bit instruction occupies two consecutive cells (where the second cell stores displacement D and increment for B), and a 48-bit instruction occupies three consecutive cells (where the second and third cells store respectively two displacements and increments). Therefore, this addressing scheme allows attachment of primary memories sizes equivalent to those used for the IBM 360/370. Consider now the organization of program storage in several memory elements.

Since the program may be stored in several memory elements, the problem that must be resolved is how to organize the sequencing of consecutive program segments stored in different memory elements. Namely, if a current task is stored in memory ME_i, and the next task is stored in memory element ME_j, one has to organize an instruction fetch from ME_j upon completion of the task stored in ME_i. These functions are performed with a special memory jump instruction "jump instruction $ME_i \rightarrow ME_j(A_d)$" considered below.

5.4.3 Instruction "Jump $ME_i \rightarrow ME_j(A_d)$"

This instruction concludes the task stored in ME_i and initiates fetches of the next task from ME_j, where A_d is its base address. It performs the following actions:

1. Since only one PE generates instruction addresses for the respective ME, PE_i stops and PE_j initiates generation of instruction addresses.
2. PE_j receives the base address A_d of a program stored in ME_j.
3. Connecting elements within the $C_i(k)$ computer are switched to new transfer modes, since program instructions will henceforth be coming from a different memory element ME_j but must still be distributed to all modules in the computer.

The "jump $ME_i \rightarrow ME_j(A_d)$" occupies two consecutive 16-bit cells, where the first cell stores position code $i^* = j$ of ME_j, and the second cell stores the base address A_d. Selection of the PE_j processor element which has to generate instruction addresses is made by comparing the code i^* stored in the instruction with the position code i stored in the MCD of the processor elements. If $i^* = i$, the address A_d is written to the counter of the respective PE_i. Concurrently, flip-flop B_0 (Fig. III.11) is set and enables L_1 to place the effective instruction address on the address bus connecting ME_i and PE_i. For the remaining PEs, $i^* \neq i$ and flip-flop B_0 is reset ($O \rightarrow B_0$), disabling the feeding of the addresses from these PEs. Since the connecting elements receive program instructions and possess the same MCD they may establish new transfer modes by comparing their own position codes i with the position code i^* they receive through the instruction. If $i < i^*$, the connecting element MSE_i is altered to the left-transfer mode. If $i \geq i^*$, MSE_i is altered to the right-transfer mode.

Example 18. Suppose that in computer $C_1(3)$, the program segment stored in ME_2 in addresses 1621 through 1867 is succeeded by a program segment stored in ME_3 and specified by the base address 1513 (Fig III.15). To perform this transition the program segment in ME_2 is terminated by the instruction "jump $ME_2 \rightarrow ME_3(1513)$," in cells 1867 and 1868 of ME_2. The

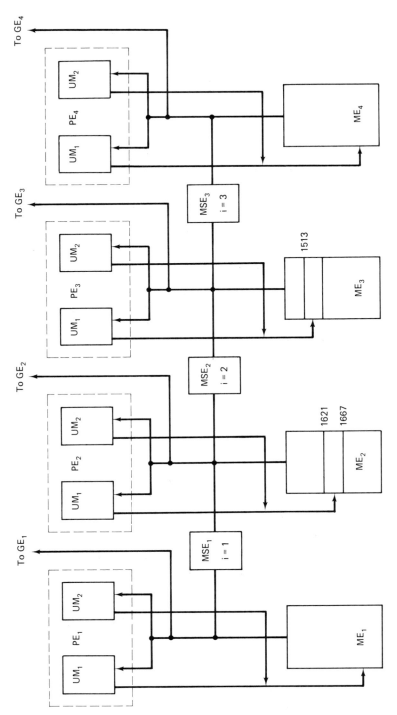

FIGURE III.15 Jump $ME_2 \rightarrow ME_3$ (1513)

instruction stores $i* = 3$ and base address 1513, and its cycle is composed of the three intervals T_1, T_2, and T_3. During T_1, the control portion of the instruction is fetched to all LSI modules of the $C_1(3)$ computer. During T_2, address 1513 is sent to the counters of all processor elements. During T_3, $i* = 3$ is compared with all the position codes in the processor elements and connecting elements. Since $C_1(3)$ includes PE_1, PE_2, and PE_3, $i* = i = 3$ in PE_3 and $1 \rightarrow B_0$ in it alone. For PE_1 and PE_2, $i* \neq i$ and $O \rightarrow B_0$. Therefore, only in PE_3 is address 1513 enabled to the address bus. As for connecting elements, $C_1(3)$ integrates MSE_1 and MSE_2. For MSE_1, storing $i = 1$, $i < i*$, so it is switched to the left-transfer mode. For MSE_2, storing $i = 2$, $i < i*$, so it is switched to the left transfer mode as well. ∎

Using this instruction one may distribute consecutive program segments arbitrarily among memory elements and thus fill all the empty spaces created in the primary memory because of variable word sizes with programs. This allows high utilization of the primary memory.

5.5 Reconfigurable Paths for the Processor Signals

Consider the problems encountered in the software formation of variable size processors. A DC-group contains n PEs and each PE may be a component of up to n different size processors ranging from 16 to $16 \cdot n$ bits. Furthermore, a PE may function as the least-significant, middle, or most-significant element in a processor.

This implies that the same PE must be provided with connections which establish correct paths for the overflow signal if this PE is the most-significant element in a processor; the carry-in, carry-out signals if it is a middle element in a processor; and the end-around carry if it is the least-significant element in a processor. Similarly, the correct paths for equality signals via software have to be established. This means that when two $16 \cdot k$-bit words are compared, the resulting equality signal should be fed only into each LSI module of the given computer and prevented from going into other concurrent computers. In addition, since each PE is sliced into several modules,† one has to establish the correct propagation of carries, overflows, and equalities within each PE. Therefore, to form a processor one has to establish reconfigurable paths for its adder and for equality signals.

There are two approaches to establishing reconfigurable paths for these signals: (1) Carries and overflows have to propagate through the memory–processor interconnection network that connects the processor resource and the memory resource [14], and (2) carries and overflows are propagated through a dedicated reconfigurable processor bus [9, 11, 13]. Since the memory–processor interconnection network introduces significant reconfiguration delays into carry-propagation times, this section will study different reconfigurable Processor busses.

†This section discusses a dynamic organization of bit-slicing, whereby the boundaries of each computer—expressed in terms of its most- and least-significant PE—may change during execution. A static organization of bit-slicing where the boundaries of the computer are fixed is implemented in bit-sliced computers. (Ed.)

5.5.1 Reconfigurable Busses for the End-Around Carry

We introduce two alternative busses: A minimal delay bus that minimizes delays of the reconfiguration logic and a minimal interconnection bus that minimizes the number of between-module interconnections.

As will be shown below, the number of pins in each UM assigned for the processor bus will depend on the type of bus. For the minimal delay processor bus, it will take $n + 1$ pins in each UM, where n is the number of CEs in the resource. For the minimal interconnection bus it will take 4 pins in the UM regardless of the number of CEs. To simplify consideration, pins assigned for the processor bus will be numbered as 1, 2, ...; these pins must be added to these of UM shown in Figs. III.11 and III.12.

5.5.1.1 Minimal delay processor bus

Assume that in a $k \cdot h$-bit processor containing k processor modules, each left processor module PE_i is more significant than its immediate neighbor on the right, PE_{i+1}. Otherwise, if no ordering among processor modules in accordance with their significance is established, the number of combinations for end-around carry paths to be implemented equals $n!$ leading to a great complexity in the reconfigurable bus. If the processor resource contains n processor modules, $PE_1, PE_2, \ldots, PE_n$, then the leftmost processor module PE_1 may be the most significant for n different computers: $C_1(1)$ which processes h-bit words; $C_1(2)$ which processes $2h$-bit words, ..., $C_1(n)$ which processes $n \cdot h$-bit words.

Example 19. Figure III.16 contains a processor resource assembled from four processor modules $PE_1, \ldots, PE_4$. The PE_1 processor module may be the most significant for computer $C_1(1)$, including only PE_1, or for $C_1(2)$ including PE_1 and PE_2, or $C_1(3)$, including PE_1, PE_2, PE_3, etc. ∎

It then follows that PE_1 may initiate n different overflow paths that have to be implemented through n dedicated connections. Each overflow fans out from PE_1 to a respective least-significant PE. In order not to introduce separate external logic which selects the path for overflow depending on how many processor modules are in $k \cdot h -$ bit processor, assume that PE_1 contains logic L_1 (Fig. III.17) which connects the most-significant bit of the adder to n different pins $O_1, O_2, \ldots, O_n$. Selection of the pin O_i is determined by the position of the least-significant processor module PE_i, so that O_1 receives overflow in $C_1(1)$ containing PE_1 only, O_2 receives overflow in $C_1(2)$ in which PE_2 is the least significant, O_3 receives overflow in $C_1(3)$ with least-significant PE_3, etc. As a result, each least significant module PE_i may receive end-around carry through the same pin O_p (Fig. III.16).

A similar distribution of overflows occurs in any other processor module

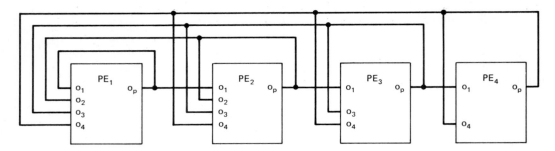

FIGURE III.16 Minimal delay processor bus

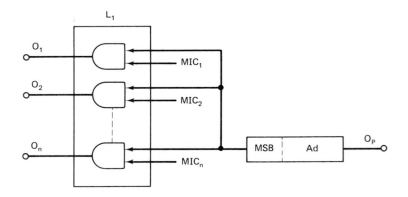

FIGURE III.17 Connection of the adder with pins $O_1, O_2, \ldots, O_n$

PE_i when it functions as the most-significant module in computer $C_i(k)$. For instance, PE_2 may serve as the most-significant module for computer $C_2(1)$ including only PE_2, and overflow is sent to pin O_2; for $C_2(2)$ including PE_2 and PE_3, overflow is sent to pin O_3; for $C_2(3)$ including PE_2, PE_3, and PE_4, overflow is sent to pin O_4, etc. Therefore, logic L_1 has to be present in all processor modules of the resource. Selective activation of the L_1 logic is performed by the microcommands $MIC_1, MIC_2, \ldots, MIC_n$ which respectively enable the overflow produced by the local adder to one of the pins $O_1, O_2, \ldots, O_n$.

Each processor module activates only one microcommand. Selectivity in this activation is determined by the following control codes stored in each PE:

1. k-code showing how many processor modules, PE, are contained in the $C_i(k)$ computer. For instance, $C_2(3)$ contains three processor modules, PE_2, PE_3, PE_4, giving code $k = 3$, and signal $k_3 = 1$.

2. i-code showing the position of the processor module PE_i among all other processor modules of the resource. For instance, for PE_1, $i = 1$ giving signal $i_1 = 1$; for PE_2, $i = 2$ and signal $i_2 = 1$; etc.

3. b-code marking the most-significant PE of the $C_i(k)$ computer. The remaining PEs store $b = 0$. For instance, for computer $C_1(4)$, the most-significant PE_1 stores $b = 1$, while the remaining PE_2, PE_3, and PE_4 have $b = 0$; for $C_2(3)$ computer PE_2 stores $b = 1$, while PE_3 and PE_4 store $b = 0$; etc.

Selective activation of microcommands may be implemented if they are realized by the following logical functions:

$$\begin{aligned} MIC_1 &= k_1 \cdot i_1 v \bar{b} \\ MIC_2 &= k_2 \cdot i_1 v k_1 \cdot i_2 \\ MIC_3 &= k_3 \cdot i_1 v k_2 \cdot i_2 v k_1 \cdot i_3 \\ MIC_4 &= k_4 \cdot i_1 v k_3 \cdot i_2 v k_2 \cdot i_3 v k_1 \cdot i_4 \\ & \quad \cdots \\ MIC_n &= k_n \cdot i_1 v k_{n-1} \cdot i_2 v \ldots v k_1 \cdot i_n \end{aligned}$$

Software formation of new end-around carries is therefore reduced to writing two new codes k and b in the processor modules that are involved in forming a new computer $C_i(k)$. This is performed at the moment of switching to a new architectural state. Thus, software reconfiguration of the path for the end-around carry requires only one level of logic L_1 in each processor module.

Example 20. Let the entire processor resource $PE_1, \ldots, PE_4$ form a $C_1(4)$ computer (Fig. III.18). Since $C_1(4)$ contains four PEs, $k = 4$; PE_1 has position code $i = 1$, PE_2 has $i = 2$; PE_3 has $i = 3$. Since PE_1 is the most significant module, it stores $b = 1$, and the remaining processor modules have $b = 0$. The following microcommands are activated in each PE: PE_1 activates MIC_4 since $k_4 \cdot i_1 = 1$. The remaining microcommands are deactivated. Therefore, the overflow signal produced by the local adder is fanned out to pin O_4. Since pin O_4 is connected with pin O_p of PE_4, the adder overflow is routed as an end-around carry to the least-significant bit of the adder in PE_4 and connected to pin O_p. For processor modules PE_2, PE_3, and PE_4, MIC_1 is activated, since $\bar{b} = 1$ and the carryouts of the local adders fan-out to pin O_1. All O_1 pins of these modules are connected with O_p pins of the next more

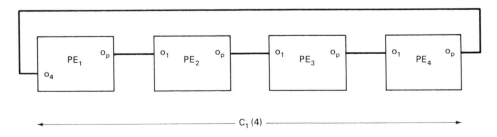

FIGURE III.18 End-around carry path for $C_1(4)$ computer

significant modules. Thus for $C_1(4)$ a correct end-around-carry path is formed.

Let the resource PE_1 through PE_4 form processors for computers $C_1(1)$ and $C_2(3)$ (Fig. III.19). For computer $C_1(1)$ containing PE_1 only, $k = 1$ and $b = 1$ are written to PE_1. Since for PE_1, $k_1 \cdot i_1 = 1$, it activates MIC_1. Since in PE_1, O_1 is connected to pin O_p, the correct end-around-carry path is formed. For $C_2(3)$, $k = 3$ is stored in PE_2, PE_3, and PE_4. For PE_2, $i = 2$ and $b = 1$; for PE_3, $i = 3$ and $b = 0$; for PE_4, $i = 4$ and $b = 0$. Thus, in PE_2, $i_2 \cdot k_3 = 1$ activates MIC_4; in PE_3 and PE_4, $\overline{b} = 1$ activates MIC_1. The resulting path of the end-around carry is shown in Fig. III.19. ∎

5.5.1.2 Minimal interconnection processor bus

This technique provides that for any number, n, of processor modules, a reconfigurable end-around carry bus can be formed with the use of four pins per module. However, the delay $\Delta RD(k)$ for $k \cdot h$-bit processor induced by this technique is twice as great as that for the minimal delay processor bus.

The minimal interconnection processor bus provides that two adjacent PEs be connected only by two connections. These establish carry and overflow paths, respectively. Thus to implement this bus takes four pins per PE and reconfiguration logic inside each PE that selectively enables local carry-out to one of the pins. The connection between pins 2 and 1 establishes the path for the carry from one PE to the next more significant PE (Fig. III.20), and the connection from pin 3 to pin 4 establishes the path for overflow from one PE to the next less significant PE.

Inside each PE the selective transfer of carries and overflows to the output pins is performed by the reconfiguration logic L_1 as shown in Fig. III.21. Each local adder carry-out is always sent to pin 2. For the processor containing only one PE (i.e., $k = 1$), it is enabled with MIC_1 to the local adder input, i.e., MIC_1 activates the path for the end-around carry. For a $k \cdot h$-bit processor containing more than one PE, the most-significant PE activates MIC_2, which routes the local carry-out to pin 3, since for this case

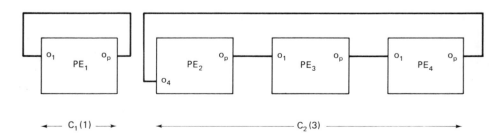

FIGURE III.19 End-around carry paths for computers $C_1(1)$ and $C_2(3)$

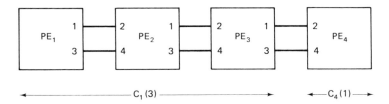

FIGURE III.20 Minimal interconnection processor bus

this adder's carry-out is actually the overflow of a $k \cdot h$-bit adder. If a PE is the least-significant module in the processor, the LSB of its adder receives the end-around carry routed from pin 4 by MIC_3. Every PE that is neither the most nor the least significant should route the overflow of the $k \cdot h$-bit adder from pin 4 to pin 3 of the PE with the use of MIC_4, and all but the least-significant one should enable, with MIC_5, the external carry-out sent to pin 1 by the next less significant PE.

Software formation of the correct path for carry-overflow is therefore reduced to selective generation of MIC_1 to MIC_5 in each PE. Activation and deactivation of these microcommands depends only on two-bit variable control code b that marks the most-significant, least-significant, and middle PE. Assume that $b = 10$, 01, and 00 recognize the most-significant, least-significant, and middle PE's, and that b_{10}, b_{01}, and b_{00} are the respective signals that decode these codes. For the $C_i(1)$ computer in which its only PE is both most and least significant, $b = 11$ and $b_{11} = 1$. Selective activation of MIC_1 through MIC_5 is performed by the following functions: $MIC_1 = b_{11}$; $MIC_2 = b_{10}$; $MIC_3 = b_{01}$, $MIC_4 = b_{00}$; and $MIC_5 = b_{00} \lor b_{10}$.

Example 21. Let the processor resource $PE_1, \ldots, PE_4$ (Fig. III.20) form computers $C_1(3)$ and $C_4(1)$. Since $C_1(3)$ contains three PEs, $k = 3$. PE_1 is the most significant and it stores $b = 10$; PE_2 stores $b = 00$ since it is a middle element; PE_3 stores $b = 01$, since it is the least significant. The following microcommands are activated in each PE: PE_1 activates MIC_2 and MIC_5, since it satisfies $b_{10} = 1$. Therefore, the overflow of the $3 \cdot h$-bit adder in the $C_1(3)$ computer is fanned out to pin 3. Likewise the external carry-out generated in the less significant PE_2 is sent to the adder input in PE_1 (Fig. III.21). For processor module PE_2, since $b_{00} = 1$, MIC_4 and MIC_5 are activated. Thus, the $3h$-bit adder overflow is passed from pin 4 to pin 3 by MIC_4, and the external carry-out generated in PE_3 is sent to the adder input in PE_2 by MIC_5. For PE_3, b_{01} activates MIC_3. It routes the $3h$-bit adder overflow from pin 4 to its LSB. Thus in $C_1(3)$ a correct end-around-carry path has been formed. For the $C_4(1)$ computer containing PE_4 only, PE_4 stores $b = 11$. Therefore it activates only $MIC_1 = b_{11}$. This forms the local end-around-carry path. None of the other MICs are activated. ∎

306   CHAPTER III

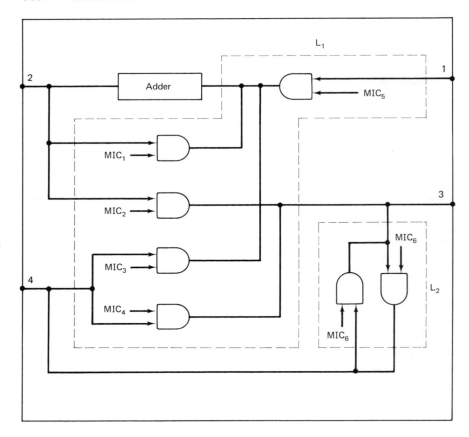

FIGURE III.21 Reconfiguration logic for the minimal interconnection processor bus

5.5.2 Reconfigurable Bus for the Equality Signal

The equality signal generated in a $k \cdot h$-bit computer during comparison of two $k \cdot h$-bit operands participates in several conditional branch operations.† Since each PE of the computer may compare only two h-bit bytes, during comparison k equality signals, one generated by each PE, may be produced. They have to be received by the most significant PE which then selects true or false program sequences dependent on the result of a conditional test. This takes $(k-1)$ pins in this PE. Since for a dynamic architecture, each PE may be the most significant in some $k \cdot h$-bit computer, and k ranges from $k = 1$ to $k = n$, each PE has to have $(n - 1)$ pins assigned for equality signals. As was shown in Sec. 5.3., this number of pins may be sharply reduced if instead of an equality signal β, one transfers the inequality, $\bar{\beta}$, because a correct equality path in a $C_i(k)$ computer may then be formed by

†The functions of equality signal during conditional branches are discussed in Chap. I, Sec. 6.2. (Ed.)

assigning only one pin per PE.‡ Indeed, if in a $C_i(k)$ computer all these pins are connected, then the presence of a $\overline{\beta}$ signal in this connection means that the two $k \cdot h$-bit words are not equal, otherwise they are equal. To prevent unwanted $\overline{\beta}$ transfers from one computer to another, two PE neighbors send their $\overline{\beta}$ signals through the same pin 5 to pins 3 and 4 of the connecting element MSE between them (Fig. III.22). In this MSE, pins 3 and 4 are tied via logic L_2 that is activated by MIC_6 (Fig. III.21). The MIC_6 is generated only if two PE neighbors belong to the same computer. Otherwise it is deactivated, disconnecting pins 3 and 4.

Consider now how activation and deactivation of MIC_6 may be accomplished. To reduce the number of connecting elements in the system, it is reasonable to establish the $\overline{\beta}$-path through connecting elements contained in the memory–processor bus. Then MIC_6 will be activated in all MSE connecting elements integrated by computer $C_i(k)$, assuming a transfer mode (right or left) when there is no data fetch. Thus each MSE which separates two computers will have a no-transfer mode, which deactivates MIC_6. This will be distinguished by a special flip-flop B_0 in each MSE which distinguishes a transfer mode ($B_0 = 0$) from the no-transfer mode ($B_0 = 1$). Note, the MIC_6 function will have as one of its arguments, the functional orientation code q_{MSE}, which will activate it only in MSE elements. That is, $MIC_6 = q_{MSE} \cdot \overline{B}_0$.

Since we are using a universal module concept, the same flip-flop B_0

‡One may assign a single pin per module to transfer the β signal, if all β-pins are connected in a form of a hard-wired AND. (Ed.)

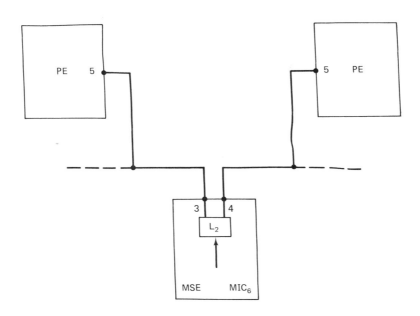

FIGURE III.22 Reconfigurable path for equality signal

should perform different functions in connecting elements, MSE, and in processor elements, PE. In connecting elements, MSE, flip-flop B_0 distinguishes when MSE is integrated into a $C_i(k)$ computer ($B_0 = 0$) and when it separates two computers ($B_0 = 1$). In processor elements, PE, the same B_0 distinguishes the mode of address generation from PE to local ME: if $B_0 = 1$, PE transfers its address to logic L_1, otherwise it does not (Fig. III.11). Both types of actions of B_0 are recognized by the functional orientation code, q.

5.6 Organization of Architectural Reconfigurations in System

Dynamic reconfiguration from one state to another is supervised by the monitor system. This section introduces several important organizations which will allow a DC-group to perform an architectural switch from one state to another.

The following problems have to be solved:

1. *Task synchronization.* For each state-successor, at least one new computer is formed from the resources sacrificed by other computer(s) which had been functioning in the predecessor state. Before giving up their resources, all such computers must finish their executions and inform the V monitor of this occasion. It then follows that the V monitor has to establish the moment when the required resource is free and the DC-group is ready for the next architectural transition. Thus the need arises to organize task synchronization.

2. *Priority analysis.* DC-group computers may compute independent programs which make conflicting calls for the next transition. The V monitor must resolve these conflicts on the basis of priorities assigned to the programs. Therefore, one must organize a priority analysis during which the V monitor allows or denies an architectural switch requested by a program.

3. *Storage of variable control codes.* For different architectural states, one has to write the new meanings of the variable control codes (k, p, m_E, and b) to all UMs belonging to a newly formed computer. These codes may acquire different meanings for UMs of the same computer as well as for concurrent computers existing in the same state. A DC-group may assume 2^{n-1} architectural states, and all the variable control codes for each state form an array which has to be stored in a memory. To speed up an architectural switch, the storage of this array should be organized in a way that would minimize the time of searching for any given item in the array.

4. *Organization of the architectural switch to a new state.* This includes finding the priority of a requesting program, testing for the availability of the requested equipment, accessing the necessary variable control codes, writing them only to UMs of new computers, and

establishing the appropriate transfer modes in the connecting elements.

Let us consider how these problems may be solved.

5.6.1 Task Synchronization and Priority Analysis

If a program is computed by a sequence of different size computers, it must be partitioned into tasks where each task is computed in a $C_i(k)$ computer. Each task is concluded by the special "Stop-N_d" instruction which sends the V monitor an n-bit code that shows the computer elements this computer consists of. Upon receipt of this code, the V monitor updates its own n-bit idle equipment code showing what computer elements of the DC-group are free. When a program makes a request for architectural transition it sends a code of requested equipment to the V monitor showing which computer elements should participate in forming new computers of the next state. If the requested resource is free and the program priority is high, the V monitor allows this architectural transition.

Example 22. Let a DC-group be in the state N_6 in which computers $C_1(2)$, $C_3(1)$, and $C_4(2)$ are function independently (Table III.5). Suppose that $C_1(2)$ finishes its task first and sends code 11000 to the V monitor showing that CE_1 and CE_2 that make up this computer are free. The second computer which finishes its task is $C_4(2)$ and it sends the code 00011 to the V monitor, showing that CE_4 and CE_5 are now free. Suppose that upon completion of the "Stop-N_6" instruction, $C_4(2)$ requests an architectural transition into state N_4. This means that in the program for $C_4(2)$, the synchronization instruction stop N_6 is immediately followed by the architectural switch instruction "$N_6 \rightarrow N_4$" requesting a DC-group transition into state N_4. This instruction sends the code 00111 to the V monitor which shows that in N_4, the DC-group has to form only one new computer, $C_3(3)$, requiring the resources of CE_3, CE_4, and CE_5. Since at the moment this code is received, only $C_1(2)$ and $C_4(2)$ have finished their executions, the V monitor stores a 11011 showing that only CE_3 is not yet free. By performing logical multiplication of 11011 and 00111, the V monitor establishes that transition into state N_4 is impossible since $11011 \wedge 00111 \neq 00111$, because $C_3(1)$ is still executing. When $C_3(1)$ finishes its task, its stop-N instruction sends the code 00100. Upon receiving this code, the V monitor allows the transition $N_6 \rightarrow N_4$, provided that this program has the highest priority in state N_6. ∎

5.6.2 Priority Analysis

Priority analysis is performed by a special control program stored in the V monitor. When a DC-group computes several independent programs, they may have conflicting calls on the next DC-group transition; i.e., two programs may ask for transitions to the N_k and N_e states respectively where $N_k \neq N_e$. Possible conflicting calls are resolved by assigning to each task of any given program a priority code. This assignment is performed as follows: Every program is specified by the program user code, NP. All tasks of this program are also assigned the same NP code. For every NP program, a page of memory in M (V) is assigned. This page contains 2^{n-1} cells, so that each DC-group state is provided with one cell. The page address is the NP code, while the relative address within the page is the N_d state code ($d = 0, \ldots, 2^{n-1} - 1$). A 16-bit cell of one NP page is called a basic cell. Each basic cell (specified by the NPN_d address) is partitioned into two zones: I (bits 1 to 10) and II (bits 11 to 16). Each bit j of zone I stores the priority of a task to perform reconfiguration from state N_d during the jth iteration of the N_d state, so that bit 1 stores the priority of the task computed in N_d when it is established first. Bit 2 stores a task priority for the second iteration of state N_d, etc.

Zone II stores two codes: x (4 bits) and z (2 bits), where the x code specifies location of a priority bit in zone I, and the z code is used to compute the address of an additional priority array, which is used if N_d is iterated more than 10 times (size of zone I). That is, if $z = 0$, a basic cell stores a current priority bit j. If $z = 1, 2, 3$, the current priority bit is in an additional priority array including three cells for each state N_d. Bit j of zone I is 1 ($j = 1$) if the respective task is denied execution of the $N_d \rightarrow N_f$ transition. Otherwise, the $N_d \rightarrow N_f$ transition is allowed. If N_d is iterated more than 10 times, priorities of the tasks for each additional iteration are stored in an additional cell specified by the address $Y + 3 \cdot 2^{n-1} \cdot NP + 3 \cdot N_d + (z - 1)$, where Y is the initial address of the additional array and NP and N_d are the decimal meanings of the NP and N_d codes. Since a two-bit z code allows three additional cells for every N_d (48 priority bits), one N_d state can be iterated in a single program up to 58 times (10 + 48).

Example 23. Let a DC-group with $n = 4$ execute five independent programs assigned codes $NP_0 = 000, \ldots, NP_4 = 100$. Suppose that the program with $NP_3 = 011$ is in the 22nd iteration of state N_2 which is to be succeeded by N_1 state if the task priority allows it. The basic array in the M(V) memory occupies $2^{n-1} \cdot 5 = 8 \cdot 5 = 40$ cells. Let the initial address of the additional array be: $Y = 50$. For the 22nd iteration, $z = 1$, $x = 12$, because after each iteration x is incremented by one and the current x for additional cells is found as $\#(i) = 10 + (z - 1) \cdot 16 + x$, where $\#(i)$ is the current number of iterations. For $\#(i) = 22$, $22 = 10 + (1 - 1) \cdot 16 + x = 10 + 12$. The V monitor specifies the address of additional cell as $50 + 3(8 \cdot 3 + 2) + 1 - 1 = 128$ and bit 12 of this cell specifies priority of this iteration. If this bit is 1, $N_2 \rightarrow N_1$ is halted; if it is 0, $N_2 \rightarrow N_1$ is allowed. ∎

5.6.3 Storage of Variable Control Codes

To form a new DC-group computer one has to write four variable control codes to all its CEs: k (computer size code), p (processor code), b (significance code), and m_E (data fetch code). As was shown above, for all UM contained in one CE the same values of the variable control codes ought to be written. To store them requires N bits, where $N = 2 \log_2 n + 2 + \log_2 S$ where $\log_2 n$ is the size of the k code; the p code also takes $\log_2 n$ bits since $p < k$ if carry look-ahead circuits are used and $p = k$ otherwise; two bits are required to store b; and St_0 is the speed of the slowest memory element. If $n \le 16$ and $S \le 16$, $N = 8 + 2 + 4 = 14$ bits; i.e., storage of all variable control codes for one CE takes one 16-bit cell.

Since a DC-group has n CEs, the control codes for an architectural state may be stored in a single $16 \cdot n$ bit cell composed of n 16-bit memory bytes where each byte stores control codes for one CE. Because a DC-group may assume 2^{n-1} states, to store the control codes for all architectural states requires an array of 2^{n-1} cells of the size $16n$-bit each. This array may be mapped either into I/O memory or primary memory. If it is stored in the initial cells of I/O memory M(GE), then the effective address of one $16 \cdot n$ bit cell may be made the code of the architectural state.

Example 24. Table III.6 lists all variable control codes for a DC-group having $n = 5$. The speeds of its five MEs are: $13t_0$ for ME_1, $12t_0$ for ME_2,

TABLE III.6

			$M(GE_1)$											
		k			p			b				m_E		
		1	2	3	4	5	6	7	8	9	10	11	12	13	14
N_0	$C_1(5)$	0	1	0	1	0	1	0	1	1	0	1	1	0	1
N_1	$C_1(4), C_5(1)$	0	1	0	0	0	1	0	0	1	0	1	1	0	1
N_2	$C_1(3), C_4(2)$	0	0	1	1	0	1	0	0	1	0	1	1	0	1
N_3	$C_1(3), C_4(1), C_5(1)$	0	0	1	1	0	1	0	0	1	0	1	1	0	1
N_4	$C_1(2), C_3(3)$	0	0	1	0	0	0	1	1	1	0	1	1	0	1
N_5	$C_1(2), C_3(2), C_5(1)$	0	0	1	0	0	0	1	1	1	0	1	1	0	1
N_6	$C_1(2), C_3(1), C_4(2)$	0	0	1	0	0	0	1	1	1	0	1	1	0	1
N_7	$C_1(2), C_3(1), C_4(1), C_5(1)$	0	0	1	0	0	0	1	1	1	0	1	1	0	1
N_8	$C_1(1), C_2(4)$	0	0	0	1	0	0	1	0	1	1	1	1	0	1
N_9	$C_1(1), C_2(3), C_3(1)$	0	0	0	1	0	0	1	0	1	1	1	1	0	1
N_{10}	$C_1(1), C_2(2), C_4(2)$	0	0	0	1	0	0	1	0	1	1	1	1	0	1
N_{11}	$C_1(1), C_2(2), C_4(1), C_5(1)$	0	0	0	1	0	0	1	0	1	1	1	1	0	1
N_{12}	$C_1(1), C_2(1), C_3(3)$	0	0	0	1	0	0	1	0	1	1	1	1	0	1
N_{13}	$C_1(1), C_2(1), C_3(2), C_5(1)$	0	0	0	1	0	0	1	0	1	1	1	1	0	1
N_{14}	$C_1(1), C_2(1), C_3(1), C_4(2)$	0	0	0	1	0	0	1	0	1	1	1	1	0	1
N_{15}	$C_1(1), C_2(1), C_3(1), C_4(1), C_5(1)$	0	0	0	1	0	0	1	0	1	1	1	1	0	1

TABLE III.6

		\multicolumn{14}{c}{M(GE₂)}

		k				p		b				m_E			
		1	2	3	4	5	6	7	8	9	10	11	12	13	14
N_0	$C_1(5)$	0	1	0	1	0	1	0	1	0	0	1	1	0	1
N_1	$C_1(4), C_5(1)$	0	1	0	0	0	1	0	0	0	0	1	1	0	1
N_2	$C_1(3), C_4(2)$	0	0	1	1	0	1	0	0	0	0	1	1	0	1
N_3	$C_1(3), C_4(1), C_5(1)$	0	0	1	1	0	1	0	0	0	0	1	1	0	1
N_4	$C_1(2), C_3(3)$	0	0	1	0	0	0	1	1	0	1	1	1	0	1
N_5	$C_1(2), C_3(2), C_5(1)$	0	0	1	0	0	0	1	1	0	1	1	1	0	1
N_6	$C_1(2), C_3(1), C_4(2)$	0	0	1	0	0	0	1	1	0	1	1	1	0	1
N_7	$C_1(2), C_3(1), C_4(1), C_5(1)$	0	0	1	0	0	0	1	1	0	1	1	1	0	1
N_8	$C_1(1), C_2(4)$	0	1	0	0	0	1	0	0	1	0	1	1	0	0
N_9	$C_1(1), C_2(3), C_3(1)$	0	0	1	1	0	1	0	0	1	0	1	1	0	0
N_{10}	$C_1(1), C_2(2), C_4(2)$	0	0	1	0	0	0	1	1	1	0	1	1	0	0
N_{11}	$C_1(1), C_2(2), C_4(1), C_5(1)$	0	0	1	0	0	0	1	1	1	0	1	1	0	0
N_{12}	$C_1(1), C_2(1), C_3(3)$	0	0	0	1	0	0	1	0	1	1	1	1	0	0
N_{13}	$C_1(1), C_2(1), C_3(2), C_5(1)$	0	0	0	1	0	0	1	0	1	1	1	1	0	0
N_{14}	$C_1(1), C_2(1), C_3(1), C_4(2)$	0	0	0	1	0	0	1	0	1	1	1	1	0	0
N_{15}	$C_1(1), C_2(1), C_3(1), C_4(1), C_5(1)$	0	0	0	1	0	0	1	0	1	1	1	1	0	0

TABLE III.6

		k				p		b				m_E			
		1	2	3	4	5	6	7	8	9	10	11	12	13	14
N_0	$C_1(5)$	0	1	0	1	0	1	0	1	0	0	1	1	0	1
N_1	$C_1(4), C_5(1)$	0	1	0	0	0	1	0	0	0	0	1	1	0	1
N_2	$C_1(3), C_4(2)$	0	0	1	1	0	1	0	0	0	1	1	1	0	1
N_3	$C_1(3), C_4(1), C_5(1)$	0	0	1	1	0	1	0	0	0	1	1	1	0	1
N_4	$C_1(2), C_3(3)$	0	0	1	1	0	1	0	0	1	0	0	0	1	1
N_5	$C_1(2), C_3(2), C_5(1)$	0	0	1	0	0	0	1	1	1	0	0	0	1	1
N_6	$C_1(2), C_3(1), C_4(2)$	0	0	0	1	0	0	1	0	1	1	0	0	1	1
N_7	$C_1(2), C_3(1), C_4(1), C_5(1)$	0	0	0	1	0	0	1	0	1	1	0	0	1	1
N_8	$C_1(1), C_2(4)$	0	1	0	0	0	1	0	0	0	0	1	1	0	0
N_9	$C_1(1), C_2(3), C_3(1)$	0	0	1	1	0	1	0	0	0	0	1	1	0	0
N_{10}	$C_1(1), C_2(2), C_4(2)$	0	0	1	0	0	0	1	1	0	1	1	1	0	0
N_{11}	$C_1(1), C_2(2), C_4(1), C_5(1)$	0	0	1	0	0	0	1	1	0	1	1	1	0	0
N_{12}	$C_1(1), C_2(1), C_3(3)$	0	0	1	1	0	1	0	0	1	0	0	0	1	1
N_{13}	$C_1(1), C_2(1), C_3(2), C_5(1)$	0	0	1	0	0	0	1	1	1	0	0	0	1	1
N_{14}	$C_1(1), C_2(1), C_3(1), C_4(2)$	0	0	0	1	0	0	1	0	1	1	0	0	1	1
N_{15}	$C_1(1), C_2(1), C_3(1), C_4(1), C_5(1)$	0	0	0	1	0	0	1	0	1	1	0	0	1	1

$3t_0$ for ME₃, $2t_0$ for ME₄, and t_0 for ME₅. Fast CLA circuits are used in the adder, where the size of one group is four bits. This yields a time t_0 of one clock period (the time of an 8-bit addition) equal to $6t_d$, where t_d is one gate delay. Consider how one can find the codes for state N_2 in which computers $C_1(3)$ and $C_4(2)$ exist. Since $C_1(3)$ consists of CE₁, CE₂, and CE₃, codes for these computer elements are stored in the first three 16-bit bytes, M(GE₁),

TABLE III.6

M(GE$_4$)

		k				p			b				m_E		
		1	2	3	4	5	6	7	8	9	10	11	12	13	14
N_0	$C_1(5)$	0	1	0	1	0	1	0	1	0	0	1	1	0	1
N_1	$C_1(4), C_5(1)$	0	1	0	0	0	1	0	0	0	1	1	1	0	1
N_2	$C_1(3), C_4(2)$	0	0	1	0	0	0	1	1	1	0	0	0	1	0
N_3	$C_1(3), C_4(1), C_5(1)$	0	0	0	1	0	0	1	0	1	1	0	0	1	0
N_4	$C_1(2), C_3(3)$	0	0	1	1	0	1	0	0	0	0	0	0	1	1
N_5	$C_1(2), C_3(2), C_5(1)$	0	0	1	0	0	0	1	1	0	1	0	0	1	1
N_6	$C_1(2), C_3(1), C_4(2)$	0	0	1	0	0	0	1	1	1	0	0	0	1	0
N_7	$C_1(2), C_3(1), C_4(1), C_5(1)$	0	0	0	1	0	0	1	0	1	1	0	0	1	0
N_8	$C_1(1), C_2(4)$	0	1	0	0	0	1	0	0	0	0	1	1	0	0
N_9	$C_1(1), C_2(3), C_3(1)$	0	0	1	1	0	1	0	0	0	1	1	1	0	0
N_{10}	$C_1(1), C_2(2), C_4(2)$	0	0	1	0	0	0	1	1	1	0	0	0	1	0
N_{11}	$C_1(1), C_2(2), C_4(1), C_5(1)$	0	0	0	1	0	0	1	0	1	1	0	0	1	0
N_{12}	$C_1(1), C_2(1), C_3(3)$	0	0	1	1	0	1	0	0	0	0	0	0	1	1
N_{13}	$C_1(1), C_2(1), C_3(2), C_5(1)$	0	0	1	0	0	0	1	1	0	1	0	0	1	1
N_{14}	$C_1(1), C_2(1), C_3(1), C_4(2)$	0	0	1	0	0	0	1	1	1	0	0	0	1	0
N_{15}	$C_1(1), C_2(1), C_3(1), C_4(1), C_5(1)$	0	0	0	1	0	0	1	0	1	1	0	0	1	0

TABLE III.6

M(GE$_5$)

		k				p			b				m_E		
		1	2	3	4	5	6	7	8	9	10	11	12	13	14
N_0	$C_1(5)$	0	1	0	1	0	1	0	1	0	1	1	1	0	1
N_1	$C_1(4), C_5(1)$	0	0	0	1	0	0	1	0	1	1	0	0	0	1
N_2	$C_1(3), C_4(2)$	0	0	1	0	0	0	1	1	0	1	0	0	1	0
N_3	$C_1(3), C_4(1), C_5(1)$	0	0	0	1	0	0	1	0	1	1	0	0	0	1
N_4	$C_1(2), C_3(3)$	0	0	1	1	0	1	0	0	0	1	0	0	1	1
N_5	$C_1(2), C_3(2), C_5(1)$	0	0	0	1	0	0	1	0	1	1	0	0	0	1
N_6	$C_1(2), C_3(1), C_4(2)$	0	0	1	0	0	0	1	1	0	1	0	0	1	0
N_7	$C_1(2), C_3(1), C_4(1), C_5(1)$	0	0	0	1	0	0	1	0	1	1	0	0	0	1
N_8	$C_1(1), C_2(4)$	0	1	0	0	0	1	0	0	0	1	1	1	0	0
N_9	$C_1(1), C_2(3), C_3(1)$	0	0	0	1	0	0	1	0	1	1	0	0	0	1
N_{10}	$C_1(1), C_2(2), C_4(2)$	0	0	1	0	0	0	1	1	0	1	0	0	1	0
N_{11}	$C_1(1), C_2(2), C_4(1), C_5(1)$	0	0	0	1	0	0	1	0	1	1	0	0	0	1
N_{12}	$C_1(1), C_2(1), C_3(3)$	0	0	1	1	0	1	0	0	0	1	0	0	1	1
N_{13}	$C_1(1), C_2(1), C_3(2), C_5(1)$	0	0	0	1	0	0	1	0	1	1	0	0	0	1
N_{14}	$C_1(1), C_2(1), C_3(1), C_4(2)$	0	0	1	0	0	0	1	1	0	1	0	0	1	0
N_{15}	$C_1(1), C_2(1), C_3(1), C_4(1), C_5(1)$	0	0	0	1	0	0	1	0	1	1	0	0	0	1

M(GE$_2$), and M(GE$_3$); the time for 48-bit addition, specifies $p = 0100$ and this p is written in the same three bytes. In the $C_1(3)$ computer, CE$_1$, CE$_2$, and CE$_3$ handle the most-, middle-, and least-significant bytes, respectively, so for CE$_1$ $b = 10$ is stored in M(GE$_1$), for CE$_2$, $b = 00$ is stored in M(GE$_2$) and for CE$_3$, $b = 01$ stored in M(GE$_3$). Since ME$_1$, with access time, $13t_0$, is the slowest memory element in $C_1(3)$, $m_E = 1101$ for all three bytes. Since the

$C_4(2)$ computer consists of CE_4 and CE_5, their respective codes are stored in bytes of $M(GE_4)$ and $M(GE_5)$, code $k = 0010$ is written in both bytes. Since the time for 32-bit addition is $3t_0$, $p = 0011$; CE_4 handles the most-significant byte ($b = 11$), and CE_5 handles the least-significant byte ($b = 01$). Since the slowest memory element ME_4 has access time $2t_0$, $m_E = 0010$ for both bytes. ∎

5.6.4 Architectural Switch to a New State

DC-group transition from one state to another is performed by a special "$N_d \to N_f$" instruction which may be executed by any computer functioning in the state N_d. This instruction initiates the V monitor, which first checks the priority of the calling program and the availability of the equipment requested. In order that the monitor execute these actions, the "$N_d \to N_f$" instruction should transfer to it the program user code NP and the codes of the requested equipment RE, showing which computer elements participate in forming the new computers of the N_f state. For instance, for the transition $N_6 \to N_4$, where $N_6 = (C_1(2), C_3(1), C_4(2))$ and $N_4 = (C_1(2), C_3(3))$, the $RE = 00111$. Indeed, CE_3, CE_4, and CE_5 form the only new computer, $C_3(3)$, of the state N_4, and $C_1(2)$ functions in both states.

If the program is denied execution of the $N_d \to N_f$ transition, the instruction stops, otherwise the monitor checks the availability of the requested equipment. Note that the instruction "$N_d \to N_f$" stops execution only if the priority of the requesting program is not high. As for deadlocks, which in this context mean that the equipment needed for reconfiguration is not yet available, the problem of deadlocks is solved as follows: the V monitor always stores the code of idle CEs and updates it with the synchronization instructions, stop-N. Each $N_d \to N_f$ instruction sends to the V monitor the code of requested equipment, RE. Upon receiving the RE code, the V monitor compares the codes of the requested and the idle equipment. If the requested computer elements are free they fetch new meanings for the variable control codes necessary for the next state, N_f. Otherwise $N_d \to N_f$ waits until the requested computer elements stop execution in the predecessor state, N_d. The control codes for the next state, N_f, are stored in a $16 \cdot n$ bit cell of the I/O memory under address N_f and the instruction sends this address via σ-bus to those GE elements that correspond to the ones positions of the RE code (Fig. III.23). As a result, only those CEs which form new computers fetch the variable control codes. For each such CE_i, the new variable control codes are fetched from memory $M(GE_i)$ and written in the Y_2 registers of all its universal modules (Fig. III.24).

The last stage of an architectural switch is establishment of new transfer modes in the connecting elements, MSE. The task of the V monitor is to identify which connecting elements should switch and which transfer mode each of them should assume. To this end the monitor uses the code, RE, of the requested computer's elements. Since every computer element, CE_i, is

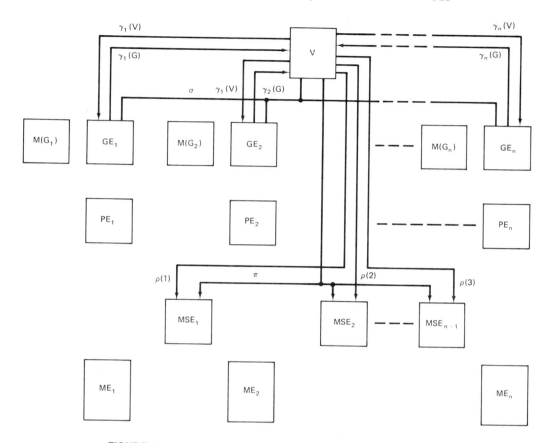

FIGURE III.23 Connection of the V monitor with DC-group elements

followed by connecting unit, MSE_i, except for CE_n, the least singificant, which has no MSE_n following it, the first $(n - 1)$ bits of the RE code form code RE^* which shows the positions of the MSEs which should switch their modes. That is, if MSE has a one in RE^*, it should switch because it corresponds to a requested CE in the RE code. If MSE has a zero in RE^*, it should be left unchanged, because the respective CE continues to be integrated into the same computer existing for both states N_d and N_f.

The V monitor uses the N_f code to find the transfer mode that should be assumed by each MSE that is to be switched. This may be done if a special state assignment technique is used. A DC-group has 2^{n-1} states. Since in each state, separation of two adjacent computers is made by the MSE between them assuming the no-transfer mode, and integration of k CEs into one $C_i(k)$ is made by $k - 1$ connecting elements assuming a transfer mode, assigning a transfer mode or no-transfer mode for the MSE_i in the ith position ($i = 1, \ldots, n - 1$) with a 0 or 1 respectively yields an $(n - 1)$-bit code.

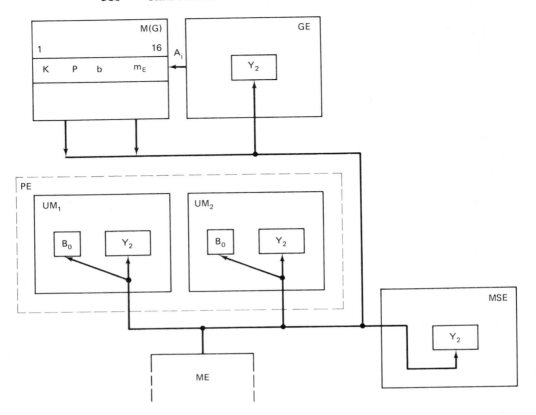

FIGURE III.24 Writing variable control codes to all LSI modules of CE

This code shows the positions and the number of independent computers functioning in this state. Furthermore, for each computer $C_i(k)$, this code shows the positions of all the CEs which it integrates.

Example 25. In Table III.6 all states of a DC-group with $n = 5$ are assigned with a four-bit code. Each state is represented by the decimal value of its code. For N_{11}, code 1011 shows three connecting elements, MSE_1, MSE_3, and MSE_4, in the no-transfer mode and one, MSE_2, in a transfer mode. Since MSE_1 stores 1, it separates CE_1 and CE_2, CE_1 functions as $C_1(1)$ and CE_2 is integrated by another computer. Since MSE_2 stores 0, it ties CE_2 and CE_3 together forming computer $C_2(2)$. Since MSE_3 and MSE_4 store 1, they separate CE_3 from CE_4 and CE_4 from CE_5, therefore CE_4 and CE_5 function respectively as $C_4(1)$ and $C_5(1)$. Thus, $N_{11} = 1011$ specifies $C_1(1)$, $C_2(2)$, $C_4(1)$, and $C_5(1)$ functioning concurrently. ■

Such a state assignment allows the V monitor to specify the mode of operation (transfer or no-transfer) that should be assumed by each MSE which should switch. To this end, it performs a bitwise comparison of RE^* and N_f which establishes:

If bit $i = 1$ for both RE^* and N_f, MSE_i switches to no-transfer mode

If $i = 1$ for RE^*, and $i = 0$ for N_f, MSE_i switches to a transfer mode

If $i = 0$ for RE^*, MSE_i remains unchanged

Establishment of the proper transfer modes in the connecting elements concludes the switch of the architecture to state N_f. As shown above, the technique considered requires that the $N_d \to N_f$ instruction store the following codes: program user code NP for priority analysis, n-bit code RE of requested equipment, $(n-1)$-bit code N_f of the next architectural state. All these codes may be mapped into one program instruction made of two 16-bit cells.

When each new computer begins execution it should be provided with the following information: The new program's base address, which memory element stores the first program segment, and the modes of transfer which should be established in all the connecting elements it integrates. In order to exclude this information from the $N_d \to N_f$ instruction, assume that each new computer begins by fetching cell 0 from its most-significant memory element. Then all connecting elements will assume the right transfer mode. Usually cell 0 stores a jump to some other location of the same memory element. Then the program may jump from one memory element to another using techniques discussed above.

Example 26. Let the program P_2 (user code 010) computed by computer $C_1(4)$ request the $N_1 \to N_3$ transition. The codes stored in the architectural switch instruction are found as follows. To form new computers $C_1(3)$ and $C_4(1)$ (Table III.5) in state N_3 takes the resources of CE_1, CE_2, CE_3, and CE_4. Therefore, $RE = 11110$. In this DC-group, $n = 5$, therefore the instruction $N_1 \to N_3$ stores $NP = 010$, $RE = 11110$, and $N_3 = 0011$. In the process of instruction execution the $RE^* = 1111$ is formed. By comparing RE^* and N_3 the V monitor establishes that both MSE_1 and MSE_2 should establish the right transfer mode because they have 1 in RE^* and 0 in N_3, and both MSE_3 and MSE_4 should establish the no-transfer mode because they have 1 in RE^* and 1 in N_3. Indeed, in N_3 MSE_1 and MSE_2 are integrated by $C_1(3)$, and MSE_3 and MSE_4 separate $C_1(3)$ from $C_4(1)$ and $C_4(1)$ from $C_5(1)$, respectively.

6. DC-GROUP WITH MINIMAL DELAY

This dynamic architecture is characterized by the minimal communication times between any two modules of the resource (PE–PE, PE–ME) which are independent of the module locations. This allows obtaining permanent and minimal times for accessing instructions and $h \cdot k$-bit data words for any variable size computer which may be formed in a DC-group. In addition, this dynamic architecture implements simple parallel communications between computers which can also be established during permanent and minimal time.

6.1 Resource Diagram

This DC-group as well as the DC-group with minimal complexity contains n computer elements, CE, and one monitor, V. One CE processes h-bit words in parallel. Assume $h = 16$. Each CE includes an h-bit PE, an h-bit wide ME, and an I/O element, GE, equipped with a small memory M(GE) (Fig. III.25).

The processor and I/O resource are connected with the memory resource via a reconfigurable memory–processor bus containing two types of connecting modules: an *address connecting element* (ASE) and a *memory connecting element* (MSE). The ASE elements connect each processor element (PE) with the memory elements (ME) and broadcast the memory addresses, read, and write signals (w_1, w_2) from this PE to all MEs. The number of ASE elements assigned to each PE is n, the number of MEs in the resource, so that ASE_i transfers an address and two signals from a specific PE to ME_i ($i = 1, \ldots, n$).

The MSE elements are connected to each ME and exchange h-bit bytes between this ME and the PEs. Each ME is assigned n MSEs, where n is the number of PEs in the resource, so that MSE_i exchanges h-bit bytes between this ME and PE_i. Both the MSE and the ASE connecting elements may be implemented on the universal module (UM) considered in Sec. 5.3.

6.2 Organization of Communications Between PE and ME

In order to satisfy the flexibility requirement each PE of the resource must have a direct access to each ME. Thus each PE–ME pair must be connected by an address path and an h-bit data path. Consider how these two paths may be organized.

6.2.1 Address Path

Any processor element, PE, may generate address and w (read or write) signals for any memory element, ME, in the resource. The PE outputs this address through the *address channel* (A-channel) which is connected to all the ASEs (Fig. III.26). Selection of the ASE that broadcasts w and this address is specified by the *selective activation code* (SEL) sent through channel B. The SEL is a $\log_2 n$-bit code where n is the number of ASE connecting elements assigned to each PE. Its value, i, is interpreted to mean the position of the ASE that is to broadcast the address (SEL = i). Since each

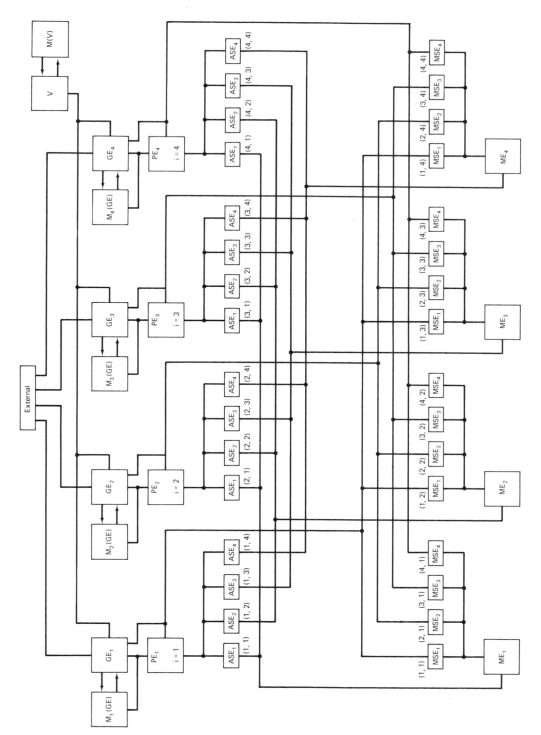

FIGURE III.25 Resource diagram for DC-group with minimal delay

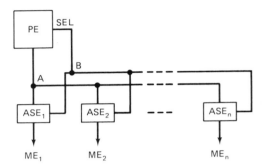

FIGURE III.26 Address broadcast from PE_i to ME_j

ASE is realized on a UM, it is equipped with the modular control organization considered in Sec. 5.2.2 that has, among other control codes, the position code i, so that ASE_1 stores $i = 1$, ASE_2 stores $i = 2$, etc.

Upon receiving code SEL, each ASE compares it with its own position code i. The ASE in which SEL $= i$ is opened for broadcasting the address to ME_i. For example, suppose that PE_1 must send an address to ME_2. It sends SEL $= 2$ through the B channel which activates ASE_2 to transferring the address to ME_2 it receives through the A channel.

6.2.2 The h-Bit Data Path

The h-bit byte exchange between a pair of PEs and MEs is accomplished through the connecting element MSE. Consider how this MSE is selected. For each ME attached to n MSEs, selective activation of MSE_j establishes an h-bit data path between this ME and PE_j (Fig. III.25). Selective activation of MSE_j is performed by a read signal (w_1) if an h-bit byte has to be read from the ME or by a write signal (w_2) if an h-byte has to be written to the ME. These two signals are obtained from the connecting element ASE that broadcasts addresses and w's to this ME. It then follows that to implement such selective activation one has to connect each ASE_j (which sends an address to ME_j) and each MSE_i (which sends an h-bit byte between PE_i and ME_j) with a two-line connection (i, j), where one line is used for signal w_1 and the other is used for signal w_2.

Therefore, to organize an h-bit data path between PE_i and ME_j one has to construct the (i, j) connection using the following rule: PE_i broadcasts an address through its ASE_j ($i \rightarrow j$), and the MSE_i assigned to ME_j broadcasts the respective h-bit byte to or from ME_j ($i \rightarrow j$). It then follows that the (i, j) connection is the connection for the w_1 and w_2 signals between ASE_j (assigned to PE_i) and MSE_i (assigned to ME_j). In each ASE and MSE the (i, j) connection takes only two pins. In order to implement the address and h-bit data paths described above, the reconfigurable memory–processor bus has to have all (i, j) connections, where $i, j = 1, \ldots, n$.

Example 27. Suppose processor element, PE_2, wants to write an h-bit byte into a cell of memory element, ME_4 (Fig. III.25). This processor element sends the address of this cell and signal w_2 through its A-channel (Fig. III.26) and the code SEL = 4 through its B-channel, thus selecting ASE_4. ASE_4 then broadcasts the address and w_2 to ME_4. Since the (2, 4) connection connects the ASE_4 assigned to PE_2 with the MSE_2 assigned to ME_4, ASE_4 also sends signal w_2 to MSE_2 activating it into a mode of h-bit byte transfer to ME_4. Therefore PE_2 broadcasts an h-bit byte to ME_4 through the MSE_2 assigned to ME_4. ∎

6.3 Communications Inside One Computer

A dynamic architecture forms the available resources into variable size computers $C_i(k)$ where k ($k = 1, \ldots, n$) shows the number of PEs and MEs integrated inside the computer, and i shows the position of the most-significant PE. For each computer $C_i(k)$ one has to organize the instruction path and the $k \cdot h$-bit data path.

6.3.1 Instruction Path

Implementation of the *program universality* concept for dynamic architectures necessitates that instructions have a size of one ME and be stored consecutively in one ME. Complex programs may be stored in several MEs so that each ME stores one program segment. The instruction fetched from one ME would then be sent concurrently to all the modules from which the given computer is assembled. It follows that if an instruction is fetched from one ME, it must be broadcast through only those MSE assigned to this memory element that connect it with the processor elements integrated into the given computer. The remaining MSE connecting elements assigned to this ME should block the instruction broadcast since they connect this ME with PEs from other computers. In order to minimize the bandwidth of the address bus, generation of addresses for instructions fetched from ME_i ought only to be performed in the PE_i storing the same position code i, so PE_1 generates instruction addresses only for ME_1, PE_2 does it only for ME_2, etc.

Let us consider how the memory–processor bus could organize such an instruction fetch. All processor elements integrated into a given computer store a special *program selection code* (PSC) that shows the position i of the memory element ME_i that contains the currently executed program segment. For instance, if PSC = 2, instructions are fetched from ME_2, etc. This code is written to all PEs of the computer by a special "memory jump" instruction that organizes the fetching of the next program segment from a new ME. (Execution of a similar $ME_i \rightarrow ME_j$ instruction was considered in Sec. 5.4.5.)

Consider now the organization of instruction fetches from memory element, ME_j, i.e., all PEs of the computer store PSC = j. Each such PE compares its own position code i with the PSC. Since these two codes only match in PE_j, the PE broadcasts the instruction address to the A-channel. Concurrently, all the PEs of the computer send the PSC to their B-channels; i.e., SEL = PSC = j. Thus in each PE, the PSC selects ASE_j. Since only PE_j generates an instruction address, it broadcasts it through its ASE_j to ME_j. The ASE_j of each PE_i sends signal w_1 through the (i, j) connection. This activates the respective MSE_i assigned to ME_j. Thus the path for broadcasting the instructions from the ME_j that stores them to each PE of the computer is established.

Example 28. Let the $C_1(3)$ computer fetch instructions from ME_2 (Fig. III.27). All the PEs integrated into this computer store PSC = 2. Since only in PE_2 does the position code $i = 2$ coincide with the PSC (i = PSC), this processor element sends the instruction address A_k to its channel-A. All the processor elements of computer $C_1(3)$ (PE_1, PE_2, PE_3) send PSC = 2 through their B-channels leading to activation of the ASE_2 element attached to each PE. PE_2 broadcasts the address and signal w_1 through its ASE_2 to ME_2. This ASE_2 sends signal w_1 through the (1, 2) connection and activates the mode of instruction broadcast from ME_2 to PE_2 in MSE_2. For PE_1, PSC = 2 activates its ASE_2. This ASE_2 sends signal w_1 through the (1, 2) connection and activates the mode of instruction broadcast from ME_2 to PE_1 in MSE_1. Similarly, the (3, 2) connection activates instruction broadcast from ME_2 to PE_3 through MSE_3. Therefore, the connecting elements MSE_1, MSE_2 and MSE_3 established the paths for instruction broadcast from ME_2 to PE_1, PE_2, and PE_3, respectively. ∎

6.3.2 Data Path

Implementation of the program universality concept also necessitates that for a dynamic architecture one $h \cdot k$-bit word processed by computer $C_i(k)$ be stored in a single parallel cell of k memory elements, all with the same address. (For instance, for computer $C_1(3)$, a $3h$-bit word is stored in all MEs under the same address A_p (Fig. III.28).) Consequently, a data array containing "d" $k \cdot h$-bit words will require d parallel cells for storage in k different memory elements. When a $k \cdot h$-bit word is accessed, each of k processor elements generates the same address, resulting in a concurrent fetch of all "k" h-bit bytes comprising the data word. Let us consider how the memory–processor bus could organize such a parallel data exchange of $k \cdot h$-bit words.

Each PE may access its own ME if the address for the ME is broadcast through the ASE with the same position. For instance, PE_1 may access ME_1 by broadcasting an address for ME_1 through its ASE_1; PE_2 may access ME_2 by broadcasting addresses through ASE_2, etc. It follows the position code, j, of PE_j must be sent to its B-channel to activate address broadcast from PE_j

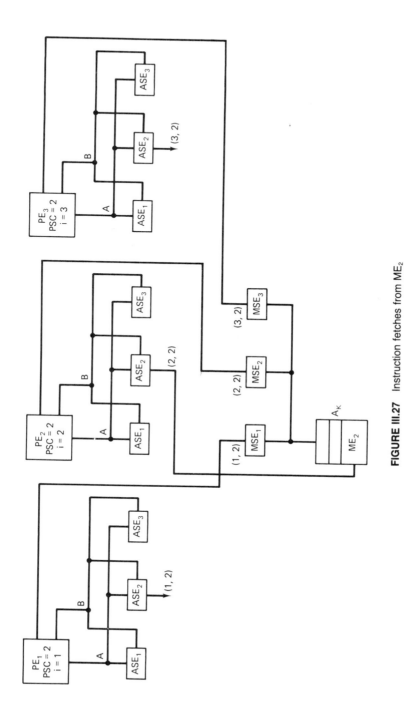

FIGURE III.27 Instruction fetches from ME_2

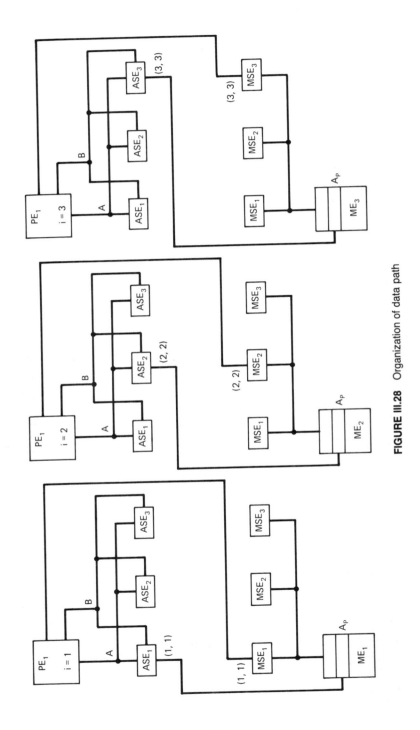

FIGURE III.28 Organization of data path

324

to ME_j. Activation of ASE_j leads to activation of MSE_j, belonging to ME_j, into an h-byte broadcast between ME_j and PE_j. Therefore, in the $C_i(k)$ computer, concurrent transfer by each PE of its own position code to the B-channel leads to broadcasting of the same address to all k MEs and parallel $h \cdot k$-bit word exchange between the processor and primary memory.

Example 29. Suppose computer $C_1(3)$ has to fetch a $3h$-bit word from cell A_p (Fig. III.28). In the interval for word fetch, PE_1 transfers its own position code, $i = 1$, to the B-channel. This allows ASE_1 to propagate the address A_p to ME_1. ASE_1 activates MSE_1 into a mode of h-bit broadcast from ME_1 to PE_1 through the (1, 1) connection. Similarly, codes $i = 2$ in PE_2 and $i = 3$ in PE_3 cause ASE_2 assigned to PE_2 and ASE_3 assigned to PE_3 to broadcast A_p addresses to ME_2 and ME_3. ASE_2 activates MSE_2 assigned to ME_2, through the (2, 2) connection and ASE_3 activates MSE_3 assigned to ME_3 through the (3, 3) connection. This establishes the path for the $3h$-bit word fetch from the primary memory to the processor of computer $C_1(3)$. ■

6.4 Communications Between Computers

The memory–processor bus just considered provides high flexibility for data exchanges between concurrent computers. It can make the following between-computer information exchanges possible:

1. "X processor–Y memory": For this exchange the X computer accesses the primary memory of the Y computer. This access may be performed without interrupting the operation of the processor in the Y computer.
2. "X processor–Y processor". Data exchange between the processors of computers X and Y.
3. "X memory–Y memory". Data exchange between primary memories of computers X and Y.

6.4.1 Information Exchange Between the Processor and the Memory of Two Different Computers

This exchange is performed by a special instruction, "X processor–Y memory," executed by the X computer. This instruction stores the following values: the base address, A_0, of the block of information stored in the primary memory of the Y computer, and a displacement constant z showing the difference in positions of the most-significant PEs contained in computers X and Y. If the position code i of the most-significant PE in the X computer is $i = j_x$ and in the Y computer $i = j_y$ then for $j_x < j_y$, $j_y = j_x + z$, and for $j_x > j_y$, $j_y = j_x - z$; i.e., to obtain j_y one has to perform a complemented addition of j_x and z.

When the instruction X processor–Y memory is fetched by all PEs of the computer, each PE adds its own position code i to the displacement constant z and sends the result, $i + z$, to the B-channel; i.e., SEL $= i + z$. This code selects element ASE_{i+z} for broadcasting address A_0 to the memory element, ME_{i+z}. ASE_{i+z} concurrently causes element MSE_i assigned to ME_{i+z} to broadbroadcast h-bit bytes from ME_{i+z} to PE_i. This establishes data communication between PE_i and ME_{i+z}. The instruction may iterate d times to access a block of data where d is the dimension of the block.

Example 30. Let the resource form two 32-bit computers, $C_1(2)$ and $C_3(2)$, and suppose computer $C_1(2)$ executes the instruction X processor–Y memory (Fig. III.29). This instruction fetches thirty 32-bit words from the primary memory of computer $C_3(2)$ starting at the base address $A_0 = 1750$, and $z = 2$ (the most-significant PEs of these computers are PE_1 and PE_3, leading to $z = 3 - 1 = 2$). When the instruction is executed in the $C_1(2)$ computer the following actions are performed: position codes i in PE_1 and PE_2 are added to z and sent to the B-bus giving SEL $= i + z$. In PE_1, $i + z = 1 + 2 = 3$ activates ASE_3, and in PE_2, $i + z = 2 + 2 = 4$ activates ASE_4 to broadcast the same address, A_0, to ME_3 and ME_4, the units that form the primary memory of computer $C_3(2)$ (Fig. III.30). Through the (1, 3) connection ASE_3 activates the MSE_1 element assigned to ME_3. This leads to establishing a data path between PE_1 and ME_3. Through the (2, 4) connection, ASE_4 activates MSE_2 assigned to ME_4. This connects PE_2 and ME_4. Therefore a data path between the 32-bit processor of $C_1(2)$ and the 32-bit memory of $C_3(2)$ is established. ∎

6.4.2 Parallel Byte Exchange Between the Processor and the Memory of Two Different Computers

The flexibility of the bus considered is such that it allows organization of parallel byte exchanges between computers. This means that the X computer may fetch not only the entire $k \cdot h$-bit word(s) but any portion of this word that is a multiple of h from the primary memory of the Y computer. Namely, the X computer may access $h, 2h, \ldots, (k - 1) \cdot h$-bit bytes of a $k \cdot h$-bit word.

Two types of X processor–Y memory byte exchanges, exist:

1. The size of computer X is smaller than that of computer Y, and computer X fetches a byte from computer Y that matches its size. Such an exchange allows a smaller computer to perform a fast access to the words that match its size from the primay memory of the larger computer.

2. Computer X fetches a byte from computer Y that is smaller than X's size. This byte exchange may be used if computer X's size exceeds

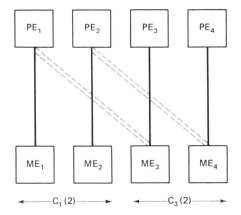

FIGURE III.29 X processor–Y memory exchange

that of computer Y or if computer X performs associative scanning of of data arrays with smaller word sizes that are stored in separate memory modules of computer Y, etc.†

†Associative scanning of data arrays is discussed in Chap. II, Sec. 3.2. (Ed.)

The byte exchange of the first type may be performed by the same X processor–Y memory instruction considered above; however, the instruction should store a modified displacement constant z that shows the difference in position codes assigned to respective ME and PE elements where ME stores the byte, PE accesses it.

Example 31. Let the available resource (Fig. III.31) be formed into three computers $C_1(2)$, $C_3(1)$, and $C_4(1)$. Suppose that $C_4(1)$ has to fetch an h-bit byte from memory cell 510 of ME_2, which belongs to the primary memory of $C_1(2)$. $C_4(1)$ executes instruction X processor–Y memory which stores $A_0 = 510$, and $z = -2$ ($z = 2 - 4 = -2$, since ME_2 is accessed by PE_4) (Fig. III.32). When this instruction is executed in $C_4(1)$, PE_4 performs $i + z = 4 - 2 = 2$ and causes ASE_2 to broadcast address A_0 to ME_2. ASE_2 activates MSE_4 assigned to ME_2 through the (4, 2) connection. As a result h bits in ME_2 are fetched by PE_4. ■

For the second type of byte exchange in which only a portion of the PEs in the X computer must access a data word stored in the primary memory of the Y computer, one may use a modified X processor–Y memory instruction. In addition to the A_0 and z values specified above, this instruction stores two more values, q and p, showing positions of processor elements PE_q and PE_p in the X computer that receive the most- and least-significant bytes of the word stored in the Y computer, respectively.

When this instruction is fetched to all PEs of the X computer, each PE

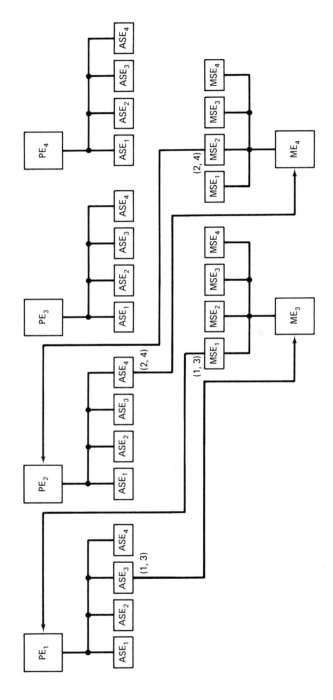

FIGURE III.30 Exchange between PE_1, PE_2 and ME_3, ME_4

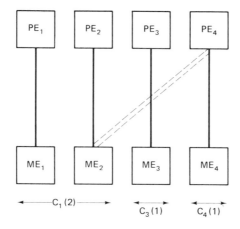

FIGURE III.31 Architectural state $[C_1(2), C_3(1), C_4(1)]$

compares the q value received through the instruction with its own position code i. If $i < q$, this PE is blocked from the word fetch. If $i \geq q$, the next test is performed. All processor elements which satisfied the first test, $i \geq q$, are then tested for $i \leq p$. If $i \leq p$, the respective PE fetches word. If $i > p$, this PE does not fetch the word. Thus, two sequential tests, $i \geq q$ and $i \leq p$, allow one to identify a consecutive portion of the PEs of the X computer that should fetch the word from the Y computer. The remainder of execution of the instruction is similar to that considered above.

Example 32. Let the resource form computers $C_1(3)$ and $C_4(4)$ and suppose that PE_5 and PE_6 in $C_4(4)$ need data words from ME_2 and ME_3 contained in $C_1(3)$ (Fig. III.33). Thus $C_4(4)$ uses a modified version of the X processor–Y memory instruction that stores the following values: displacement $z = 2 - 5 = -3$ since ME_2 stores the word and PE_5 accesses it, and $q = 5$ and $p = 6$ since PE_5 and PE_6 receive the most- and least-significant bytes of the $2h$-bit word, respectively. In the $C_4(4)$ computer, PE_4 is blocked from fetching since PE_4 does not pass the first test $i \geq q$ ($i = 4 < q = 5$). The remaining PE_5, PE_6, and PE_7 continue the next test: $i \leq p$, where $p = 6$. This test is satisfied only in PE_5 and PE_6 because $i = 5 \leq 6$ and $i = 6 \leq 6$. PE_7 is blocked from the data fetch since $i = 7 > 6$. Thus a fetch is performed only by PE_5 and PE_6. Each of these PEs calculates code $SEL = i + z$, where i is the position code and z is the displacement value. For PE_5, $SEL = 5 - 3 = 2$ activates ASE_2. For PE_6, $SEL = 6 - 3 = 3$ activates ASE_3. This establishes the data path between PE_5 and ME_2 and PE_6 and ME_3 and results in a data fetch from ME_2 and ME_3 to PE_5 and PE_6, respectively. ∎

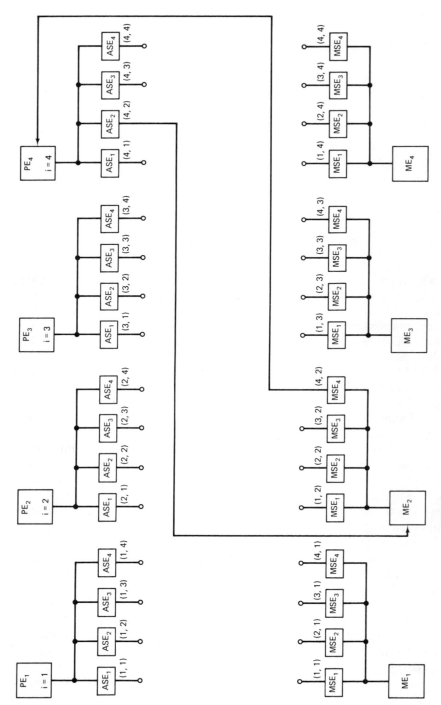

FIGURE III.32 Execution of X processor–Y memory instruction in $C_4(1)$ computer

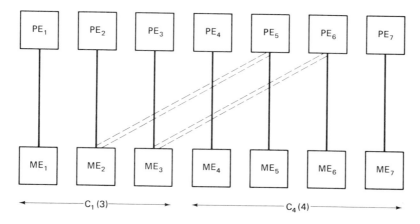

FIGURE III.33 X processor–Y memory byte exchange

Note that the byte exchange instruction considered above may perform a concurrent fetch-shift operation; i.e., a word fetched from primary memory of computer Y may be shifted before being written to the processor of com-computer X. To do this one has to change the meaning of the q and p values.

6.4.3 Information Exchange Between the Processors of Two Different Computers

To perform a parallel exchange "X processor–Y processor," each PE of the X processor has to be connected in parallel with a matching PE of the Y processor. The bus being considered may establish each such connection between two PEs through the path made of a pair of two connecting elements MSE (assigned to one memory element ME) that may pass an h-bit byte in opposite directions.

For each memory element ME, its n MSE connecting elements connect the ME with n PEs of the processor resource. Therefore, to establish an h-bit path between PE_a and PE_d, one may use the two MSE connecting elements MSE_a and MSE_d that connect this ME with PE_a and PE_d, respectively. In addition it is necessary that MSE_a be activated by the write signal w_2 so that it will pass the h-bit byte from PE_a to the memory input, and MSE_d has to be be activated by the read signal w_1 so that it will pass the h-bit byte from the memory input to PE_d.

Such parallel interconnections between two processor modules may generally be organized through the pair of MSEs assigned to any memory element. However, since the instruction performing this exchange is executed by the X computer, an organization must be introduced which performs no transfer of any of the codes stored in this instruction from computer X to computer Y. It then follows that any PE of the Y computer receives no displacement z which may change the value of the SEL code it generates. Each PE of the Y computer may therefore only form SEL $= i$, where i is its own position code. This will activate the ASE_i with matching position which then opens a data path between that PE_i and ME_i. For instance in PE_3, SEL $= 3$ selects ASE_3 which activates MSE_3 assigned to ME_3 through the (3, 3) connection. This establishes the data path between PE_3 and ME_3 (Fig. III.34).

To eliminate the transfer of displacement z from computer X to computer puter Y and to form a path between each PE_a and each PE_d where PE_a belongs to computer X and PE_d belongs to computer Y, one has to select a pair of MSE_a and MSE_d connecting elements that are assigned to ME_d in computer Y. Activation of MSE_a assigned to ME_d may then be done by PE_a of computer computer X when it forms SEL $= a + z = d$ where a is the value of its position code and z is the displacement constant stored in the instruction. Activation of the MSE_d, the second element of this pair, is made by PE_d of computer Y when it sends SEL $= d$ equal to its position code. This organization requires that the instruction X processor–Y processor store the displacement $z = d - a$ which equals the difference in positions between the communicating PEs. Note that all PEs of the X computer may use the same displacement z, because for the next pair of communicating PEs (PE_{a+1}, PE_{d+1}), $z = (d + 1) - (a + 1) = d - a$.

Example 33. Let the resource form two computers, $C_1(2)$ and $C_3(2)$ (Fig. III.34). In order to establish a parallel connection between the processors of these computers, $C_1(2)$ executes the instruction X processor–Y processor. This instruction stores displacement $z = 2$ since PE_1 communicates with PE_3 ($z = 3 - 1$) and PE_2 communicates with PE_4. When the instruction is fetched, PE_1 forms SEL $= 1 + 2 = 3$. This activates ASE_3 causing activation of the MSE_1 of ME_3. For PE_2 SEL $= 2 + 2 = 4$. This activates ASE_4 and the MSE_2 of ME_4. In the $C_3(2)$ computer, both PE_3 and PE_4 form SELs coinciding with the values of their position codes. That is, for PE_3, SEL $= 3$ activates ASE_3, activating the MSE_3 of ME_3. For PE_4, SEL $= 4$ activates ASE_4, connected with the MSE_4 of ME_4. Therefore, PE_1 is connected with PE_3 through the MSE_1 and MSE_3 assigned to ME_3; PE_2 is connected with PE_4 through the MSE_2 and MSE_4 assigned to ME_4. This sets up a parallel $2h$-bit path between the processors of the two computers.

Note that the instruction X processor–Y processor executed by PE_1 and PE_2 does not transfer memory addresses through connecting elements ASE_3 and ASE_4, respectively. Rather PE_1 uses its ASE_3 in order to activate MSE_1 of ME_3. Likewise, PE_2 uses its ASE_4 to activate MSE_2 of ME_4. ∎

6.4.4 Parallel Byte Exchanges Between Two Processors

The bus being considered may organize an h-bit data path between any two PEs of the resource. This allows organization of parallel byte exchanges

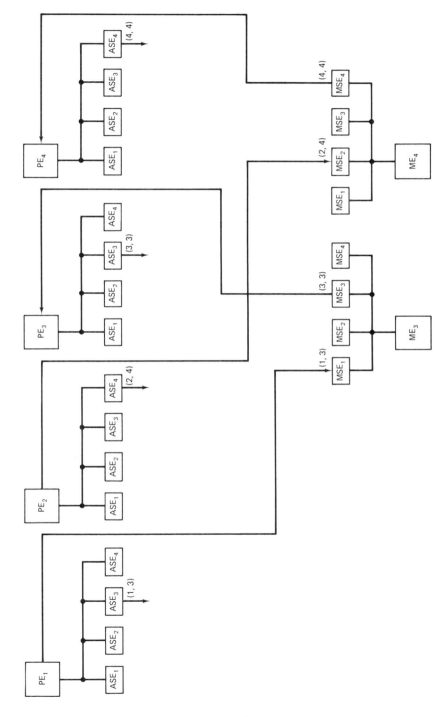

FIGURE III.34 Execution of X processor–Y processor instruction in $C_1(2)$ computer

between computers whenever processor X sends not the entire word but any portion that is a multiple of h to processor Y. This permits communication between different size processors and allows a smaller processor Y to receive a word from a larger processor X. Organization of byte exchanges between processors are similar to those considered for X processor–Y memory exchanges.

6.4.5 Parallel Shifts Inside One Computer

Existing limitations on the number of pins in LSI modules make it quite difficult to organize various types of parallel shifts of $k \cdot h$-bit words since any m-bit shift takes m pins in each PE. for $f \cdot h$-bit shifts; (where $f \geq 1$) i.e., shifts that are a multiple of h, it is expedient to use the memory–processor bus since these shifts require no additional interconnections between PE elements. The $f \cdot h$-bit shift of a word is reduced to parallel transfer of each h-bit byte of this word from PE_i to PE_j where $j = i + f$. The direction of the shift depends on the sign of f, i.e., for a left shift $j = i - f$, and for a right shift $j = i + f$.

To organize an h-bit path from PE_i to PE_j one may use the same serial connection of two connecting elements MSE_i and MSE_j assigned to the memory element ME_j that was used for conventional communication between different PEs considered earlier. For shifts however, some PEs have to perform two actions: Send their own byte to a destination PE, and receive a shifted byte from another PE. Since each PE is connected via an h-bit bus with MSE elements, it may perform these two actions only in two clock periods. During the first clock period, PE_i issues h bits to a register of MSE_i assigned to ME_j. Activation of this MSE_i is performed through the ASE_j assigned to PE_i when PE_i sends $SEL = i + f = j$, where i is its position code and f is the shifting constant stored in the instruction. At the second clock period, each PE sends a SEL equal to its own position code. For each PE_j (which has to receive a shifted byte from PE_i) $SEL = j$ selects the local ASE_j, and then the MSE_j assigned to ME_j copies the shifted byte from MSE_i to the PE_j.

In the organization of an $f \cdot h$-bit shift, one has to consider that the meaning of $SEL = i + f$ has to identify position of those PEs which actually belong to the computer performing this shift. This means that for a left shift, the f most-significant PEs should not form $SEL = i - f$, otherwise, each of them will transfer an h-bit byte to a PE contained in the left neighboring computer. Likewise for a right shift, the f least-significant PEs should not form $SEL = i + f$, for otherwise, each of them will transfer an h-bit byte to a PE contained in the computer to the right. Blocking unwanted lefthand and righthand PEs from forming $SEL = i + f$ uses the same constants q and p introduced above for the byte exchange X processor–Y memory. For $f \cdot h$-bit shifts the test $i > q$ shows positions of PEs that may participate in a left shift, and the test $i < p$ specifies positions of PEs that may participate in a right shift. Using both constants q and p, the instruction may perform an $f \cdot h$-bit shift of a portion of a word restricted by PE_q on the left ane PE_p on the right.

Example 34. Let the resource be formed into a $C_1(4)$ computer, and a $2h$-bit word stored in PE_1 and PE_2, have to be shifted by $2h$ bits to the right. Thus $f = 2$, $q = 0$, and $p = 3$, since no left PEs and the right PE_3 and PE_4 should be blocked from shifting. Show that $2h$-bit shift may use the same communication path between two $2h$ bit processors that was considered in Example 33 (Fig. III.34). When the shift instruction is fetched, all PEs of computer $C_1(4)$ perform two sequential tests: $i > q = 0$ and $i < p = 3$ where i is the position code of each PE. These two tests determine that PE_1 and PE_2 each store a word to be shifted. During the first clock period of shifting, PE_1 forms $SEL = 1 + 2 = 3$. This selects ASE_3 and h bits from PE_1 are written to the MSE_1 assigned to ME_3. For PE_2, $SEL = 2 + 2 = 4$ sends h bits from PE_2 to the MSE_2 assigned to ME_4. During the second clock period each PE of the computer generates a SEL equal to its own position code. For PE_3, $SEL = 3$ activates ASE_3 and MSE_3 assigned to ME_3. Therefore the h bit byte stored in the MSE_1 of ME_3 is now broadcast through MSE_3 to PE_3. Similar actions are performed by PE_4. Since for PE_1 and PE_2 no shifted byte was received by the MSEs assigned to their MEs, generation of each $SEL = i$ in these PEs does not lead to the fetching of a shifted byte. ∎

6.4.6 Information Exchanges Between the Memories of Two Computers

The bus being considered allows for simple exchanges of data arrays stored in the primary memories of two computers. In order to speed up this exchange and not interrupt the work of the processor in the one computer, the generation of addresses for both memories will be performed by the computer that needs this array. Consider how such exchange may be organized.

Suppose that the X computer needs that a data array stored in the memory of Y computer be transferred to the primary memory of X computer. Since each PE may form only one address at a time, the generation of two addresses, A_x for memory X and A_y for memory Y, is performed in two clock periods.

The exchange is performed with a special instruction "X memory–Y memory," which is executed in computer X. The following values are used for this exchange: two base addresses A_x and A_y for memories X and Y, the displacement z showing the difference in positions of the most significant MEs contained in memories X and Y, and the array's dimension. When the instruction is fetched by all PEs of computer X, each PE_i forms $SEL = i + z$, $SEL = i + z$, and sends the A_y address to the ASE_{i+z} which stores it during one clock period. During the second clock period, the following parallel actions are performed: Each ASE_{i+z} element assigned to PE_i broadcasts the A_y A_y address to memory element ME_{i+z}; each PE_i forms a second $SEL = i$. This This leads to accessing the local memory element ME_i. The ASE_i assigned to PE_i activates the MSE_i assigned to ME_i; ASE_{i+z} assigned to PE_i activates the MSE_i assigned to ME_{i+z}. Therefore, an h-bit data path is established between ME_i and ME_{i+z}. Similar data paths are established between all other pairs of MEs contained in memory X and Y.

The bus considered allows byte exchanges between the memories of two computers. The organizations for such exchanges are similar to those considered above.

Example 35. Let the resource form two computers, $C_1(2)$ and $C_3(2)$ and let the instruction X memory–Y memory executed in $C_1(2)$ fetch 100 data words from $C_3(2)$ (Fig. III.35). The following values are used in execution of this instruction: $z = 2$ since ME_1 and ME_3 are two most-significant ME's, two base addresses, A_x and A_y, for memories X and Y, and array dimension $d = 100$. When the instruction is fetched by PE_1 and PE_2, PE_1 forms $SEL = 1 + 2 = 3$, selects the local ASE_3, and sends address A_y to ASE_3. Through the $(1, 3)$ connection, ASE_3 activates the MSE_1 assigned to ME_3 into an h-bit broadcast from ME_3, during the next clock period, PE_1 forms $SEL = 1$ and broadcasts A_x through the ASE_1 to ME_1. ASE_3 concurrently sends address A_y to ME_3. Through connection $(1, 1)$, ASE_1 activates the MSE_1 assigned to ME_1 into h-bit reception. Therefore, two MSE_1's assigned to ME_1 and ME_3, respectively, establish an h-bit path for word broadcast from ME_3 to ME_1. Similar actions are performed by PE_2 in establishing the path from ME_4 to ME_2. ∎

6.5 Reconfigurable Paths for Processor Signals

These remain the same as for the DC-group with the minimal complexity, because both DC-groups use dedicated connections for these busses. As was shown in Sec. 5.4 the use of dedicated busses for the processor signals (carry and overflow) minimizes the carry propagation times which leads to a much faster rate of information processing.

The alternative organization considered by Lipovski and Tripathi in [14] provides that carries and overflows propagate through the processor–memory bus interconnecting processor and memory resources. This leads to large delays introduced by reconfiguration logic into carry propagation times.

6.6 Organization of Architectural Reconfiguration in a System

As was shown in Sec. 5.6., this part of the architectural reconfiguration of a dynamic architecture includes two actions:

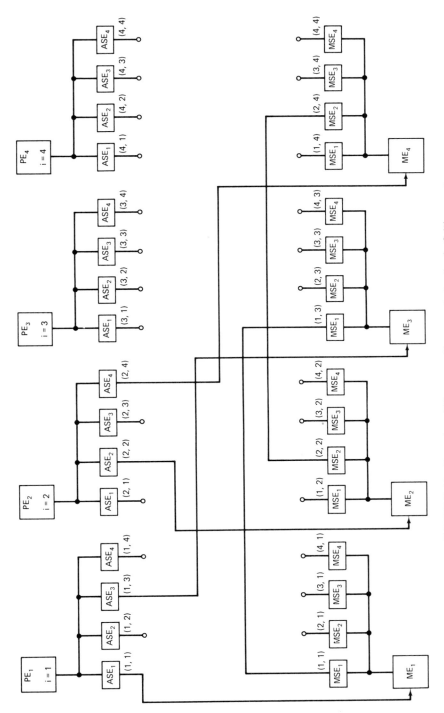

FIGURE III.35 Execution of X memory–Y memory instruction in $C_1(2)$ computer

1. Search and write new control codes to modules of the resource.
2. Switch all the reconfigurable busses that must separate the computers of the next architectural state.

Organization of searching and writing new control codes remains the same as in the DC-group with minimal complexity, since it does not depend on the structure of the reconfigurable busses. On the other hand, since different dynamic architectures are distinguished by the reconfigurable memory–processor busses they use, organization of switching reconfigurable busses varies for different dynamic architectures.

For DC-groups with minimal complexity, switching the memory–processor bus (Sec. 5.6.) takes two steps: During the first step, the monitor performs a bitwise comparison of two codes (requested equipment, RE^*, and the next architectural state, N_f) and establishes the no-transfer mode in those connecting elements which have to separate the new computers of the next architectural state. During the second step, the monitor establishes the right transfer mode in all connecting elements that are integrated into new computers.

For the DC-group with minimal delay it takes no time to switch the memory–processor bus during a system's transition into a new state. The reason for this is that all the information paths in the bus are established when pertinent PEs generate SEL codes. These codes are available in the ASE elements during the time of communication between each pair of exchanging resource modules PE and ME.

This bus therefore minimizes the time of architectural reconfiguration since this time now includes only the time to search and write new control codes.

PERFORMANCE OF DYNAMIC ARCHITECTURES

7. INTRODUCTION

This part assesses two dynamic architectures—the DC-group with minimal complexity and the DC-group with minimal delay—from the standpoint of the major performance factors considered below.

Each dynamic architecture may be evaluated from two viewpoints: *adaptation* and *performance*. The adaptation aspect looks at how well a particular dynamic architecture is adapted to an algorithm to be executed. As was shown in [8], this aspect of dynamic architecture is characterized by such parameters as:

1. Adaptation to program streams. This gives the percentage of the resource used to compute a given set of concurrent programs.

2. Adaptation to program structures. This examines execution speed-up arising from the program being computed by a particular instruction set.
3. Array adaptations. This shows the percentage of redundant equipment created during array (vector) computations.
4. Bit size adaptation. This measures the delay in the execution of the entire task (program) because of selection of a permanent computer size for executing all of its instructions; etc.

The performance aspect characterizes each dynamic architecture from the viewpoint of time of computation and of complexity of the respective hardware resource.

Performance of dynamic architectures is affected by the following major factors:

1. The time required for architectural reconfiguration.
2. Delays introduced into signal propagation by reconfigurable busses.
3. Time required for between-computer communications.
4. Modular expansion of existing systems with new hardware components.
5. Cost of realization.

Let us consider these factors.

8. TIME REQUIRED FOR ARCHITECTURAL RECONFIGURATION

In a system with a dynamic architecture one may distinguish two modes of computations:

Mode 1. An algorithm is computed by a single architectural state that is established before computation.

Mode 2. An algorithm is computed by a sequence of architectural states. The algorithm is partitioned into several tasks, and for each task the architectural state is found that minimizes its computation time.

Complex real-time algorithms require mode 2 in most cases, because as a rule, these algorithms are characterized by a changeable number of information streams and changeable processing requirements [2, 3]. It follows that such algorithms require the architecture to reconfigure from one state to another during computation. The time for architectural reconfiguration includes the following actions:

1. The time required to search, fetch, and write all the necessary control codes to those modules of the resource that are formed into new computers in the next architectural state.
2. The time required to switch the reconfigurable busses that must separate computers of the next architectural state and provide correct information exchanges between them.

Since each time that an algorithm requires an architectural reconfiguration the computers affected stop, the time of architectural reconfiguration has to be added to the time of algorithm computation. It then follows that the time of reconfiguration has to be minimized in order to not offset speed advantages gained by dynamic architectures due to their adaptations to the executed algorithm.

9. DELAYS INTRODUCED BY RECONFIGURABLE BUSSES

A dynamic architecture is characterized by delays introduced by reconfigurable interconnections. If one assumes that a dedicated connection, $A \to B$, between modules A and B introduces no delays, then a reconfigurable interconnection between modules A and B, C, or D in which either interconnection $A \to B$ or $A \to C$ or $A \to D$ is activated selectively requires additional logic.† This introduces additional delays into the path of signal propagation‡ that slow down the rate of information processing.

One may distinguish the following signal paths that are delayed due to the reconfiguration logic:

†Organization of reconfigurable interconnections is discussed in Chap. I, Sec. 24.2. (Ed.)

‡A technique for computing delays of logical circuits is discussed in Chap. I, Secs. 3, 4, and 5. (Ed.)

1. *Instruction path*. In dynamic architectures the available resource may be partitioned into a variable number of concurrent computers. The instruction bus must separate the instruction streams computed by independent computers. It follows that the instruction path is reconfigurable, and the reconfiguration logic adds additional delay to the instruction fetch time.
2. *Data exchange path*. Since each $h \cdot k$-bit computer must maintain a parallel data exchange with $h \cdot k$-bit words, and several computers of different sizes may be formed in a single state, the data bus must separate concurrent data streams processed by independent computers in one state. Thus the data path is also reconfigurable, and additional delays introduced by the reconfiguration logic are added to the data fetch times.
3. *End-around-carry path*. Since the available processor resource, containing n h-bit processor modules, may be arranged into $h \cdot k$-bit processors where k can range from 1 to n, a dynamic architecture

must form a reconfigurable path for the end-around carries. This means that in addition to the usual delays that characterize a static end-around-carry path, there is an additional delay due to reconfiguration in this path.

4. *Paths for equality signals.* The equality signal produced during the comparison of two $h \cdot k$-bit operands participates in several conditional branch operations, and must be routed to the most significant processor module of the respective $h \cdot k$-bit computer. Since each processor module in a dynamic architecture may be the most significant one in some $h \cdot k$-bit computer, the path for equality signals must also be reconfigurable. Additional delay arises while propagating the equality signal(s) through the reconfigurable equality path due to reconfiguration logic.

Since delays introduced by each of the paths listed above affect selection of the period(s)† of the synchronization sequence(s), then minimizing delays induced by the logical circuits performing reconfiguration in each of these paths causes a faster rate of information processing.

†All factors that affect selection of clock periods are discussed in Chap. I, Sec. 5. (Ed.)

10. MODULAR EXPANSION

As was indicated in Sec. 2.2, modular expansion is restricted by the following factors:

1. Technological constraints on the number of admissible interconnections.
2. Increase in delays introduced by the reconfiguration logic.
3. Ability of the system to maintain a specified level of realiability.

11. COST OF REALIZATION

The low cost of microprocessors advertised by manufacturers has led to a popular misconception that the cost of hardware for a complex parallel system is also indefinitely low. Current experience with LSI implementation of complex parallel mainframes indicates, however, that each such system requires fabrication of tens of different LSI module types [26]. Since production of each module type may cost up to $100,000 and take 1 year of work, producing a single copy of a complex parallel system requires enormous investments.

On the other hand, attempts to use off-the shelf modules for fabrication of a complex system do not solve the problem since only 30 to 40 percent of all the necessary circuitry may be mapped onto LSI modules. The remaining 60 to 70 percent of the hardware has a conventional (say TTL) implementation [32].

Thus the existing experience with LSI implementation of complex parallel systems contests the indiscriminate use of the thesis that LSI technology allows one to obtain arbitrarily cheap hardware for complex systems. This hardware may become cheap only if the architecture can be assembled from LSI modules selected from a minimal number of module types and have mostly pin-to-pin connections among modules; i.e., the number of external circuits connecting LSI modules should be restricted.

Thus in evaluating different dynamic architecture one has to consider:

1. The number of different module types used in assembling the system.
2. The overall percentage of circuits mapped onto LSI modules in logical portion of the system.

12. PERFORMANCE EVALUATION

The five factors listed in Secs. 7 through 11 show that for a dynamic architecture, performance improvement may be achieved if one perfects the organizations of reconfigurable interconnections among modules and reduces the number both of different module types and of the logical circuits that interconnect LSI modules.

As was shown, to implement a dynamic architecture one must use two types of busses:

1. A reconfigurable memory–processor bus which performs variations in the number of instruction and data streams.
2. A reconfigurable processor bus which implements variable $16k$-bit processors.

The architectural realization of these busses affects both the speed of computations and complexity of the resources organized into a particular dynamic architecture.

Since a DC-group employs two types of busses we will find the meanings of five performance factors assumed for these busses. These will be found for the DC-group with minimal delay and the DC-group with minimal complexity.

12.1 Assessment of the Minimal Delay Memory–Processor Bus

This bus has several important properties which greatly improve the performance of dynamic architecture. The major advantages of this bus are due to the fact that it reduces to a minimum all delays due to reconfigurable interconnections and achieves minimal communication times between various modules of the resource. A reconfigurable path between any two modules, PE and ME, of the resource requires the time of signal propagation through three consecutively connected AND gates. This delay is permanent for all pairs, PE and ME, and thus independent of the module positions in the resource.

Organization of between-module reconfigurations is also very simple. They are reduced to a change in each PE of only one code of selective activation, SEL, which then selects a new pair of connecting elements ASE and MSE, and thus establishes communication between this PE with a new module in the resource. The cost of this bus is also very low, since each ASE and MSE element may be implemented as copies of a universal module (LSI module). The bus does not require any external logic, and all connections between ASE, MSE, PE, and ME are pin-to-pin. The bus allows for significant modular expansion, since additional attachment of new resource modules, PE, ME, MSE, and ASE, requires practically no additional pins in each module of the resource. For instance, if one more than doubles the numer of PE and ME modules from 15 to 30 then each PE, MSE, and ASE will need only one additional pin, since code SEL now increases from four bits ($\log_2 15$) to five bits ($\log_2 30$). Therefore a linear increase in the number of resource modules PE and ME causes the number of pins required in each PE, ASE, and MSE to increase logarithmically.

Note that one of the major advantages of this bus is ease of establishing between-computer communications. Such between computer exchanges as X processor–Y memory, X processor–Y processor, and X memory–Y memory open new dimensions in speeding up the execution of complex algorithms requiring high interaction between parallel information streams.

Therefore, the use of a minimal delay memory–processor bus in a dynamic architectue significantly improves all five of the performance factors considered above that will be used in the evaluation. Let us assess the meanings of the five performance parameters as they apply to this bus.

12.1.1 Time Required for Architectural Reconfiguration

This bus minimizes the time of architectural reconfiguration. Indeed, as was shown in Sec. 8, for any dynamic architecture, architectural reconfiguration includes two actions:

1. Searching and writing new control ideas into all modules of requested resource.
2. Switching connecting elements of a reconfigurable memory–processor bus into new transfer modes determined by a new partitioning of the resource units into a set of concurrent computers and/or arrays.

For the given bus the reconfiguration time includes the time to search and write new control codes to modules of the requested resource, inasmuch as any switch of a connecting element to new transfer modes takes no time at all since it is performed dynamically by sending a new selective activation code (SEL) to the A-channel of the respective processor element, PE. On the other hand, reference [11] introduced one organization for storage control codes that allows one to have the time of search and write control codes equivalent to the time of 64-bit addition. Accordingly, for the bus, the time of architectural reconfiguration is miniml since it take the minimal time of action, (a), only.

12.1.2 Delays Introduced by Reconfigurable Interconnections

These delays are minimal. Indeed, a 16-bit path between a PE and a ME goes through two reconfiguration circuits in the ASE and the MSE where address broadcast through ASE takes $2t_d$, and 16-bit broadcast through MSE takes t_d where t_d is one gate delay. Therefore reconfiguration delay, $\Delta RD = 3t_d$ for any path between PE and ME.

Note that for parallel exchange between two processors $\Delta RD = 2t_d$ since a path between two PEs uses a pair of consecutively connected MSEs. The important feature of the bus is that the ΔRD time is constant and independent of the locations of PE and ME in the resource.

12.1.3 Time Required for Information Exchanges Between Independent Computers

This time is minimal. Indeed, a parallel data path X processor–Y memory can be established in one step, taking time $3t_d$, since for each path of communication between PE and ME, it requires activation of only two connecting elements, ASE and MSE. The parallel data paths for exchanges X processor–Y processor and X memory–Y memory are established in two steps, each taking the time, $3t_d$, since each PE has to activate two ASE and a PE may activate only one ASE at a time.

12.1.4 Modular Expansion

The bus is amenable to a significant modular expansion in spite of the fact that all the resource modules may have an LSI implementation, which as a rule, is very sensitive to a pin restriction. Indeed, if the existing system has new PE and ME modules added, in each PE and ASE the only pins that are affected by modular expansion are those for broadcasting code SEL. However, since the size of SEL grows logarithmically, the increase in pins is insignificant.

12.1.5 Cost of Realization

The cost of realization for this bus is low, because:

1. The use of copies of the universal module for implementing all modules of the resource (PE, MSE, and ASE).
2. Total exclusion of external logic in the bus. This allows organization of only pin-to-pin connections among the PE, ASE, MSE, and ME modules.

12.2 Assessment of the Minimal Complexity Memory–Processor Bus

Let us examine the behavior of the five performance parameters assumed for this bus.

12.2.1 Time Required for Architectral Reconfiguration

This is non-minimal, since during architectural transition, the bus requires that connecting elements MSE switch their transfer modes. As was shown in Sec. 5.6.4, this may be performed in two steps taking $4t_d$ and $2t_d$, respectively.

12.2.2 Delays Introduced by Reconfigurable Interconnections

This is non-minimal for instructions and minimal for $h \cdot k$-bit words. For instructions, the reconfiguration delay is $\Delta RD(k)=(k-1)t_d$. A $h \cdot k$-bit word is fetched with no reconfiguration delay.

12.2.3 Time Required for Information Exchanges Between Independent Computers

This is non-minimal. Ths bus allows time-shared transfer to h-bit bytes from one computer to another; i.e., a $h \cdot k$-bit word can be transferred during k clock periods. Another mode of between-computer exchange is by merging computers X and Y into a single computer. If computer X needs data stored in the memory of computer Y, then these two may merge into a single new computer which has access to the memory units storing the necessary data. But this requires reconfiguration into a new architectural state. Computers may also exchange words through I/O channels.

12.2.4 Modular Expansion

As far as pins are concerned modular expansion is practically unlimited, since the PE, MSE, and ME elements have a permanent number of pins that are independent of the number of modules, PE and ME. For the monitor unit, the number of communication pins grows logarithmically with an increase in the number of resource units. However, an arbitrary growth in the number K of PE and ME in one computer may be restricted by reconfiguration delay ΔRD, since for instruction fetch $\Delta RD = (k - 1)t_d$ and an increase in k leads to an increase in instruction fetch time.

12.2.5 Cost of Realization

This is minimal. Since not more than n independent computers may be formed concurrently, it takes $(n - 1)$ connecting elements to separate them. Thus the bus contains the minimal number of connecting elements. All MSE may be copies of a universal module. The bus requires no external logic and all connections between resource modules are pin-to-pin.

12.3 Assessment of Processor Busses

Each DC-group may use one of the two dedicated processor busses considered above: minimal delay or minimal interconnection (Sec. 5.5.1). Let us find performance factors for each of these busses.

12.3.1 Minimal Delay Processor Bus

12.3.1.1 Time required for architectural reconfiguration

This time is very insignificant. Accordingly, this bus introduces no additional time into the overall time of architectural reconfiguration for the system.

Indeed, the time to reconfigure this bus equals that of searching and writing two new codes—the significance code, b, and the computer size code, k—into the modules of the requested resource. However since codes b and k are among the other control codes which are fetched in parallel during reconfiguration and the time to search and write these codes is small, the reconfiguration of this bus requires no additional delay.

12.3.1.2 Delays introduced by reconfigurable interconnections

These delays are minimal. Indeed, for each processor element, PE, that may be dynamically connected to a new $16k$-bit processor, the time of reconfiguration delay, ΔRD is that of one logical circuit L_1 (Fig. III.17): i.e., $\Delta RD = t_d$ is one gate delay. For a $16k$-bit processor assembled from k PEs, $\Delta RD(k) = k \cdot t_d$. This value achieves an absolute minimum since each L_1 introduces a minimal delay, and the number k is determined by the word size of an $h \cdot k$-bit processor.

12.3.1.3 Time required for information exchanges between computers

This criterion is not applicable in the evaluation of processor busses.

12.3.1.4 Modular expansion

For the minimal delay processor bus, modular expansion is limited. Indeed, since each PE of this bus takes $n + 1$ pins, where n is the number of PEs in the resource, attachment of new PEs to the system is restricted by the current pin count per module.

12.3.1.5 Cost of realization

The cost of this bus is very low. Indeed, this bus supports the concept of a unique PE, since each processor module contains the same logical circuit L_1 selectively activated by the different codes k and b written to all modules of requested resource. The bus requires no external logic and all interconnections among modules are pin-to-pin.

12.3.2 Minimal Interconnection Processor Bus

12.3.2.1 Time required for architectural reconfiguration

Since to reconfigure this bus one has to write only one new code b to each PE, the time of bus reconfiguration is insignificant and does not affect the time of a system's transition to a new state.

12.3.2.2 Delays introduced by reconfigurable interconnections

This time is twice as great as that of the minimal delay processor bus. Indeed, each module, PE, introduces a reconfiguration delay $\Delta RD = 2t_d$, where one t_d is for carry and the other is for overflow. For a $h \cdot k$-bit processor assembled from k PEs, $\Delta RD(k) = 2k \cdot t_d$.

12.3.2.3 Modular expansion

This bus allows unlimited modular expansion. Indeed, the bus takes only four pins in each module regardless of the number of processor modules in the resource. This number of pins is minimal for $n \geq 4$. Indeed, if the processor resource contains two PEs ($n = 2$), it takes two pins in each module to organize the reconfigurable end-around-carry path. For $n = 3$, it takes 3 pins in each module (one for carry-in, two for 2 overflows). For n PEs, PEs, it takes n pins in each module (one for carry-in, $n - 1$ for $n - 1$ overflows, respectively). However, for the minimal interconnections bus, only four pins are required in each PE regardless of the number, n, of PEs. It then follows that for $n \geq 4$, the minimal interconnection bus takes the minimal number of pins in each PE.

12.3.2.4 Cost of realization

The cost of this bus is low, inasmuch as the bus may be assembled from identical PEs. Each PE contains the same reconfiguration logic connected with the same pins in each module and selectively activated by codes k and b. The interconnections among PEs are also pin-to-pin since the bus requires no external logic to connect the PEs of the resource.

13. SUMMARY

In improving the performance of a dynamic architecture a major choice is in selection of the reconfigurable busses that perform the redistribution of the hardware resources among the independent computers. Above we analyzed four such busses:

1. Minimal delay memory–processor bus
2. Minimal complexity memory–processor bus
3. Minimal delay processor bus
4. Minimal interconnection processor bus

These busses allow simple modular expansion and they support flexible information exchanges between modules of the resource integrated in the same or different computers. Another feature of these busses is that they allow the use of a universal LSI module type which serves as a unique building block for all functional units of dynamic architecture (processors, I/O, monitors, connecting units). The organizations introduced provide a total exclusion of external logic which connect LSI modules: All resource LSI modules are connected only via pin-to-pin connections.

PROGRAMMING SYSTEMS WITH DYNAMIC ARCHITECTURE

14. INTRODUCTION

To realize the capability of dynamic architectures to increase the throughput using the available resources, the following problem has to be solved. For each user program one has to find a sequence of minimal size computers which may execute it. Next, the available hardware resources of the multicomputer system have to be assigned among user programs, each of which is executed by a sequence of the minimal size computers found earlier. The hardware resource assignment is then reduced to finding a flow chart of architectural states which gives the maximal concurrency in execution of the given set of user programs.

Overall assignment of processor and memory resources for dynamic architectures is performed by the adaptation system. This system includes two subsystems: *assignment* and *monitor*. The first system, assignment, assigns the hardware resources to programs and constructs a flow chart of architectural transitions. The monitor subsystem supervises the correct execution of the flow chart of architectural transitions. Since basic concepts for the monitor system were discussed in Sec. 5.6., here we will focus on the assignment system only.

Since most user programs are written in a high-level language, the assignment subsystem begins the hardware resource assignment by finding, for each high-level statement, maximal data word sizes and sizes of data arrays for computed results. Then for each user program, it finds the diagram of computer sizes which are used in computations of this program [12]. Since a multicomputer system executes several independent programs, their computer sizes diagrams then are merged into a diagram showing the assignment of the resource among user programs. This diagram serves as a basis for the construction of an architectural flow chart; i.e., the assignment subsystem constructs the flow chart of transitions showing what sequence of architectural states is formed and which program is allowed to perform each transition. Using this flow chart, the assignment subsystem performs memory scheduling among instruction and data arrays.

For algorithms computed in real-time which are specified by response restrictions, the resource assignment is done on the basis of statistical meanings of data words and loop iterations. Indeed, for each such algorithm, the most probable (statistical) ranges of initial data words and loop iterations are known in advance. The techniques that are presented here allow one to estimate bit sizes and array volumes of computed results. Using these data, the resource is dynamically partitioned into a sequence of concurrent computer sets which maximize the number of executed tasks. If, during any one computation, a particular data word size exceeds that of a computer, this data size is reduced via transformation into floating-point form. However, for dynamic architectures the number of such transformations is minimal since the techniques presented below allow preliminary evaluation in bit sizes of computed results and construction of a flow-chart diagram that takes into account this evaluation.

We now elaborate on the basic principles of designing assignment system. Techniques of program analysis aimed at assigning available

15. FINDING THE HARDWARE RESOURCES FOR A USER'S PROGRAM

resources—both processor and memory—and constructing a flow chart of architectural transitions are presented.

For each program written in a high-level language, we have to find word sizes for a sequence of computers that could execute this program, i.e., the program should be partitioned into tasks, each of which is executed by a new size computer. To perform this partitioning it is necessary for each statement in the high-level language to find the maximal bit sizes of all *computed variables* executed in this statement. This will give a bit size for the computer which can execute this statement. Finding word sizes for computed variables is greatly simplified if we construct a *program graph*† in which each node includes one or more statements and where the arrows show sequencing. For each node, we find a minimal computer size in which this node may be executed. Next, we construct a diagram of bit sizes for all graph nodes which serves as a basis for finding a diagram of the computer sizes required for executing a given user program. This diagram will partition the program into tasks where each task is composed of the sequence of nodes requiring the same computer size.

In order to perform efficient assignment of the hardware resources, it is necessary to estimate the duration for each task. This will allow us to perform a resources assignment among tasks that will minimize the time of idleness of executed programs. Section 16 introduces a technique for finding the number of clock periods, t_0, required for the execution of each task. Since all computers with dynamic architectures use the same clock period, t_0 [9], expressing the duration of all tasks in terms of the same t_0 shows the relative timing in their executions. The analysis of each user program ends with the construction of its resource diagram which shows computer sizes and memory volumes requested by evey task in this program.

†See Chap. IV, Sec. 3, where a similar model is used for finding infinite loops in programs. This section also performs an interesting survey of various program graphs representations used to represent an algorithm (program). (Ed.)

15.1 Bit Sizes of Computed Variables

A program, P, written in a high-level language is transformed into a P *program graph* in which each node is assigned one or several statements and arrows show sequencing of nodes. There are two types of nodes in the graph: *simple* and *complex*. A *simple node* has one outgoing arrow, which means an an unconditional transition to its immediate successor. A *complex node* has several outgoing arrows. Selection of its immediate successor depends on a conditional test executed by a control statement in this mode.

Nodes in the graph may be *iterative* and *noniterative*. Each *iterative node* is specified with the parameter, Z, which is the maximal number of iterations it may go through in a program loop. Parameter Z may be given as follows:

1. It can be presented explicitly in the program.
2. It can be defined via analysis of a control statement which specifies an exit from the program loop, or,
3. In case it can not be defined via 1. or 2. its statistical value can be given based on experience with past computations of the same program.

A program's transformation into a program graph is formalized and specified in the following section.

15.1.1 Algorithm for Constructing a Program Graph (Algorithm 1)

The objective of this algorithm is to assign high-level statements to graph nodes and to determine sequencing among those nodes. The algorithm assumes that the program is stripped of all the unnecessary references and that all referencing is done for conditional or unconditional branching or for looping. This algorithm includes several separate sequences of steps aimed at handling the following types of statements:

Type 1. A non-control statement that is not referenced by any other statement.
Type 2. A statement referenced by another control statement.
Type 3. All control statements (except the DO statement).
Type 4. The DO statement.
Type 5. The DO reference statement or the DO object.

The flow chart of the algorithm is shown in Fig. III.36. Let us show how this algorithm handles the five types of statements mentioned above.

Let $A*$ be the statement under analysis.

Type 1 Statements

If $A*$ is a statement of Type 1, it is assigned to the same node that includes the preceding statement, B, unless B is a control statement.

If B is a control statement, $A*$ is assigned a separate node. This procedure is established in Steps 2, 3, 4, 5, 25, and 6. The tests in Steps 2 and 3 identify statement $A*$ as Type 1; Steps 4 and 5 find whether or not the preceding statement, B, is a control statement; Step 25 creates a new node for $A*$ if B is a control statement; and, if B is not, Step 6 includes $A*$ into the same node that includes B.

352 CHAPTER III

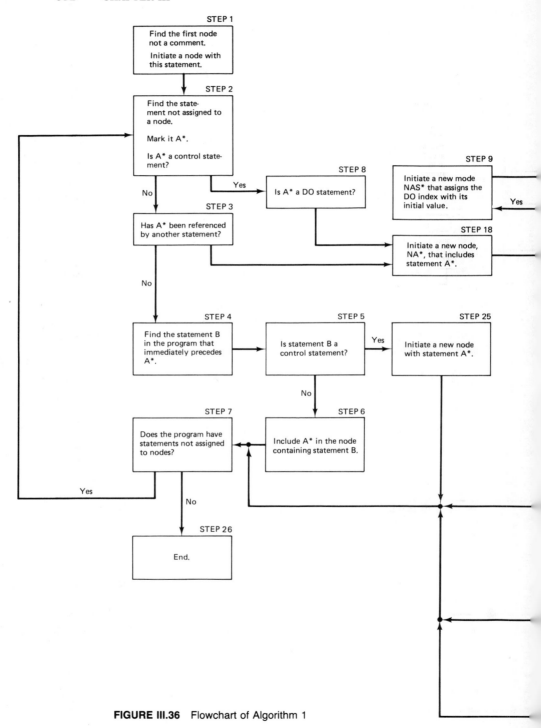

FIGURE III.36 Flowchart of Algorithm 1

Designing and Programming Supersystems with Dynamic Architectures 353

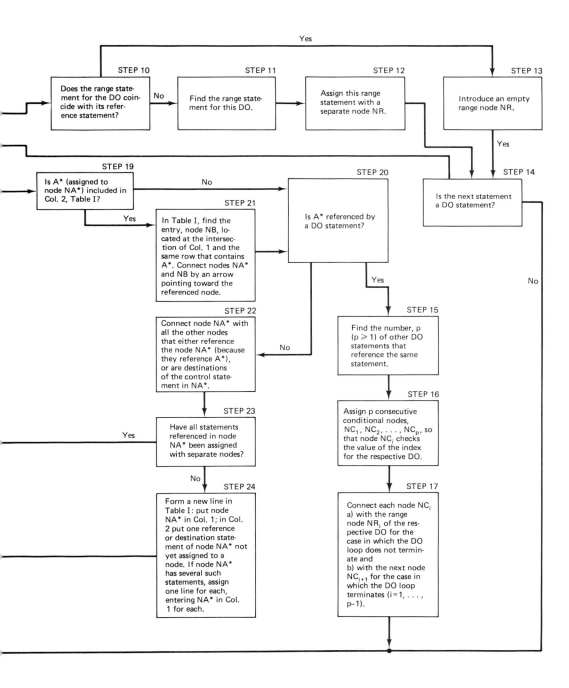

Type 2 Statements

If A^* is a Type 2 Statement, it must be assigned a separate node, NA^*, since it is the destination of one or more control statements. This is done in Steps 2, 3, and 18.

If a control statement, B, that references A^* has been assigned to the node, NB, nodes NB and NA^* should be connected by the arc, $NB \rightarrow NA^*$ (Step 22).

If the control statement B has not yet been assigned to a node since it follows (not necessarily immediately) statement A^* in the program, the $NB \rightarrow NA^*$ connection between the current node NA^* and the not yet created node NB should be established in the future, during the formation of node NB. Therefore, this connection must be remembered in a special table, Table 1, comprising two columns: Column 1 contains graph nodes already assigned with statements; and Column 2 contains source or destination statements not yet formed into nodes (Steps 24). (Here, by a *source statement* we mean a statement that references another statement. By a *destination statement* we mean a statement that is referenced by another statement.) Therefore, each time a new node, NA^*, is formed that includes a statement A^* of Type 2, a test is made to check if A^* is in Table 1, Column 2 (Step 20). If it is, the newly formed node NA^* is connected to the source node(s), NB of Column 1 that were formed previously and reference A^* (Step 21). All these actions that handle Type 2 statements are performed by Steps 2, 3, 18, 20, 19, 21, 22, 23, 24.

Type 3 Statements

If A^* is a Type 3 Statement, it must be assigned to a separate node, NA^* (Steps 2, 8, 18), that will be connected with all other nodes that are destinations of NA^* (Step 22). Also, A^* must be tested for inclusion in Table 1, Column 2. If A^* is in Column 2, it references node NB, formed previously, where node NB is located in Column 1, Table 1. Thus the $NA^* \rightarrow NB$ connection must be formed (Steps 20 and 21).

When node NA^* is formed, some of its destinations may not be assigned to separate nodes, so those that are not yet assigned must be put in the Column 2 entries of Table 1, where the Column 1 entries contain node NA^* (Step 24).

Type 4 and Type 5 Statements

If A^* is a DO statement, it must be treated as follows.

Each DO must be assigned a separate node, NAS, that assigns the DO index to its initial value (Step 9). The NAS node must be followed by the range node NR, which includes the DO's range statement(s). If the range statement for the DO coincides with its reference statement, then the range node should be an empty node. Therefore, steps 10, 11, 12, and 13 are aimed at finding out what type of node should be assigned as the DO range—empty or non-empty.

Example 36. In the program sequence below the following assignment of *NAS* and *NR* nodes is performed:

$$DO\ 2\quad I = 1,\ 100$$
$$DO\ 1\quad J = 1,\ 100$$
$$1\quad A[I,J] = 0$$
$$2\quad A[I,I] = 1$$

First we form the assignment node *NAS* 2 for the statement DO 2 that performs the initialization $I = 1$ (Fig. III.37). *NAS* 2 is followed by an empty range node *NR* 2, since the range and reference statements for this DO coincide. Next, during a second iteration of Step 9 we form a second initial assignment node *NAS* 1 for the statement DO 1 that sets $J = 1$. The range node *NR* 1 is also empty since the reference and range statements for this DO match as well. ∎

Example 37. The sequence

$$DO\ 17\ I = 1,\ 11,\ 2$$
$$X = I - 1$$
$$DO\ 17\ J = 1,\ 5$$
$$Y = J - 3$$
$$17\ \text{Print},\ X,\ Y,\ Z = X^{**}2 + Y^{**}2$$

has two non-empty range nodes; *NR* 1 containing $X = I - 1$, and *NR* 2 containing $Y = J - 3$. These follow assignment nodes *NAS* 1 and *NAS* 2, respectively, where *NAS* 1 sets $I = 1$ and *NAS* 2 sets $J = 1$.

Following the formation of the range nodes, the algorithm begins a new loop in Step 2 aimed at locating the DO reference and assigning it to a separate node, *NA** (Steps 18 and 20). Node *NA** may be referenced by p other DO's, where $p \geq 1$. In this case it must be succeeded by p conditional nodes NC_1, $NC_2, \ldots, NC_p$, each of which checks the index of the respective DO associated with it. For the case that the DO loop terminates, NC_i is connected with NC_{i+1}. If the DO does not terminate, NC_i is connected with its range node, NR_i. ∎

Example 38. Let us illustrate the entire procedure for the program sequence of Example 36. Following the formation of assignment nodes *NAS* 2, *NAS* 1 and the empty range nodes *NR* 2 and *NR* 1, the algorithm creates

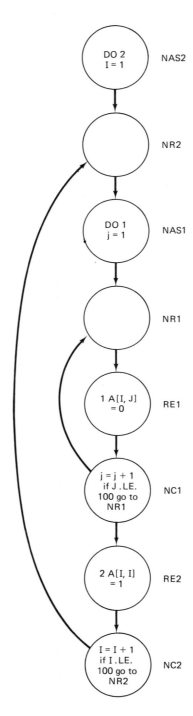

FIGURE III.37

the reference node $RE1$ for the statement DO 1 and assigns to it $RE1 = 1$ $A[I,J] = 0$ (Step 18). It next follows $RE1$ by the conditional node $NC1$ which increments the DO index J and checks whether its value terminates the DO loop. Thus $NC1$ is connected with its range node $NR1$ and with the reference node for the statement DO 2 $I = 1, 100$; etc. ∎

Example 39. Consider a program to tabulate the function $B = (C^2 + 3C + 2)(C - 8)$ for $C = 0, 4, \ldots, 16$, but B skips the value when $C - 8 = 0$. Figure III.38 shows an expression of this program in Fortran. Fig. III.39 shows the respective program graph built in accordance with the Algorithm 1. The Algorithm assigns node 1 to the first IF. Statements 2 and 3 referenced by IF are also assigned separate nodes 2 and 3. Statement Print, B, C is included into node 2, since it is not a control statement nor is it referenced by other nodes. Statement IF $(C - 16), 1, 1, 4$ is assigned node 4, since it is a control statement. It is followed by statement Stop, assigned to node 5, since it is referenced by node 4. ∎

15.1.2 Finding the Number of Iterations in the Loop

Each variable computed in a node of the program graph ought to be analyzed to obtain its maximal bit size and the dimension of the data array required for its storage.

To this end, one first specifies for each node its *parameter of iteration*, Z, showing the maximal number of iterations this node can execute. To find Z, one has to mark all nodes in the graph which initiate program loops. For each such node, i, one finds the maximal number of iterations, Z_i, of the respective loop, and this number is assigned to all nodes included into the loop. If node b belongs to several program loops which iterate $Z_a, Z_c, \ldots, Z_n$ times, respectively, then the parameter of iteration, Z_b, of this node is:

$$Z_b = Z_a \cdot Z_c \cdot \ldots \cdot Z_n \qquad (1)$$

```
        C = 0
  1     IF(C – 8.)2, 3, 2

  2     B = (C * C + 3. * C + 2.) * (C – 8.)
        PRINT, B, C

  3     C = C + 4
        IF(C – 16.)1, 1, 4

  4     STOP
        END
```

FIGURE III.38 Expression of program in FORTRAN

358 CHAPTER III

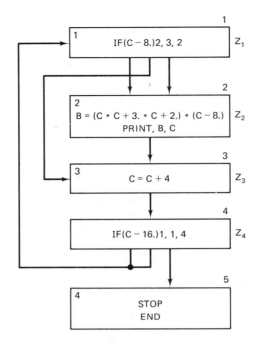

FIGURE III.39 Program graph constructed in accordance with algorithm 1

Typically, for a program loop, the number of its iterations, Z, may be given either explicitly or determined via analysis of a control statement IF which specifies an exit from the loop. Usually this statement performs a conditional test between two variables X and Y: $X > Y$, $X < Y$, $X \geq Y$, $X \leq Y$, $X = Y$, $X \neq Y$, etc. X and Y may be computed variables which are changed during each iteration. (If one variable is a 0 or a constant, it is a particular case of a more general condition that X and Y change.)

Let Δ_1 and Δ_2 be additive increments for X and Y, i.e., $X_i = X_{i-1} + \Delta_1$ and $Y_i = Y_{i-1} + \Delta_2$. Assume that the control statement specifying exit from the loop is: IF $(X . \text{GT}. Y)$ GO TO N. (Similar techniques may be used for other types of IF). Since the loop has to iterate Z times until $X > Y$, if $Z > 0$ then in the beginning of the iterative process (during the first iteration) $X_0 < Y_0$. Also in order that X be greater than Y in the future (after Z iterations), $\Delta_1 > \Delta_2$. After Z iterations, the following statement will become true for the first time:

$$X_0 + Z \cdot \Delta_1 > Y_0 + Z \cdot \Delta_2 \qquad (2)$$

If X and Y are changed via multiplication, i.e., $X_I = X_{i-1} \cdot \Delta_1$ and $Y_i = Y_{i-1} \cdot \Delta_2$, then after Z iterations

$$X_0 \cdot \Delta_1^Z > Y_o \cdot \Delta_2^Z \tag{3}$$

It follows from Eq. (2), that after $(Z-1)$ iterations $X_0 + (Z-1) \cdot \Delta_1 \leq Y_0 + (Z-1) \cdot \Delta_2$, or

$$Z \leq \frac{Y_0 - X_0}{\Delta_1 - \Delta_2} + 1 \tag{4}$$

If X and Y are changed via multiplication, after $(Z-1)$ iterations

$$X_0 \cdot \Delta_1^{Z-1} \leq Y_0 \cdot \Delta_2^{Z-1}, \quad \left(\frac{\Delta_1}{\Delta_2}\right)^{Z-1} \leq \frac{Y_0}{X_0}; \quad \text{and}$$
$$(Z-1) \cdot (\log_2 \Delta_1 - \log_2 \Delta_2) \leq \log_2 Y_0 - \log_2 X_0.$$

$$Z \leq \frac{\log_2 Y_0 - \log_2 X_0}{\log_2 \Delta_1 - \log_2 \Delta_2} + 1 \tag{5}$$

For both Eqs. (4) and (5), X_0 and Y_0 are the initial values of variables X and Y in the control statement, IF $(X.\text{GT}.Y)$ GO TO N, and Δ_1 and Δ_2 are their increments during iteration. If Y is constant, the $\Delta_2 = 0$ and instead of Eq. (4) one obtains

$$Z \leq \frac{Y_o - X_0}{\Delta_1} + 1 \tag{6}$$

and instead of Eq. (5) one obtains

$$Z \leq \frac{\log_2 Y_0 - \log_2 X_0}{\log_2 \Delta_1} + 1 \tag{7}$$

Similar formulas can be developed for other IF statements and for other changes in the X and Y variables during iteration.

Example 40. Assume that in IF $(X.\text{GT}.Y)$, X and Y change via addition, $X = X + \Delta_1$ and $Y = Y + \Delta_2$ and $X_0 = 4$, $Y_0 = 16$, $\Delta_1 = 2$ and $\Delta_2 = 1$. Applying Eq. (4) one obtains $Z = \frac{16 - 4}{2 - 1} + 1 = 13$. Indeed, condition $X.\text{GT}.Y$ is true after 13 iterations because the value of X is $X = X_0 + \Delta_1 \cdot Z = 4 + 2 \cdot 13 = 30$ and $Y = Y_0 + \Delta_2 \cdot Z = 16 + 1 \cdot 13 = 29$; i.e., $X > Y$. After 12 iterations $X > Y$ is false because $X = 4 + 2 \cdot 12 = 28$ and $Y = 16 + 1 \cdot 12 = 28$; i.e., $X = Y$. ∎

Example 41. Let X and Y change via multiplication and $X_0 = 3$, $Y_0 = 35$, $\Delta_1 = 5$, $\Delta_2 = 2$. Applying Eq. (5), one obtains

$$Z \leq \frac{\log_2 35 - \log_2 3}{\log_2 5 - \log_2 2} + 1 = \frac{6 - 3}{3 - 1} + 1 = \frac{4}{2} + 1 = 3$$

Indeed, after three iterations $X = 3 \cdot 5^3 = 375$ and $Y = 35 \cdot 2^3 = 280$ and

$X > Y$ is true. After two iterations $X = 3 \cdot 5^2 = 75$, $Y = 35 \cdot 2^2 = 140$; i.e., $X > Y$ is false.

Handling Exceptions

There are instances when the parameter of iteration, Z, is not given explicitly, nor can it be determined analytically.

This is true for the case whan a loop containing node b operates with initial variables, X_0 and Y_0, that are computed in another loop. All such cases can be handled by assuming that Z is a maximum of the statistical values of Z observed during previous computations of the algorithm. Also if a loop, L_2, uses an initial variable A_0, computed in another loop, L_1, then one first finds the bit size of the variable A_0 via analysis of the loop L_1 and then assumes that A_0 is the maximal word that can be stored in the bit size that was found. This word can then be taken as an initial variable for loop L_2. One can treat a sequence of loops, $L_1, L_2, \ldots, L_k$, in which loop L_i computes initial variables for loop L_{i+1} in a similar way.

If the actual number of iterations is larger than its statistical estimates, then the actual bit size, BS, of node b may exceed the analytical bit size, BS', obtained during bit size analysis. Since a selected computer size may coincide with BS', the actual bit size, BS, must be reduced to BS' using a special floating-point operator discussed in Sec. 3.2.

15.1.3 Two Types of Bit Sizes

After finding the iteration parameter of each node, one examines all nodes which contain computed variables; i.e., those which are assigned by the program to changeable values. For each computed variable, X, belonging to a node, one defines two parameters:

1. *Its maximal bit size, $BS(X)$*, which shows the maximal number of bits required to store X.
2. *Its maximal intermediary bit size, $BS(X^*)$*, which shows the maximal bit size of all the intermediate results, which the computer forms in calculation X.

Note that for complex arithmetic expressions, the bit size of intermediary results, $BS(X^*)$, may exceed that of the final result, $BS(X)$. For this case $BS(X^*)$ identifies the bit size of the computer which may compute the corresponding computed variable X. $BS(X)$ is increased when the computer executes addition, multiplication, and shifts toward the most-significant bit in its computation (squares, cubes, etc., are executed by the computer as iterative multiplications), whereas divisions and subtractions reduce the bit size of computed variables.

15.1.4 Bit Sizes of Integers

We determine the bit sizes of integers involved in operations leading to an increase in bit sizes (additions and multiplications). The next paragraph will extend this analysis to real and floating-point variables.

15.1.4.1 Addition

We distinguish two possible options for addition, $X = A + B$.

1. *Nonrecursive addition.* X is assigned neither to A nor B; i.e., the addition is not of the form $X = X + A$ or $X = X + B$; then

$$BS(X) \leq \max [BS(A), BS(B)] + 1 \qquad (8)$$

where $BS(A)$ and $BS(B)$ are maximal bit sizes of the computed variables A and B, respectively, obtained after Z iterations.

2. *Recursive addition.* X was assigned either A or B; i.e., $X = X + A$ (or $X = X + B$) where this addition iterates Z times.
The bit size of a computed variable X is upper bounded as

$$BS(X) \leq \max [BS(X_0), BS(A)] + \log_2 (1 + Z) \qquad (9)$$

where $BS(X_0)$ is the bit size of X variable before iterative process; if A is a constant, $BS(A)$ is its bit size; if A is a computed variable, then it is its final bit size found after iterations.

Equation (9) can be obtained as follows: Assume that A is a constant, i.e., it does not change during iterations, and find $BS(X)$ for this case. The $BS(X)$ for a changeable A will be obtained from $BS(X)$ found for a constant A. If A is constant, after the first iteration $X_1 = X_0 + A$, after the second iteration, $X_2 = X_1 + A = X_0 + 2A$; ..., after the Zth iteration, $X_z = X_0 + Z \cdot A$.

Find the maximal value assumed by X. Since $X_0 \leq 2^{BS(X_0)} - 1$ and $A \leq 2^{BS(A)} - 1$, then X is upper bounded as $X \leq (2^{BS(X_0)} - 1) + Z \cdot (2^{BS(A)} - 1)$. Let $M = \max [BS(X_0), BS(A)]$. Then $X \leq (2^M - 1) + Z \cdot (2^M - 1) = (2^M - 1) \cdot (1 + Z)$. It then follows that $BS(X) \leq M + \log_2 (1 + Z) = \max [BS(X_0), BS(A)] + \log_2 (1 + Z)$. Thus one obtains Eq. (9).

Example 42. Let $X_0 = 5$, $A = 9$, and $Z = 14$; find $BS(X)$ after 14 iterations. Using Eq. (9), we find that $BS(X) \leq \max [BS(5), BS(9)] + \log_2 (1 + 14) = 4 + 4 = 8$ bits. Actual value of X after 14 iterations is $X = 5 + 14 \cdot 9 = 131$. Its bit size is 8. Therefore a correct bit size was found. ∎

If for recursive addition, $X = X + A$, A also changes during iterations, then before finding $BS(X)$, one finds $BS(A)$ and substitutes the value found to the Eq. (9).

Example 43. Let $X + A$ iterate $Z = 16$ times and during each iteration A also changes as $A = A + B$. Assume that in the beginning $X_0 = 5$, $A_0 = 4$, $B = 9$; i.e.

After the first iteration: $X_1 = X_0 + A_0 = 9$

After the second iteration:
$$X_2 = X_1 + A_1 = X_0 + A_0 + A_0 + B = X_0 + 2A_0 + B$$

After the third iteration: $X_3 = X_2 + A_2 = X_0 + 3A_0 + 3B$

After the Z-th iteration:
$$X_z = X_{z-1} + A_{z-1} = X_0 + Z \cdot A_0 + [1 + 2 + \cdots + (Z-1)] \cdot B$$

Using Eq. (9), we first find $BS(A) = \max[BS(9), BS(4)] + \log_2(1 + 16) = \max(4, 3) + \log_2(1 + 16) = 4 + 5 = 9$ bits. Then $BS(X) = \max[BS(5), 9] + \log_2(1 + 16) = 9 + 5 = 14$ bits. The actual value of X is $X = 5 + 16 \cdot 4 + (1 + 2 + \ldots + 15) \cdot 9 = 1149$. Its bit size is 12. Therefore a correct upper bound was found. ∎

15.1.4.2 Multiplication

Consider three possible options for multiplication $X = A * B$

1. *Nonrecursive multiplication with complex changeable factors A and B.* At least one of the values A or B is described by expressions containing more than one arithmetic operation. (For instance, $A = C^2 - D + E$, or $B = K * D + c$, etc.) For this case

$$BS(X) \leq BS(A) + BS(B) \tag{10}$$

where $BS(A)$ and $BS(B)$ are maximal bit sizes of the computed variables A and B respectively obtained after Z iterations.

2. *Nonrecursive multiplication with simple changeable factors.* In $X = A * B$, X assumes neither A nor B and A and B are described by arithmetic expressions containing not more than one operation.
Suppose that A and B are variables involved in bit increasing operations—addition and multiplication.
   a. *A and B are changed via addition.* That is, $X = (A + \Delta_1)*(B + \Delta_2)$ where Δ_1 and Δ_2 are A and B increments during each interation. For this case:

$$BS(X) \leq 2BS(\max[A_0, B_0, \Delta_1, \Delta_2]) + 2\log_2(1 + Z) \tag{11}$$

where $\max[A_0, B_0, \Delta_1, \Delta_2]$ is the maximal value among four meanings A_0 and B_0 (before iteractive process) and Δ_1 and Δ_2.
   Let us show that Eq. (11) is valid. After the first iteration, $X_1 = A_1 * B_1 = (A_0 + \Delta_1)*(B_0 + \Delta_2)$; after the second iteration,

$X_2 = A_2 * B_2 = (A_1 + \Delta_1) * (B_1 + \Delta_2) = (A_0 + 2\Delta_1) * (B_0 + 2\Delta_2), \ldots,$ after the Zth iteration $X_z = A_z * B_z = (A_0 + Z \cdot \Delta_1) * (B_0 + Z \cdot \Delta_2) = A_0 * B_0 + A_0 * \Delta_2 * Z + B_0 * \Delta_1 * Z + \Delta_1 * \Delta_2 * Z^2$. Let us evaluate the bit size of this expression. First, find the maximal value $A_m * B_m = \max[(A_0 * B_0), (A_0 * \Delta_2), (B_0 * \Delta_1), (\Delta_1 * \Delta_2)]$. Then after Z iterations, $X \leq A_m * B_m * (1 + 2Z + Z^2) = A_m * B_m * (1 + Z)^2$ and $BS(X) \leq BS(A_m) + BS(B_m) + 2 \log_2 (1 + Z) \leq 2BS(A_m) + 2 \log_2 (1 + Z)$, where $A_m = \max(A_0, B_0, \Delta_1, \Delta_2)$.

Example 44a. Let initial factors A_0 and B_0 be $A_o = 14$, $B_0 = 7$, their increments $\Delta_1 = 4$, $\Delta_2 = 6$ and the number of interations $Z = 19$. Using Eq. (11), find that $BS(X) \leq 2BS(\max [14, 7, 4, 6]) + 2 \log_2 (1 + 19) = 2BS(14) + 2 \log_2 20 = 2 \cdot 4 + 2 \cdot 5 = 18$ bits. ∎

b. *A and B are changed via multiplication.* That is, $X = (A * \Delta_1) * (B * \Delta_2)$, where Δ_1 and Δ_2 are multiplicative factors which change A and B during each iteration.

The bit size of the product is upper bounded as

$$BS(X) \leq BS(A_0) + BS(B_o) + Z \cdot [BS(\Delta_1) + BS(\Delta_2)] \qquad (12)$$

Indeed, after the first iteration, $X_1 = A_1 * B_1 = A_0 * B_0 * \Delta_1 * \Delta_2$; after the second iteration, $X_2 = A_2 * B_2 = A_0 * B_0 * (\Delta_1 * \Delta_2)^2; \ldots$; after the Zth iteration, $X_z = A_z * B_z = A_0 * B_0 * (\Delta_1 * \Delta_2)^z$. Taking the bit size of this expression leads to Eq. (12).

Example 44b. Let $A_0 = 13$, $B_0 = 7$, $\Delta_1 = 2$, $\Delta_2 = 3$, $Z = 4$. Using Eq. (12), find $BS(X)$: $BS(X) \leq BS(13) + BS(7) + 4 [BS(2) + BS(3)] = 4 + 3 + 4(2 + 2) = 7 + 16 = 23$ bits. ∎

Note that Eqs. (11) and (12) assume that in the product $(X) = Z * B$ both values A and B change via bit increasing operations. If only one value, say, A changes while B remains constant, then $\Delta_1 \neq 0$, $\Delta_2 = 0$ $\Delta_2 = 0$ and Eq. (11) is transformed into

$$BS(X) \leq 2BS(\max A_0, B_0, \Delta_1) + \log_2 (1 + Z) \qquad (13)$$

Likewise Eq. (12) is transformed into

$$BS(X) \leq BS(X_0) + BS(B_0) + Z \, BS(\Delta_1) \qquad (14)$$

Equations (13) and (14) can be used in case product $X = A * B$ grows, grows, although Δ_1 is a bit increasing increment (addend or multiplier) and Δ_2 is a bit decreasing increment (subtrahend or divisor). For this case we may assume that $\Delta_2 = 0$; i.e., the value of B is a constant. Then Eqs. (13) and (14) will give a correct upper bound. If both A and B are changed via division or subtraction during Z

iterations, then the bit size of the product $BS(X)$ after Z iterations will be smaller than that before iterations. As a result there is no need to specify $BS(X)$ at all, since we will have a bit decreasing operation.

3. *Recursive multiplication.* X was assigned either A or B; i.e., $X = X*A$ (or $X = X*B$), and multiplication iterates Z times. Then the bit size of the X variable is upper bounded as

$$BS(X) \leq BS(X_0) + Z \cdot BS(A) \tag{15}$$

If A is a constant, then after the first iteration, $X_1 = X_0*A$; after the second iteration, $X_2 = X_1*A = X_0*A^2$; ... ; after the Zth iteration, $X_Z = X_0*A^Z$. Then taking the bit size of this equation leads to Eq. (15). If A changes during the iterations then one treats A as a computed variable and finds $BS(A)$ using one of the equations introduced above. Then $BS(X)$ is found by substituting $BS(A)$ to Eq. (15).

Example 45. Let $X = X*A$ iterate $Z = 3$ times and during each iteration A also change as $A = A + B$. Assume that in the beginning $X_0 = 3$, $A_0 = 7$, and $B = 3$. First find $BS(A)$ after three interations. Since A changes via recursive addition with constant addend B, we use Eq. (9) to find $BS(A) \leq \max [BS(A_0), BS(B)] + \log_2 (1 + 3) = \max [3, 2] + \log_2 (1 + 3) = 3 + 2 = 5$ bits. Then using Eq. (15) and substituting $BS(A)$ into it, we obtain $BS(X) \leq BS(3) + 3 \cdot 5 = 2 + 15 = 17$ bits. The actual value of product is 2730. To store it takes 13 bits. ■

Above we considered bit sizes of integers participating in bit increasing operations—additions and multiplications. Similar formulas can be obtained for integers involved in bit decreasing operations—subtractions and divisions. For instance, for subtractions we have to subtract $\log_2 (1 + Z)$ in Eq. (9), and Eq. (8) is transformed into $BS(X) \leq \max [BS(A), BS(B)]$, where $X = A - B$. For divisions $X = A \div B$, and Eq. (10) is transformed to $BS(X) \leq BS(A) - BS(B)$. Similar transforms can be extended to Eqs. (11) to (14).

15.1.5 Bit Sizes of Real and Floating-point Variables

Let us show that the techniques introduced above can be extended to real and floating-point variables.

1. *Real variables.* The bit size of a real variable, BSR, is

$$BSR = BSI + BSF \tag{16}$$

where BSI and BSF are the bit sizes of the integer and fractional portions. BSI may be obtained via the same techniques considered above for integers; BSF is determined in advance by the user, since for most fractions there exists no exact conversions to the binary form. Therefore the number of bits a fractional number will take for its storage is determined by required accuracy of computation. It then follows that *BSF* must be given in advance. Having found *BSI*, and substituting it into Eq. (16) one finds the bit size of a real variable.

Henceforth, the analysis presented above can be extended not only to integers but also to real numbers.

2. *Floating-point variables*. The bit size of a floating-point number is determined as follows. Since mantissa is a fraction, its bit size is specified in advance by required accuracy of computation; as for exponents they may be treated as integers and their bit sizes may be determined using the formulas developed above for operations over integers.

For instance, for multiplication of two floating-point numbers,† the bit size of the mantissa-product remains the same; the bit size of the exponent of the result is determined using the formulas which give the bit size of a sum of two integers, because during multiplication, exponents of operands are added.

†An algorithm of multiplication for two floating-point numbers is discussed in Chap. I, Sec. 8.2. (Ed.)

Example 46. Suppose that for a recursive floating-point multiplication $X = X*A$ where the product $X = XE \times XM$, XE is the exponent, and XM is mantissa. Similarly, $A = AE \times AM$. Let the required accuracy of the mantissa operation be 10 bits, the multiplication iterates $Z = 10$ times, the exponent AE of the multiplier is constant and $AE = 5$, and the initial value XE_0 of the XE exponent is $XE_0 = 3$. Since $XE_i = XE_{i-1} + AE$, $(i=1,\ldots,Z)$, the exponent operation is reduced to recursive addition. Thus, using Eq. (9), we may find the bit size of the exponent result, XE. $BS(XE) \leq \max [BS(XE_0), BS(AE)] + \log_2(1 + Z) = \max[BS(3), BS(5)] + \log_2(1 + 10) = \max(2, 3) + 4 = 7$ bits. The actual value of XE is found as follows. After the first iteration, $XE_1 = XE_0 + AE = 8$; after the second iteration $XE_2 = XE_1 + AE = XE_0 + 2AE = 13$; ... ; after the tenth iteration $XE_{10} = XE_0 + 10 \cdot AE = 3 + 50 = 53$. To store $XE = 53$ takes 6 bits. Therefore, a correct upper bound is found. The overall size of the floating-point number $X = XE \times XM$ after 10 iterations is 16 bits. Our analysis gives it as 17 bits, since we obtained 7 bits for XE and the bit size of mantissa is fixed and takes 10 bits. ∎

15.1.6 Bit Sizes of Variables Computed in Graph Nodes

For each variable X computed in the node d of the program graph one finds $BS(X)$ and $BS(X^*)$ using techniques outlined in previous paragraphs. Then the bit size of a node d, BSN_d is obtained as a maximum of all bit sizes assumed by variables computed in this node; i.e., $BSN_d = \max [BS(X)]$. Likewise $BSN_d^* = \max [BS(X^*)]$. Therefore each node, d, is specified by two bit parameters, BSN_d and BSN_d^*, which will be used for finding the hardware resource required by the P program.

Example 47. Let us find the bit size of the computed variable, B, given in Example 36. Since variable B changes during iterative process, first we find the number of iterations, Z_2 of the node 2 which computes B. Node 2 is contained in the iterative loop made up of nodes 1, 2, 3, and 4. Since the exit of this loop is specified by the control statement IF $(C - 16.)$ 1, 1, 4 that is assigned to node 4, $Z_2 = Z_4$. To find Z_4, we use Eq. (6) since the value of Y $(Y = 16)$ is not changed during iterations. Thus $Y_0 = 16$, $X_0 = 0$, $\Delta_1 = 4$, $\Delta_2 = 0$ and $Z_4 \leq (Y_0 - X_0)/\Delta_1 + 1 = (16 - 0)/4 + 1 = 5$. Therefore, the program loop of nodes 1, 2, 3, and 4 iterates $Z_4 = 5$ times. Since the expression for B uses no bit size decreasing operations which reduce $BS(B)$ as compared to $BS(B^*)$ we obtain $BS(B) \geq BS(B^*)$. Thus there is no need to find $BS(B^*)$. Variable B is obtained as a result of multiplication $B = D*E$ with complex changeable factors D and E, where $D = (C*C + 3.*C + 2.)$ and $E = (C - 8.)$, because to obtain D takes four arithmetic operations; D is changed via bit increasing operations; and E is changed via bit decreasing operation. First, find $BS(D)$ after Z iterations. Since D contains a computed variable C involved in $C*C$ and $3*C$, to find $BS(D)$, one has to find the bit size of a temporary result $BS(D^*)$, where $BS(D^*) \leq \max [BS(C*C), BS(3*C)] + 1$; $BS(C*C)$ is found via Eq. (11), as $BS(C*C) \leq 2BS [\max(C_0, \Delta)] + 2 \log_2 (1 + Z)$, where $C_0 = 0$, $\Delta = 4$, and $Z = 5$. Therefore, $BS(C*C) \leq 2BS \max (0, 4) + 2 \log_2 (1 + 5) = 2BS(4) + 2 \cdot \log_2 6 = 2 \cdot 3 + 2 \cdot 3 = 12$. $BS(3*C)$ is found via Eq. (13), since the multiplicand 3 is not changed. $BS(3*C) \leq 2BS \max (3, C_0, \Delta) + \log_2 (1 + Z) = 2BS [\max (3, 0, 4)] + \log_2 (1 + 5) = 2BS(4) + \log_2 6 = 2 \cdot 3 + 3 = 9$. $BS(D^*) \leq \max [BS(C*C), BS(3*C)] + 1 = \max (12, 9) + 1 = 13$ bits and $BS(D) \leq \max [BS(D^*), BS(2)] + 1 = \max (13, 2) + 1 = 14$ bits.

Let us find $BS(E)$, where $E = (C - 8.)$ is the multiplier of $B = D*E$. $BS(E) = \max [BS(C = C + 4), BS(8)]$ where C is involved in the recursive addition $C = C + 4$. $BS(C = C + 4)$ is obtained via Eq. (9). $BS(C = C + 4) \leq \max [BS(C_0), BS(4)] + \log_2 (1 + Z) = \max [BS(0), BS(4)] + \log_2 (1 + Z) = 3 + \log_2 (1 + 5) = 6$ bits. Thus, $BS(E) = \max [6, 4] = 6$ bits.

Next, since we found the upper bounds on bit sizes assumed by D and E with due regard to iterative process we may use Eq. (10) to find $BS(B)$: $BS(B) \leq BS(D) + BS(E) = 14 + 6 = 20$ bits.

Thus for node 2, $BSN_2 = BSN_2^* = 20$ bits.

Nodes, 1, 3, and 4 use variable C involved in the recursive addition $C = C + 4$. Since $BS (C = C + 4) \leq 6$, for nodes 1, 3, and 4, $BSN = BSN^* = 6$ bits. ■

As shown above, finding the bit sizes for the computed variables is a simple process which uses bit sizes of the variables before iteration, the number of iterations, and the bit sizes of other variables which were computed previously.

15.1.7 Array Dimensions of Variables Computed in Graph Nodes

During the analysis of each graph node, one has to specify the dimensions of data arrays, DA, for all computed variables. To this end, one finds a statement at a node which provides for the storage of a computed variable or variables. Let this statement specify that v computed variables, $X_1, X_2, \ldots, X_v$, be stored and the iteration parameter for the node be Z. Then the required dimension of the array to store variables $X_1, X_2, \ldots, X_v$ is $DA = v \times Z$. The dimension of the data array for all variables computed in the node, DAN, is the sum of the DAs obtained for all statements: $DAN = \Sigma DA$.

Example 48. Using the program graph of Fig. III.36, we find that node 2 prints B and C variables; i.e., $v = 2$. Since node 2 iterates $Z_2 = 5$ times, the dimension of the data array which contains all B and C variables is $DA = v \times Z = 2 \cdot 5 = 10$ data words. Since no other variables of node 2 have to be stored, $DAN_2 = 10$. ∎

As a result of the analysis introduced above each node is characterized by the following parameters: iteration parameter Z, two bit size parameters BSN and BSN^*, and the data array dimension parameter DAN.

15.2 Diagram of the Hardware Resources

Analysis of the program nodes is followed by the construction of the diagram of the hardware resources required by a program, or *P-resource diagram*. The P-resource diagram is made as follows:

15.2.1 Bit Size Diagram of a Program Graph

Having found bit sizes for all nodes in a program graph, we construct a bit size diagram in which the horizontal axis specifies the nodes of the program graph, and the vertical axis shows the two bit size parameters, BSN and BSN^* for each node. Figure III.40(a) shows a bit size diagram built for a program graph containing seven nodes. Vertical dotted and solid lines show the respective BSN^* and BSN parameters for the case $BSN^* > BSN$. For example, for node 7, $BSN_7^* = 64$ bits and $BSN_7 = 32$ bits. The bit size diagram serves as a basis for finding the computer size diagram required by the program.

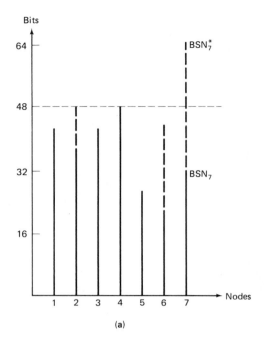

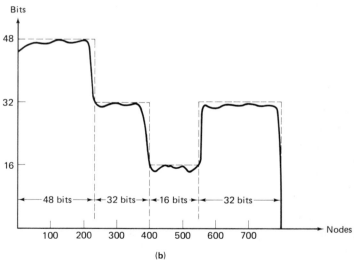

FIGURE III.40 (a) Bit-size diagram; (b) Computer size diagram

15.2.2 Alignment of the Bit Size Diagram

The bit size diagram is aligned in order to exclude excessive changes in computer sizes. This means that for a sequence of nodes, requiring approximately the same bit size, a dominant bit size is found. This is then accepted as the size of the computer that will compute this sequence. If a node in the sequence requires an exceptionally large bit size for fixed-point operation, then it is necessary to replace it using a floating-point operation. For instance, Fig. III.40(a) shows a bit size diagram for 7 nodes. The dominant size for this sequence is 48 bits. However for node 7, the maximal intermediary bit size $BSN^* = 64$ bits. Suppose that in analyzing node 7, we find that this size is obtained for a fixed-point multiplication operation. Reduction to 48 bits may be accomplished by replacing the fixed-point operation with a floating-point multiplication. We introduce some techniques that allow this replacement.

15.2.2.1 Alignment in bit sizes

Let the bit size of the node for a fixed-point operation (FI bits) have to be reduced to a smaller bit size for a floating-point operation (FL bits). First we find a fixed-point variable, X, computed in the node that requires FI bits; i.e., $BS(X) = FI$. When X is transformed into floating-point form, it includes exponent, XE, and mantissa, XM; i.e., $X = XE \times XM$. Since XE and XM have to be stored in a new and smaller bit size FL, $FL = BS(XE) + BS(XM)$.

Let us find the number of bits, $BS(XE)$, to be assigned to XE. When X is a fixed-point number, it cannot be larger than $2^{FI} - 1$. Thus $X \leq 2^{FI} - 1$ and $XE \leq FI$. Consequently, $BS(XE) \leq BS(FI)$, and $FL \leq BS(FI) + BS(XM)$.

Therefore, for a floating-point number, $X = XE \times XM$, stored in FL bits, one has to assign to the exponent, $BS(FI)$ bits, where FI is the bit size of the same number in fixed-point form. Then the remaining bits may be assigned for mantissa XM, $BS(XM) = FL - BS(FI)$.

Example 49. Let a fixed-point form of a variable X take 64 bits. Suppose that X has to be converted into a floating-point form taking 48 bits. Thus $FI = 64$ and $FL = 48$. Since a 64-bit number cannot be larger than $2^{64} - 1$, the exponent XE of X is bounded above as $XE \leq 64$ and $BS(XE) \leq BS(64) = 7$ bits. Therefore in a floating-point register taking $FL = 48$ bits, we have to assign $BS(64) = 7$ bits for XE. The remaining bits may be assigned for mantissa: $BS(XM) = 48 - 7 = 41$ bits. ■

If a node in the graph is specified with a smaller number of bits than the dominant bit size, its size is increased. This increase may be performed by introducing additional precision bits if all operations in the node are fixed-point, or by converting floating-point numbers taking FL bits into a fixed-point numbers taking FI bits, where $FL \leq FI$. Both types of alignment are described by simple and well-known procedures.

Alignment of all graph nodes to dominant bit sizes results in finding the computer size diagram which specifies the sequence of computer sizes required for computation of the program. This diagram partitions the program into tasks, where each task is composed of nodes requiring the same computer size. Figure III.40(b) shows the computer size diagram (dotted lines) obtained from the bit size diagram (curved line) and built for a program containing 800 nodes. This diagram partitions the program into four tasks computed by 48-, 32-, 16-, and 32-bit computers, respectively.

15.2.3 Finding the Time for Computing Each Task

The adaptive assignment of the resource of dynamic architecture among independent programs must be based on the execution time of each task in a given size computer. Since a system with dynamic architecture may compute several concurrent programs, one architectural state may feature several independent tasks computed in concurrent computers. The problem is in deciding which task should switch the architecture into the next state. If the switching is triggered by the task requiring the longest execution time, the remaining tasks will finish their executions earlier and will wait for the end of the longest task. Therefore, performance will deteriorate. An expedient solution would have the longest task executed in several states using the same computer. In this case, the shorter tasks may switch architectures and continue their execution without waiting for the end of the longest task. The system will perform correctly because the resource of the longest task will not be requested for reconfiguration. Therefore, if one assigns the hardware resources on the basis of time, the idle time of the computed programs will be minimized.

However, the diagram of computer sizes has been built for program nodes. That is why it is necessary to find a tentative time for executing a task in a given size computer. Since all computers of dynamic architecture use the same clock period, t_0, (Sec. 5.2.2.), the time to execute a task may be expressed in terms of t_0. Thus, we first find the number of clock periods, t_0, required for each node included into the task. This is done with due regard to iterative execution as specified by the iteration parameter Z of the node.

Having found the time for each node, we find the time of the task as the sum of times for the nodes included into the task. Therefore, to find the time of one task, we have to find the time to execute one node. This may be accomplished if we find the time to execute one machine instruction since each statement in the node may be interpreted as a sequence of machine instructions.

15.2.3.1 Time to execute one instruction

As we saw in Sec. 5.2.2, the sequencing of the intervals for one instruction is defined by an interval sequencer which is included in the MCD of every LSI module. In this sequencer, each interval is identified with one state, and a sequence of states is selected by the instruction op code. For each interval, variation of its duration is achieved with the *operation sequencer*. This sequencer is controlled by one of the three codes (p, m_E, or m_I) considered in Sec. 5.3.2. Variation of operation times may be accomplished if the sequencer executes a loop which lasts either p, m_I, or m_E clock periods, t_0. If the interval sequencer initiates the operation sequencer, the interval sequencer retains its state (interval) until the operation sequencer completes its operation. Therefore, the interval duration is a function of one of three codes stored in an LSI module of the computer.

When a new computer is formed, it is sufficient to write new values of codes p, m_E, and m_I into all its LSI modules for the control unit to start generation of new time intervals.

Example 50. Let us find the number of clock periods, t_0, required for the addition instruction. This instruction is executed during the following time intervals: $T_1 = m_I t_0$ performs the instruction fetch, where m_I is the fetch instruction code; $T_2 = t_0$ performs instruction decoding; $T_3 = m_E t_0$ fetches the operand, where m_E is fetch data code; and $T_4 = 2t_0 \cdot p$ performs the addition where p is the processor code. Assume that $m_I = m_E = 2$, and for a 16-bit computer, $p = 1$; for a 32-bit computer, $p = 2$; for a 48-bit computer, $p = 3$, etc. Then, for a 16-bit computer, the time to execute this instruction is: $T = T_1 + T_2 + T_3 + T_4 = 2t_0 + t_0 + 2t_0 + 2t_0 \cdot 1 = 7t_0$. In a 32-bit computer this instruction takes time $T = 2t_0 + t_0 + 2t_0 + 2t_0 \cdot 2 = 9t_0$ to execute.

15.2.3.2 Time to execute one graph node

Having found the times to execute all the instructions of the instruction set, we construct a special table in which rows correspond to machine instructions. An entry into a row shows the number of clock periods required for the execution of that instruction. A processor-dependent instruction is assigned n rows, each showing the instruction time in a compter of given size ranging from 16 to $16 \cdot n$-bits, respectively. A processor-independent instruction is assigned a single row since its execution time does not depend on the computer size. Therefore, if a statement is interpreted as f machine instructions, then its time is found as the sum of the times of all its instructions. If this statement belongs to a node with iteration parameter Z, then the numer of clock periods thus obtained is multiplied by Z. Having obtained the time of each statement, one may find the time of the node and the time of the task.

Example 51. For the program graph of Fig. III.39, let us find the time to execute node 2. This node computes $B = (C*C + 3.*C + 2.) + (C - 8.)$ and prints B and C variables. Execution of B takes 4 machine instructions: (1) The addition instruction, I_+, which fetches one operand from the memory; (2) the modified addition instruction, I_+^*, which adds two operands stored in processor registers; (3) the multiplication instruction, I_x, which fetches one operand from the memeory; and (4) the modified multiplication instruction, I_x^*, which multiplies two operands stored in processor registers.

Since for computation of B we need the C and $C - 8$ variables used in preceding node 1, assume that the current C and $C - 8$ are stored in registers $R1$ and $R2$ of the processor. Then the following instruction sequence computes B:

1. I_x^* computes $C \times C \rightarrow R3$ (i.e., the product is sent to a double precision register $R3$).
2. I_x computes $3. \times C \rightarrow R4$.
3. I_+^* finds $R3 + R4 \rightarrow R5$ ($R5$ stores $C*C + 3.*C$).
4. I_+ finds $R5 + 2 \rightarrow R6$ ($R6$ stores $C*C + 3.*C + 2.$)
5. I_x^* computes $R6 \times R2 \rightarrow R7$ ($R7$ stores a current variable B). Statement "Print, B, C" is interpreted as two print instructions, I_p.
6. I_p sends the $R7$ register contents to a cache register of an I/O device.
7. I_p also sends the $R1$ register contents to the I/O device. ■

In Example 45, we found that the bit size of node 2, $BSN_2 = 20$ bits. Assume that this node is executed in a 32-bit computer storing the following control codes: processor code $p = 2$; instruction fetch code $m_I = 2$; and data fetch code $m_E = 3$. Using the technique illustrated in Example 48, we find that the time to execute I_+ in a 32-bit computer is $T(I_+) = 10t_0$; for the modified addition instruction, $T(I_+^*) = 7t_0$, since the time of operand fetch ($T_3 = 3t_0$) is eliminated.

Let us find the time $T(I_x)$ of multiplication instruction in a 32-bit computer. For simplicity assume that no fast multiplication schemes are used and the multiplication operation takes the time of 32 additions (each taking the time $2t_0 \cdot p = 4t_0$) and 32 shifts (one t_0 per shift). $T(I_x) = T_1 + T_2 + T_3 + T_4$, where $T_1 = m_i \cdot t_0 = 2t_0$ fetches an instruction; $T_2 = t_0$ decodes I_x; $T_3 = m_E t_0 = 3t_0$ fetches the multiplicand; and $T_4 = 32 \cdot 2t_0 \cdot p + 32t_0 = 160t_0$ performs the multiplication as 32 additions and 32 shifts. Therefore $T(I_x) = 2t_0 + t_0 + 3t_0 + 160t_0 = 166t_0$.

For a modified multiplication instruction, $T(I_x^*) = 163t_0$, since the time for operand fetch is eliminated. For a print instruction, $T(I_p) = T_1 +$

$T_2 + T_3$, where $T_1 = m_l \cdot t_0 = 2t_0$, $T_2 = t_0$, and $T_3 = 2t_0$, since T_3 sends the result of $R3$ to a local cache of an I/O device. Therefore, $T(I_p) = 2t_0 + t_0 + 2t_0 = 5t_0$.

Find the time of one iteration of node 2: $T_0(N_2) = T(I_x^*) + T(I_x) + T(I_x^*) + T(I_+) + T(I_x^*) + T(I_p) + T(I_p) = 163t_0 + 166t_0 + 7t_0 + 10t_0 + 163t_0 + 5t_0 + 5t_0 = 519t_0$.

In the Example 45, we found that node 2 iterates $Z = 5$ times. Therefore the total time to execute node 2 during Z iterates is $T_Z(N_2) = 519 \cdot t_0 \cdot Z = 519 \cdot 5t_0 = 2595t_0$.

15.2.4 P-resource Diagram

After finding the time to execute each task in a given size computer, one constructs the hardware resource diagram (or *P-resource diagram*) for the entire program, P. In this diagram the horizontal axis shows the time each task functions, and the vertical axis contains two parts: the upper portion showing computer sizes, and the lower portion showing the dimensions of data array. Figure III.41 shows the P-resource diagram constructed on the basis of an aligned diagram of Fig. III.40. The horizontal axis indicates the number of clock periods, t_0, that execute each task in a given size computer. In accordance with Fig. III. 41, the first and second tasks (executed by 48-bit and 32-bit computers, respectively) take 220 and 180 nodes, respectively; however, the first task takes around $4 \times 10^5 t_0$, whereas the second one is executed in about $2 \times 10^5 t_0$. Although the two tasks use approximately the same number of nodes, the first one requires twice as long as the second one to execute.

Each task above was specified by a unique computer size. To find the space required by its data, we add all the *DAN* dimensions found earlier for the task nodes to the dimension of the array of initial data which lists all the initial words used in computation of all the task nodes.

Having constructed P-resource diagram for each program P, we perform the overall assignment of the system resource among user programs. This assignment takes into account specific attributes of dynamic architectures considered above.

16. ASSIGNMENT OF THE SYSTEM RESOURCES AMONG PROGRAMS

To find a sequence of states to be assumed by a dynamic architecture computing a set of user programs, one has to perform a hardware resource assignment among tasks to identify which computer elements should be assigned to each task of any given program, P. This may be done by using all the P-resource diagrams obtained in Sec. 15. The assignment system

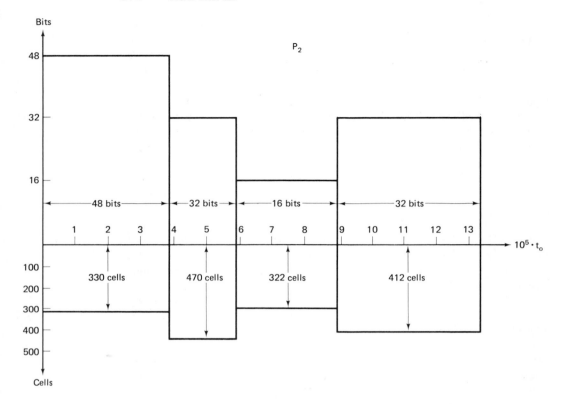

FIGURE III.41 P resource diagram

unifies all such diagrams into the following two diagrams showing the distribution of the DC-group resources:

1. A *CE resource diagram* which charts the computer elements assigned to each task and defines the flow chart of architectural transitions.
2. A *ME resource diagram* which indicates the allocation of the DC-group primary memory among various user programs.

16.1 CE Resource Diagram

To construct a CE resource diagram one has to use all the P-resource diagrams. Their computer size portions merge into a CE resource diagram in which the vertical axis shows the maximal bit size ($h \cdot n$-bits) for one DC-group made of n CEs, and the horizontal axis shows the time for executing each task. For simplicity of mapping assume that $h = 16$. Mapping each P-resource diagram onto the CE diagram is made in accordance with the priority assigned to that program. This means that the request of the highest priority program for resources is satisfied first. The tasks of all other programs are mapped onto the resources remaining; i.e., if a DC group having n CE cannot satisfy the resource needs of all its programs, then the resources used by programs with lower priorities can be taken over by programs with higher priorities. Thus the programs with lower priorities may be interrupted if their resources are requested by programs with higher priority. (Fig. III.42).

The vertical axis of the CE diagram also shows the positions of all the CEs. Since a DC-group may use memory elements with differing speeds, the CE diagram shows the positions of all fast and slow MEs. Consequently, if a program requires that some of its tasks be executed faster, it selects those MEs whose speed is maximal. This implies that the resource requirements of each task may be moved up and down on the vertical axis to meet its speed requirements.

Another reason for such up and down movements is to minimize the time of data transfer from one computer to another. Since in a DC-group with minimal complexity a $C_i(k)$ computer may have parallel data exchanges only with data words stored in its own primary memory, all between-computer information exchanges are performed in a time-shared mode. If one computer needs an array of $16 \cdot k$-bit data words stored in the memory of another computer, it takes k time intervals to fetch all the bytes of each $16 \cdot k$-bit data word. To minimize the number of time-shared exchanges, it is desirable to adopt the following strategy: if the program P is computed by a sequence of different size computers, it is necessary that all smaller size computers be formed from the less significant CEs included into the largest size computer of this sequence. Then all larger size computers may perform parallel fetches of data words computed by smaller size computers, since these words will be stored in the least-significant portions of their primary memory. This will minimize the number of time-shared data exchanges between computers. Fetches of such words may be performed by a special instruction, "fetch-nonstandard", which is considered in [9], and which organizes economical storage of short words in the primary memory of a given size computer.

Example 52. Let a program P be computed by a sequence of 32- and 48-bit computers. Suppose that the 48-bit computer of this sequence is assigned computer elements CE_1, CE_2, and CE_3. Its primary memory consists of ME_1, ME_2, and ME_3. For the 32-bit computer one must assign computer elements CE_2 and CE_3. The primary memory of the 32-bit computer will then contain ME_2 and ME_3. Thus, the 48-bit computer will have a parallel access to 32-bit data words computed by the 32-bit computer since they will be stored in the least-significant portion of its primary memory (ME_2 and ME_3). ∎

The construction of a CE resource diagram is equivalent to finding the sequence of DC-group transitions specified by the requirements of the exe-

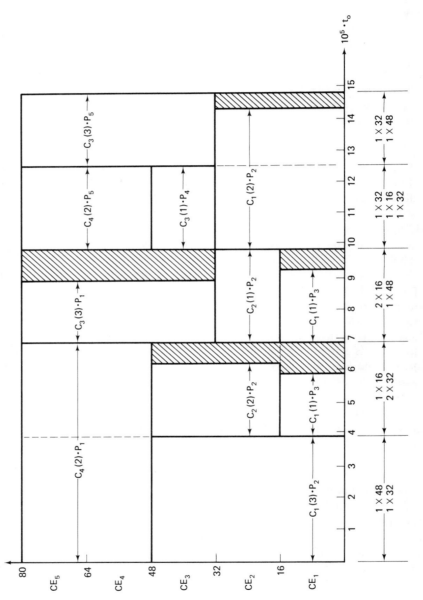

FIGURE III.42 CE resource diagram

cuted programs on the CE resource. Indeed, DC-group states are found by merging all concurrent computers into a single state. A transition from one state to another is performed by a program which requests a resource and which has a priority higher than those of other requesting programs.

Example 53. Find a unified CE resource diagram for five programs $P_1 - P_5$ given by their P resource diagrams (Figs. III.41 and III.43). Let P_1 have the highest priority, so its resource requirements will be satisfied first. The first task of P_1, (requiring a 32-bit computer) is assigned CE_4 and CE_5 (Fig. III.42). The next task of P_1 (requiring a 48-bit computer) is assigned CE_3, CE_4, and CE_5. This task may use all the 32-bit constants of the 32-bit computer, C_4 (2), because they are stored in ME_5 and ME_4. The next in the priority hierarchy is program P_2, whose P_2 resource diagram is shown in Fig. III.41. First, P_2 requires a 48-bit computer which is assigned with the available resources of CE_1, CE_2, and CE_3. Next it needs a 32-bit computer assigned to CE_2 and CE_3. The released resource of CE_1 is used for computation of P_3.

This procedure is continued until all resources are not assigned to all programs. As a result one obtains a CE resource diagram which specifies the sequence of architectural states. Indeed, the first state of the DC-group is characterized by one 48-bit and one 32-bit computer, i.e., 1×48, 1×32. For the next state, one 16-bit and two 32-bit computers are working concurrently. As can be seen from Fig. III.42, transition from the first state to the second is performed by P_2, since it is the first high priority program to require a new computer belonging to the second state. Therefore, although the priority of P_1 is higher than that of P_2, transition to the next state is performed by P_2, since P_1 does not need a new resource and computer C_4 (2) in which it is run exists for both states. A transition from state (1×16, 2×32) to state (2×16, 1×48) is performed by P_1, since it is the highest priority program. P_2 finishes its task in the C_2 (2) computer earlier than P_1 and it waits until P_1 ends its task in the 32-bit computer. (Shaded areas mean waiting time.) Similarly, a waiting period occurs for P_3 since its priority is even lower than that for P_2. The transition from state (2×16, 1×48) to state (1×32, 1×16, 1×32) is performed by P_2, because P_1 has ended its execution and among the remaining programs P_2 is assigned since it has the highest priority. Since the resource taken before by P_1 is now available, it is used for computing new programs P_5 and P_4 in computers C_4 (2) and C_3 (1), respectively. ∎

16.2 DC-Group Flow Chart

We have shown above that the CE resource diagram specifies the DC-group flow chart. This is a directed graph in which nodes mean architectural states. Each arrow is marked by the program which is allowed to execute this

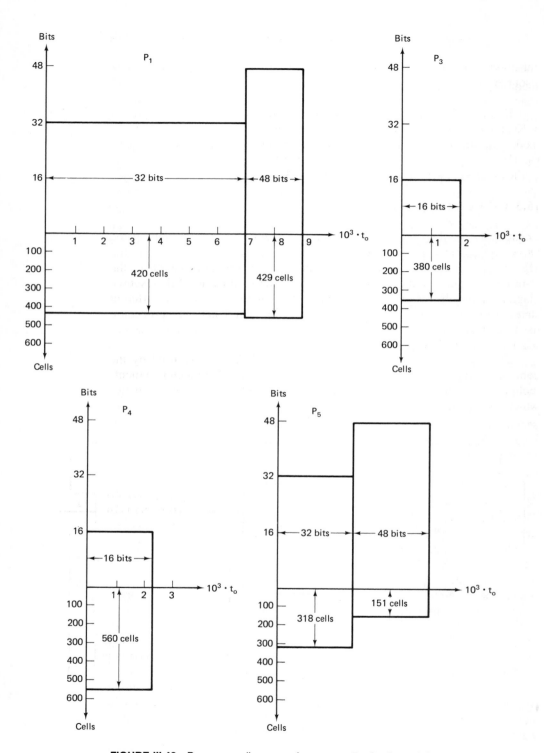

FIGURE III.43 P resource diagrams of programs P_1, P_3, P_4, and P_5

transition. This program is the highest priority program that needs a new resource.

Example 54. The flow chart for the CE resource diagram constructed in Example 51 is shown in Fig. III.44. As seen from this figure, in state N_2 (node 1), computer C_4 (2) executes program P_1 and computer C_1 (3) executes P_2. The arrow from node N_2 to node N_{10} is marked by the program P_2, which is allowed to execute this transition, etc. ∎

16.3 ME Resource Diagram

Construction of the ME resource diagram is equivalent to distributing the DC-group primary memory among various user programs. To construct the ME resource diagram, one has to use the memory size portions of all the P-resource diagrams and the DC-group flow chart. Since the ME resource diagram shows the scheduling of primary memory for accessing in different states of the DC-group flow chart, for each architectural state, the procedure has to find two memory spaces which store the data and instruction arrays of one user program, respectively.

The primary memory is first assigned for all data arrays used by the computers functioning in different states. The positions of computer elements included in each C_i (k) computer give positions of the memory elements which should store the data array used by this computer. For instance, since

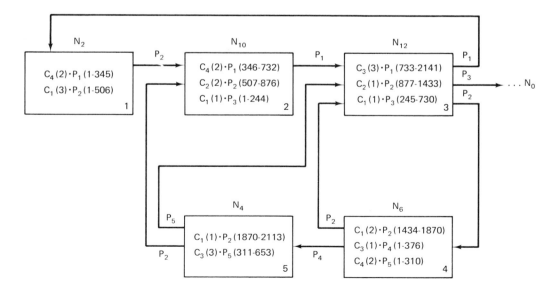

FIGURE III.44 DC-group flowchart

computer C_1 (2) consists of CE_1 and CE_2, its data array, made up of 32-bit words, should be stored in ME_1 and ME_2. Following the allocation of all data arrays, we perform the allocation of programs. Inasmuch as each C_i (k) computer may store its program in k memory elements, the smaller the k, the smaller the number of memory elements which may store the program for C_i (k). For instance, a 16-bit computer, C_3 (1), may store its program only in ME_3 whereas a 32-bit computer, C_3 (2), may store its program in ME_3 and ME_4. That is why one must allocate programs for small size computers first, since they may be stored in a limited number of MEs.

In allocating programs one first begins to fill the empty spaces created by variable size data arrays. This will allow primary memory to be completely filled. If a program, P, is computed in several architectural states by new computers, then each task of the program executed in state N_d by computer C_i (k) is concluded by the synchronization instruction Stop-N_d, considered in Sec. 5.6.1. If in the flow chart, the program P is on the arrow connecting states N_d and N_f, then the Stop-N_d instruction of this program is succeeded by an $N_d \to N_f$ instruction, which performs an architectural reconfiguration (Sec. 5.6.4). Computer sequencing of consecutive program segments of its program that are stored in memory elements ME_i and ME_j respectively, is performed by the instruction, "Jump $ME_i \to ME_j$," which performs a program jump from ME_i to ME_j (Part II, Sec. 2.4.5.).

Example 55. The scheduling of memory in accordance with the flow chart of Fig. III.44 is shown in Fig. III.45. In all five MEs the first 100 cells are assigned for storage of constants, addresses, etc. Thus data arrays for computers C_4 (2) and C_1 (3), which function in state N_2 (Fig. III.44), begin with cell 101. Since C_4 (2) consists of CE_4 and CE_5, its data array for program P_1 is stored in the parallel cells of ME_4 and ME_5 (cells 101–520). Likewise the data array for C_1 (3), made up of CE_1, CE_2, and CE_3 is stored in parallel cells of ME_1, ME_2, and ME_3 (cells 101–430). The next state, N_{10}, forms two new computers, C_2 (2) and C_1 (1), while C_4 (2) continues to function in both states; C_2 (2) stores its data array in parallel cells of ME_2 and ME_3 (cells 431–900), and C_1 (1) only stores its 16-bit data array in ME_1 (cells 431–710). This procedure is continued until all data arrays of the given flow chart are allocated.

The next stage is finding the empty spaces created in the primary memory by variable size data arrays. In our case, ME_1 has an empty space from cell 711 to cell 1222, and ME_4 and ME_5 have empty 32-bit cells from 521 to 900 and from 1648 to 1889. First, we fill the empty space formed in ME_1. We use it for program P_3 computed by computer C_1 (1) in state N_{10}, since P_3 may be stored only in ME_1. Thus cells 711–1222 store one program segment. For better illustration, the first cell of each task of program P computed in one architectural state is marked with symbol "Begin P," which does not take an additional instruction. So cell 711 of ME_1 is marked with "Begin P_3," mean-

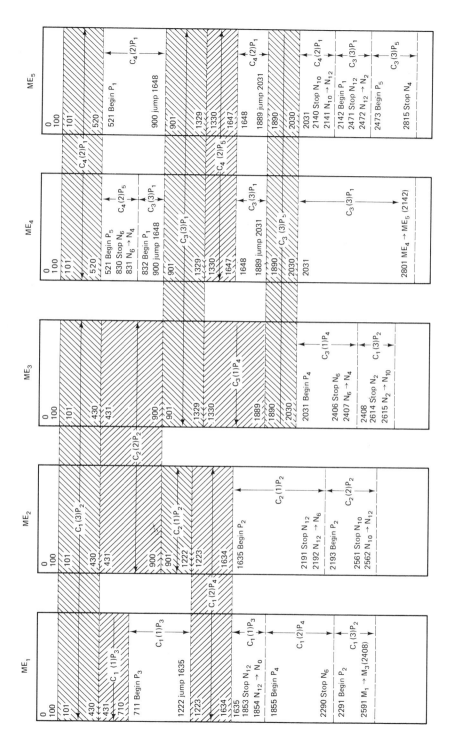

FIGURE III.45 Memory scheduling in accordance with the flowchart of Fig. III.41

ing that it contains the first instruction of P_3 when it is executed in state N_{10}. Next, P_3 performs "jump (1635)" within the same ME_1. Note that P_3 is executed in states N_{10} and N_{12} by the same C_1 (1) computer, thus no synchronization instruction, Stop-N_{10}, is introduced because the C_1 (1) resource is not requested. In state N_{12}, P_3 is concluded by instruction Stop-N_{12} which is then followed by instruction $N_{12} \rightarrow N_0$.

Consider now the filling of the empty spaces created in ME_5. In ME_5, we start to allocate P_1, executed by C_4 (2) in state N_2 (cells 521–900). In cell 900, P_1 jumps to cell 1648 of ME_5. This continuous sequence is maintained up to the cell 1889 of ME_5. In this location P_1 jumps to location 2031 of ME_5. Next it is maintained up to cell 2140 and concluded by synchronization instruction Stop-N_{10}, because it must perform transition $N_{10} \rightarrow N_{12}$ (cell 2141). Thereafter P_1 continues to be executed by C_3 (3) which fetches it from cells 2142 to 2470. Then again it is concluded by Stop-N_{12} and $N_{12} \rightarrow N_2$, which returns it again to state N_2 and computer C_4 (2), which fetches P_1 from ME_5. If a program is stored in several ME_i, in each ME_i it is concluded by the Jump $ME_i \rightarrow ME_j$ instruction which begins program fetches from ME_j. For instance, the P_2 program computed by the C_1 (3) computer in state N_2 is stored in two memory elements ME_1 and ME_3. Its first portion, stored in ME_1 (cells 2291–2590), is concluded with the Jump $ME_1 \rightarrow ME_3$ (2408) instruction which begins instruction fetches from ME_3 beginning address 2408. ∎

17. SUMMARY

This part discussed some important software problems arising in dynamic architectures. It introduced analysis techniques for a user program written in a high-level language which are aimed at finding the minimal sizes of computers which may execute this program. Next, it considered techniques for assigning the DC-group hardware resources among several concurrent programs, thus allowing one to increase the execution parallelism obtained on the hardware.

The presented methodology was shown to be simple and straightforward, requiring no complex computations. As a result, resource assignment among user programs can be performed quickly and efficiently.

CONCLUSIONS

During the past few years, major problems and basic approaches toward developing a very complex parallel system working in real-time have been formulated in the scientific literature. The most essential requirements to such systems (Supersystems) are as follows:

First, the Supersystem must adapt its architecture to new algorithms or modified versions of old algorithms in order to compute them during the limited time intervals these algorithms may require.

Second, Supersystems may need to augment their throughput continuously during their lifetimes in order to be capable of computing in real-time a continuously increasing number of concurrent programs.

Third, most of the real-time algorithms for Supersystems are characterized by a variable number of information streams (instructions and data). The number of information streams may be in the hundreds or thousands. This necessitates enormously complex hardware resources because each stream has to be processed in real-time and thus requires a separate processor. But as a system's complexity goes up, its reliability goes down. Thus a Sypersystem may become unsuitable for some important applications.

Specific requirements on Supersystems create a need for new computer architectures that may adapt via software to various computational requirements of an algorithm(s). This chapter demonstrated for one type of architecture (dynamic), all of these requirements may be satisfied.

A system with dynamic architecture is capable of establishing not a single, but many architectures, each of which is dedicated to a given class of application. Also it may switch from one architecture to another during the algorithm's running time, so that it allows creation of a match between the architecture and a complex real-time algorithm or a portion of such an algorithm. Thus, if a new or modified algorithm appears, it is possible to select an architecture or sequence of architectures which minimizes the disparity between the architecture and the algorithm.

A system with dynamic architecture is capable of partitioning its resources into a variable number of computers to accommodate the required number of programs it needs to handle. Therefore, if the system has to process an increased number of programs, it may partition its resources into a larger number of computers with smaller word sizes so that each program is run on a separate computer. This allows all programs to be tracked in real-time, although the precision of computations may suffer. In case a particular program requires a more precise computation, a larger size computer may be formed for its handling so the accuracy of computations will be maintained at the required level.

Finally, since a dynamic architecture should compute each program in a minimal size computer, it will partition the available resources into a maximal number of concurrent computers. Therefore, it will maximize the number of instruction and data streams computed by the available resources. As a consequence, a given throughput may be achieved with less complexity of the resources than in conventional systems.

REFERENCES

[1] F. R. BAILEY, "Computational Aerodynamics—Illiac IV and Beyond," *Digest of Papers,* Spring CompCon '77, pp. 8–11.

[2] C. R. VICK, "Research and Development in Computer Technology, How Do We Follow the Last Act," Keynote Speech, *Proc. of the International Conference on Parallel Processing.* pp. 1–5, 1978.

[3] ———, "A Dynamically Reconfigurable Distributed Computing System," Doctoral Disseratation, The Graduate Faculty of Auburn University, Alabama, 1979.

[4] C. R. VICK, J. E. SCALF, and W. C. MCDONALD, "Distributed Data Processing for Real-time Applications," *Proc. of the Sixth Texas Conference on Computing Systems,* pp. 174–191, 1977.

[5] H. FITZGIBBON, B. BUCKLES, and J. SCALF, "Distributed Data Processing Design Evaluation Through Emulation," *Proc. Computer Software and Applications Conference (CompSac),* pp. 364–369, 1978.

[6] J. VON NEUMANN, "Probabilistic Logic and the Synthesis of Reliable Organism from Unreliable Components," *Automata Studies,* C. E. SHANNON and J. MCCARTHY, Eds., Princeton University Press, Princeton, N.J., 1956, pp. 48–98.

[7] J. L. BAER, "Multiprocessing Systems," *IEEE Trans. on Computers,* **C-25,** pp. 1271–1277, December 1976.

[8] S. I. KARTASHEV, S. P. KARTASHEV, and C. V. RAMAMOORTHY, "Adaptation Properties for Dynamic Architectures," 1979 National Computer Conference, *AFIPS Conference Proc.,* AFIPS Press, **48,** pp. 543–556, 1979.

[9] S. I. KARTASHEV and S. P. KARTASHEV, "A Multicomputer System with Software Reconfiguration of the Architecture," *Proc. of the Eighth International Conference on Computer Performance,* SIGMETRICS CMG VIII, Washington, D.C., pp. 271–286, 1977.

[10] S. I. KARTASHEV and S. P. KARTASHEV, "Dynamic Architectures: Problems and Solutions," *Computer,* **11,** pp. 26–40, July 1978.

[11] ———, "Multicomputer System with Dynamic Architecture," *IEEE Trans. on Computers,* **C-28,** *10,* pp. 704–721, October 1979.

[12] ———, "Software Problems for Dynamic Architecture: Adaptive Assignment of Hardware Resources," *Proc. Computer Software and Applications Conference (CompSac),* pp. 775–780, 1978.

[13] ———, "Performance of Reconfigurable Busses for Dynamic Architectures," *Proc. of the 1st International Conference on Distributed Computing Systems,* pp. 261–273, Huntsville, Alabama, 1979.

[14] G. J. LIPOVSKI and A. TRIPATHI, "A Reconfigurable Varistructured Array Processor," *Proc. International Conference on Parallel Processing,* pp. 165–174, 1977.

[15] D. W. ANDERSON, F. J. SPARACIO, and R. M. TOMASULO, "IBM System 360 Model 91, Machine Philosophy and Instruction Handling," *IBM Journal of Research and Development,* pp. 8–24, January 1967.

[16] S. F. ANDERSON, J. G. EARLE, R. E. GOLDSCHMIDT and D. M. POWERS, "The IBM System 360, Model 91: Floating-point Execution Unit," *IBM Journal of Research and Development,* pp. 34–53, January 1967.

[17] W. J. WATSON, " The TI ASC—A Highly Modular and Flexible Super Computer Architecture," *AFIPS 1972 Fall Joint Computer Conf.,* AFIPS Press, 1972, pp. 221–228.

[18] R. M. RUSSELL, "The CRAY-1 Computer System," *Communications ACM,* **21,** pp. 63–72, January 1978.

[19] R. N. IBBETT and P. C. CAPON, "The Development of the MU5 Computer System," *Communications ACM*, **21**, pp. 13–24, January 1978.

[20] M. J. IRWIN, "Reconfigurable Pipeline Systems," *Proc. 1978 ACM Annual Conference*, **1**, pp. 86–92.

[21] S. P. KARTASHEV and S. I. KARTASHEV, "Adaptable Pipeline System with Dynamic Architecture," *Proc. of the 1979 International Conference on Parallel Processing*, pp. 222–230.

[22] ———, "Designing LSI Metacomputer System with Dynamic Architecture," DCA Association, Lincoln, Neb, 1974.

[23] ———, "Designing of LSI Metacomputer System with Dynamic Architecture Made of Microcomputers," *Proc. 3rd Annual International Symposium on Mini- and Microcomputers and their Applications*, pp. 88–93, Zurich, Switzerland, 1977.

[24] ———, "A Microprocessor with Modular Control as a Universal Building Block for Complex Computers," *Proc. 3rd Euromicro Symposium on Microprocessing and Microprogramming*, pp. 210–216, Amsterdam, Holland, 1977.

[25] ———, "Designing of LSI Modular Computers and Systems (Modern State of the Art)", Key Presentation at the MIMI '77, Montreal, *Proc. of the International Symposium on Mini- and Microcomputers*, Montreal, Canada, 1977.

[26] B. R. BORGERSON, "The Viability of Multimicroprocessor Systems," *Computer*, **9**, 1, pp. 26–30, January 1976.

[27] S. I. KARTASHEV and S. P. KARTASHEV, "Selection of the Control Organization for a Multicomputer System with Dynamic Architecture," *Proc. 4th Euromicro Symposium on Microprocesing and Microprogramming*, Munich, Germany, 1978. pp. 346–357.

[28] ———, "LSI Modular Computers, Systems and Networks," *Computer*, **11**, pp. 7–15, July 1978.

[29] G. M. AMDAHL, G. A. BLAAUW, and F. P. BROOKS, "Architecture of the IBM System/360," *IBM Journal of Research and Development*, **8**, pp. 86–101, April 1964.

[30] *IBM System/370 Principles of Operation*, publ. A22-6821-3, International Business Machines Corp., White Plains, N.Y., 1974.

[31] L. R. GOKE, "Connecting Networks for Partitioning Polymorphic Systems," Doctoral Dissertation, Dept. of Electrical Engineering, University of Florida, 1976.

[32] B. R. BORGERSON, G. S. TJADEN, and M. L. HANSON, "Mainframe Implementation with Off-the-Shelf LSE Modules, *Computer*, **11**, pp. 42–48, July 1978.

Chapter IV

Verification of Complex Programs and Microprograms

R. Negrini
M. G. Sami
R. Stefanelli

PREVIEW

The reliability of software and firmware is a problem that has become more important recently, since the difficulty of finding a solution increases steeply with the complexity of (micro) programs. This problem can constitute a real bottleneck in the design and realization of data processing systems.

Complex modular architectures constitute a class of systems where software reliability is critical. Here are a few points that illustrate this assertion:

- Such architectures execute very complex and sophisticated programs, whose reliability is therefore difficult to assess.
- Modules are often specialized through microprogramming, and errors in firmware would affect the machine reliability itself.
- In a dynamic architecture, software must be proved reliable whatever the architecture that executes it.

- Finally, but certainly not least, such architectures are often used for such highly critical applications as aerospace and nuclear plant control applications.

The existing techniques for assessing or improving software reliability often are very costly, involve long and complex computations, sophisticated interaction with programmers, etc. Techniques for verifying complete correctness, even on the level of semantics are only a particular example of this fact and have been actually developed only for a limited number of programs with simple semantical structures.

In this chapter we present a methodology for analyzing a property of particular importance, *termination*. Our aim is to identify the existence of possible infinite loops, an obviously relevant error in a program. We propose to verify termination characteristics by means of an algorithm as simple (and fast) as possible, requiring little preprocessing of the program (and no further preprocessing than that necessary for allocating a program on a dynamic architecture), and also requiring no specialized information or interaction on the programmer's part. While the characteristics thus defined give only partial insight in a program's termination, they point out possible problem areas in a complex program, thus directing the programmer's intervention in the debugging phase.

We will show how all the properties and algorithms presented apply to microprograms as well as to programs; moreover, problems such as the relative timing of operations and synchronization can be in most cases accounted for without introducing further complexity in the processing.

We will also see that programs executed upon distributed, reconfigurable architectures can be analyzed, requiring no reference to the actual machine structure that executes them.

A simple overview of our presentation:

- Section 1 provides an overview of techniques for assessing software reliability.
- Section 2 defines the particular problem discussed in this chapter, and illustrates it with some examples.
- Section 3 presents the models used for determining termination properties, in the following steps:
  3.1 Basic models for conventional programs.
  3.2 Extensions for microprograms.
  3.3 Modeling techniques for timing and synchronization problems; extension to multiprocessor systems.

Thus, while subsection 3.1 is absolutely necessary for subsequent reading, subsections 3.2 and 3.3 provide information for more complex problems.

- Section 4 presents the conditions which guarantee the termination properties here discussed and therefore outlines the corresponding algorithm for a very broad class of programs defined here as *series-parallel* and obtained by structured programming.
- Section 5 and 6 extend the results of Sec. 4 to programs with unconstricted structures; again, Sec. 4 is absolutely necessary for pointing out what are the characteristics of our method; Secs. 5 and 6 provide an extension.
- Section 7 gives some brief conclusions and indicates further research areas.

1. INTRODUCTION

The problem of obtaining reliable software and assessing software reliability, has acquired growing importance over the years; the difficulty of its solution increases very steeply with the increasing complexity of programs, and it can become a real bottleneck in the design of a data processing system.

The same considerations hold when microprograms are considered. In conventional systems microprogram complexity was usually rather low, and it was reasonable to assume complete testing and validation of microprograms. However, the last few years have seen a growing trend towards complex microprograms. This is both because of firmware implementation of sections of such programs as operating systems and compilers and because of the introduction of general-purpose (universal) modules that become dedicated via microprogramming in multiprocessor systems. Microprogram reliability is a vital issue, since it directly affects also the correct operation of a possibly fault-tolerant hardware.†

Multiprocessor architectures‡ are a typical instance of computer systems specifically designed for the execution of very large programs, and they easily involve sophisticated microprogramming. Moreover, their application fields are often critical ones, as far as reliability is concerned (e.g., defense systems, aerospace applications, etc.). Thus, the problem of software (and firmware) reliability can be particularly important.

Extensive preprocessing** is already required in order to find an optimal allocation of the program† using the various processors; it would be obviously quite useful to exploit this same preprocessing for reliability purposes. It is also important that the reliability-affecting parameters thus obtained depend upon the *semantics* of the program, and not upon its allocation upon parallel processors, so that characteristics identified be kept valid independent of the dynamic reconfiguration of a modular system.

Program-processing–oriented scheduling and performance prediction of programs in multiprocessor computing systems was developed in the 1960s.

†A comprehensive insight on fault-tolerant hardware is presented in Chap. VI, Secs. 2 and 4, where Sec. 2 treats different techniques of a fault-tolerant hardware design attained via hardware redundancy, spare modules, or graceful degradation; and Sec. 4 considers diagnosis techniques and their role in improving the fault-tolerance of a modular system. (Ed.)

‡General prerequisite information on multiprocessor architectures is given in Chap. I, Sec. 14.2. (Ed.)

**Chapter V discusses such program preprocessing that minimizes the complexity of interconnection network that connects all available resources together. (Ed.)

†For more on optimal allocation of the program(s), see Chapter V, Sec. 2.3. (Ed.)

The representation of programs by means of graph models and the processing of such models for specific purposes is extensively discussed in [1]. There, operations performed upon the graph lead to the definition of a lower and an upper bound on the number of processors required for maximum parallelism in a homogeneous processor system.

In a similar way, graph models were introduced for the representation of microprograms and were used for horizontal microprograms optimization in [2].

New problems, and new approaches to graph model processing, have been introduced with the definition of dynamic architectures as defined in [3] to [6]. *For each user program*, it now becomes necessary to "find a sequence of minimal size computers which execute it; next, available hardware resources of the multicomputer system have to be assigned among user programs, each of which is executed by a sequence of the minimal size computers found earlier."[5]‡ A simple analysis of a control-flow representation is not sufficient for this purpose; program preprocessing must involve an analysis of variables introduced in each statement and of relations among such variables. Such information allows us to determine not only the bit-size for the different modules, but even to choose their "native" instruction set [6].

‡This type of preprocessing is discussed in Chap. III, Secs. 14 to 17, that considers analysis of programs aimed at finding their requirements in terms of processor sizes and memory volumes. (Ed.)

As it will be seen in this chapter, the preprocessing gathers also all information necessary to verify a property we term *structural termination* and which gives useful insight for the (micro)program's correct operation and for its reliability. In fact, it will be seen that this property can be identified directly at compilation, by means of very simple computations and algorithms, and without requiring any interaction with the programmer. This is related with the (micro)program's semantics, and is not affected by reconfiguration.

Improvement of software reliability has involved such different areas as programming methodology, language design, and software validation.† While the first two research areas aim to decrease the probability of errors in a program—however complex— by means of suitable techniques used in design and implementation, software validation aims at identifying errors present in an existing program. See for instance [7]. The two approaches are thus complementary rather than mutually exclusive. Here, we are concerned with software validation.

†An overall organization of fault-tolerant software is considered in Chap. VI, Sec. 5; that discusses sequencing of all steps aimed at effecting fault-free computation of programs. (Ed.)

A number of techniques have been suggested for program validation; we shall examine briefly the meaning of such techniques as *dynamic analysis*, *interpretative analysis*, and *static analysis*. Testing is increasingly seen as a powerful methodology for software reliability assessment. The definition of structural termination can be seen as bridging the gap between static analysis and structural testing; although belonging to the first area, structural termination analysis is useful for pointing out possible problem areas and directing subsequent test strategies.

One of the approaches to dynamic analysis considered in [8] consists of monitoring the program behavior during the run and gathering information concerning one or more executions. Its main drawback is its dependence on input data. Moreover, when dynamic architectures are considered, dynamic analysis can become also reconfiguration-dependent.

Data dependence does not affect interpretive analysis methods such as symbolic execution. With this technique, a particular path in the program is chosen and validated by assigning values to the input data and then executing the program path with reference to the symbolic values. When the execution has been completed, the final symbolic representation obtained can be analyzed to determine path correctness, giving data-independent results [9].

The obvious problem with symbolic execution arises with long, complex programs; final symbolic representations can easily become so long and complex as to become unmanageable and practically meaningless. Moreover, selection of a path involves also the specification of the number of iterations for the loops.† In the case where the number of loop iterations in this path is not given explicitly nor can be determined analytically, it is impossible to determine structural termination of this path. As a result, program analysis becomes incomplete.

†See also Chap. III, Sec. 15, where the number of loop iterations is evaluated to find bit sizes of computed variables. (Ed.)

Finally, static analysis comprises such different techniques as *information gethering*, *data flow analysis*, and *program verification*. The first two are now being introduced in optimizing compilers; they lead to the accumulation of such information as variable usage tables and routine control graphs, they identify instances of illegal data usage such as references to undefined variables, etc. Typically, these techniques are aimed at achieving the following objectives: (a) each instruction of a program can be reached from the program entry point; (b) from any instruction, the end of the program can be reached; and (c) all variable are defined prior to usage, etc. In the rest of the chapter we will assume that such controls have been performed upon the program prior to structural termination analysis, or—in our words— that the program is "formally correct."

Now consider program verification or proof of correctness. This consists of "demonstrating the consistency between (1) a computer program and (2) specifications or assertions describing what the program is supposed to do."[10] The most widely known method of verification is the "inductive assertion" or Floyd method, based on work by Floyd [11], later formalized by Hoare [12]. To apply this method, the programmer has to provide in addition to the program itself and to its I/O specifications, a set of "assertions" at specific points.† Each assertion states a relation which should be true whenever program control reaches it; successful verification aims at proving that each assertion is in fact true. Thus, if all the assertions inserted in the program are shown to be true and *if the program terminates*, the assertion at the end of the program will also be true. Actually, the statement "if the program terminates" introduces the great difficulty of this method. In a program having a complicated loop structure, it is necessary to place the

†See more on checking points and predicates that perform program testing in Chap. VI, Sec. 2.2. (Ed.)

assertions in the loops in such a way as to have any closed path cut by at least one assertion. Furthermore, the nature of these assertions must be such as to provide one with the definite answer on the termination conditions of the program analyzed.

Generation of the appropriate assertions is recognized as one of the major problems in using the inductive assertion method; the problem becomes increasingly difficult if it is sought also to verify program termination.

Reference [13] states that "Because of the limitations of automatic theorem proving and the necessity to specify completely the program and to describe explicitly the input, output, and loop invariants, program verification requires considerably more human interaction than the other validation methods. Verification often uncovers program errors (or specification errors), but *because of the complex human interactions, it is usually only applied to small segments of a program*" (our italics). The programmer is, in fact, required to describe again, in a different language, what the program is expected to do. Of course, with the added information there is also added possibility of errors.

The same criteria of formal verification can be obviously applied also to microprograms.† In [14] a method explicitly related to microprograms is presented: Having defined an architectural-level machine M and a register-transfer-level machine μM, the authors proceed to verify that an instruction sequence on M is correctly emulated by an instruction sequence on μM. Termination of instruction sequences on M is always assumed implicitly.‡

Testing, as opposed to program verification, aims not to guaranteeing correctness, but at identifying a number (hopefully as near as possible to all) of errors in the software (or in the firmware). Consider in particular *structural* testing. The control flow of the program, as represented by a suitable program graph model, and the data representation are analyzed. Tests chosen are as elementary as possible. They are aimed at covering as many as possible faults in the program requiring no additional information concerning its behavior. Clearly this does not guarantee that the program is correct even for complete coverage of its paths. Therefore, incomplete fault coverage is achieved at the expense of reduced complexity of the technique.

Conventional testing criteria do not allow a general identification of program terminations. However, this is quite obviously a critical characteristic and it can lead to potentially dangerous situations during a run. For example, for dynamic architectures it is necessary to identify the maximum number of iterations of each loop as one of the necessary parameters for assigning hardware resources to program tasks [5]; nontermination of a loop would create extensive malfunctions in the whole system. The aim of the work described in this chapter is to introduce such techniques of structural termination that

†See Chap. I, Secs. 1.6 and 2.4 for prerequisite material on microprograms. (Ed.)

‡More information on the architectural level and the register transfer level of representation for a microprogrammed computer is given in Chap. 1, Sec. 10. (Ed.)

1. Require no additional information besides that provided by the (micro)program itself.

2. Require no details concerning semantics or data dependence. Only the boolean data dependence, i.e., existence or lack of dependence from a given variable will be considered.

In fact, it will be seen that the property termed *structural termination* can be identified by very simple and fast computations and requires no additional preprocessing besides that already necessary for adapting the program to a multiprocessor computing system. While structural termination does not guarantee that a loop will, or will not, terminate for any data set, it provides insight in the loop's behavior, since it points out possible nonterminating loops (i.e., the ones for which there is at least one data set leading to nontermination) and the behavior during a run of variables upon which termination depends. Structural termination thus points out possible error areas and prepares the program for test definition that must be performed next. Our analysis will associate the loops with three-valued matrices whose dimensions do not depend on the complexity of statements and instruction sequences (as it happens, for instance, in symbolic execution) but only from the number of variables appearing in the loop.

The first definitions (in Secs. 2 and 3.1) will be made with reference to programs; the particular problems introduced by microprograms and the solutions provided will then be analyzed. Models and formal representations used for determining termination properties in the two cases will be introduced later in Sec. 3.

First the conditions of structural termination will be presented for programs and microprograms that are characterized by graph models with rather "simple" structures, i.e., series-parallel structures. Actually, such models of the control flow of a (micro)program† are typically associated with *structured* programs, and therefore they constitute a very important class. We will also see subsequently, how additional, but still simple, preprocessing allows us to reduce formally any program graph model to a series-parallel structure, not for computation purposes but just for the determination of structural termination.

†More on the control flow of microprograms represented by directed graphs is in Chap. I, Sec. 1.6. (Ed.)

This chapter refers mainly to ideal execution, in a sense that problems created by timing and by synchronization of asynchronous processes and/or processing elements are not considered. When discussing horizontal microprograms, we will see that such timing and synchronization aspects which can be reduced to either fixed or computable delays can be readily accounted for. Totally asynchronous cases; i.e., cases in which the relative delays between processing elements cannot be evaluated a priori are not considered here; they would require greater model complexity and extended operation upon the models than the ones used here. A solution to this last problem, possibly keeping the simplicity of computations involved, constitutes the theme for subsequent studies.

2. DEFINITION OF THE PROBLEM

In this section, we define the concept and meaning of *structural termination*; criteria for determining it will be described in the remainder of the chapter.

We propose to analyze at compilation only a conventional user-written program, without requiring either additional information inserted or subsequent interaction with the programmer. It is easy to conclude that within such bounds the only property related to program correctness (and different from "formal" or "syntactical" correctness) which can be examined is termination of program loops. Since loopless programs terminate by definition, the termination analysis for them is not required. It is required only for those programs that contain at least one loop. For complex programs which certainly contain loops, termination analysis becomes essential, since otherwise it becomes prohibitively difficult to determine whether a complex program contains infinite loops or not.

Termination analysis will be performed on the basis of reduced information extracted from the program: for each executable statement of the program, only the *existence* of interdependences among variables will be considered, while the nature of such dependences will be ignored. Thus, a statement such as

$$v_3 = (v_1 - 3v_2)/v_4$$

will be reduced to the simpler form

$$v_3 = f(v_1, v_2, v_4)$$

for subsequent processing.

Since only the "structure" of statements is examined, and no attention is given to further semantic characteristics, we feel justified to denote the termination property we are looking for as *structural termination*.

In order to give a clear statement of structural termination, we refer to three simple examples analyzed in detail.

Prior to this analysis, we notice that termination constitutes a problem only if the number of iterations for a loop is not predetermined. Thus, the following examples, although simple, have been chosen so as to illustrate the possible problems of a loop whose exit condition depends upon variable(s) computed inside the loop itself.

Example 1. Consider the following program segment:

.
.
.

Repeat

$$v_1 = 2v_2$$

$$v_2 = v_3 - 1$$

$$v_3 = v_4 + 1$$

$$v_4 = v_4 + 2$$

Until $(v_1 \geq 8)$

.
.
.

Assume the following initial values: $v_2 = 1$, $v_3 = 1$, and $v_4 = 1$. Refer to Fig. IV.1(a), in which the values of the four variables involved are tabulated for the different iterations of the loop. The initial conditions correspond to time t_0, i.e., prior to any execution of the loop. The value of v_1 is immaterial, while v_2, v_3, and v_4 are assigned constant values. Define time i ($i > 0$) as immediately preceding computation of the closing predicate until $(v_1 \geq 8)$ in the ith iteration of the loop. Thus, at time t_1 the variables have acquired the values

$$v_1 = 2, \quad v_2 = 0, \quad v_3 = 2, \quad v_4 = 3$$

Evaluation of the closing predicate causes iteration of the loop. At time t_2 the variables have become

$$v_1 = 0, \quad v_2 = 1, \quad v_3 = 4, \quad v_4 = 5$$

The loop is again iterated until, at time t_5, it is

$$v_1 = 10, \quad v_2 = 7, \quad v_3 = 10, \quad v_4 = 11$$

At this time, the loop terminates. The closing predicate depends on (or "is dominated by") variable v_1 whose value is updated during each iteration; if the values assigned to the variables before entering the loop had been different, the number of iterations before exit would also be different. The same is true if the closing predicate, although still depending on v_1, had a different expression; in this case, a predicate such as "until $(v_1 \geq -1)$" for initial values $v_4 = v_3 = v_2 = 1$ would lead to nontermination of the loop. ∎

The continued modification of variable v_1 guarantees that the number of iterations of the loop is ultimately data-dependent. For the class of loops that we propose to examine, in which the number of iterations is undetermined and must be data-dependent, a condition such as the one met in this example is *necessary* in order to have correct implementation of the loop. Therefore, we define the program as *characterized by structural termination*. As we noted, the condition itself does not ensure termination for any possible initial data set or closing predicate. In other words, the condition is *not necessary* for termination.

396 CHAPTER IV

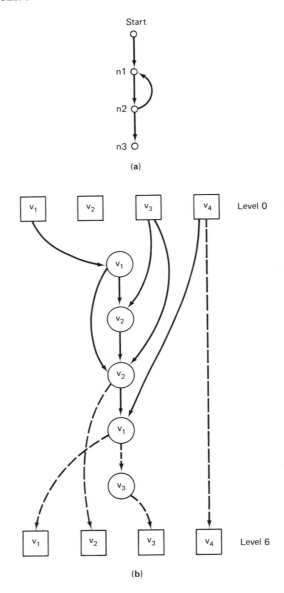

FIGURE IV.1 (a) Tabulated values of variables in Example 1 during subsequent iterations; (b) the same for Example 2; (c) the same for Example 3

Example 2. Consider the following program segment:

.
.
.

Repeat

$$v_1 = 2v_2$$
$$v_2 = v_3 - 1$$
$$v_3 = v_4 + 1$$
$$\text{Until } v_1 \geq 8$$

Assume again initial conditions $v_4 = v_3 = v_2 = 1$. As before, we tabulate the values assumed by the different variables in the successive iterations of the loop (Fig. IV.1(b)). We see that after a *transient* consisting of times t_0, t_1, and t_2, all variables acquire constant values. In particular, v_1 (the variable that dominates the closing predicate) becomes constant with a value of 2, and *the condition for termination is never reached*. Not only that: Whatever the initial data, either the program terminates in the course of the transient (whose length $k_o = 3$ is *not* data-dependent) or else it never terminates. This happens with different closing predicates, as long as they are dominated by v_1. ∎

Summing up, we can observe that in this case no real data dependence is achieved for the number of iterations. On the contrary, it is quite probable that the program will not terminate. As a consequence, we will consider this to be one of the cases where structural termination *is not* achieved.

Example 3. We consider now a program with a slightly more complex structure.

Repeat
$$v_3 = v_4 - 1$$
$$\text{if } (v_1 \geq 0) \text{ do}$$
$$\qquad v_1 = v_2 - 1$$
$$\qquad v_2 = v_3 - 1$$
else
$$\qquad v_1 = v_2 + 1$$
$$\qquad v_2 = v_3$$
$$\text{Until } (v_1 < -2)$$

In this case, the loop encloses two alternative instruction sequences whose choice depends on the evaluation of the first predicate *if* $(v_1 \geq 0)$. Assume the following initial conditions:

$$v_4 = 2, \quad v_2 = 2, \quad v_1 = 1$$

(the initial value of v_3 is immaterial) and tabulate, as before, values of the variables at end of each iteration (Fig. IV.1(c)).

During the first iteration $v_1 = 1$, thus the first alternative instruction sequence is executed; at t_1, values are, respectively

$$v_1 = 1, \quad v_2 = 0, \quad v_3 = 1, \quad v_4 = 2$$

The same instructions are executed during the second iteration: at time t_2,

$$v_1 = -1, \quad v_2 = 0, \quad v_3 = 1, \quad v_4 = 2$$

(v_4 never changes its value, actually); at t_3, the second alternative instruction sequence is executed, with results

$$v_1 = 1, \quad v_2 = 1, \quad v_3 = 1, \quad v_4 = 2$$

Noted that once variable v_1, the variable that dominates the predicate closing the loop, after an initial transient of $k_o = 1$ cycle, keeps assuming only values in a finite set (values 1, -1, 0), the period length being $k_1 = 3$. As in the previous example, the program loop never terminates. Here, different initial data or even a different closing predicate would have led either to termination inside the first $k_o + k_1 = 4$ cycles (with k_o, k_1 not data-dependent) or else to nonterminations.

The same considerations made when discussing Example 2 lead us to consider this program as *not* structurally terminating. ∎

From the three examples analyzed above we can finally deduce that, given a program with a loop whose closing predicate P_e is dominated by a variable v_i, the following three cases may occur:

1. In each iteration of the loop, the value of v_i is modified: there may exist one or more initial conditions such that v_i assumes a value leading to exit from the loop after a number k of iterations, where k cannot be determined and in fact, depends upon the initial data.†

2. After a finite (usually small) number k_o of iterations, v_i becomes constant. k_o depends on the instruction sequence inside the loop, not on the input data.

3. After the first k_o iterations, v_i changes its value with a periodical law, cyclically assuming a finite, fixed set of values: the period length is k_1. Again, k_o and k_1 depend on the instruction sequence, not on the input data.

†Some techniques for finding the number of iterations for terminating loops that are dependent on their initial data set are discussed in Chap. III, Sec. 15.1. (Ed.)

In case 1, the number k of iterations after which the program terminates is generally data-dependent. As we alereadly stated, while nontermination also is possible, a first, necessary condition for correct implementation of data-dependent iterative computation has been satisfied. We refer to this condition as *structural termination*.

On the contrary, in cases 2 and 3 the number k of iterations is either infinite or else if it is finite, it is data-independent and bound by a value k_0 (case 2) or $k_0 + k_1$ (case 3) a priori computable.

In the remainder of this chapter, structural termination will be determined first with reference to programs with a very simple structure; afterwards, the problem will be extended to a much larger class of programs defined as series-parallel, i.e. programs whose representative graph models have series-parallel structure. It will be seen how all these considerations apply to microprograms.

Recall that a program built following the rules of structured programming and, in particular, avoiding all *goto* statements, leads to series-parallel structure; therefore, the class of (micro)programs to which our basic method can be applied is not only a large one, but a growing one with the increasing application of structured programming techniques.

We shall examine also the general case of (micro)programs with nonseries-parallel structure. A set of transforms will be performed upon the program graph models so as to obtain series-parallel *structures* "equivalent" to those given as far as the problem of structural termination is concerned. (That is, we are not interested in deriving a *semantically equivalent program*, but just in finding how structural termination criteria developed for series-parallel programs can be applied to the general case).

3. PROGRAM GRAPH MODELS AND REPRESENTATION OF THE RELATIONS AMONG VARIABLES

3.1. Program Graph Models

The problem of representing program structure and relations among variables in a way suitable for determining structural termination will be examined first for the simplest case, i.e., a program executed on a conventional single processor or in a multiprocessor environment without any problem of synchronizations and resource sharing concerning other program sections. Such a program will be denoted from now on as a *sequential program*. After assessing a basic modeling methodology, microprograms also will be considered (subsection 3.2) and suitable extensions will be made. Finally (subsection 3.3), timing and synchronization problems will be examined, and extensions to parallel processing will be considered.

A widely known program graph model, presented in [18], defines an "instruction sequence" as a program segment characterized by:

- One entry point
- One exit point
- At most one branching, coinciding with the exit point.

The model (here referred to as the Estrin model) is derived from the program as follows:

- Each node represents an instruction sequence
- Control flow from instruction sequence $s\alpha$ to instruction sequence $s\beta$ is represented by an oriented edge going from node α to node β.†

†See also Chap. III, Sec. 15.1.1 which uses a similar graph for finding bit sizes of computed variables. (Ed.)

Obviously, there is considerable freedom in defining instruction sequences; i.e., the model's nodes, given a sequel of instructions without any conditional branching, we can arbitrarily associate it with one single node rather than with a sequence of series-connected nodes.

Refer to Example 3; we can label the *repeat* statement n_1, the *if* statement n_2, and *until* statement n_3; that is, we label each "branching" or "meeting" instruction. We have the following instruction sequences:

$s_1: v_3 = v_4 - 1;$ evaluation of *if* statement

$s_2: v_1 = v_2 - 1; v_2 = v_3 - 1$

$s_3: v_1 = v_2 + 1; v_2 = v_3$

$s_4:$ evaluation of *until* $(v_1 < -2)$

The graph model in Fig. IV.2(a) is thus obtained. In general, the simplest way to derive an Estrin program graph model consists in associating an entry point with

1. Each labelled statement
2. Each statement immediately following a branching instruction
3. Each statement reached as a consequence of a branching

Exit points are then either branching points or else are defined by default with respect to entry points, since each instruction sequence is a group of consecutive statements starting with an entry point and ending with the (nearest) exit point. Edges connect nodes in the model whenever control flow connects the corresponding instruction sequences; thus, in Fig. IV.2(a) from node s_1 (the "beginning" node) there are two outgoing edges (conditional branching) reaching s_2 and s_3 respectively. These, in turn, have one outgoing edge only, reaching node s_4. From node s_4, evaluation of the predicate represented by *until* statements either leads back to s_1 edge from s_4 to s_1 or else leads to subsequent program segments, not explicitly represented here (arrow leading to dot sequence). (It may be noted at once that the same model has been used in [19] for microprograms.)

An alternative representation has been proposed in [20]; it can be obtained from the previous (or Estrin) model as follows:

(a)

	v_1	v_2	v_3	v_4
t_0	–	1	1	1
t_1	2	0	2	3
t_2	0	1	4	5
t_3	2	3	6	7
t_4	6	5	8	9
t_5	10	7	10	11
		Exit		

(b)

	v_1	v_2	v_3	v_4
t_0	–	1	1	1
t_1	2	0	2	1
t_2	0	1	2	1
t_3	2	1	2	1
t_4	2	1	2	1
t_5	2	1	2	1

(c)

	v_1	v_2	v_3	v_4
t_0	1	2	–	2
t_1	1	0	1	2
t_2	–1	0	1	2
t_3	1	1	1	2
t_4	0	0	1	2
t_5	–1	0	1	2
t_6	1	1	1	2
t_7	0	0	1	2
t_8	–1	0	1	2

FIGURE IV.2 (a) Estrin graph model for Example 3; (b) an intermediate step for obtaining the model used in this chapter; (c) graph model of type used in this chapter

- Each node in the Estrin model is substituted with an oriented edge termed the *instruction edge* and oriented from entry to exit of the corresponding instruction sequence (thick arrows in Fig. IV.2(b)).
- Each edge in the Estrin model is kept as a *control edge* (thin arrows in Fig. IV.2(b)); control edges are always oriented from exit of an instruction edge top entry of a (possibly different) instruction edge.

All series connections can be subsequently simplified through merging into a single edge (merging of an instruction edge and a control edge results in an instruction edge). In the resulting graph (Fig. IV.2(c)) edges represent at once instruction sequences *and* control flow; nodes simply denote *singular points* in the control flow; that is, either merging points or branching points.†

---

†Although Fig. IV.2(c) has kept the distinction between thick arrows and thin ones, this distinction is immaterial for determining structural termination. Subsequent examples, therefore, will avoid it and consider just one type of edge in the graph model.

This second type of graph has proved to lend itself very easily to various preprocessing actions useful for such operating system policies as task allocation, hierarchical memory management, etc. For example, it allows straightforward computation of "probable activities" of instruction sequences both prior to and during a run [20]. Note that in this graph only two types of nodes appear, that is

- "Type-i" nodes, in which one or more arrows merge while there is only one outgoing arrow: in graph-theory terms, a type-i node has input-degree $i \geq 1$, output-degree $\sigma = 1$. With reference to programming practice, a type-i node is the entry point of an instruction sequence which can be referenced by more than one control statement: thus, considering again Example 3 and the related Fig. IV.2(c) node labeled n_2 (entry point of instruction sequence s_4) is a type-i node, since it is reached both by sequence s_2 and by sequence s_3.
- "Type-σ" nodes, having only one incoming arrow while one or more arrows depart; in graph-theory terms, a type-σ node has input-degree $i = 1$, output-degree $\sigma \geq 1$. With reference to programming, a type-σ node with $\sigma > 1$ stands for a conditional branch statement that specifies all exits from a given instruction sequence. Obviously, a type-σ node can be the exit node of one instruction sequence only (just as type-i node can be entry node of one instruction sequence only).

From the above definition, it follows that each type-σ node having $\sigma > 1$ is associated with a predicate corresponding to the branch condition. For any given result of this predicate, during a run, one and only one of the out-going edges will be followed. The evaluation of the predicate will in general depend from a set of variables $v_1, v_2, \ldots$ that will be said to *dominate* the predicate itself. Consider for instance the very simple case

$$\text{if } (v_1) \text{ then } e_1 \text{ else } e_2$$

The predicate is simply based upon evaluation of a boolean variable v_1; if $v_1 = true$ the predicate is true and control is passed to statment e_1, otherwise the predicate is false and e_2 is executed. v_1 is then the variable dominating the predicate.

From the existence of type-i and type-σ nodes only, it follows that no self-loops will exist in the graph model. Actually, existence of a self-loop would coincide with an error situation; that is, an instruction sequence without at least one exit path towards the program "stop."

Actually, the program graph model now described, which will be used in this chapter, can be also derived directly from the program itself. Consider the following sample program (Example 4), given with an abstract notation which keeps all information necessary for subsequent processing: there, n_k denotes a label, v_i denotes a variable, $f(v_i, v_j \ldots)$ is a proper function of variables $v_i, v_j \ldots$ (i.e., f is not a simple assignment), $P(v_i)$ is a predicate dominated by variable v_i.

Example 4

$$\begin{aligned}
&\textit{Start}\\
&n_1: v_1 = f(v_1)\\
&\quad v_2 = f(v_1, v_3)\\
&\quad v_2 = f(v_1, v_2, v_3)\\
&\quad v_1 = f(v_2, v_4)\\
&\quad v_3 = v_1\\
&n_2: \textit{if } P(v_1) \textit{ then } n_1 \textit{ else}\\
&n_3: \text{stop}
\end{aligned}$$

The representative graph model for this program is given in Fig. IV.3. ∎

We will now assume that any program considered is formally correct following the definition given in Sec. 1. *Formal* correctness refers in fact to syntactic properties of the program; its validation is already performed by many compilers, but it gives no insight into the semantics of the program, nor does it point out any problem areas excepting obvious errors such as

- References to undefined variables
- Statements which cannot be reached from any point in the program
- Branching to undefined labels

On the graph model, formal correctness means that given any edge there is at least one oriented path from the Start node to the Stop node traversing it.[†] Thus, the graph model in Fig. IV.3 is formally correct.

The characteristics to be examined in order to identify structural termination are

- The program graph model, giving the structure of the control flow (from now on called the *control graph* or *C-graph*)
- The set V of all variables appearing in the program
- The relations among the variables introduced in each edge of the C-graph.

---

†We assume that any program has only one Start and one Stop. This assumption causes no loss of generality, since any program can be very simply reduced to this case.

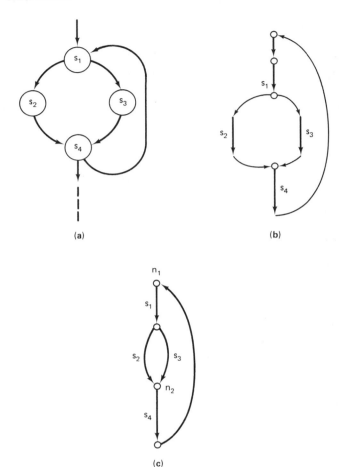

FIGURE IV.3 (a) Program graph model for Example 4; (b) graph representing I/O relationships for Example 4; (c) V-matrix for Example 4

A particular problem is introduced by indexed or otherwise structured variables (e.g., arrays) which would required us to account not only for the relations directly involving *either* the variables *or* their indices, but also for the one between the variables *and* its indices. Actually, the same problem will appear when microprograms are considered, concerning variables referenced through addressing (e.g., primary storage words, etc.).

It is not easy to find a representation at the same time simple and fully meaningful; in fact, it would be necessary to go into far greater detail of the program semantics than is done for determining structural termination. A partial solution will be presented when discussing microprograms†; there, the case is relatively easier since it is reasonable to consider only linearly address-

†Different techniques for representing microprograms are discussed in Chap. I, Secs. 1.6. and 2. (Ed.)

able storage. However, given the complex data structuring that can be introduced within high-level language programming, we limit ourselves here to programs referencing only nonindexed or nonstructured variables.

We have already seen how the C-graph is derived from the program; we consider now how the other information is derived and organized.

The set V consists of all variables appearing in assignment statements, plus one additional "dummy" variable representing all constants appearing in simple assignment statements in the program. Since structural termination is analyzed locally for the various loops, it is useful to examine local and thus restricted, subsets of variables.

With each edge of the C-graph derived from an instruction sequence is associated an ordered set of relations among variables; each relation is derived from program statements by the following rules:

- The ordering of relations is the same as the ordering of program statements; labels are kept unchanged.
- All variable names are kept unchanged, except for the dummy one marking constants.
- Simple assignment statements, i.e., statements such as "$v_i = v_j$", are kept unchanged.
- Relations derived from proper functions simply point out that a dependence exists between variables on the left and on the right, without defining the nature of such dependence (they are thus defined as abstract relations). In real time programs, data acquisition statements are seen as proper functions which modify the input data. This is justified by our philosophy of structural termination as depending on variables which undergo continued modification (without any periodical law). In fact, information acquired in real time can be defined just in such way; a loop whose iteration number depended upon it would certainly have data-dependent behavior.
- Branching statements (which are represented by type-σ nodes in the C-graph) are kept unchanged as far as the control flow modification is concerned; i.e., *goto*, *if—then—else*, etc., appear unchanged in the abstract relation. If a conditional branching instruction is complex, with its predicate depending on the evaluation of a statement rather than simply on a number of variables, the instruction itself will be split into
  - Proper functions and/or assignment statements corresponding to the statements evaluated (possibly with introduction of further temporary variables).
  - Predicates dominated by the variables thus evaluated.

The condition expressed in the branching is then abstractly represented by a predicate P defined as above, with which no assignment is associated. Thus, a closing statement of a "Repeat" instruction expressed as *until* ($v_4 + 2v_3 - v_1 \leq v_1 + 1$) would be represented by abstract relation sequence:

$$v_{d1} = f(v_4, v_3, v_1)$$
$$v_{d2} = f(v_1)$$
$$\text{until } P(v_{d1}, v_{d2})$$

These abstract relations express all information necessary for the subsequent assessment of structural termination, since they allow us to identify modifications performed on variables and dependences between variables. From the definition of structural termination given in Sec. 2, we already know that continued modification of predicate-dominating variables is the condition for structural termination; we see in the Sec. 4 how the existence of such a condition is determined.

Processing required for this last purpose is best performed with reference to matrices and graphs; this is true both because of the compact representation allowed and because of the very simple algorithms possible. We will first represent a sequence of abstract relations by means of an immediately derived graph; in order to describe clearly the rules for building it, we refer again to Example 4. The set of variables appearing in Example 4 consists of $\{v_1, v_2, v_3, v_4\}$. No dummy variable needs to be introduced, since there are no simple assignments to constant values. It is assumed that v_1, v_2, v_3, and v_4 have been previously defined in a preceding program edge. Thus, at the beginning of Example 4 they have already been given input values. We then can introduce a set of four "input" nodes in the graph; they are drawn as square nodes to show that they represent an "interface" with the rest of the program, and they are each associated with one variable (see Fig. IV.3(b)).

All input nodes are drawn on the same line or "level"; the input level is assigned value 0. Then, the first abstract relation of Example 4 is processed: the variable on the left is v_1. A node (representing the new value assigned to v_1) is drawn on the line immediately below level 0; the level is assigned ordering number 1, and the node is marked with v_1 (drawn as a circular node to distinguish it from "interface" nodes). The abstract relation represents a proper function in which the new value of v_1 depends upon the previous value of v_1 itself; this "previous level" is the input one, and the proper function is represented by a full-line oriented edge going from input node v_1 to level-1 node v_1. This edge is called a *strong edge*. Next, the second abstract relation is processed; a node v_2 is inserted on a still lower line; that is assigned level 2. The new value of v_2 depends upon the previous values of v_1 and v_3; v_1 has been updated in the previous statement, so that the dependence exists with respect to the *last value* of v_1 (node at level 1). As for v_3, it still retains its initial value; no new nodes marked v_3 have as yet been inserted after level 0. Oriented edges are drawn from the "nearest" nodes marked v_1 and v_3 to node

v_2; in general, at any given level and for an abstract relation depending upon a variable v_k, the node marked v_k at level i nearest j ($i < j$) is searched for and used as a "source" of the edge representing the relation. In the same way, the third and fourth abstract relations are processed; they all represent proper functions, so strong edges are inserted in every case. The fifth abstract relation, on the other hand, is a simple assignment $v_3 = v_1$; a node v_3 is inserted at level 5 and to represent the simple transfer of a value a dotted arrow or *weak edge* is drawn from the nearest v_1 node to the new node.

The sixth statement is a predicate in which no assignment is made; therefore, no corresponding addition is made to the graph. At this point, the whole instruction sequence has been processed and all variables have acquired "output" values which will constitute the "interface" with subsequent instruction sequences. Another interface level—the output level or level 6—is then introduced, in which all variables are represented by as many square nodes. The output values are assigned to these nodes by means of *weak* edges, i.e., simple assignments, coming for each output node from the nearest node marked with the same name. In the particular case of v_4, which did not appear on the left in any of the abstract relations, the nearest node is the input one.

Actually, this graph is redundant with respect to what is required in order to determine structural termination; for this last purpose, it is sufficient to know *input-output relationships*, i.e., to know if the output value of any given variable v_i depends from the input values of any variable v_j in set V and if this dependence is strong, that is, if it involves at least one proper function, or else weak. Such information is just part of the information given by a graph such as the one in Fig. IV.3(b) (and which is useful, for example, in the preprocessing necessary for reconfiguration of a dynamic architecture).† In fact, the existence of an I/O relationship between an input value v_i and an output value v_j can be derived simply from the existence in the graph of an oriented path going from input node v_i to output node v_j. (Nevertheless, for simplicity, from now on we refer to such graphs as I/O relationship graphs).

Recall that in Sec. 2 the focal point was defined as the continued modification of variables dominating the closing predicate of a loop. The particular sequence of actions through which this modification is performed is not essential. Passing to the graph derived from abstract relations, the above considerations mean that the particular sequence of nodes on a given path from input node v_i to output node v_j is immaterial for further processing.

We define now a path from input node v_i to output node v_j to be strong if it contains at least one strong edge, and weak otherwise. (Note that, from the definition of strong and weak edges and paths, it follows that there may exist more than one strong path from v_i to v_j; on the contrary, if there exists a weak path from v_i to v_j it is the only possible path between these nodes, since since two or more simple assignments cannot "coexist" at the same time on the same output variable). Thus, we see in Fig. IV.3(b) that

†See Chap. III, Secs. 14 to 17 for a description of some aspects of such preprocessing. (Ed.)

- There are strong paths from input node v_1 to output nodes $v_1, v_2,$ and v_3.
- There are strong paths from input node v_3 to output nodes $v_1, v_2,$ and v_3.
- There are strong paths from input node v_4 to output nodes v_1 and v_3.
- There is a weak path from input node v_4 to output node v_4.

We represent in a very compact way all these I/O relationships by means of a matrix, called the *V-matrix*, whose *columns* represent *input values* of the variables, and whose *rows* represent *output values* of the variables. (It is therefore a square matrix.) Any given element $v(j,i)$ of the V-matrix contains information on existence and nature of a path from input node v_i to output node v_j. More precisely, each such element can be assigned one and only one of three values, $+1$, -1, or 0, according to the following rules:

- $v(j,i) = +1$ if (and only if) there is a *strong* path from input v_i to output v_j
- $v(j,i) = -1$ if (and only if) there is a *weak* path from input v_i to output v_j
- $v(j,i) = 0$ if (and only if) there is *no* path from input v_i to output v_j

Refer again to Example 4 and to the graph in Fig. IV.3(b); its V-matrix can be built column by column as follows:

- Column v_1: recalling the strong paths from input v_1 to output v_1, v_2, v_3, entries $+1$ are inserted in rows v_1, v_2, v_3, while it is $v(4,1) = 0$.
- Column v_2: all entries are 0.
- Column v_3: entries $+1$ in rows v_1, v_2, v_3, entry 0 in row v_4.
- Column v_4: entry $v(1,4) = +1$, entry, $v(2,4) = 0$, entry $v(3,4) = 0$, entry $v(4,4) = -1$.

The V-matrix thus completed is shown in Fig. IV.3(c). Again, we may notice that due to the meaning of a weak path, if *in a row* there is a -1 entry, all other entries in the row are 0. In fact, if the output value of a variable v_j is obtained through a simple assignment from the input value of a variable v_i (possibly with $i = j$) it cannot at the same time depend on the input values of other variables.

For subsequent operations, it will now be sufficient to process the V-matrices associated with the edges of the C-graph; operations to be performed will be seen to be very simple. Also note that the storage space itself necessary for the V-matrices is quite limited. Moreover, such matrices will often be rather sparse, i.e., they will contain few nonzero entries—the more so with growing dimensions of V. Suitable compact representations for sparse matrices have been studied; see, for example, [21].

3.2 Extension to Microprograms

Let us look now at microprograms. While it is immediately possible (and, in fact, usual) to keep unmodified the C-graph representation, it is necessary to consider a number of problems prior to representing abstract relations and I/O relationships. These problems are

- Horizontal microprogramming (i.e., microinstructions consisting of several different microorders, concurrently executed).†
- Definition of the variable set.
- Definition of abstract relations.

First consider the case of *horizontal microprograms*. Assume that a suitable definition has been found for *variables* and *abstract relations*. A horizontal microinstruction can be represented by means of a set of concurrent abstract relations, each representing one microorder,‡ provided we accept that all microorders in a microinstruction require the same time (or alternatively, that the longest microorder time is fixed as microinstruction time).* Microcompilers and microassemblers perform verification of the formal correctness that was introduced in Sec 1. For horizontal microprograms, verification of formal correctness must also guarantee that no variable appears on the left in more than one microorder of the same microinstruction as this would lead to conflict for access to and modification of the variable.

Assignments, whether they are simple or involve execution of the same functions, are obviously made here also with reference to previously computed values; i.e., it is assumed that if in one microinstruction there appear two microorders, the one defining variable v_j as a function of variable v_i, the other one assigning a new value to v_i, then the new value of v_j is computed before updating of v_i, i.e., computation of v_j precedes updating of v_i. (Thus, "race" conditions between microorders are avoided.) When drawing the graph representation of I/O relationships, concurrence of microorders in one microinstruction is represented by inserting all nodes associated with leftside variables at the same level. Consider the following "abstract" microprogram:

Example 5. The microprogram sequence consists of three horizontal microinstructions μ_1, μ_2, and μ_3. It is

$$\mu_1: v_1 = f(v_1,v_2);\ v_3 = f(v_1);\ v_4 = v_1$$
$$\mu_2: v_2 = f(v_1,v_3);\ v_3 = f(v_4)$$
$$\mu_3: v_4 = f(v_2)$$

†See Chap. I, Secs. 10.2 and 10.3, that discusses concurrency issues in microoperation execution that arise in horizontal microprogramming. (Ed.)

‡In Chap. I, a microorder is called a microcommand. (Ed.)

*Concurrent and phased execution of (horizontal) microinstructions is discussed in Chap. I, Sec. 10.3.2. (Ed.)

410 CHAPTER IV

The set of variables V is made up by v_1, v_2, v_3, and v_4. The corresponding input nodes are then inserted at level 0 in the graph (Fig. IV.4(a)).

Microinstruction μ_1 consists of three concurrent microorders. Thus, at level 1 three nodes are inserted, marked v_1, v_3, and v_4. The functional relation $v_1 = f(v_1, v_2)$ is represented by strong edges going to v_1 from the nodes marked v_1 and v_2 at level 0; the same is repeated for microorder $v_3 = f(v_1)$. Last,

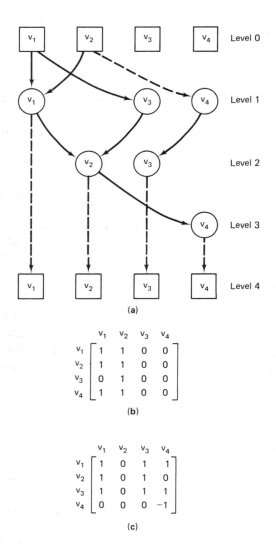

FIGURE IV.4 (a) Graph representing I/O relationships for Example 5 (horizontal microprogram); (b) V-matrix for Example 5

$v_4 = v_1$ is represented by a weak edge going to node v_4 at level 1 from node v_1 at level 0.

In the same way, microinstruction μ_2 leads to insertion of two nodes (marked v_2 and v_3) at level 2; edges are again drawn from the nearest nodes representing variables which appear on the right side in the abstract relations.

Microinstruction μ_3 leads to the insertion of just one node, v_4, at level 3; finally, the four output nodes are inserted at level 4 and weak edges connect each of them with the nearest node marked with the same variable. It is now possible to state that

- There are strong paths from input node v_1 to output nodes v_1, v_2, and v_4.
- There are strong paths from input node v_2 to output nodes v_1, v_2, v_3, and v_4.
- There are no paths from input nodes v_3 and v_4 to any output node.

The V-matrix in Fig. IV.4(b) is thus completed. ∎

Let us get back to the other two problems; i.e., definition of "variable set" and of "abstract relations."

Variables have to be defined with reference to machine architecture. A variable is an information unit capable of being independently accessed, processed, read, or written. In this context, an n-bit register R appears as an n-bit variable, a carry flag C is a one-bit variable, etc. Summing up, the set of variables V comprises the names of

- All registers identifiable by a unique name (this allows also for I/O ports) and definable as a collection of bits connected via identical logical circuits with other registers or logical circuits.
- In accordance with what has been done for programs, a "dummy" representing all constants.

Relations can be

- *Simple transfers* of an information unit from a "source" variable to an "object" variable. These correspond to simple assignments in programs, and will be represented in the graph by weak edges.
- *Transfers through functional units* (functional transfers), involving some processing of the source variable(s). They correspond to *proper functions* in programs, and will be represented in the graph by strong edges.† It remains to consider *read* transfers. From an architectural point of view these do not differ from a register-to-register transfer. However, if we refer to the focal point of structural termination, that

†See Chap. I, Secs. 1.2 and 1.3, that discuss microoperations implementing simple and functional transfers. (Ed.)

is, a continued data dependence, it can be said that an external read, in microprograms just as in real-time programs, leads to a non-determinable value of the variables thus read. It seems therefore reasonable, in this context, to interpret a read as a functional transfer.

We must still consider the case of registers defined through the contents of a different, addressing, register, and of accesses made through addressing. Suitable definitions have to be given in terms of variable and abstract relations.

Actually, as stated in Sec. 3.1, the problem is not different from the one met in a program with indexed or "structured" variables such as arrays, etc. Indices are in fact an addressing mechanism. We present here a partial solution for the case of machine architectures and of microprograms, which is justifiable thanks to the limited number of addressing modes and of addressable structures in comparison with the complex structures possible in programs. A more complex and general solution is being sought.

First assume R_S to be the source register, and R_O the object register involved in an abstract relation.† It is possible that R_S and R_O have different lengths. For instance, it may happen that R_S is smaller than R_O and that its content is transferred into a segment of R_O (or, on the contrary, that a segment of R_S is transferred into a smaller R_O). If the information stored into R_S is not otherwise processed during the transfer, the operation will still be considered a simple transfer and will be represented by a weak edge. It might happen although it is unusual that simultaneous transfers occur from different (smaller) sources to one object register. In this case, separate logic and bussing would be necessary to perform the various operations, and the various segments of R_O need to be separately addressed. It becomes therefore quite reasonable to consider R_O as made up of several parallel working units and to represent it by as many variables; the same holds for a source R_S whose segments can be transferred independently and in parallel to different object registers. Of course, when the complete register is then involved in other microinstructions, it will be necessary to represent it again as a set of variables rather than as a single variable and to insert a group of concurrent microorders instead than a single one operating upon it.

Consider now the case where the particular segment of the larger register involved is defined, each time, through the value of a third variable R_a. Assume, for instance, that R_S is the smaller register and R_O the larger one. It is necessary to recognize

- Whether the transfer is simple or functional.
- What are the relationships of R_S and R_O with R_a and how they can be represented.

For this purpose, it is useful to refer back to the definition of structural termination and to the related considerations introduced in Sec. 2. Let us

†See reference on page 411. (Ed.)

assume that the (micro)instruction sequence being examined is enclosed in a simple loop ending with a predicated $P(v_i)$ dominated by a variable v_i. (For instance, referring to Example 5 let instruction μ_3 be followed by the conditional branching

$$\text{if } P(v_3) \text{ then goto } \mu_1$$

Should only simple transfers occur among the variables in the whole sequence, then it is obvious that, after an initial transient which depends on the values the variables had when the loop was entered, variable v_i will either assume one constant value or else permute cyclically among the values of a finite set of constants. Both such instances were discussed in detail in Sec. 2 (where they constituted cases 2 and 3), where an analysis of some numerical examples strengthened our assertions that in such cases *structural termination does not hold*. Simple transfers have already been associated with weak edges.

Assume now the content of R_S to be kept constant, while the content of R_a varies with any given rule—obviously inside a finite set of values. As a consequence, the contents of R_O after an initial transient will permute cyclically among a finite and predetermined set of values following a known rule. Thus, the same instance, which was seen to lead to lack of structural termination, considered above occurs. We consider this type of transfer as a *simple* one, represented in the graph by a weak edge.

No conceptual difference exists between the case we are now examining and that of a random-access memory or of a register file accessed through an addressing register. Let M denote a random access memory, MAR be its memory addressing register, and let MB be the associated memory buffer register. As a consequence of the previous discussions, a microoperation of the type

$$(M(MAR)) \rightarrow MB$$

will be considered in the sequel as a simple transfer from register M to register MB; the addressing register MAR will be represented only in such abstract relations which directly transfer or process its contents.

Finally, we can deduce that the set of variables V will be extended to comprise as many variables as there are *register files* or *storage areas* accessed each through a single addressing register.

Consider the following example of vertical microprogram (i.e., one in which each instruction comprises one microorder only), corresponding to a segment of the execute phase of an instruction such as "increment a word in memory until it reaches value 0" (e.g., an instruction allowing implementation of a software counter, a programmed delay, etc.). Let M, MB, and MAR have the meanings just assigned.

Example 6.

$$\mu_1: MB = (M(MAR))$$

μ_2: $MB = (MB) + 1$

μ_3: $M(MAR) = (MB)$

μ_4: if $(MB.NE\cdot 0)$ then μ_1 else ....

The abstract representation of this microinstruction sequence becomes

μ_1: $MB = M$

μ_2: $MB = f(MB)$

μ_3: $M = MB$

μ_4: if $P(MB)$ then μ_1 else ....

It is now possible to draw the graph representing I/O relationships following the by now well-known rules (Fig. IV.5(a)). At level 0, input nodes

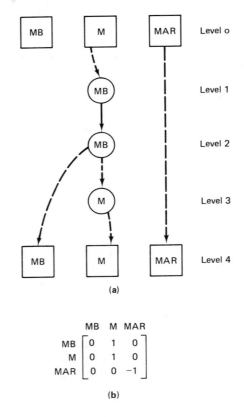

FIGURE IV.5 (a) I/O relationships graph for Example 6; (b) V-matrix for Example 6

MB, M, and MAR represent the input values of variables involved. At level 1, microinstruction μ_1 leads to insertion of a node marked MB and connected by means of a weak edge with node M at level 0, etc.

Microinstruction μ_4, which does not affect the value of any variable, but rather marks the end of the microinstruction sequence, leads to ending the graph with the set of output nodes at level 4; since the contents of MAR have not been affected during the whole sequence, a weak edge is simply drawn from input node MAR to output node MAR. The usual examination of paths in the graph leads to the complete V-matrix in Fig. IV.5(b). The input value of MB was lost (i.e., no path connects input node MB to any output node) thus column MB of the V-matrix only has zero entries. The strong paths from input node M to output nodes MB and M, respectively, lead to $+1$ entries in column M, row MB and column M, row M; the single weak path from input node MAR to output node MAR leads to a -1 entry in column MAR, row MAR. (All entries not explicitly recalled are 0.) ∎

3.3. Modeling Timing and Synchronization Characteristics

Up to now, all considerations have been made with reference to an ideal environment where no delays are present or where alternatively each (micro)instruction in a (micro)program, or each microorder in a horizontal microinstruction, take up the same time and absolute synchronization is a priori guaranteed.†

This is the usual hypothesis when program correctness is examined. In fact, validation of sequential programs does not involve timing or synchronization considerations. The same simplifying hypothesis is not always as acceptable when microprograms are considered; the no delay assumption reduces to assuming that any (functional) transfer in the machine can be performed in one same "basic cycle time," an hypothesis largely denied by actual architectures. In fact, these automatic synchronization assumptions can be acceptable for strictly vertical microprograms (which do not differ from sequential programs, from this point of view) or whenever some kind of hardware-implemented lock disables any further processing until the slowest microorder has been completed.

Similar problems arise when multiprocessor systems are considered. If there are no shared resources, in particular, no shared memory areas, and if distribution of functions among processors is performed on the basis of (possibly complex) tasks, it is possible to consider single tasks as sequential programs as far as structural termination is concerned. In, on the other extreme, a step-by-step parallelism is kept between processors executing different instructions (or instruction sequences) of one program, the case gets very near to that of horizontal microprograms.

Let us consider how these facts can be modeled in the representation of I/O relationships and how they influence analysis of structural termination.

†Technique for finding a microoperation delay is considered in Chap. I, Sec. 1.3. Microinstruction delay is discussed in Sec. 10.3.2, of the same chapter. (Ed.)

†Phased execution of microinstructions is discussed in Chap. I, Sec. 10.3.2. (Ed.)

Consider the following first instance, exemplified with reference to microprogramming. A given (functional) transfer, leading to an updated value of variable v_i, requires a known number n_i of basic cycle times, whereas the simplest microorders require one basic cycle.† For optimizing the computer performance, after the transfer has been "launched" it is possible to start execution of subsequent microinstructions, until it becomes necessary to use the updated value of v_i. If we adopt in the graph representing I/O relationships the convention of associating transition from level j to level $j + 1$ with the basic machine cycle, the fixed delay of n_i cycles is quite simply modeled by inserting the node representing the updated value of v_i a number n_i of levels lower than the one deriving from the previous microinstruction, and by verifying that this relative positioning of nodes is respected by all abstract relations. Clearly, no new problem arises as far as structural termination is concerned. Consider the following.

Example 7. A segment of horizontal microprogram performs the following set of operations.

The content of the accumulator AC is incremented by one, squared, normalized, and divided by the content of a memory word addressed through MAR. In the meantime, the content of an index register IX is incremented and transferred into MAR. The following timing characteristics are given: A memory read requires three basic cycles; the address must be kept stable for one cycle after memory read command; operation $(AC)^2$ requires two basic cycles; all other (functional) transfers require one basic cycle. The microprogram can be written as follows:

μ_1: $MB = (M(MAR))$

μ_2: $AC = (AC)^2$; $IX = (IX) + 1$

μ_3: $AC = \text{norm}(AC)$; $MAR = (IX)$

μ_4: $AC = (AC)/(MB)$

and the abstract relations sequence is

μ_1: $MB = M$

μ_2: $AC = f(AC)$; $IX = f(IX)$

μ_3: $AC = f(AC)$; $MAR = IX$

μ_4: $AC = f(AC, MB)$

The timing in execution of this microprogram is as follows:

cycle 1: memory read command (μ_1)

cycle 2: operation commands on AC, IX (μ_2)

cycle 3: transfer from *IX* to *MAR* (the previous operation on *AC* completes at the end of this cycle)

cycle 4: normalization of *AC* (thus, μ_3 is split up into two cycles)

cycle 5: operation command on *AC* and *MB*.

The graph in Fig. IV.6 is thus obtained. As usual, level 0 corresponds to the input nodes, while each level from 1 to 5 corresponds to a basic machine cycle. Since a node at level *i* represents the *value already obtained* by the corresponding variable, nodes in the graph will be inserted at the level representing not the time at which the command is given, but the one at which the updated value is available. Thus, no node is inserted at level 1 (the result of microinstruction μ_1 will be available only at level 3, where node *MB* is inserted). At level 2, only node *IX* is inserted; the result of microorder $(AC)^2 \to AC$ will be available only at level 3, where node *AC* is inserted. At level 4, a new node *AC* can be introduced (normalization only requires one cycle), and the same holds at level 5.

In a second instance, n_i may not be known a priori but an upper limit may be set for its value, and the (micro)program is written so as to guarantee that v_i is not accessed before this upper limit has elapsed. This case reduces to the previous one, simply by substituting n_i with the upper-limit value. ■

While previous cases can still be considered as "synchronous," a more difficult and interesting instance is represented by asynchronous delays; i.e.,

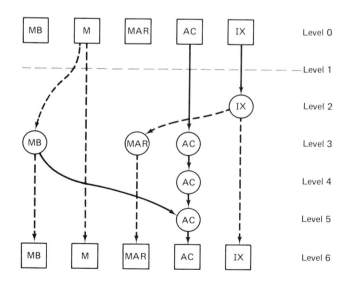

FIGURE IV.6 Input-output relationships graph for Example 6

delays whose value can in no way be predicted. Typically, these instances will arise not from technological constraints (as the previous ones did) but from interaction between asynchronous units.

Assume first that synchronization—in our case, correct access to the updated variable(s)—is performed through the use of other variables acting as "flags." Consider the case of a program $c1$ which can continue its execution only if flag v_S (set by a concurrent process $c2$) validates acquisition of an external information v_2. The instruction sequences involved constitute

Example 8.

$$
\begin{array}{ll}
c1 & c2 \\
n_1: \textit{if } (v_S) \textit{ then} & v_S = \textit{false} \\
n_2: v_1 = f(v_1, v_2) & l_1: \textit{read } v_2 \\
\quad \textit{else } n_1 & l_2: \textit{if } (v_2 \cdot \text{NE} \cdot 0) \textit{ then} \\
n_3: \ldots & l_3: v_S = \textit{true} \\
& \quad \textit{else } l_1
\end{array}
$$

The two program segments are represented by the C-graph in Fig. IV.7(a). Although we have two separate C-graphs, termination of the loop in $c2$ is actually the condition which dominates the loop in $c1$, and v_s is the variable synchronizing execution of $c1$ with the operations performed in $c2$. We can therefore consider execution of $c2$ to "precede" execution of $c1$; since both processes access common variables, it is correct to use the output interface values of $c2$ as input interface values of $c1$. We come to a composition of two graphs representing I/O relationships, built up as follows. (See Fig. IV.7(b).) First, input nodes concerning $c2$ are inserted at level 0 (variable v_1 is added in order to keep the one-to-one correspondence between variables used in $c2$ and in $c1$). At level 1, a node v_2 is inserted and connected via a strong edge with the input value of v_2. Now, v_S is assigned a constant value *depending* upon the value of v_2. We represent this fact by drawing a strong edge *from* v_2 to v_S. At this point (level 2) c_1 has been completely processed and output nodes may be inserted. These values constitute the basis for computation of statement n_2. Again, a dependence of v_1 from v_S (through the synchronizing clause) is explicitly inserted. The graph is completed by representing the passing of values from the outputs of c_2 to the inputs of c_1, by the insertion of node v_1 as depending from v_1, v_2 *and* v_S and by the final output nodes (Fig. IV.7(b)). ∎

In general, we may state that if V_S is the set of variables acting as synchronization mechanism and controlling access to a variable v_i, dependence of v_i from V_S must be stated. This is performed by considering the synchronization mechanisms as proper functions, and represented in the graph by

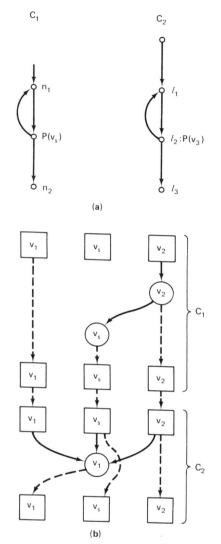

FIGURE IV.7 (a) C-graph for Example 7; (b) I/O relationships graph for Example 7

strong edges going from "controlling" to "controlled" nodes. The problem is thus reduced to a simple examination of structural termination characteristics. (The predicate upon the flag will need to be examined.)

In principle, it remains to consider the "totally asynchronous" instance, in which no evaluation of delays is possible and no synchronization mechanism occurs. Actually, access to a variable v_i involved in such an operation

420   CHAPTER IV

mode gives completely random results, since there is no way of controlling the relative timing between updating and access. This is therefore a case of noncorrect system design and it seems meaningless to discuss program validation for such an instance.

4. TERMINATION OF SERIES-PARALLEL CONTROL GRAPHS

In this section we will see how simple computations on V-matrices allow us to determine the structural termination of a class of programs characterized by series-parallel C-graphs, from the simplest to the most general case.

In order to perform such computations, an algebra has to be introduced; in particular, it is necessary to define basic operations on V-matrices.

Let us consider first actual operations involving different instruction sequences during a run. Whatever the structure of the control graph (in terms of branchings, meeting points, etc.) execution of a (micro)program reduces to execution of a first instruction sequence a followed by execution of a second instruction sequence b, etc. In other words, during a run we have a number of instruction sequences executed in series to each other. It is necessary to see how such a series connection is translated to the V-matrices.

Consider the simple C-graph corresponding to the catenation of two abstract relation sequences.

Example 9.

$$\left.\begin{array}{l} n_1: v_1 = f(v_1, v_2) \\ v_2 = v_3 \\ v_3 = f(v_1, v_4) \\ v_4 = f(v_2) \end{array}\right\} a$$

$$\left.\begin{array}{l} n_2: v_2 = f(v_1) \\ v_1 = v_4 \\ v_4 = f(v_2, v_3) \\ n_3: v_3 = f(v_1) \end{array}\right\} b$$

where the first sequence (a) consists of the first four abstract relations, and the second one (b) comprises relations from n_2 and n_3. The original instruction sequences are executed one directly following the other; the C-graph is the series connection of two edges (Fig. IV.8(a)).

Let us now represent the I/O relationships concerning the whole series connection. Output values from the first relation sequence constitute the input

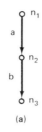

(a)

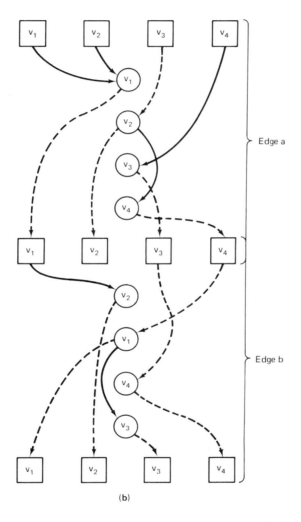

(b)

FIGURE IV.8 (a) C-graph for Example 8; (b) Graph representing I/O relationships for edge a followed by edge b (Example 8); (c) V-matrices for edges a, b; (d) I/O relationships graph for equivalent edge; (e) V-matrix for equivalent edge

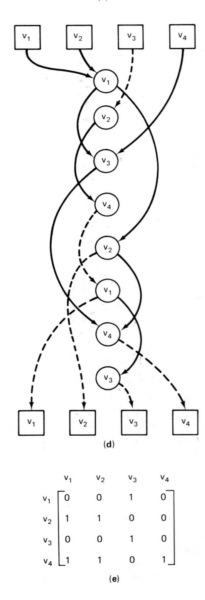

FIGURE IV.8 (continued)

values for the second one, and thus there is just one set of interface values between the two sequences.

We start by drawing the graph representing I/O relationships for sequence a, following the usual rules. The output nodes of this first graph section constitute the input nodes of a second graph section, concerning sequence b. Thus, the complete graph of Fig. IV.8(b) is obtained. The V-matrices concerning instruction sequences a and b, denoted as M_a and M_b, respectively are given in Fig. IV.8(c). They are built following criteria by now well-known, and representing the following features

- In the graph associated with a: strong paths from v_1 to v_1 and v_3; from v_2 to v_1 and v_3; from v_3 to v_4; from v_4 to v_3; a weak path from v_3 to v_2.
- In the graph associated with b: strong paths from v_1 to v_2 and v_4; from v_4 to v_3; weak paths from v_3 to v_4, from v_4 to v_1. ∎

The series connection of two instruction sequences may well be considered as one single sequence. It is therefore possible to draw one graph representing I/O relationships, omitting the interface section (Fig. IV.8(d)). We associate the V-matrix M_e (Fig. IV.8(e)) with the graph representing I/O relationships of the "equivalent" edge.

It seems natural that M_e could be derived through suitable operations on M_a and M_b. In fact, let us apply the usual rules of matrix multiplication upon $M_b \cdot M_a$, assuming that the two basic operations of addition and multiplication are defined as follows:

Addition	Multiplication
$0 + 0 = 0$	$0 \times 0 = 0$
$0 + 1 = 1$	$0 \times 1 = 0$
$0 + (-1) = -1$	$0 \times (-1) = -0$
$1 + 1 = 1$	$1 \times 1 = 1$
$1 + (-1) = 1$	$1 \times (-1) = 1$
$-1 + (-1) = -1$	$-1 \times (-1) = -1$

(both operations are commutative).

Then, we compute the various entries of M_e.

$$M_e(1, 1) = \sum_{k=1}^{4} M_b(1, k) \times M_a(k, 1)$$
$$= 0 \times 1 + 0 \times 0 + 0 \times 1 + (-1) \times 0 = 0$$

$$M_e(1, 2) = \sum_{k=1}^{4} M_b(1, k) \times M_a(k, 2)$$
$$= 0 \times 1 + 0 \times 0 + 0 \times 1 + 0 \times (-1) = 0$$

$$M_e(1, 3) = \sum_{k=1}^{4} M_b(1, k) \times M_a(k, 3)$$
$$= 0 \times 0 + 0 \times (-1) + 0 \times 0 + (-1) \times 1 = 1$$

We can easily verify that by adopting the two operators just defined and the basic rules of matrix multiplication, the V-matrix representing I/O relationships of two series-connected edges can be obtained by multiplying the V-matrix of the second, or successor, edge (left-hand factor) by the V-matrix of the first, or predecessor, edge (right-hand factor).

The above rule can be obviously extended to a series connection of any number of C-graph edges.

We see now how V-matrices can be used for determining structural termination; by application of the algebra introduced above. We shall examine first the simplest C-graph structure: a condition *guaranteeing* structural termination (a "necessary and sufficient condition") will be found. Later on, general series-parallel C-graphs will be examined.

The simplest C-graph structure which is interesting for structural termination is the "simple loop," defined and studied in the following subsection.

4.1. Simple Loop

A *simple loop* is defined as a C-graph (or, better, a segment of a C-graph) in which only one loop exists, and there is only one possible path inside the loop. The C-graph in Fig. IV.3(a) (referring to Example 4) is a simple loop; on the other hand, Example 3 is not a simple loop since its C-graph (Fig. IV.2(c)) shows two alternative paths inside the loop. If there is just one path from the node where the loop is entered to the closing predicate, this path is either a single edge or else it is the series connection of various edges (which can be immediately replaced by a single equivalent edge). From now on, we shall consider therefore a simple loop as having always the structure given in Fig. IV.3(a).

Let M be the V-matrix of the instruction edge enclosed in the loop. During a run, after k iterations the edge has been executed k consecutive times, just as if k identical edges had been connected in series. Thus, the relationships between variables' values at the moment of entering the loop and output values after k iterations can be summarized by the kth power of the V-matrix associated with the edge.

Now, given the meaning attached to possible values of V-matrix entries, we know that a V-matrix whose entries only are 0 or -1 represents an instance in which either no variable is modified or else only simple assignments are performed. This is to say that execution of the instruction sequence with which such a V-matrix is associated either causes no variable to change its value, or else it simply causes a permutation in the assignment of a fixed set of values to a fixed set of variables.

If the kth power of the V-matrix associated with the edge enclosed in a simple loop has the above characteristics, we know that all variables, and in particular the one(s) that dominate the closing predicate, will not undergo the continued modification required for structural termination. Of course, in principle it would seem necessary to examine all possible powers of the V-matrix in order to verify whether such an instance occurs. Therefore, it is possible to derive the presence or absence of structural termination by a very simple examination of V-matrix M. A more intuitive approach to this proof can be obtained if a new graph is introduced, in fact, just a new way of presenting information already gathered by means of previous preprocessing.

This new graph is called the V-graph and it is simply a pictorial way of representing information summarized by the V-matrix. It is more compact than the I/O relationships graph—just as the V-matrix is—since it only shows whether an I/O relationship exists and what its nature is. The V-graph is derived from the V-matrix as follows:

- There is one node in the V-graph for each variable appearing in the V-matrix. Thus, refer back to Example 4, and to the related V-matrix (Fig. IV.3(c)). In order to build the V-graph, four nodes marked v_1, v_2, v_3, and v_4 are introduced.
- Each entry $v(i,j) \neq 0$ in the V-matrix is represented in the V-graph by an oriented edge going from node v_i to node v_j. The edge is strong (full arrow) if the entry has value $+1$, it is weak (dotted arrow) if the entry has value -1.
- No other edges appear in the graph.

(Note that, in this way, the V-matrix is the incidence matrix of the V-graph.)

Thus, in the V-matrix of Fig. IV.3(c) there are $+1$ entries in column v_1 for rows v_1, v_2, and v_3. In the V-graph we draw full arrows from node v_1 to node v_1 itself (self-loops are possible in the V-graph), to node v_2 and to node v_3. Column v_2 only contains 0 entries; no arrows depart from node v_2 in the V-graph. Column v_3 has $+1$ entries for rows v_1, v_2, and v_3; full arrows go from node v_3 to nodes v_1, v_2, and v_3. Column v_4 has $+1$ entries for rows v_1 and v_3 and a -1 entry for row v_4. In the V-graph, we insert full arrows from v_4 to v_1 and v_3, and a dotted arrow from v_4 to v_4 (Fig. IV.9).

It is possible to identify a path in the V-graph from node v_i to node v_j as a sequence of oriented edges from node v_i to node v_k, from v_k to v_{k+1} and so on until v_j is reached (if such a path exists). Again, we say that a path is strong if it contains at least one strong edge, and weak otherwise. Thus, the path from v_1 to v_2 is strong, the one from v_4 to v_4 is weak. It is possible to have more than one strong path from v_i to v_j comprising different sets of nodes, if any exists at all. Thus, in Fig. IV.9 we see that v_3 can be connected to v_2 by such different strong paths as v_3-v_2 or $v_3-v_1-v_2$ (existence of loops and self-

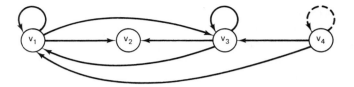

FIGURE IV.9 V-graph for Example 4

loops may then bring to an infinite number the number of paths). In fact, a row of the V-matrix may have any number of $+1$ entries. This means that any number of strong edges may enter a node in the V-graph.

On the other hand, if a weak edge goes from one node v_j to node v_i it is is the only (possibly cyclic) path between them. If a row in the V-matrix has a -1 entry, the fact that all other entries in the row are 0 guarantees that if a weak edge enters a node in the V-graph no other edge does. A weak path is a sequence of weak edges. For any node on such a sequence, the path is the only one entering it, so that ultimately it is also the only node from its initial to its final node. In Fig. IV.9, the only weak path is the self-loop on v_4; it is also the only path reaching node v_4.

Hereafter, a looplike path in the V-graph—that is, a path whose initial and final nodes coincide—will be called a *cycle*.

It is now possible to prove the basic theorem concerning structural termination (formal proof is given in [15]:

Theorem 1. Consider a simple loop with a closing predicate P dominated by a set of variables $v_p = \{v_{1p}, v_{2p}, \ldots, v_{nP}\}$. Refer to the associated V-graph. If for at least one of the variables in V_p, v_{ip}, it happens that

1. Either v_{ip} belongs to a strong cycle, or else
2. v_{ip} does not belong to a strong cycle but there is (at least) one path going from a node on a strong cycle to v_{ip} (in which case it is said that v_{ip} "depends" from a strong cycle)

then structural termination of the loop is guaranteed. On the other hand, if the above condition is not satisfied the program will *not* be characterized by structural termination (i.e., the condition is necessary and sufficient).

While a formal proof of this theorem is given in [15], it is possible to illustrate it in a more intuitive way. Refer to the simplest case of a closing predicate dominated by a single variable. Consider for instance Example 4, whose closing predicate is dominated by v_1. A strong edge in the V-graph from v_j to v_i means that after execution of the instruction sequence, the output value of v_i depends upon a proper function from the input value of v_j. In general, a strong path from v_j to v_i implies this kind of functional dependence through as many iterations of the loop as there are edges in the path. Now, if the path from v_j to v_i is a cycle, its length is infinite. This means that functional dependence of v_i upon v_j is continued; that is, that at each iteration v_i is modified through a functional dependence and assumes a value different from the previous one. This exactly corresponds to the condition for structural termination. Example 4 is associated with a V-graph where variable v_1 actually appears in two strong cycles. Example 4 is structurally terminating.

The same considerations seen above hold if the variable dominating the closing predicate appears on a path which originates on a strong cycle. This means that the variable depends from a set of variables undergoing continued functional modification, and that the previous instance is met again. Consider for instance Example 1, which was used to illustrate the meaning of structural termination. The program is a simple loop, whose V-matrix is given in Fig. IV.10(a).† The V-graph is given in Fig. IV.10(b). The only strong cycle in the V-graph is the self-loop on node v_4, from which node v_1 depends through an oriented path.

On the other hand, assume that the condition in Theorem 1 is not satisfied. Assume, for instance, that the V-graph is acyclic. Such a graph has "sources," i.e., nodes into which no edge enters. A source in the V-graph represents a variable whose value is never modified. After one iteration of the loop, all nodes connected to sources by an edge correspond to variables whose values also has become constant; after two iterations the same happens for nodes depending from sources through paths consisting of two edges (or "at distance 2 from the sources"), etc. After a number of iterations equal to the maximum length of a path in the V-graph, all variables have assumed constant values. If, in particular, the variable which dominates the closing predicate is on such a path, we find that the program does not satisfy structural termination (case 2, Sec. 2).

Actually we have no completely acyclic V-graphs; the rules for identi-

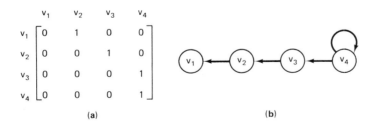

FIGURE IV.10 (a) V-matrix for Example 1; (b) V-graph for Example 1

†The reader is invited to derive this V-matrix from the program.

fying I/O relationships and then building V-matrices lead us to insert a weak self-loop on all variables which are in no way modified in the course of the program. Obviously, this representation gives a "weak cycle". Again, Theorem 1 is not satisfied and the program does not exhibit structural termination. Refer to Example 2; the associated V-matrix and V-graph are given in Figs. IV.11(a) and (b), respectively. The only cycle in the V-graph is the weak self-loop on v_4; v_1 is connected by a path (it does not matter if it is a strong one) to this cycle. The program, as we know, is not structurally terminating since after three iterations (notice that three is the length of the path from the weak cycle to v_1) the value of v_1 becomes constant.

The above considerations are extended to the case in which the variable dominating the predicate appears in any weak cycle, or depends from such a cycle. In this instance, there is a periodical cycling of the variable through the values in a finite set, as in case 3 of Sec. 2.

While identification of strong cycles in the V-graph is a straightforward affair for manual analysis of structural termination, we need a way to identify the same condition by automatic processing. Clearly, this processing will be much easier upon V-matrices than upon V-graphs; the information content is the same, but the representation is much more computer-oriented. Thus, we want to identify existence of strong cycles by examination of V-matrices. To this end, we define first of all a "submatrix of a V-matrix M over a subset $\bar{I}$ of the index set I" as a matrix obtained from M by deleting all rows and columns marked with indices not belonging to $\bar{I}$. With any submatrix we can immediately associate a subgraph of the V-graph consisting of all nodes marked with indices in $\bar{I}$, connected with all edges corresponding to entries in the submatrix.

For example, refer to the V-matrix in Fig. IV.10(a) and let $\bar{I} = \{v_3, v_4\}$. The corresponding submatrix is made up of entries (v_3, v_3), (v_3, v_4), (v_4, v_3), and (v_4, v_4) (see Fig. IV.12(a).) The subgraph represented by this submatrix is shown in Fig. IV.12(b).

We define now a *strong submatrix S* of a matrix M as a submatrix over $\bar{I}$ characterized by the following features:

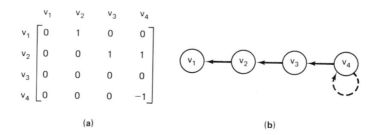

FIGURE IV.11 (a) V-matrix for Example 2; (b) V-graph for Example 2

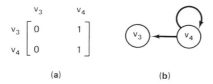

FIGURE IV.12 (a) Submatrix of V-matrix in Fig. IV.10(a); (b) subgraph corresponding to submatrix in Fig. IV.12(a)

- There is at least one nonzero entry in each column.
- There is at least one $+1$ entry.

For Example 4, submatrix S over $\bar{I} = \{v_1, v_3, v_4\}$ is a strong submatrix (Fig. IV.13(a); the associated subgraph is shown in Fig. IV.13(b)). We see now how the existence of strong submatrices of a V-matrix guarantees the existence of strong cycles in the V-graph; i.e., the first condition for structural termination. This is particularly interesting because finding the largest strong submatrix of a given V-matrix (i.e., the one with the largest set of indices, and which clearly contains all smaller submatrices) is a very simple operation; it is sufficient to

- Delete from I all indices marking columns whose entries are all zeroes and examining the corresponding submatrix.
- Iterate the previous step until either a strong submatrix is found, a zero-dimension submatrix is reached, or the submatrix has only 0 and -1 entries.

Thus, consider V-matrix in Fig. IV.10(a). First, we delete index v_1 and obtain the submatrix in Fig. IV.14(a). [Since column v_2 is now an all-zero column, index v_2 is removed (Fig. IV.14(b).] Again, column v_3 has only 0 entries. Index v_3 is removed. The residue submatrix of dimension 1 (Fig. IV.14(c)) is a strong one.

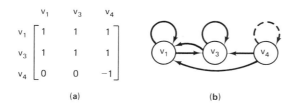

FIGURE IV.13 (a) Strong submatrix for Example 4; (b) strong subgraph for Example 4

$$
\begin{array}{c}
\begin{array}{ccc}v_2 & v_3 & v_4\end{array} \\
\begin{array}{c}v_2 \\ v_3 \\ v_4\end{array}\left[\begin{array}{ccc}0 & 1 & 0 \\ 0 & 0 & 1 \\ 0 & 0 & 1\end{array}\right]
\end{array}
\qquad
\begin{array}{c}
\begin{array}{cc}v_3 & v_4\end{array} \\
\begin{array}{c}v_3 \\ v_4\end{array}\left[\begin{array}{cc}0 & 1 \\ 0 & 1\end{array}\right]
\end{array}
\qquad
\begin{array}{c}
\begin{array}{c}v_4\end{array} \\
v_4\left[\begin{array}{c}1\end{array}\right]
\end{array}
$$

(a)   (b)   (c)

FIGURE IV.14 (a) First step for obtaining strong submatrix of Fig. IV.10(a); (b) second step; (c) strong submatrix of Fig. IV.10(a)

Consider now V-matrix in Fig. IV.11(a). Again, column v_1 has only 0 entries and index v_1 is removed (Fig. IV.15(a)). Column v_2 of the submatrix has only 0 entries. The submatrix of Fig. IV.15(b) is obtained. Finally, the submatrix of Fig. IV.15(c) is reached, its only entry is -1. The given V-matrix has no strong submatrix (just as the corresponding V-graph has no strong cycle).

Having thus motivated our interest in strong submatrices, we define their most important properties. (Again, full proofs are to be found in [15].)

Theorem 2. Let M_a and M_b be two submatrices having each a strong submatrix (denoted respectively as S_a and S_b) over the same subset $\bar{I}$. The product matrices $M_a \cdot M_b$ and $M_b \cdot M_a$ also contain a strong submatrix over $\bar{I}$.

Refer back to the V-matrices in Fig. IV.8(c); they both have a strong submatrix over $\bar{I} = \{v_1, v_3, v_4\}$ (S_b is also the largest strong submatrix of M_b; the largest strong submatrix of M_a is M_a itself.) S_a, S_b are given in Fig. IV.16(a). Now, examine matrices $M_b \cdot M_a$ and $M_a \cdot M_b$ (Fig. IV.16(b)). We can immediately verify that they have strong submatrices over $\bar{I}$, denoted as S_{ba} and S_{ab}, respectively, (Fig. IV.16(c)).

The justification of Theorem 2 can be easily derived from our definition of V-matrix algebra.

An interesting particular case of V-matrix multiplication is—as we already verified—multiplication of a V-matrix by itself, since the result repre-

$$
\begin{array}{c}
\begin{array}{ccc}v_2 & v_3 & v_4\end{array} \\
\begin{array}{c}v_2 \\ v_3 \\ v_4\end{array}\left[\begin{array}{ccc}0 & 1 & 1 \\ 0 & 0 & 0 \\ 0 & 0 & -1\end{array}\right]
\end{array}
\qquad
\begin{array}{c}
\begin{array}{cc}v_3 & v_4\end{array} \\
\begin{array}{c}v_3 \\ v_4\end{array}\left[\begin{array}{cc}0 & 0 \\ 0 & -1\end{array}\right]
\end{array}
\qquad
\begin{array}{c}
\begin{array}{c}v_4\end{array} \\
v_4\left[\begin{array}{c}-1\end{array}\right]
\end{array}
$$

(a)   (b)   (c)

FIGURE IV.15 (a) First step in search of strong submatrix of Fig. IV.11(a); (b) second step; (c) final step. No strong submatrix exists

$$S_a = \begin{array}{c} \\ v_1 \\ v_3 \\ v_4 \end{array} \begin{array}{ccc} v_1 & v_3 & v_4 \\ \begin{bmatrix} 1 & 0 & 0 \\ 1 & 0 & 1 \\ 0 & 1 & 0 \end{bmatrix} \end{array} \qquad S_b = \begin{array}{c} \\ v_1 \\ v_3 \\ v_4 \end{array} \begin{array}{ccc} v_1 & v_3 & v_4 \\ \begin{bmatrix} 0 & 0 & -1 \\ 0 & 0 & 1 \\ 1 & -1 & 0 \end{bmatrix} \end{array}$$

(a)

$$M_b \cdot M_a = \begin{array}{c} \\ v_1 \\ v_2 \\ v_3 \\ v_4 \end{array} \begin{array}{cccc} v_1 & v_2 & v_3 & v_4 \\ \begin{bmatrix} 0 & 0 & 1 & 0 \\ 1 & 1 & 0 & 0 \\ 0 & 0 & 1 & 0 \\ 1 & 1 & 0 & 1 \end{bmatrix} \end{array} \qquad M_a \cdot M_b = \begin{array}{c} \\ v_1 \\ v_2 \\ v_3 \\ v_4 \end{array} \begin{array}{cccc} v_1 & v_2 & v_3 & v_4 \\ \begin{bmatrix} 1 & 0 & 0 & 1 \\ 0 & 0 & 0 & 1 \\ 1 & 0 & 1 & 1 \\ 0 & 0 & 0 & 1 \end{bmatrix} \end{array}$$

(b)

$$S_{ba} = \begin{array}{c} \\ v_1 \\ v_3 \\ v_4 \end{array} \begin{array}{ccc} v_1 & v_3 & v_4 \\ \begin{bmatrix} 0 & 1 & 0 \\ 0 & 1 & 0 \\ 1 & 0 & 1 \end{bmatrix} \end{array} \qquad S_{ab} = \begin{array}{c} \\ v_1 \\ v_3 \\ v_4 \end{array} \begin{array}{ccc} v_1 & v_3 & v_4 \\ \begin{bmatrix} 1 & 0 & 1 \\ 1 & 1 & 1 \\ 0 & 0 & 1 \end{bmatrix} \end{array}$$

(c)

FIGURE IV.16 (a) Strong submatrices of Fig. IV.8(c); (b) product matrices $M_b \cdot M_a$, $M_a \cdot M_b$ for Fig. IV.8(c); (c) strong submatrices of product matrices

sents I/O relationships for iterated executions of the instruction sequence. In fact, it can be stated that:

If a V-matrix M contains a strong submatrix over $\bar{I}$, then all the powers of M contain a strong submatrix over $\bar{I}$.

At last, we reach the basic (to our purposes) characteristic of V-matrices.

Theorem 3. If a V-matrix M contains a strong submatrix S over an index subset $\bar{I}$, the corresponding V-graph contains at least one strong cycle. Nodes corresponding to indices in $\bar{I}$ either belong to a strong cycle or belong to paths leading to a strong cycle.

Thus, given a C-graph with a simple loop, in order to verify structural termination, it is sufficient to

1. Verify whether the V-matrix associated with the loop has a strong submatrix.
2. Verify whether at least one of the variables which dominate the closing predicate belongs to a strong cycle or is connected with a strong cycle.

We shall not examine in detail an algorithm for step 2, since there are many algorithms of the kind presented in classical graph theory. We can just briefly outline a simple procedure for identifying which of the nodes with indices in $\bar{I}$ belong to strong cycles. Refer to Fig. IV.13(a); from Fig. IV.13(b) we see that $\bar{I} = \{v_1, v_3, v_4\}$ involves two nodes belonging to strong cycles (v_1 and v_3) while strong edges depart from v_4 leading to the strong cycle.

This last fact appears quite clearly in the strong submatrix of Fig. IV.14(a). There are nonzero entries in column v_4, i.e., there are strong edges going from v_4 to the strong cycle or to a path leading to the strong cycle. But in row v_4 the only nonzero entry appears in position (v_4, v_4) and has value -1; that is, no edge goes from any other of the nodes in $\bar{I}$ to v_4, and the self-loop on v_4 is a weak one. Therefore, v_4 cannot belong to one of the strong cycles in the V-graph.

Having examined the problem of structural termination for a simple loop, we think that a few rather important considerations are in order.

The determination of structural termination is made on the V-matrix *of the simple loop only*; it is not necessary to consider all variables not appearing in any statement inside the loop. This allows us, in general, to keep the V-matrix rather small. The analysis can be performed on a local basis. Moreover, it is quite reasonable to assume that large V-matrices will be rather sparse; i.e., they have relatively few nonzero entries. Suitable methods have been developed [21] for representing and processing sparse matrices so as to minimize storage occupation and processing time.

The simple-loop analysis can be applied to programs whose C-graphs contain any number of simple loops; structural termination is determined separately for each of them. We consider in the next subsection the class of series-parallel C-graphs which cannot be solved by examining cascades of simple loops.

4.2 Complex Series-Parallel Control Graphs

By *complex series-parallel C-graphs* we mean those C-graphs in which at least one loop is not a simple one, i.e., at least one loop encloses another loop or else loop encloses two or more alternative paths, as Example 3 does (see Fig. IV.2(c)). Refer to the following examples.

Example 10.

n_1: $v_1 = f(v_2)$
n_2: if $P_h(v_1)$ do
$\quad v_3 = v_1$
$\quad v_1 = f(v_1)$ ⎫
$\quad v_2 = f(v_1)$ enddo ⎬ a
else do
$\quad v_3 = f(v_2, v_3)$ ⎫
$\quad v_2 = f(v_2)$ enddo ⎬ b
endif
n_3: $v_3 = f(v_1, v_2)$
n_4: if $P_e(v_2)$ then n_1
$\quad$ else. . . .

The corresponding C-graph is given in Fig. IV.17(a). The loop closed by predicate $P_e(v_2)$ encloses two alternative paths. A predicate P_h decides whether instruction sequence a or b will be executed. ∎

Example 11.

n_1: Repeat
$\quad v_1 = v_3$
$\quad v_3 = f(v_3)$
n_2: Repeat
$\quad v_2 = f(v_1, v_3)$
$\quad v_1 = f(v_1)$
Until $P_1(v_1)$
$\quad v_2 = f(v_3, v_1)$
n_3: Until $P_2(v_3)$

The corresponding C-graph is given in Fig. IV.17(b). The internal loop, with closing predicate P_1, is enclosed by the external loop with closing predicate P_2. ∎

We can immediately verify that, together with simple loops, these two instances comprise all types of loops that can appear in series-parallel C-graphs (suitably repeated). We will prove, first, that both such instances actually reduce to the case of a loop enclosing two or more alternative paths, as far as structural termination is concerned. Then, some conditions guaranteeing strucutral termination for such a loop will be presented. The problem is clearly far more complex than the simple loop and a general solution would

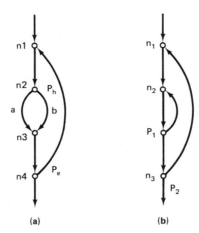

FIGURE IV.17 (a) *C*-graph for Example 10; (b) *C*-graph for Example 11

require complex and long enumerative procedures. In keeping with our purpose of using algorithms and computations as simple as possible, we limit ourselves to solution of those cases which satisfy easily identifiable conditions.

Consider any given V-matrix M, compute its powers $M^2, M^3, \ldots, M^k, \ldots$. Owing to V-matrix algebra, we can easily verify that after an initial sequence of λ_0 powers (a transient of length λ_0) all powers of M repeat following a cyclical pattern with period length λ. Thus, refer to the V-matrix in Fig. IV.3(c) (repeated in Fig. IV.18(a) for the convenience of the readers); the initial transient has length $\lambda_0 = 1$, and after that all power of M from M^2 onwards are identical, i.e., $\lambda = 1$ (Figs. IV.18(b) and (c)).

Now consider the case of an edge enclosed in a simple loop. We are interested in I/O relationships after any number of iterations of the loop. The above considerations tell us that I/O relationships also repeat themselves following a periodical pattern, after the first λ_0 iterations; if λ_0 and λ have

$$\begin{bmatrix} 1 & 0 & 1 & 1 \\ 1 & 0 & 1 & 0 \\ 1 & 0 & 1 & 1 \\ 0 & 0 & 0 & -1 \end{bmatrix} \quad \begin{bmatrix} 1 & 0 & 1 & 1 \\ 1 & 0 & 1 & 1 \\ 1 & 0 & 1 & 1 \\ 0 & 0 & 0 & -1 \end{bmatrix} \quad \begin{bmatrix} 1 & 0 & 1 & 1 \\ 1 & 0 & 1 & 1 \\ 1 & 0 & 1 & 1 \\ 0 & 0 & 0 & -1 \end{bmatrix}$$

$$\quad M \qquad\qquad\qquad M^2 \qquad\qquad\qquad M^3$$
$$\quad (a) \qquad\qquad\qquad (b) \qquad\qquad\qquad (c)$$

FIGURE IV.18 (a) V-matrix given in Fig. IV.3(c). (b), (c) second and third powers of V-matrix

already been computed, we can immediately assert that the I/O relationships after n iterations are defined by the power of M or order l, l being given by expression

$$\ell = (n - \lambda_0)_{mod \; \lambda} + \lambda_0 + 1$$

Thus, in Example 4 the I/O relationships after any number $n \geq 2$ of iterations are given by matrix M^2.

Consider now Example 1, whose V-matrix (Fig. IV.10(a)) is again reproduced in Fig. IV.19(a). Computation of its powers leads to our finding $\lambda_0 = 2$, $\lambda = 1$ (Figs. from IV.19(b) to IV.19(d)). Thus, after $n = 6$ iterations the power of M is:

$$\ell = (6-2-1)_{mod \; 1} + 2 + 1 = 0 + 2 + 1 = 3$$

NOTE: Since $\lambda = 1$ is a divisor of any number, ℓ becomes independent of the number of iterations. Thus it assumes permanent value $\ell = \lambda_0 + 1 = 3$. Again, consider the following example.

Example 12.

$$\textit{Repeat}$$
$$v_4 = v_3$$
$$v_3 = v_2$$

$$M: \begin{array}{c|cccc} & v_1 & v_2 & v_3 & v_4 \\ \hline v_1 & 0 & 1 & 0 & 0 \\ v_2 & 0 & 0 & 1 & 0 \\ v_3 & 0 & 0 & 0 & 1 \\ v_4 & 0 & 0 & 0 & 1 \end{array}$$

(a)

$$M^2: \begin{array}{c|cccc} & v_1 & v_2 & v_3 & v_4 \\ \hline v_1 & 0 & 0 & 1 & 0 \\ v_2 & 0 & 0 & 0 & 1 \\ v_3 & 0 & 0 & 0 & 1 \\ v_4 & 0 & 0 & 0 & 1 \end{array}$$

(b)

$$M^3: \begin{array}{c|cccc} & v_1 & v_2 & v_3 & v_4 \\ \hline v_1 & 0 & 0 & 0 & 1 \\ v_2 & 0 & 0 & 0 & 1 \\ v_3 & 0 & 0 & 0 & 1 \\ v_4 & 0 & 0 & 0 & 1 \end{array}$$

(c)

$$M^4: \begin{array}{c|cccc} & v_1 & v_2 & v_3 & v_4 \\ \hline v_1 & 0 & 0 & 0 & 1 \\ v_2 & 0 & 0 & 0 & 1 \\ v_3 & 0 & 0 & 0 & 1 \\ v_4 & 0 & 0 & 0 & 1 \end{array}$$

(d)

FIGURE IV.19 (a) V-matrix for Example 1; (b) to (d) powers of V-matrix

$$v_2 = f(v_1)$$
$$v_1 = f(v_3)$$
$$v_3 = f(v_1)$$
$$v_4 = f(v_4)$$
Until $P(v_1)$

The V-matrix M and its powers from 2 to 5 are given in Fig. IV.20(a) to (e). There, we see that after the initial transient of length $\lambda_0 = 1$, powers of M appear in a periodical pattern with $\lambda = 2$. Thus, after $n = 5$ iterations the input-output relationships are given by the power of M of order

$$l = (5-1-1)_{mod\ 2} + 1 + 1 = 1 + 1 + 1 = 3 \quad \blacksquare$$

This is a very important result in light of our particular purpose. In fact, consider a C-graph such as the one in Fig. IV.17(b), with an outer loop enclosing an inner loop. Assume that structural termination of the inner loop has been proved. For subsequent analysis, the particular semantics of the program segment enclosed in the outer loop are not vital. Rather, we are interested in determination of possible I/O relationships concerning the outer loop. The set of powers of the inner loop V-matrix gives us just that; a set of alternative I/O relationships. Moreover, the set itself is finite (in our examples, its dimensions are quite small). As far as subsequent analysis of structural termination is concerned, the inner loop may be substituted by a set of λ alternatives, equivalent edges, whose I/O relationships are given by one of the periodical power of M respectively. In practice, we again have a C-graph similar to that in Fig. IV.17(a), where alternative edges are chosen on the base of an *input predicate* P_i dominated not by a variable but by the number n (mod λ) of iterations of the inner loop. Obviously, in all this analysis we assume that termination does not occur in the first λ_0 iterations. If a complete analysis were sought, it would be sufficient to examine separately two equivalent C-graphs, the first one concerning the transient and enclosing λ_0 alternative equivalent edges, the second one enclosing λ equivalent edges.

Let us consider, in general, the second case. (It is quite reasonable to assme $n > \lambda_0$, and no loss of generality is risked by this.) In Example 4, the loop could be substituted for by just one equivalent edge, whose I/O relationships are given by M^2. In Example 12, the loop after the first iteration can be substituted for by two equivalent edges, whose I/O relationships are given respectively by M^2 and M^3, the choice between the two being performed on the basis of $n = 2 \cdot i$ or $n = 2 \cdot i + 1$, n being the number of iterations and i being any integer greater than 0.

Thus, the examination of any series-parallel C-graph with the purpose of determining structural termination requires analysis of

- Either simple loops
- Or loops enclosing two or more alternative equivalent edges, as in Fig. IV.17(a).

$$M = \begin{array}{c} \\ v_1 \\ v_2 \\ v_3 \\ v_4 \end{array} \begin{array}{cccc} v_1 & v_2 & v_3 & v_4 \\ \left[\begin{array}{cccc} 0 & 1 & 0 & 0 \\ 1 & 0 & 0 & 0 \\ 0 & 1 & 0 & 0 \\ 0 & 0 & 1 & 0 \end{array}\right] \end{array}$$

(a)

$$M^2 = \begin{array}{c} \\ v_1 \\ v_2 \\ v_3 \\ v_4 \end{array} \begin{array}{cccc} v_1 & v_2 & v_3 & v_4 \\ \left[\begin{array}{cccc} 1 & 0 & 0 & 0 \\ 0 & 1 & 0 & 0 \\ 1 & 0 & 0 & 0 \\ 0 & 1 & 0 & 0 \end{array}\right] \end{array}$$

(b)

$$M^3 = \begin{array}{c} \\ v_1 \\ v_2 \\ v_3 \\ v_4 \end{array} \begin{array}{cccc} v_1 & v_2 & v_3 & v_4 \\ \left[\begin{array}{cccc} 0 & 1 & 0 & 0 \\ 1 & 0 & 0 & 0 \\ 0 & 1 & 0 & 0 \\ 1 & 0 & 0 & 0 \end{array}\right] \end{array}$$

(c)

$$M^4 = \begin{array}{c} \\ v_1 \\ v_2 \\ v_3 \\ v_4 \end{array} \begin{array}{cccc} v_1 & v_2 & v_3 & v_4 \\ \left[\begin{array}{cccc} 1 & 0 & 0 & 0 \\ 0 & 1 & 0 & 0 \\ 1 & 0 & 0 & 0 \\ 0 & 1 & 0 & 0 \end{array}\right] \end{array}$$

(d)

$$M^5 = \begin{array}{c} \\ v_1 \\ v_2 \\ v_3 \\ v_4 \end{array} \begin{array}{cccc} v_1 & v_2 & v_3 & v_4 \\ \left[\begin{array}{cccc} 0 & 1 & 0 & 0 \\ 1 & 0 & 0 & 0 \\ 0 & 1 & 0 & 0 \\ 1 & 0 & 0 & 0 \end{array}\right] \end{array}$$

(e)

FIGURE IV.20 (a) to (e) V-matrix for Example 12 and powers of it from 2 to 5

Consider, in particular, Example 11 (the C-graph representation in Fig. IV.17(b)). The V-matrix M_1 of the inner loop is given in Fig. IV.21(a); the condition for structural termination is satisfied. It is $\lambda_0 = 1$, $\lambda = 1$. (See Figs. IV.21(b) and (c).) After the first iteration, the inner loop can be substituted for by just one equivalent edge whose I/O relationships are given by M_1^2.

Thus, we limit our study to *loops enclosing two or more parallel alternative edges*. Let $l_1, l_2, \ldots, l_m$ be the alternative paths internal to the loop and $M_1, M_2, \ldots, M_m$ be the associated V-matrices. The acutal choice of a

$$M_1 = \begin{bmatrix} & v_1 & v_2 & v_3 \\ v_1 & 1 & 0 & 0 \\ v_2 & 1 & 0 & 1 \\ v_3 & 0 & 0 & -1 \end{bmatrix} \quad M_1^2 = \begin{bmatrix} & v_1 & v_2 & v_3 \\ v_1 & 1 & 0 & 1 \\ v_2 & 1 & 0 & 1 \\ v_3 & 0 & 0 & -1 \end{bmatrix} \quad M_1^3 = \begin{bmatrix} & v_1 & v_2 & v_3 \\ v_1 & 1 & 0 & 1 \\ v_2 & 1 & 0 & 1 \\ v_3 & 0 & 0 & -1 \end{bmatrix}$$

(a)   (b)   (c)

FIGURE IV.21 (a) V-matrix for Example 11; (b), (c) second and third powers of V-matrix

path is effected by a head predicate P_h where P_e is the predicate closing the loop. Structural termination must be independent of the exit from P_h, therefore it must be verified for cycles holding any of the edges $l_1, l_2, \ldots, l_m$ or any of their sequences. Refer to Example 10 (Fig. IV.17(a)). The loop must show structural termination whether an alternative path comprising a only is always chosen, an alternative path comprising b only, or also for any possible sequence of paths involving a or b. Obviously, we could not accept the loop as "structurally terminating" if the property was not valid for a succession of iterations involving a sequence a,a,a,b,a.

In general, verifying the complete condition would involve the examination of a very high number of matrices (corresponding to all possible sequences leading to different products of V-matrices) and therefore it would be impractical. We present in this section

1. Two conditions *sufficient* for structural termination (Lemmas 1 and 3). That is, whenever one of these conditions is satisfied, structural termination is guaranteed (although we cannot assert that if the condition is not satisfied, the program will lack structural termination).
2. A condition sufficient to exclude structural termination (Lemma 2). Whenever this is satisfied, the program is not structurally terminating.

Actually, the condition presented in Lemma 1 can be considered as a particular instance of Lemma 3. Since verifying it involves a far simpler procedure than the one required by Lemma 3, we present it separately.

The simplest condition for structural termination is based, once again, upon the existence of strong submatrices and upon their properties.

Lemma 1. Let $l_1, l_2, \ldots, l_m$ all satisfy, one by one, the condition for structural termination; i.e., all V-matrices have a strong submatrix, and the variables dominating the closing predicate are connected with strong cycles in the known way. Determine now a subset $\bar{I}$ of I by the following procedure:

- Delete from I all indices corresponding to columns in any of M_1, $M_2, \ldots, M_m$ consisting only of 0 entries.
- Repeat the previous step upon submatrices obtained until either $\bar{I}$ is a void set or else no further deletions are possible.

If

1. $\bar{I}$ is nonvoid.
2. All submatrices $S_1, S_2, \ldots, S_m$ over $\bar{I}$ obtained from $M_1, M_2, \ldots, M_m$ are strong.
3. At least one of the variables which dominate the closing predicate either belongs to a strong cycle as determined by submatrices S_1, $S_2, \ldots, S_m$ or else is connected with such a cycle by an oriented path (depends from such a cycle)

then structural termination of the loop is guaranteed.

The proof follows from Theorem 2. In fact, any possible product over the given matrices will exhibit a strong submatrix over $\bar{I}$, and will therefore present structural termination.

Consider Example 10. Acutally, the two alternative paths to be considered are the two complete sequences:

$$v_1 = f(v_2) \qquad v_1 = f(v_2)$$
$$v_3 = v_1 \qquad v_3 = f(v_2, v_3)$$
$$v_1 = f(v_1) \qquad v_2 = f(v_2)$$
$$v_2 = f(v_1) \qquad v_3 = f(v_1, v_2)$$
$$v_3 = f(v_1, v_2)$$

In Fig. IV.22(a) we give the respective V-matrices M_a and M_b and in Fig. IV.22(b) the V-graphs (each of which separately satisfies structural termination for the closing predicate dominated by v_2). Both matrices share a strong submatrix of dimension 1 over $\bar{I} = v_2$.

Now, we can easily verify that any power of any of the two matrices, any product of whatever sequence of such powers shares a strong submatrix which denotes a strong cycle over v_2, i.e., the variable dominating the closing predicate. As a result, whatever the choice determined by predicate P_h, the loop will exhibit structural termination.

Now, it is interesting to see an opposite condition, that is, one that is sufficient for the *lack* of structural termination. We consider this for the simplest case, i.e., a loop enclosing just two alternative paths; it can be immediately extended to any number of paths.

Lemma 2. Consider a loop enclosing two alternative paths a, b: let

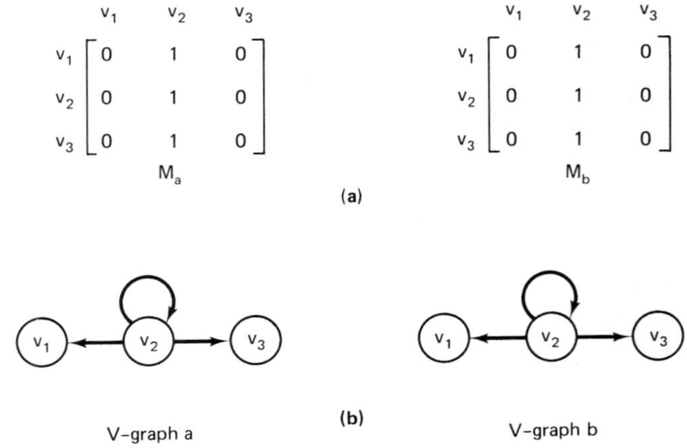

$$
\begin{array}{c c}
\begin{array}{c}
\;\;v_1\;\;v_2\;\;v_3 \\
\begin{array}{c} v_1 \\ v_2 \\ v_3 \end{array}\!\!
\left[\begin{array}{ccc} 0 & 1 & 0 \\ 0 & 1 & 0 \\ 0 & 1 & 0 \end{array}\right] \\
M_a
\end{array}
&
\begin{array}{c}
\;\;v_1\;\;v_2\;\;v_3 \\
\begin{array}{c} v_1 \\ v_2 \\ v_3 \end{array}\!\!
\left[\begin{array}{ccc} 0 & 1 & 0 \\ 0 & 1 & 0 \\ 0 & 1 & 0 \end{array}\right] \\
M_b
\end{array}
\end{array}
$$

(a)

V-graph a   (b)   V-graph b

FIGURE IV.22 (a) V-matrices for alternative paths in Example 10; (b) V-graphs for alternative paths in Example 10

M_a, M_b be the respective V-matrices, such that each of them, separately, satisfies the condition for structural termination but that they do not together satisfy Lemma 1. Now, if it happens that

- For each row in M_a containing at least one $+1$ entry, the column in M_b marked with the same index has only 0 entries (or vice versa, i.e., row of M_b, column of M_a).
- For each $+1$ entry in M_b marked as $M_b(h,k)$ the column of M_a marked with index h only contains 0 entries (or vice versa),

then the loop lacks structural termination.

The proof given in [22] is based upon the algebra of V-matrices; we are interested here in considering an example.

Example 13. The program is:

$$
\left.\begin{array}{l}
\textit{Repeat} \\
\quad \textit{if } P(v_1) \textit{ then do} \\
\qquad v_1 = v_3 \\
\qquad v_3 = v_2 \\
\qquad v_2 = f(v_2, v_1) \textit{ enddo}
\end{array}\right\}a
$$

$$
\left.\begin{array}{l}
\textit{else do} \\
\qquad v_2 = v_1 \\
\qquad v_1 = f(v_1) \textit{ enddo}
\end{array}\right\}b
$$

$$\textit{Until } P(v_2)$$

The V-matrices M_a, M_b for the two alternative paths are represented in Fig. IV.23(a); each of them exhibits structural termination with respect to the closing predicate. On the other hand, Lemma 1 is not satisfied (The index subset I obtained following its rules is void.) while Lemma 2 is satisfied. The $+1$ entries of M_a are in row v_2, while column v_2 of M_b only has 0 entries, the only $+1$ entry of M_b is in row v_1 and column v_1 of M_a has only 0 entries. Consider now the product matrix $M_a \cdot M_b$ (Fig. IV.23(b)); it has no strong submatrix. In other words, if the head predicate $P(v_1)$ causes execution first of instruction sequence b, then a, then again $b, a \ldots$ etc, the loop will not exhibit structural termination.

$$M_a = \begin{array}{c} \\ v_1 \\ v_2 \\ v_3 \end{array} \begin{array}{ccc} v_1 & v_2 & v_3 \\ \left[\begin{array}{ccc} 0 & 0 & -1 \\ 0 & 1 & 1 \\ 0 & -1 & 0 \end{array}\right] \end{array} \qquad M_b = \begin{array}{c} \\ v_1 \\ v_2 \\ v_3 \end{array} \begin{array}{ccc} v_1 & v_2 & v_3 \\ \left[\begin{array}{ccc} 1 & 0 & 0 \\ -1 & 0 & 0 \\ 0 & 0 & -1 \end{array}\right] \end{array}$$

(a)

$$M_a \cdot M_b = \begin{array}{c} \\ v_1 \\ v_2 \\ v_3 \end{array} \begin{array}{ccc} v_1 & v_2 & v_3 \\ \left[\begin{array}{ccc} 0 & 0 & -1 \\ 1 & 0 & -1 \\ -1 & 0 & 0 \end{array}\right] \end{array}$$

(b)

FIGURE IV.23 (a) V-matrices for Example 13; (b) product matrix $M_b \cdot M_a$

Thus, by a very simple procedure (in practice, the algorithm only requires us to compare the patterns of entries in the V-matrices) programs which lack structural termination can be readily identified. Considerations made in Sec. 2 allow us to define such loops as probable "problem areas" and therefore it is useful to signal their presence to the programmer.

Finally, we give a further condition which, if satisfied, guarantees structural termination of a loop enclosing two or more alternative paths. It is rather more complex than Lemma 1, both in statement and in the algorithm that verifies it. For simplicity, we present it only for the case of two alternative paths (it can be easily extended to any number of paths). A likely procedure would be to

- Verify if Lemma 1 is satisfied. If so, structural termination is guaranteed. Otherwise

- Verify whether Lemma 2 is satisfied. If so, the loop is not structurally terminating and the error is communicated to the programmer. Otherwise
- Verify the following:

Lemma 3. Consider a loop enclosing two alternative paths a and b whose V-matrices M_a and M_b both singly satisfy structural termination. If for any nonzero entry in M_a, $M_a(i,j)$(and, respectively, $M_b(h,k)$) it happens that the row of M_a (respectively, of M_b) with index i (h) and the column of M_b (M_a) with index j (k) have at least one nonzero entry in a position with the same ordering number (that is, their vector product gives nonzero result) then, in order to verify structural termination of the loop, it is sufficient to examine a very limited set of sequences represented by matrix products $M_b^l \cdot M_a^n$ and $M_a^n \cdot M_b^l$, where l and n may assume all values from 1 to the period length of the powers, respectively, of M_a and M_b.

The complete proof of Lemma 3 is given in [22]; there is an interesting *corollary* stating that the analysis above can be performed simply upon the strong submatrices S_a and S_b of M_a and M_b. Rather than examine the proof in in detail, let us consider the following.

Example 14.

> *Repeat*
> *if* $p_h(v_2)$ *then do*
> $\quad v_3 = v_1$
> $\quad v_1 = f(v_1)$
> $\quad v_2 = f(v_1)$ *enddo*
> *else do*
> $\quad v_3 = f(v_2, v_3)$
> $\quad v_2 = f(v_2)$ *enddo*
> *Until* $P_e(v_1)$

The graphs representing I/O relationships are given in Fig. IV.24(a) while V-matrices are in Fig. IV.24(b). This example falls under the conditions of Lemma 3. Each path singly exhibits structural termination, Lemma 1 is not satisfied; Lemma 3 is satisfied. Moreover, the period length of the powers of M_a is 2 (with $\lambda_0 = 0$) and M_a^2 is given in Fig. IV.24(c). The period length of the powers of M_b is 1 (with $\lambda_0 = 0$). Thus, it is sufficient to examine the structural termination of the sequences $M_b \cdot M_a$ and $M_b \cdot M_a^2$. Both have a strong submatrix (both over v_1), thus the program is characterized by structural termination (Fig. IV.24(d)).

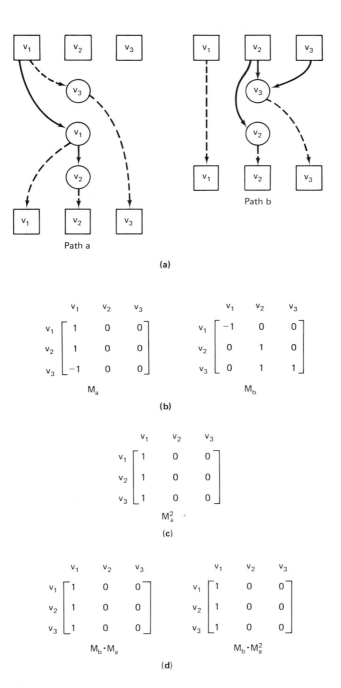

FIGURE IV.24 (a) Input-output relationships graph for Example 14; (b) V-matrices for Example 14; (c) Second power of V-matrix M_a; (d) product matrices $M_b \cdot M_a$ and $M_b \cdot M_a^2$.

5. FORMAL EQUIVALENCE BETWEEN C-GRAPHS AND TRANSITION GRAPHS

We propose now to show how the methodology derived for series-parallel C-graphs can be applied also to C-graphs that have any structure. Obviously, we are interested only in the structures involved in loops. Although the procedure is general, it best applies to loop sections.

An oriented graph with one initial node, one terminal node, and such that any of its edges is traversed by at least one path from its initial node to its terminal node can formally be considered as the transition graph of a finite-state recognizer automaton. If the symbols of a finite alphabet are associated with the graph edges, the set of all paths from initial to terminal nodes constitute a regular expression.

A C-graph responds to the above conditions. In our approach to the study of structural termination, we are not interested in the semantics of instruction sequences, but simply in their presence; thus, it is possible to associate with any edge of the C-graph a symbol from a finite alphabet,† allowing the null strong ϵ if the edge simply stands for a control transfer.

The *sequencing* of two instruction sequences is then representable by means of the *catenation* of the corresponding symbols. "Execution of instruction sequence a is followed by execution of sequence b" is represented as "symbol a followed by symbol b." The *predicate* associated with a type-o node may be seen as originating the *union* of the symbols associated with the outgoing edges. Thus, the "sequence a or sequence b is executed" (depending on the outcome of predicate P) is represented as "symbol a or symbol b is accepted," i.e., "$a \cup b$ is accepted". Consider the C-graph in Fig. IV.25 (for simplicity, consider a loopless segment of a C-graph). The possible paths in the program are bd, ae, and acd. This set of paths can be represented formally by the regular expression $bd \cup ae \cup acd$.

A *loop* enclosing an instruction sequence S (with the closing edge associated with the null string ϵ) corresponds to the possibility of having during a run a number $1 \leq n < \infty$ of iterations of S, and thus can be represented by $S \cup SS \cup SSS \cup \ldots = S^*$. *Closure* is then introduced.

No other operators can be identified. Moreover, since the C-graph has finite dimensions, it follows that the operators can be applied only a finite number of times.‡

Conversely, we know that each operator applied to regular expressions can be translated by a basic series-parallel subgraph. Thus

- Catenation of two regular expressions S_1 and S_2 can be seen as the series connection of the two associated edges (Fig. IV.26(a)).

---

†Having as many different symbols as there are instruction sequences.

‡Note that this assertion does *not* imply termination, i.e., finite run length of the program, since closure operators can introduce an infinite number of iterations.

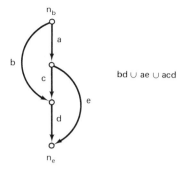

FIGURE IV.25 Sample C-graph

- Union $S_1 \cup S_2$ of two regular expressions can be seen as the parallel connection of two equivalent edges (Fig. IV.26(b)).
- Closure S* can be represented by a simple loop, with the insertion of suitable ϵ edges (Fig. IV.26(c)).

This, allowing only for series-parallel connections, leads us to conclude that for any regular expression a series-parallel transition graph can be built.

From the above considerations we can conclude that, given any C-graph, a possible procedure to obtain an equivalent series-parallel C-graph would be to first determine the equivalent regular expression and then to build the series-parallel transition graph for the regular expression. This will not constitute the C-graph of an actual program, but will still represent all the possible paths in the original program. To this end, union predictes have to be associated with equivalent predicates.

The argument below is aimed at establishing a complete equivalence between paths in the original graph and paths in the series-parallel graph. However, we also note that by using a global procedure we make it very

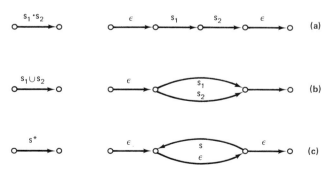

FIGURE IV.26 (a) C-graph corresponding to catenation; (b) C-graph corresponding to union; (c) C-graph corresponding to closure

difficult to determine the equivalent predicates in the series-parallel structure and therefore the study structural termination is hindered. Thus, a step-by-step identification and reduction to series-parallel structure will be suggested.

First, consider the problem of deriving the regular expression from a given transition graph. We briefly recall the algorithm presented in [17].†

1. Label the nodes in the transition graph with $s_1, \ldots, s_n$ where s_1 is the entry node and s_n the exit node. For each pair s_i, s_j $1 \leq i, j \leq n$, $i \neq j$, if there are two or more parallel equivalent edges connecting them, substitute for them with an edge labelled with the union α_{ij}^1 of the labels of the single edges. If no edge connects s_i and s_j, put entry $\alpha_{ij}^1 = \emptyset$.
2. Then perform the following steps:
   a. Initially put $k = 2$.
   b. Delete s_k from the graph and substitute α_{ij}^{k-1} with

$$\alpha_{ij}^k = \alpha_{ij}^{k-1} \cup \alpha_{ik}^{k-1} \left(\alpha_{kk}^{k-1} \right)^* \alpha_{kj}^{k-1}$$

   c. If $k < n - 1$, let $k = k + 1$ and repeat step b. If $k = n - 1$, the regular expression corresponding to the graph is $\alpha_{1,n}^{n-1}$.

Since the labelling of nodes $s_2, \ldots, s_{n-1}$ is arbitrary, up to $(n - 2)!$ formally different regular expressions can be found, all sharing the same ambiguities as the original graph. In our case the original graph is essentially nonambiguous, and so are all the equivalent regular expressions derived from it. We may note that the *reduced graphs* that are obtained in intermediate steps of the algorithm may be *mixed-type* nodes. This is inessential in that it just happens on an intermediate representation.

Conversely, starting from the regular expressions obtained, it is now possible for us to build a series-parallel graph recognizing all the strings in the regular expression. This allows to state an equivalence between *paths* in the original graph and *paths* in the series-parallel one.

While the instruction sequences denoted by the "symbols" are not affected by this procedure, predicates on the equivalent series-parallel graph will be different both for number and for positioning along the paths. (In other words, there is not a correspondence between the *nodes* in the two graphs.) We state in Sec. 6 how the equivalent predicates can be actually determined; we refer first to the equivalence of *actions* along corresponding paths.

Denote by ω_i the value of the state vector (i.e., the vector consisting of all variables in the program) at a given step of computation. Start from a given initial value ω_o.

†Hereafter, we denote by ϵ the null sequence, by $\emptyset$ the void set, and assume the regular expression algebra to be known.

We denote as the *initial sequence of computation* a sequence of actions performed from Start to a given instruction sequence, following a predetermined path in the C-graph; when each variable in this sequence assumes only one value, i.e., it is not modified due to loop iterations. To this initial sequence of computation there corresponds an initial substring of a string recognized by the transition graph. It is therefore possible to identify the same initial sequence of computation both on the given C-graph G and on the equivalent series-parallel graph G_{sp}.

Since, as we said above, the actions performed by initial sequences of computation and therefore by the strings are not modified, the two values computed following the paths in the two graphs starting with the same ω_o are identical.

The problem of determining whether a given path identified in G actually will be followed at run time also in G_{sp}, given the same ω_o, still exists. This question can be answered only by the proper definition of the predicates in series-parallel graph G_{sp}. Since our objective is to extend a structural termination analysis to any C-graph G, our purpose will be to study the behavior of sets of variables in the equivalent predicates from G and G_{sp} rather than the predicates themselves.

The procedure below is aimed at (a) finding an equivalent regular expression for an arbitrary C-graph, (b) constructing a series parallel graph for this expression, and (c) proving an equivalence of both graphs from the viewpoint of structural termination.

6. REDUCTION PROCEDURE

The algorithm proposed in [17] for finding the regular expression associated with a transition graph essentially proceeds by successive elimination of nodes. Our purpose here is to show intuitively—inasmuch as formal proof becomes too complex—that the inductive steps of the algorithm aimed at gradual replacement of graph nodes with variables do not affect structural termination properties of the program.

On the original C-graph, two basic instances may occur when node n_j is deleted

1. Node n_j is a type-i node; its elimination causes the transform exemplified in Fig. IV.27. It is evident that all paths are kept unchanged and that no predicate is affected. Therefore, structural termination properties also are not affected. Node n_m becomes a mixed-type node, but such nodes are accepted by the reduction procedure.
2. Node n_j is a type-σ node, associated with a predicate $P_j(V_j)$, $V_j = \{v_j^1, \ldots, v_j^s\}$. Its deletion causes the transform exemplified in Fig. IV.28; in this case, the paths are kept unchanged but the predi-

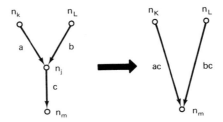

FIGURE IV.27 Transform upon type-i node in the C-graph

cate is transferred to a different node. P_k^1 will depend upon all variables appearing in P_j; moreover, any $v_j^i \in V_j$ may be redefined in a by a sequence of statements

$$v_a^2 = f(v_a^1)$$
$$v_a^3 = f(v_a^2)$$
$$\cdot$$
$$\cdot$$
$$\cdot$$
$$v_j^i = f(v_a^h)$$

All variables $v_a^1, \ldots, v_a^h$ shall appear in the set of variables associated with P_k. We use the notation $P_k'(v_j^1, \ldots, v_j^s, V_{aj})$, where V_{aj} is the set of these new variables directly or indirectly affecting the possible assignments of any of the $v_j^1, \ldots, v_j^s$ in sequence a.

Assuming that there is no undefined variable in the program (a reasonable assumption, since many good compilers already detect this type of error) it is formally possible to anticipate the decision of P_j to node n_k, since in n_k there will be at least one defined variable whose value directly or indirectly affects the $v_j^1, \ldots, v_j^s$ and thus determines the decision of P_j. Therefore, determinism is kept. We prove now that structural termination properties are also kept.

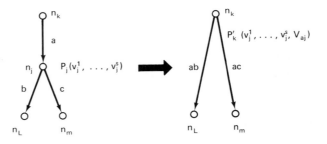

FIGURE IV.28 Transform upon type-σ node in the C-graph

Structural termination can be affected only if P_j dominates the termination of a cycle. Assume this to be the case: obviously edge a also belongs to the cycle. Two different instances may occur: variables affecting P_j are modified only in the section of cycle outside a, or also in a. Since termination is verified by a V-matrix corresponding to the whole loop, the variability of $v_j^1, \ldots, v_j^s$ is in both cases considered. The extension of the set of variables for P_k', moreover, introduces only variables already present in the loop, and therefore already in the V-matrix. Thus, no new condition for the modification of any variable is introduced, nor are any of the pre-existing ones destroyed.

Note that, if node n_k is a type-σ node, and as such associated with a predicate P_k, P_k is kept unchanged and both P_k and P_k' must be independently examined in subsequent structural termination analysis.

Considering now the intermediate steps of the reduction procedure. When mixed-type nodes appear, no new instances actually occur. In fact, we could consider that such a node is subject to two successive reductions, the first being a splitting into type-i nodes, the second one a reduction of these type-i nodes. We must note that in this case the single predicate P_j is transfored into r predicates (if r were the inputs) over *different* sets of variables.

It is now possible to describe the step-by-step reduction procedure. Consider again the single reduction step of algorithm [17]. In the most general case, this causes the introduction of

- A parallel connection of two edges.
- A series connection.
- A self-loop.

The self-loop can be immediately transformed as in Fig. IV.29(a) and (b). (Note that the ϵ-transition represents a control transfer here.) The simple loop thus found can be examined for structural termination and if the examination is positive, substituted for with its equivalent edges (Fig. IV.29(c)). P_i still exists here because another choice among different paths is possible.

We may then outline the reduction procedure in the following way

- Simplify all preexisting series-parallel subgraphs. This involves verification of all simple loops or complex series-parallel loops and substitution with equivalent edges. For the purpose of subsequent steps, substitute each parallel connection of equivalent edges a, $b, \ldots, h$ with one edge labelled $a \cup b \cup \ldots \cup h$.
- Apply the reduction step to a node randomly chosen. If a self-loop is generated, verify its termination as outlined above and substitute the equivalent parallel structure. Reduce all series connections originated, and verify the termination of any simple or complex loop now found.

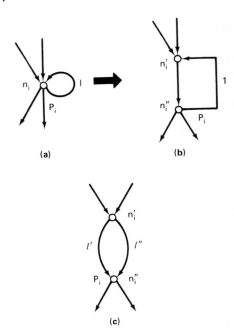

FIGURE IV.29 (a), (b) transform of a self-loop; (c) transform after verification of the loop in (b)

If the verification is positive substitute for the loop with equivalent edges; otherwise stop the procedure since structural termination does not hold.

- Repeat the above step until complete reduction of the program has been achieved. Note that all structures realizable at the end comprise only loops (simple or enclosing parallel edges).

We apply now the procedure presented above to the following example.

Example 15.

$$n_0: \ldots$$
$$\vdots$$
$$n_1: \left.\begin{matrix} v_3 = f(v_1) \\ v_4 = f(v_2) \end{matrix}\right\} b$$
$$n_2: \text{if } P_2(v_3, v_4) \text{ then } c \text{ else}$$
$$\left.\begin{matrix} v_2 = f(v_3) \\ v_1 = f(v_4) \end{matrix}\right\} d$$

n_3: **if** $P_3(v_1, v_2)$ **then** f **else**
$\left.\begin{array}{l} v_4 = f(v_1) \\ v_2 = f(v_3) \end{array}\right\} e$

goto n_1

c: . . .
   .
   .
   .

f: . . .
   .
   .
   .

g: . . .
   .
   .
   .

The corresponding C-graph is shown in Fig. IV.30(a). No series-parallel simplification is possible. Apply the procedure to node n_3. The graph of Fig. IV.30(b) is obtained; the new predicate P_2' in n_2 derives from P_3 and depends on variables v_1, v_2, v_3, and v_4 (v_3 and v_4 constituting set V_{d3}). At this point, a series-parallel graph is obtained, with only one loop consisting of sequence bde; structural termination can be immediately verified. It can be easily seen that the V-graph for bde has a strong cycle over v_1 and v_2 and that v_3 and v_4 depend on this cycle. Therefore, structural termination is ensured (Fig. IV.30(e). ■

7. CONCLUSIONS

The concept of *structural termination* has been introduced, both for programs and for microprograms. While basic theory has been developed for *sequential programs*, i.e., programs executed on single-processor machines, we have shown how the modeling approach (and the subsequent theory) can be applied to horizontal microprograms and also to parallel processing machines. Modeling of timing and synchronization problems have also been discussed.

Structural termination is a property which can be verified directly at compilation; the preprocessing already performed for programs to be distributed on multiprocessor systems (and in particular for reconfigurable architectures) gives most of the information necessary for determining structural termination. Subsequent specialized algorithms are straightforward and require quite simple computations.

Structural termination, moreover, is examined on a local basis, separately for the various loops. When nested loops are involved, the inner one is analyzed first, and only upon successful analysis does the testing proceed

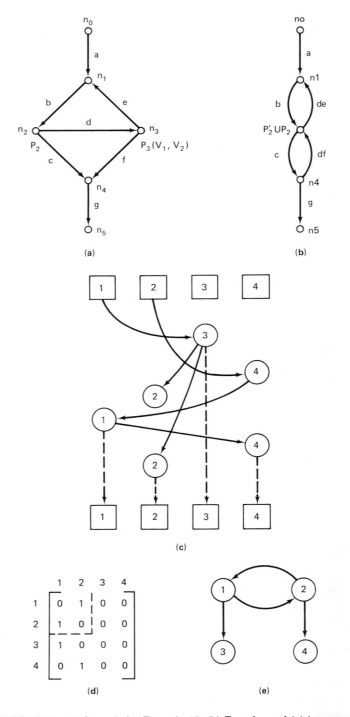

FIGURE IV.30 (a) *C*-graph for Example 15; (b) Transform of (a) by operating upon node n_3. (c) to (e): Verification of structural termination

outwards. We showed how all program structures are reduced (for our purposes) to a few basic structures, and only these need to be examined in detail.

While some insight was provided in Sec. 3 for the case of parallel processor systems, these, and in particular dynamic reconfigurable architectures, constitute a very important subject for further studies. Anyway, a few remarks can be made.

If distribution of the program upon the various processors is done at macroscopic level, i.e., at task level, and it excludes access to common memory areas (unless possibly, where block transfers or data sets are involved) no new considerations need to be made. Structural termination concerns the various tasks at a local level.

If on the other hand distribution has been performed at an elementary, single-instruction level (with a shared memory as an obvious necessity) the problem is quite near to that of horizontal microprograms.

Reconfiguration, even dynamic reconfiguration, from a number, N, of processing modules to a number of modules, M, may influence structural termination only in indirect way; i.e., if synchronization mechanisms (both hardware and software) are implemented in such way that their effect varies when computation of instruction sequences is shifted from one processing module to another one. In this case, reconfiguration involving a loop might lead to different results also where the reconfiguration is concerned. We should note that this is an intrinsic problem of process/processor synchronization, *not* of structural termination. In fact, correct synchronization should allow to consider structural termination as independent of processor parallelism.

REFERENCES

[1] J. L. BAER, "Graph Models of Computations in Computer Systems" UCLA 10P14/51-, report N.68-46, 1968.

[2] C. V. RAMAMOORTHY, and M. TSUCHIYA, "A High-Level Language for Horizontal Microprogramming," *IEEE Trans. on Computers,* **C23,** 8, pp. 791–801, August 1974.

[3] S. I. KARTASHEV and S. P. KARTASHEV, "A Multicomputer System with Software Reconfiguration of the Architecture," *Proceedings of the Eighth International Conference on Computer Performance, SIGMETRICS,* pp. 271–286, Washington, D.C., 1977.

[4] ———, "Dynamic Architectures: Problems and Solutions," *Computer,* **11,** pp. 26–40, July 1978.

[5] ———, "Software Problems for Dynamic Architecture: Adaptive Assignment of Hardware Resources," *Proc. Computer Software and Applications Conference (COMPSAC),* pp. 775–780.

[6] S. I. KARTASHEV, S. P. KARTASHEV, and C. V. RAMAMOORTHY, "Adaptation Properties for Dynamic Architectures," *Proc. National Computer Conference,* pp. 543–556, 1979.

[7] B. LISKOV, "Introduction to CLU," *Proc. Computing Systems Reliability,* Toulouse, France, September 1979.

[8] R. Fairley, "An Experimental Program-Testing Facility," *IEEE Trans. Software Engineering* **SE-1, 4,** pp. 350–357, 1975.

[9] W. E. Howden, "Symbolic Testing and the DISSECT Symbolic Evaluation System," *IEEE Trans. Software Engineering,* **SE-3, 4,** pp. 266–279, 1977.

[10] R. L. London, "A View of Program Verification," *Proc. 1975 International Conference on Reliable Software,* pp. 534–545.

[11] R. W. Floyd, "Assigning Meanings to Programs" *Proc. Symposium on Applied Mathematics,* **19,** pp. 19–32, 1967.

[12] C. A. R. Hoare, "An Axiomatic Basis of Computer Programming," *Comm. ACM,* **12,** pp. 576–580, 583, October 1969.

[13] L. A. Clarke, "Testing: Achievements and Frustrations," *Proc. Computer Software and Applications Conference COMPSAC*, pp. 310–314, 1978.

[14] G. B. Leman, W. C. Carter, and A. Birman, "Some Techniques for Microprogram Validation," *Proc. IFIP Conference on Information Processing,* 1974.

[15] M. G. Sami, and R. Stefanelli, "On Structural Termination of Programs and Microprograms," *Proc. First EUROMICRO Symposium on Microprocessing and Microprogramming,* Nice, France, 1975.

[16] C. Böhm, and G. Jacopini, "Flow Diagrams, Turing Machines and Languages with Only Two Formation Rules," *Comm. ACM,* **9,** *5,* pp. 366–371, May 1966.

[17] R. Book, S. Even, S. Greibach, and G. Otto, "Ambiguity in Graphs and Expressions," *IEEE Trans. Comp.,* **C-20,** *2,* pp. 149–153, February 1971.

[18] B. Martin, and G. Estrin, "Models of Computations and Systems: Evauation of Vertex Probabilities in Graph Models of Computations," *Journal ACM,* **14,** *2,* pp. 281–299, April 1967.

[19] R. L. Kleir, and C. V. Ramamoorthy, "Optimization Strategies for Microprograms," *IEEE Trans. Comp.,* **C-20,** *7,* pp. 783–794, July 1971.

[20] M. G. Sami, and R. Stefanneli, "On the Determination of Probable Activities of Instruction Sequences in Program Schemes" Politecnico di Milano, IEEE LC Internal report 72–12.

[21] D. J. Rose, and R. A. Willoughby, "Sparse Matrices and Their Applications" *Proc. Symposium on Sparse Matrices and Their Applications,* Yorktown Heights, N.Y., September 1971. 1971.

[22] R. Negrini, and M. G. Sami, "An Extension of Structural Termination to Programs with Arbitrary Graph Models" Politecnico di Milano, IEELC Internal Report 76–12.

Chapter V

Requirements Engineering for Modular Computer Systems

David F. Palmer

PREVIEW

This chapter describes an approach to solving the following type of problem:

1. We are given an application, or computing job, for which a modular computer system is to be developed. The application is assumed to be of a specific type, such as real time control of a given type of aircraft or chemical plant, rather than a general need to process various user jobs. The application is defined in terms of functional requirements (necessary functions, actions, transformations), performance requirements (response time, data rate, storage capacity, accuracy), operational requirements (reliability, flexibility for changes, etc.), and constraints (costs, development time, size, weight, etc.). The initial requirements are not necessarily complete, self-consistent, and unambiguous, and they are subject to change.

456 CHAPTER V

2. We are given guidelines and constraints with respect to the types of computer modules and interconnecting networks which may be used. Internal details of modules and details of network links and switches are not needed. Modules generally are to be characterized by: instruction rates in relation to modes of potential internal parallelism, memory access and input output (I/O) rates, I/O overhead, high-speed storage capacity, word size, reliability, costs, size, weight, etc. Other types of characteristics will be important later in design, e.g., instruction sets and operating system software, but we do not explicitly treat such detailed characteristics here.

3. We are to develop a preliminary design for a modular computer system that "best" satisfies the application, evaluated with respect to all user requirements and constraints. The preliminary design specifies the numbers and characteristics (speeds and capacities) of computing modules to be employed, the logical interconnection topology (message paths between modules), definition of applications software modules to be coded and mapped onto given computer modules, and requirements for control and communication (which are the base requirements for module operating system design). This task thus includes determining requirements for hardware, applications software, and systems software at a preliminary design level.

Approaches to similar problems in conventional software development have been called requirements engineering [1], software architecture design [2], and simply analysis and design [3]. We call the approach requirements engineering or high-level design synonymously to [4]. The exact point of transition to detailed design is a moot question. Indeed the subsequent internal design of a computing module might be considered as a repetition of requirements engineering at a lower level of detail. (A multilevel application of the approach for general distributed computing systems is described in [5].) The important distinction is that we are mainly concerned with structuring of requirements rather than implementation details. This high-level structuring has been identified as the primary step in achieving cost effective solutions of complex problems.

The approach described has four basic activities: analysis, partitioning, allocation, and synthesis. The analysis activity is a problem-definition phase, using techniques similar to those found in modern structured design [3]. Partitioning and allocation, which we sometimes group as a single activity, determines the computer module types and configuration and maps software requirements onto the modules. The synthesis activity explicitly defines distributed data base and intermodule communications requirements. Overall control requirements are determined by resulting interactions between modules.

Two important aspects of the approach are the method of representing requirements (i.e., as graphs or with special languages) and the use of computing aids (i.e., software packages to help organize design information, evaluate design decisions, etc.) Graphical and language representations for modular systems are presented. Two types of computer programs for complex decision making are described with respect to partitioning and allocation. A completely developed set of languages and computing aids is not currently available, however. Future applications of the approach are expected to provide impetus for language and computing aid development, which, in turn, will help refine details of the approach.

1. INTRODUCTION

1.1 Motivation

Requirements engineering, or high-level design, is the initial step in satisfying computer system user requirements. The immediate concern is to produce a system which performs required functions at an acceptable cost and within an acceptable time. Achieving this goal in the design of complex systems is seldom straightforward. Historically we see that similar problems have been addressed in architecture [6] and general systems [7] studies. The major results of these studies were development of design methodologies based on a natural, systematic translation of user requirements into an implementation specification (or blueprint). Computer scientists have embellished this concept with special design languages and computer aids for conventional system design [2, 8].

Software engineering specialists have recognized that software should be designed to perform well over its total *life cycle* (including production, operation, and maintenance), rather than just the immediate development phase [9]. In evaluating a design a user should consider the total cost of the system, including: the cost of a system not being useful (not fulfilling its functional requirement); the cost of testing and documentation during development; and the cost of maintenance (continued debugging) and modification. Analyses of costs in conventional computing systems have shown that by far the largest components are the *people-costs* involved with design, testing, maintaining, etc. Hardware costs are comparatively small. The cost relationship usually quoted is that 80% of a data processing budget typically involves people-costs, compared to 20% for computing hardware [9]. The cost ratio for modular systems would be expected to be greater because test and analyses are much more difficult and because cheaper hardware usually is used. Testing and analyses are more difficult because of the nonrepeatability caused by asynchronous, concurrent operation; results depend on relative timings of events. And testing, according to Brooks, typically consumes half of a development schedule [10].

People-costs are a function of the apparent complexity of a system; a system which is easy to understand requires much less effort to develop, test, maintain, modify, and use. Also, simple systems invite fewer errors. Since some problems are inherently large and complex, techniques for reducing the complexity of solution systems have received much attention. The current most powerful method of reducing complexity is *structured* design [3]. The goal of structured design is to divide a system into individually understandable modules.† This involves the following considerations:

†Another direction in reducing a system complexity is in equipping them with adaptable architectures that increase throughputs of the available resources. More on this topic is in Chapter III, Sec. 3. (Ed.)

1. Size: Software modules must be small enough to be fully comprehended by a person (usually 50 to 100 lines of high-level code).
2. Functional cohesiveness: A module should address a single function or purpose with respect to the problem being solved by the system.
3. Visibility: The functions and structure within a module should be clear to an examiner, without requiring review of other modules and external documentation.
4. Clear interfaces: Interactions between modules must be restricted to clearly defined *ports*; modules must not be allowed to "reach" into one another's inner data areas and control sequences.†

†Other problems of interconnections among modules for complex parallel systems are discussed in the last part of Chap. I, Sec. 24, and Chapter III, Sec. 9. These are related to designing such reconfigurable busses interconnecting resource modules that minimize communication times between any pair of modules involved in data exchanges. (Ed.)

Although these concepts appear simple, subtle problems often occur in application; experience and discipline are important to proper structuring [3]. A systematic, structured design approach is even more important for modular systems because of the additional complexity of concurrently operating modules. We particularly recommend a *top-down* structured design approach, which we will outline below.

When structured design approaches are applied in conventional systems, the application designer usually has a good notion of the target system hardware and operating system capabilities. The designer has only to structure the application software and map it onto the system. In contrast, the modular system designer typically must help define both the hardware and operating system along with the application software. Indeed, the basic concept of modular systems is that the hardware, application software, and operating system be structured to match application requirements.‡ Thus the definition of computer modules, interconnections, intermodule communications, and operating system augmentations is part of the "application" designer's responsibility. In effect, requirements engineering for modular computing systems must define requirements for the computing system in addition to requirements for applications software.

‡More information on adaptation properties of a modular architecture in in Chap. I, Secs. 23 to 26. (Ed.)

The expanded role of requirements engineering for modular systems creates a number of design decisions and structuring alternatives not found in conventional systems. This increases the problem complexity and challenges our ability to obtain simple, easily understandable solutions necessary for reduced people-costs. We need to expand conventional design approaches to handle the new complexities if modular computing systems are to be truly cost effective.

The purpose of this chapter is to describe an approach to requirements engineering which treats the additional complexities of modular computing systems.† The approach can be manually applied to any size problem. To be completely effective for medium to large problems, however, we need the support of computer aids and design guidelines, or rules of thumb. Computer aids which evaluate complex decisions are described in this chapter. Additional computer aids are very desirable to help create, catalog, and check the design; construct and exercise simulations; standardize representations; control configurations; and communicate results. Some such aids are operational and generally available [11, 12], but presently none are completely adequate because they do not treat concurrency at the requirements engineering level. Rules of thumb, such as those enjoyed in other engineering disciplines, are very valuable in simplifying decisions. For example electronics engineers know which types of circuit configurations and ratios of components lend good temperature stability, linearity of response, etc. They need not explore infinite possibilities for promising solutions. We require further experiences with modular systems to develop powerful rules of thumb.

†The origin of these additional complexities that become possible in modern modular systems is discussed in Chap. I, Sec. 17. (Ed.)

1.2 Scope

The scope of requirements engineering for modular systems is implied in the discussion above. We describe the scope more explicitly here.

Figure V-1 illustrates the design transformation to be accomplished during requirements engineering. Given the overall computer system requirements, the tasks of requirements engineering are:

1. Identify and refine application both functions to be performed and data relationships.
2. Define performance requirements and qualities/attributes (e.g., reliability) for the functions.
3. Determine the number of computer modules and their characteristics: instruction rates, I/O rates, memory rates, data storage capacities, and word size.
4. Define software modules to be executed on each computer module and their data and control interactions.
5. Define data and control interactions between computer modules and any noncomputer components.

460 CHAPTER V

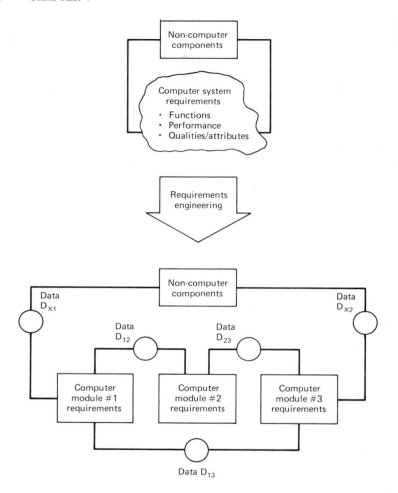

FIGURE V.1 Illustration of Requirements Engineering for Modular Computer Systems

Figure V-1 shows physical computing modules linked by data interactions, which form a logical communications network. Intermodule control is considered to be effected by similar communications; the notion of interrupts is considered to be a lower level implementation detail. Physical interconnections and a possible switching network are defined during lower level design (after requirements engineering), when specific LSI chips and board layouts are defined.

The relationships of application functions, algorithms, and software modules requires further clarification. An intitial step in requirements engineering is to identify major applications functions, e.g., to control an airplane. During following steps each function is redefined as combinations of more detailed (sub)functions; e.g., control of an airplane is decomposed into functions dealing with sensor measurements, pilot inputs, and actuator control. The results of each subsequent decomposition are usually represented by a hierarchical tree diagram. For example, Fig. V-2 shows that a function called "check for dangerous condition" decomposed into two functions, "evaluate temperature" and "evaluate pressure." "Evaluate pressure," in turn, became a set of functions, "get pressure sensor reading," etc. Note that each function may be viewed as an abstract definition of associated lower level functions. At some low level the set of functions need only be translated to a computer language to be an executable algorithm. Requirements engineering thus defines abstract algorithms, although truly executable algorithms are not developed.† Low level algorithms are not developed because their design depends on intramodule architectures and vice versa. Algorithm

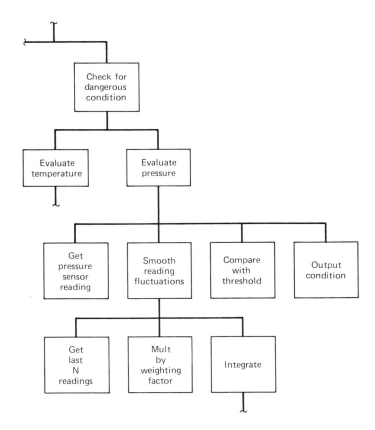

FIGURE V.2 Typical Function/Algorithm Evolution

†Requirements engineering may develop candidate executable algorithms for analysis purposes, but the candidates may be discarded during lower level design.

design should be completed in conjunction with internal design of computing modules at lower levels to better match algorithms and architectures. Requirements engineering should decompose functions only to the degree necessary to enable good estimates of functional and data requirements.

Software modules are groups of subfunctions that are to be coded as a unit. The modules are formed on the basis of size, functional cohesiveness, visibility, interface simplicity, and possibly executional characteristics (e.g., scheduling frequency or type of computation). In the example of Fig. V-2, the function "check for dangerous condition" and all underlying components would likely be a single software module.

1.3 Overview of Approach

The design approach is formulated as four major activities, as shown in Fig. V-3. The activities are further defined as a sequence of steps. Some iteration among activities and steps is almost always necessary. Detailed discussions and examples of all steps are given in the next section. As an introduction, a brief overview is given here:

1. *Analysis* The objective of the analysis activity is to assimilate and further refine user requirements in preparation for following design activities. This activity begins with the identification of relevant requirements and selection of a suitable representational form. This establishes a basis for further elaboration, which entails further func-

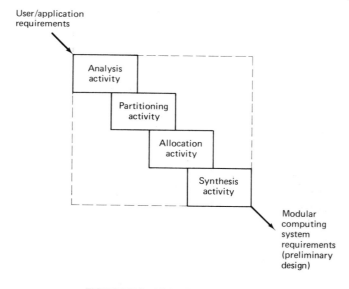

FIGURE V.3 Major Design Activities

tional decomposition of the requirements, estimations of sizing, assessment of feasibility, and assessment of performance characteristics (e.g., timing and loading). Simulations usually are used to aid and verify these tasks.

2. *Partitioning* The distribution of processing requirements to computers is achieved by the second and third activities, partitioning and allocation. Partitioning groups processing requirements defined during analysis on the basis of intrinsic commonalities, such as grouping functions that use common data. The groups may be allocated as units to computers, thereby reducing the complexity of allocation. Alternatively, group membership may be used as part of the allocation criteria. Partitioning is thus concerned with establishing and quantifying criteria for grouping, defining and comparing candidate groupings, and selecting one or more groupings to be passed on to allocation.

3. *Allocation* Allocation begins by selecting a candidate set of computers, based on characteristics of partitions (such as estimated instruction rates and storage requirements). The activity then proceeds in a manner similar to partitioning: establishing and quantifying criteria for allocation, defining and comparing the results of applying different criteria, and selecting one particular allocation. The procedure is usually iterated for a number of selections of candidate computers.

4. *Synthesis* The synthesis activity explicitly examines requirements for intermodule communications. It forms new interfacing functions and buffers to represent communications overhead in terms of delays and storage. This provides a framework for lower level design tradeoffs during intramodule and interconnecting network design.

2. DETAILS AND EXAMPLES

2.1 Analysis

Our analysis activity is similar to the *structured analysis* and *structured design* activities of conventional system design [3]. The goal of analysis is to identify and structure functional requirements of the application. In conventional systems a hierarchical control structure typically is produced, whereas in modular systems we anticipate development of concurrent control structures.

Consider the following very simple application: "The computing system is to input values of (a, b, c) corresponding to the coefficients of the quadratic equation, $ax^2 + bx + c = 0$, and output values of the two roots (r_1,

r_2) of the equation." The first step is to represent the processing requirements identified in the application problem statement. We then proceed to elaborate (add detail to) the requirements representations.

We prefer graphical representations in initial stages of requirements analysis, but we switch to a language representation when graphs become too large; also, a language representation usually is needed to apply computer aids to recording and checking the design. The most important concern of representation is to depict data flow rather than control flow. Early representation of control flow is avoided in conventional structured design because it hinders development of hierarchical structures of functions. (Figure V-2 is an example of a hierarchical structure of functions.) In modular system design, avoidance of specifying control flow is additionally important because it obscures alternative mappings of functional units onto concurrently operating computer modules.

We use data access graphs for initial representations of application requirements. Figure V-4 is the top level graph for the quadratic equation example. In data access graphs, data is represented by circular nodes and functions are represented by rectangular nodes.† An arrow from a data node to a function node represents a read-only type access of the data by the function. Write-only and read-write accesses are represented by an arrow from a function node to a data node and a double headed arrow, respectively. No physical aspects of data accessing are assumed in the graphs, i.e., so-called reads and writes may actually be accomplished by message passing. Details of the accesses are not provided at this level. For example, assuming that data node C in Fig. V-4 can contain multiple instances of (a, b, c) values, the function F may access any number of instances at a time. In fact we conceive the possibility of multiple copies of the function accessing multiple instances of data concurrently. Control of accessing also is not specified. We conceptually assume that functions read data values when they are available and write data values when calculated with no other control specified. (As a reminder that we are dealing with conceptual function and data relationships, we often call the nodes of data access graphs functional and data *entities*.) Finally, no control precedence is assumed in the graphs. In Fig. V-4, for example, the function F may concurrently read an instance of C while writing an instance of R (computed from a previous instance of C). Likewise one copy of function F may be writing an instance of R while another copy is reading an instance of C. This concept gives us flexibility in developing concurrent implementations.

The next step is to elaborate the initial data access graph. One way of elaborating is to decompose existing functions and data into component functions and data. Figure V-5 is obtained from Fig. V-4 by decomposition. In this first elaboration, interface functions ($F1$, $F4$) and the major subfunctions of the problem (calculating the two roots) are represented. The interface functions are important for isolating the computing system from

†Comprehensive treatment of other program graph models is given in Chap. IV, Sec. 3. (Ed.)

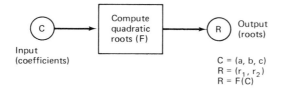

FIGURE V.4 Top Level Data Access Graph for Quadratic Equation Example

external changes. For example, the types of terminals or I/O formats may need to be changed after the system is constructed. In some cases of computers embedded in other systems, we might not know exactly what the real interface will look like because interfacing systems are being designed at the same time as the computer system. The interface functions allows us to solidify internal interfaces ($D1$, $D2$, and $D3$ in Fig. V-5) although external interfaces (C and R) may change. Our only assumptions are that C will contain sufficient information to enable us to produce $D1$, and that ($D2$, $D3$) have sufficient information to produce R. This concept is further discussed by Parnas [13].

New functional and data entities may also be introduced during elaboration, although this may be contrary to true top-down design. Two situations occur where the introduction of new entities is appropriate: (1) in some cases new requirements are "discovered" as complicated problems are being decomposed; and (2) in some cases we want to treat separately requirements having different purposes or importance. In the second case we ignore secondary requirements during initial graph construction and add them later. For example, suppose in the quadratic root problem that we had the secondary requirement of printing an error message instead of the normal calculations whenever ($a = 0$) was obtained. We might perform a few decompositions of the primary requirements (finding the roots r_1, r_2) and then add the secondary

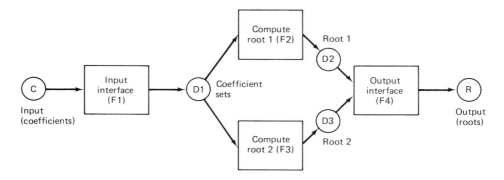

FIGURE V.5 Second Level Data Access Graph for Quadratic Equation Example

requirement. Figure V-6 illustrates adding the secondary requirement to the graph of Fig. V-5.

Figure V-7 is a further decomposition of functional entities $(F2)$ and $(F3)$ of Fig. V-6. This decomposition is based on additional information about the nature of $(F2)$ and $(F3)$ assumed to be obtained from the user or other sources. Specifically, we indicate that different calculations are to be performed depending on whether the discriminant $(D = b^2 - 4ac)$ is less than zero, equal to zero, or greater than zero.

A design issue must be discussed with respect to Fig. V-7. Suppose that the output of "discrim evaluation" $(F6)$ is the data set $D6 = (a, b, c, D)$, where D is the value of the discriminant. It appears that every root calculation function $(F8, F9, F10)$ must access $(D6)$ and then only perform calculations if D has a properly corresponding value; e.g., $(F8)$ calculates roots only if $D < 0$. Alternatively we could have shown $(F6)$ writing three separate data entities exclusively containing data sets with $D < 0$, $D = 0$, and $D > 0$. Then $(F8)$, $(F9)$, and $(F10)$ would only need to read data with properly related discriminant values. This alternative is to be avoided, however, because it would require the discrim evaluation function $(F6)$ to know about the existence of $(F8, F9,$ and $F10)$ and thus violate modularity considerations. Figure V-7, in its present form, admits concurrent implementation of $(F8)$, $(F9)$, and $(F10)$, with each function making its own decisions about whether to calculate roots. The functions $(F8, F9,$ and $F10)$ are conceptually like "guarded commands" [14].

A number of additional points are to be made with respect to the quadratic root example shown in Figs. V-4 to V-7:

1. Elaboration is a creative act. The designer must extract the nature of necessary transformations from the user and possibly other sources. The user seldom provides all important details in the initial statement of the problem. In more realistic applications, the user must be available for interaction and review during analysis. Requirements "identified" by the designer must be checked with the user. The major difficulties are assuring that *all* requirements are surfaced and that both the designer and user unambiguously understand the requirements and their interactions. As elaboration proceeds, the graph structure (alternating function and data nodes) forces explicit consideration of how functions obtain necessary information.

2. The stopping point of elaboration is not well defined. (Our guideline is that each functional entity must be sufficiently detailed to suggest candidate implementations for feasibility and sizing analyses.) This lack of a firm stopping point leads to dangers of premature algorithm design. For example, the reason for evaluating the discriminant in Fig. V-7 is to avoid trying to take the square root of a negative number. This would not be a problem in implementations where the

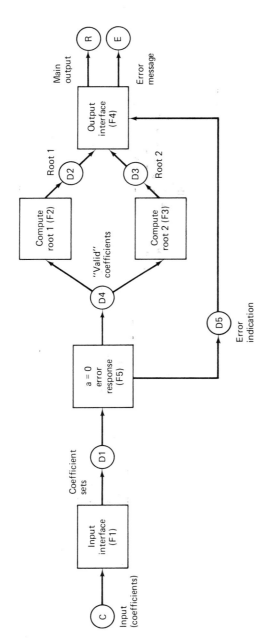

FIGURE V.6 Quadratic Equation Example with Error Message Capability

468 CHAPTER V

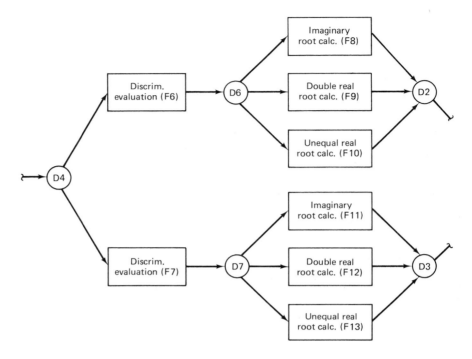

FIGURE V.7 Decomposition of Functional Entities (F2) and (F3) of Quadratic Equation Example

control system can be coded to substitute $[-\text{SQRT}(D)]$ for $[\text{SQRT}(-D)]$.

3. Data transformations and flows are not completely described in data access graphs. For example, in Fig. V-6, $(F5)$ does not read a value of $(D1)$ and then write a value of $(D4)$ *and* $(D5)$. It writes either $(D4)$ *or* $(D5)$. Similarly $(F4)$ must read *both* a value of $(D2)$ and a corresponding value of $(D3)$ to produce a value of R, and it produces E only when it reads $(D5)$. Additional markings of such *or* and *and* conditions may be used; e.g., see Yourdon and Constantine's data flow graphs [15]. (We state this information in an additional language form of representation, discussed later, instead of marking the graphs.)

4. Only data entities accessed by two or more functional entities are shown in the example graphs thus far. A subsequent example will indicate that functional entities may retain unshared data that needs to be diagrammed because retained data affects potential parallelism of replicated functions.

A more complex example is next considered: the typical feedback control system shown in Fig. V-8; see Athans for further discussion of this type of control system [16]. A digital compensator is to be developed to control a physical "process," which is composed of actuators, a plant, and sensors. (The plant most often consists of nondigital, electromechanical components interacting with a physical environment.) The compensator is to control the plant to produce a desired response or product, and to quickly shutdown the plant when dangerous conditions occur. The compensator determines appropriate control actions based on observations of the plant's state fed back from the sensors. Complete state information is usually unavailable from observations, however, and the observations that are obtained usually are not completely accurate because of noise and uncertainties of sensor action. The compensator must therefore estimate the plant's state to compute desired control signals.

Figure V-9 is a high-level data access graph for the application. The components of Fig. V-9 are derived from the textual description of the problem, plus the interface functions to isolate the design from changes in the actuators, plant, and sensors. In Fig. V-9 we represent both the major processing functions (estimate, control, and shutdown) and the major electromechanical components (actuators, plant, and sensors). This approach is helpful for developing and testing simulations. Digital simulations of the system may be constructed by substituting digital models for the electromechanical components in the diagram.

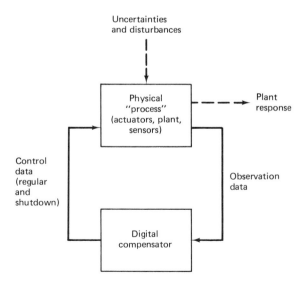

FIGURE V.8 Example Design Problem

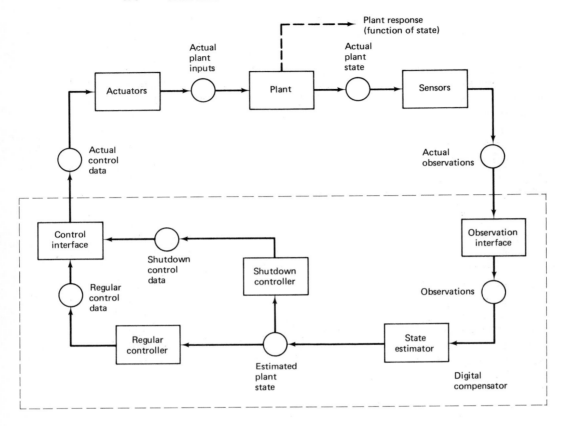

FIGURE V.9 High Level Data Access Graph

The high-level graph of Fig. V-9 is next translated through a number of iterations of elaboration. For instance, Fig. V-10 is the next elaboration of the state estimator and Fig. V-11 elaborates one functional entity, the plant model. Review of Figs. V-9 to V-11 shows that much additional information beyond the original problem statement was applied. Once again we see the need for strong user interaction during analysis.

Figure V-11 shows a case where retained data, the "next plant state," is indicated. Suppose we were to implement multiple copies of the plant model with different copies concurrently processing different instances of input data ($D2$) and ($D7$). The copy which processes the ith instance of input data needs the retained state data calculated for the $(i - 1)$st instance, which could be produced by a different copy. This otherwise internal state data consequently must become shared data when multiple copies of the plant model are implemented. The additional communications and communications overhead needs to be considered in design tradeoffs. Retained data thus must be identified

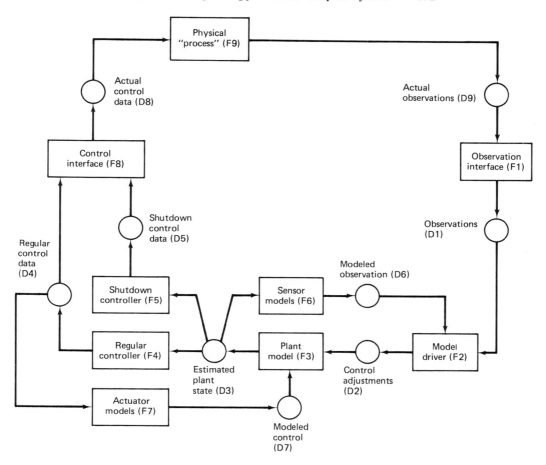

FIGURE V.10 Elaboration of the State Estimator

whenever multiple copies of functions are to be considered. Also, data storage requirements must be assessed in making design decisions.

In structured design of conventional systems, a hierarchical structure chart showing the results of elaboration usually is prepared. Figure V-12 shows the results of the present example. If this design were to be implemented as a single program on a single sequential computer, we would likely write a main program, "digital compensator," which would handle data I/O and "call" the three next lower modules. Similarly, the "main transformation" module would input/output data and call the next lower level modules. Carrying the procedure further, we would end up with a hierarchical program structure in which the most subordinate modules would perform the actual computing work, and the higher modules would simply provide branching logic and data movement.

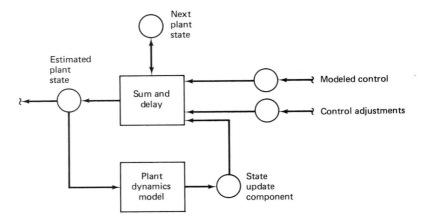

FIGURE V.11 Elaboration of Plant Model

Suppose the example design were to be mapped onto two computers. One reasonable mapping is postulated to be:

Computer 1: observation interface
state estimator

Computer 2: regular controller
shutdown controller
control interface

where we assume that each module contains all lower "son" modules in the hierarchy. A question then occurs of whether "main transformation" and "digital compensator" should appear in code and on which computer. Placing a module on one computer which controls a module on another computer requires extra communication and could reduce achievable concurrency. In addition, for reliability, we do not want to implement centralized controllers because they introduce single point failure modes.

If hierarchical structures must be implemented in modular computer systems, many high-level modules should exist in only a very restricted form; e.g., they should execute only during initiation and termination phases. Thus upon initiation the digital compensator would initiate execution of lower level modules and then become inactive until termination. Such high-level modules would not "contain" their subordinates as in conventional structured systems.

Modular system design and implementation deviates from a centralized design at this point. Instead of dealing with a hierarchy of modules, we deal with a set of communicating hierarchies. For instance, based on a number of characteristics (such as module size, data storage, data rates, type of algorithm, anticipated types of computers, etc.) we might choose as our fundamental modules:

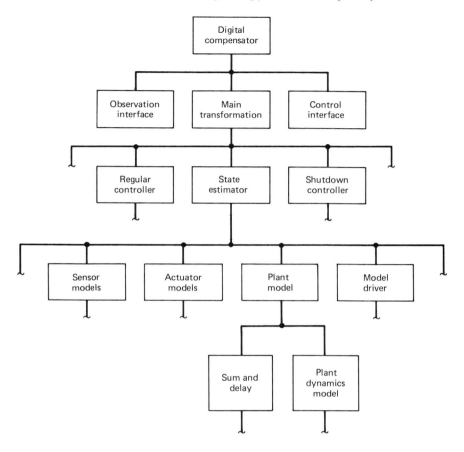

FIGURE V.12 Module Hierarchy for Example Problem

1. observation interface
2. control interface
3. regular controller
4. shutdown controller
5. sensor models
6. actuator models
7. plant model
8. model driver

Each of these modules is considered to be an individual process without a parent controlling process. Vestiges of the digital compensator, main transformation, and state estimator may be kept in initialization and termination routines, but they no longer need to be carried along in the main design effort.

Requirements structuring also is influenced by the size of identified components (numbers of instructions and amount of storage) and performance requirements. In addition, the analysis activity must prepare sizing and performance requirements for use in guiding partitioning, allocation, and synthesis. The first step to sizing is to estimate the number of instructions required for each functional entity. The number of instructions of course depends on the particular algorithm to be used and the physical computer on which the algorithm is to be implemented, both of which are not known at this point. To obtain a benchmark in the absence of other information, we assume a single, sequential computer implementation of each application module. We then estimate instructions either by comparison with existing codes for similar functions or by constructing a candidate algorithm.

To help define and investigate candidate algorithms, we find it useful to express the design in a language form. A language form also facilitates generating simulations, which are usually needed to assist and verify design steps; simulations are discussed shortly. In addition, a language form is desirable because the graphs do not contain all design information; for instance the absence of conditional data flow information was noted earlier. Graphs quickly become cluttered as more detail is added, whereas a language representation can more clearly and fully detail the design.

In translating a data access graph into a language representation, we assume that all functional entities are *data driven* processes. In other words a functional entity is to be realized by a process which executes upon arrival of necessary input data. This assumption allows us to avoid making control decisions relating to sequencing of processes. The resulting data driven design is very consistent with concepts of modular computer systems: The processes are conceived to execute asynchronously on dedicated modules when data arrives, without an external controller. We then are free to decide which processes should share modules and which processes should be replicated on more than one module. These decisions are made during partitioning and allocation.

The language form we recommend for requirements engineering is a Pascal-like pseudocode, as discussed in [17]. Figure V-13 illustrates pseudocode for the plant model in Fig. V-11. This pseudocode has no rigid rules of syntax and incorporates natural language where helpful. At the same time it permits describing control and data structures in as much detail as desired. (No data structures are defined in Fig. V-13, although some messages and records are assumed to exist.) A sequential control structure is indicated in Fig. V-13. Concurrent structures can be represented by using concurrent language constructs. For example we can use (FORK, JOIN) or (COBEGIN, COEND) type operations and may express mutual exclusion of data accessing, waiting and signaling conditions, etc. [18, 19]. A pseudocode program easily yields a commented executable Pascal program by "commenting" natural language statements and adding real code to perform the described tasks. (Executable Pascal is actually one form of pseudocode.)

```
START plant_model
DO forever
RECEIVE input message
IF modeled_control message THEN decode and store message
IF control_adjustment message THEN decode and store message
IF current control messages of both types have been received THEN
        BEGIN apply sum_and_delay
                SEND estimated_plant_state message
                apply plant_dynamics_model
                clear input message stores
        END
END plant_model
```

```
START sum_and_delay
SET estimated_plant_state to next_plant_state
SET next_plant_state to sum of modeled_control,
        control_adjustment, and state_update_component
END sum_and_delay
```

```
START plant_dynamics_model
GET current plant A-matrix
        matrix multiply A X estimated_plant_state
SET state_update_component to matrix product
END plant_dynamics_model
```

FIGURE V.13 Example Pseudocode for the Plant Model

Candidate algorithms can be expressed in pseudocode at a number of levels of detail, ranging from the abstract form in Fig. V-12 to real, executable Pascal. For sizing estimation, the code usually does not need much more detail than shown in Fig. V-12. For each line of pseudocode we estimate the equivalent number of instructions in a high-order programming language like Pascal or FORTRAN. The line estimates are then summed to obtain an estimate for the whole function. For very simple functions (no nested loops and only one type of input data) the resulting estimate will have the form

$$I = I_0 + I_1 N_d \qquad (1)$$

where I_0 is the estimated number of non data dependent (overhead) instructions,

I_1 is the estimated number of instructions per instance of input data, and N_d is the number of instances of input data to be processed at each execution.

Estimates for functions with nested loops will have the form

$$I = I_0 + I_1 N_d + I_2 N_d^2 + \cdots \qquad (2)$$

Estimates for functions with N_{d1} instances of data ($D\,1$) and N_{d2} instances of data ($D\,2$) might take the form

$$I = I_0 + I_{11} N_{d1} + I_{12} N_{d2} + I_{22} N_{d1} N_{d2} + \cdots \qquad (3)$$

In some cases the instruction count is complicated by data dependent branching within a functional entity. For example, if we wanted to estimate the number of instructions per instance of "compute root 1" ($F\,2$) in Fig. V-4, we would find that the count would depend on whether $D < 0$, $D = 0$, or $D > 0$. An average count can be obtained if we know the relative frequencies of the different cases. When relative frequencies cannot be predicted analytically, a simulation is needed.

Storage estimates are obtained through the same procedure as instruction estimates. The data storage requirements for each line of pseudocode is estimated and then summed. The resulting equations are analogous to Eqs. (1–3). For example

$$S = S_0 + S_1 N_d \qquad (4)$$

where S_0 and S_1 are overhead and data dependent storage requirements, respectively.

Instruction rate estimates are obtained by combining instruction estimates with performance requirements. For example a system may be required to accommodate a sustained input data rate of N_{di} instances per interval T_d. If a function has an instruction estimate given by Eq. (1), then its required processing rate is

$$R_{ap} = (I_0 + I_1 N_{di})/T_d \qquad (5)$$
$$= I_0/T_d + I_1 R_{di} \qquad (6)$$

where R_{di} is the input instance rate, N_{di}/T_d. Equation (5) is useful near the input port of the computer system, where N_{di} is specified. The determination of input data rates for functions which are internal to the system often requires simulation of the design.

A modular computer dedicated to this application function would need a minimum processing rate (in high-order language instructions) of

$$R_p = R_{p0} + R_{ap} \qquad (7)$$

where R_{p0} allows for overhead for control, data handling, and communications. If (n) functions shared the module the rate requirement would be

$$R_p = \sum_{i=1}^{n} (R_{p0i} + R_{api}) + R_{ps}(n) \qquad (8)$$

where R_{p0i} and R_{api} are the rate requirements of function (i) and $R_{ps}(n)$ is the additional overhead of (n) processes sharing the module. (Sharing tradeoffs are considered during partitioning and allocation.) The overhead rate R_{p0} in Eq. (7) depends on the internal hardware and control or operating system of a module. In the absence of further information at this point, we would assume a safety factor of, say, $R_{p0} = 0.25 R_{ap}$. The overhead rate $R_{ps}(n)$ is similarly estimated during partitioning and allocation, when sharing decisions are made.

Sustained throughput is not the only timing requirement commonly given. Delay is an important performance measure in most real time and interactive systems. Delay is more difficult to analyze because it is more dependent on particular algorithms and architectures.† For example, consider the alternative ways of processing N_d instances of data on a simple sequential computer. In general, algorithms may be devised to completely process a single instance at a time. An alternate algorithm could carry all instances to completion at nearly the same time using a sequence of small loops. Even if both algorithms had the same sustained throughput rate, the second algorithm would delay some data instances more than the first algorithm would. Likewise an architecture which requires setting up an array of data for concurrent processing would delay some data instances more than would a sequential architecture with the same throughput.‡ Total input to output delays of some data instances can be unacceptable even when average throughput and average delay are within acceptance.

To obtain a baseline apportionment of delay to functional entities, we assume that allowable delay should be proportioned to a function's individual processing rate requirement R_p, i.e., functions with high R_p requirements should be allowed greater delays. Thus if the total allowable delay across a series† of functions $(F1)$ and $(F2)$ is T_{12}, then the allowable delays across each function are

$$T_1 = \frac{R_{p1}}{R_{p1} + R_{p2}} T_{12} \qquad (9)$$

and

$$T_2 = \frac{R_{p2}}{R_{p1} + R_{p2}} T_{12} \qquad (10)$$

where R_{p1} and R_{p2} are the processing rate requirements for $(F1)$ and $(F2)$, respectively, calculated from Eq. (7) or (8). If $(F1)$ and $(F2)$ are in parallel, then the delay would be proportional to the $\max[R_{p1}, R_{p2}]$, assuming that both functions must complete before processing can continue. More complex

†Chapter IV, Sec. 3 presents other techniques for modeling delays and synchronization requirements for computed functions (microprograms). (Ed.)

‡The principle of sequentiality of computations implemented in a sequential architecture is discussed in Chap. I. Sec. 11. (Ed.)

---

†Series and parallel structures are identified with respect to paths taken by a single instance of input data.

structures are treated by decomposing them into series and parallel components.

Sizing estimates should be compared to module capacity limitations during analysis. If a functional entity requires a throughput or storage capacity above about 50% of a candidate module's capacity, further decomposition is suggested. A single functional entity should not require more than 50% of a module's capacity at this point in design because requirement estimates almost always increase as design progresses, and system costs typically increase considerably as capacity limits are approached [9]. If decomposition is ineffective, then explicit replication of functions to treat parallel data streams is suggested. If function replication fails to bring requirements within expected module capacity, the feasibility of solving the problem should be examined.

The rate capacities of modular computer components are usually expressed in machine instructions per second, rather than source instructions (lines of high-order language code) per second. A study has indicated that the ratio of source instructions to machine instructions varies between about 1.5 to 5, depending mainly on the richness of the computer's instruction set [20]. (Smaller ratios correlate with smaller instruction sets.) The ratio was found to be nearly independent of application and programmer and has little correlation with source language. A ratio of 3 is suggested for comparing source instruction estimates with computer capacities. Consequently a computer module with processing rate capacity

$$C_p \geq 3R_p \tag{11}$$

is needed for the estimated instruction rate R_p as given by Eq. (7) or (8).

The need for simulations has been noted above. Simulations also are needed to verify design decisions and gather information needed for partitioning, allocation, and synthesis. Usually a *functional level* simulation is adequate. In a functional simulation the overall data flow structure is examined without detailing algorithms. No computational details are provided beyond those necessary to enable proper decision branching. Of course, time is simulated also.

In constructing a functional simulation of the compensator system, for example, we initially would assume that the eight functional entities were implemented as separate processes on separate computers. The processes would be data driven, and they would communicate via messages corresponding to the data entities identified. Since in a functional simulation the simulated processes do not have real algorithms, the messages cannot have realistic data. Messages are simulated. For example the control adjustment message ($D2$) would contain:

1. Time tag: The simulated time that the message is modeled as arriving at the destination ($F3$), computed by ($F2$).

2. Size of the true message being simulated: Words or instances of data sets.
3. Modeling variables: Special simulation variables to cause proper branching or sequencing of operations. (In the present case, an "out of bounds" indicator from the physical process ($F9$) model is used to activate the shutdown controller ($F5$).)

Pseudocode is a useful starting point for simulating processes. For example, a plant model ($F3$) simulation could be generated from the pseudo code of Fig. V-12. With the following types of modification, the pseudocode could be executed, assuming a Pascal compiler is available:

1. Change English text portions to comments.
2. Add executable code as necessary to provide required functioning of the simulation, i.e., to pass messages and allow branching.
3. Provide each process with a "logical clock" to simulate time passage.
4. Add code to record data accessing, execution activity, and storage requirements versus time. (Sizing estimates described above are used here.)

The logical clock is important for simulating concurrency. A logical clock is simply a local variable which is advanced to simulate processing and waiting times. Processing time is computed using an equation like:

$$\text{PTIME} = [I_0 + I_1*(\text{SIZE OF MESSAGE})]/\text{RATE} \qquad (12)$$

where I_0 is the estimated number of nonmessage dependent instructions necessary to support the process,

I_1 is the estimated number of instructions per message component,

and RATE is the average instruction rate assumed for a candidate computer on which the process may be implemented. (RATE is considered a design parameter and is given candidate values during analysis.)

A simulated process advances its logical clock by PTIME everytime it simulates the processing of a message. A process is assumed to have spent time waiting if a message arrives with a time tag which is later than the logical clock. The clock is then advanced to the value of the time tag before processing the message. If a time tag of an arriving message is earlier than the logical clock of the receiving process, we assume that the process was busy when the message arrived—causing a delay in processing the message. If in simulation such delays become unacceptable, then RATE must be increased or multiple copies of the processor may be assumed. Timing simulation and process synchronization becomes more complex when different message streams must be merged in time order and message arrival is not predictable [21, 22].

The simulation should provide data accessing, instructional rate, and data storage requirements. Accesses A_{ij} of data D_j by function F_i are easily recorded as a function of logical time. Average and peak numbers of accesses then can be computed. Data storage is calculated for messages and local data.

In functional simulations important events usually are simulated stochastically. For example the physical process simulation may use sample values of a random variable to generate a proper number of "out of bounds" events in a given time. If the statistics of such external events are not known, a higher fidelity simulation of the system external to the computer system often is required.

In the current example we assume that the major simulation results are the numbers of data accesses per unit time. Figure V-14 shows numbers of access displayed as a *data access matrix*. The entries in the matrix are the number of accesses A_{ij} of data D_j by function F_i. The types of access (reads, writes, and read/writes) are also indicated. The data access matrix is a major input to partitioning and allocation activities.

We next need to select the number and types of computers and determine the mapping rules for the eight functional entities; note that we do not preclude realizing any entity with hardware rather than software. The selection of computers and functional mapping is the combined result of the partitioning and allocation activities.

	D1	D2	D3	D4	D5	D6	D7	D8	D9
F1	1000 (W)								1000 (R)
F2	1000 (R)	1500 (W)				1000 (R)			
F3		1500 (R)	1500 (W)				1500 (R)		
F4			1500 (R)	600 (W)					
F5			500 (R)		1 (W)				
F6			1500 (R)			1000 (W)			
F7				600 (R)			1500 (W)		
F8				600 (R)	1 (R)			600 (W)	

FIGURE V.14 Example Data Access Matrix (Numbers Show Relative Amounts of Accessing, R = READ, W = WRITE)

2.2 Partitioning

Computer selection and functional mapping is very complex for most problems because of the large number of alternatives and many interactions. For instance, if we had five *identical* computers for our example problem and our only limitation was that each computer must execute at least one function, we would have 1050 distinct ways of mapping the eight functions onto the five computers. In general we do not know how many computers to use a priori. If the number of computers must also be determined, say, in the interval [1, 8], we would have a total of 4140 alternatives to consider. The difficulty increases rapidly if we let computers differ in type, interconnection characteristics, physical location, etc. For instance, there are 10,080 ways of mapping eight functions onto five *different* computers. These numbers explode rapidly as the number of functions increases; e.g., there are more than 2×10^8 ways of mapping 15 functions onto five identical computers. Furthermore, we often must deal with additional requirements and constraints, such as the desire for functional redundancy (for reliability or load balancing) and constraints on computer and network capacities.

Because of the complexity of computer selection and functional mapping, we usually accomplish the tasks as shown in Fig. V-15. The first major step, partitioning, lowers the complexity of following steps by grouping functions according to intrinsic commonalities, e.g., common data accessing patterns. Computer characteristics are mostly ignored during partitioning, except that we may constrain sizes of partitions in terms of instructional rates and data storage requirements. Partitioning is similar to the forming of *load modules* or *tasks* in conventional software design; see for instance Yourdon and Constantine's discussion of *packaging* [24].

The selection of candidate *resources* (modular computers in our case) in Fig. V-15 usually is done simply by exhaustive enumeration, i.e., we try every feasible and reasonable alternative. In our example we would likely select N identical computers, letting N range from 1 to 8, and then determine a specific value for N based on evaluation of the ensuing allocation.† Some characteristics (throughput, storage, etc.) of the computers might be optimized in the process.

Allocation is similar to partitioning. The main difference is that allocation relates characteristics of partitions to characteristics of resources, whereas partitioning looks at commonalities of processing entities with only incidental concern for potential resource characteristics. Allocation binds partitions to physical resources.* In general a partition may be bound dynamically to different resources at different times. The results of the allocation are evaluated with respect to given constraints and criteria, and iterations of alternative resource selections and/or partitionings are possibly made.

†Chapter III, pp 348–382, treat such allocation of hardware resources (memory and processor) that takes into account program needs on computer sizes and memory volumes for storing data words. (Ed.)

*The numbers of combinations are tabulated as Stirling numbers of the second kind [23].

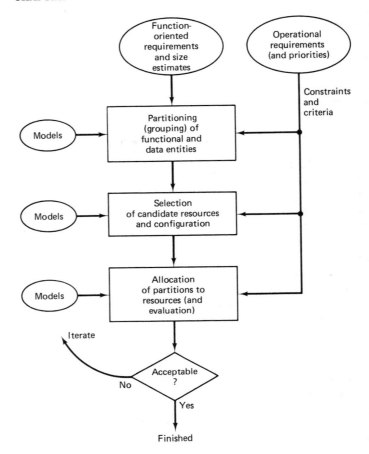

FIGURE V.15 Partitioning/Allocation Procedure

Figure V-16 lists common criteria and constraints used during partitioning and allocation. The performance criteria (item 1) and locational constraints (item 7) are usually directly quantifiable in terms of functional and data entity and physical resource characteristics. The other criteria (changeability, testability, etc.) are harder to quantify and they are the main contributors to people-costs, which are our dominant concern. All important criteria and constraints must be related to partitioning/allocation decisions via mathematical models, which we describe shortly.

The partitioning–selection–allocation procedure has been noted as a problem in constrained optimization. An attempt toward an analytic solution has met incomplete success [25]. Numerical techniques have been successful. We will outline a "fuzzy clustering" approach to partitioning [26] and an "implicit enumeration" (or "branch and bound") approach to allocation [27].

1. Performance
   - data access commonality
   - data and program storage
   - control relations (sequentiality, exclusive branches)
   - resource demands
   - architectural characteristics
   - communications delay

2. Changeability
   - localize changes

3. Testability
   - individually and meaningfully testable
   - "complete" functions
   - "loose" coupling

4. Maintainability
   - accessibility
   - testing
   - repair limitations

5. Reliability/fault tolerance
   - fault detection, isolation
   - resource uniformity
   - communications loss susceptibility
   - decentralize control and data base

6. Deployability
   - available building blocks

7. Locational constraints and computing capabilities
   - size, weight, power, operating environment
   - availability of computing hardware
   - type of processing

FIGURE V.16 Partition Formation and Allocation Criteria

Variations in techniques and decision guiding algorithms are described by Morgan and Levin [28, 29] and Boorstyn and Frank [30]. Another important approach uses "flow algorithms" [31].

A partitioning approach based on fuzzy clustering is described in the remainder of this section. A branch and bound approach is described later for allocation. Some of the referenced techniques do not treat partitioning and allocation as distinct activities, and, certainly, partitioning may be subsumed by an allocation algorithm. The primary benefit of the partitioning approach described here is that it lowers the complexity of allocation (usually drastically) with application of a very simple and stable algorithm. Also, as a distinct step toward allocation, partitioning provides decision data that might not otherwise be observable by designers.

The goal of partitioning is to group functional and data entities according to criteria like those listed in Fig. V-16. A simple example of partitioning on the basis of data access similarity follows.

Consider the system shown in Fig. V-17. Assume that a simulation of the system during the analysis activity produced frequencies of data accessing shown in Fig. V-18(a), which is a data access matrix. In this example we will

484   CHAPTER V

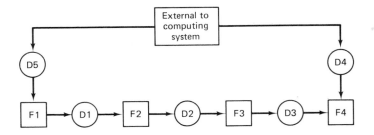

FIGURE V.17 Example System to Introduce Partitioning Approach

calculate data accessing similarities for functional entities. Data entities will be assigned to partitions according to a rule to be given shortly.

We first calculate the pairwise, or first order, similarity of functions F_i and F_j by:

$$s_1(i, j) = s_1(j, i) = \frac{\sum_k [\min(A_{ik}, A_{jk})]}{\sum_k [A_{ik} + A_{jk}]/2} \qquad (13)$$

where

A_{ik} = number of accesses of data D_k by function F_i, which is the data access matric entry in [row (i), column (k)], Fig. V-18(a)

The numerator of $s_1(i, j)$ is seen to be the total mutual accessing of data, while the denominator is the total average accessing of data by functions F_i and F_j, i.e., the sum of all data accesses by F_i and all data accesses by F_j, divided by 2. For instance, in the present example F_1 accesses D_1 with $A_{11} = 10$ and D_5 with $A_{15} = 10$, and F_2 accesses D_1 with $A_{21} = 10$ and D_2 with $A_{22} = 20$, so:

$$\sum_k [A_{ik} + A_{jk}]/2 = (10 + 10 + 10 + 20)/2 = 25$$

F_1 and F_2 only access D_1 in common, so:

$$\sum_k [\min(A_{ik}, A_{jk})] = \min(A_{11}, A_{21}) = 10$$

Therefore

$$s_1(1, 1) = 10/25 = 0.40$$

Completing the evaluation of $s_1(i, j)$ for all functions in Fig. V-18(a) yields the results shown in Fig. V-18(b).

The *direct* data accessing similarity [Eq. (13)] of F_i and F_j is shown in

	D1	D2	D3	D4	D5
F1	10				10
F2	10	20			
F3		20	10		
F4			10	10	

(a) Data access matrix (values of A_{ij})

	F1	F2	F3	F4
F1	1	0.4	0.4	0
F2	0.4	1	0.67	0.4
F3	0.4	0.67	1	0.4
F4	0	0.4	0.4	1

(c) S_2 matrix (2nd order similarity)

	F1	F2	F3	F4
F1	1	0.4	0	0
F2	0.4	1	0.67	0
F3	0	0.67	1	0.4
F4	0	0	0.4	1

(b) S_1 matrix (1st order similarity)

	F1	F2	F3	F4
F1	1	0.4	0.4	0.4
F2	0.4	1	0.67	0.4
F3	0.4	0.67	1	0.4
F4	0.4	0.4	0.4	1

(d) S_3 matrix (3rd order similarity)

FIGURE V.18 Data Access and Similarity Matrices for Example of Figure V.17

location (i, j) of the first order similarity matrix S_1, Fig. V-18(b). S_1 only shows direct, pairwise relations, however. For instance, F_1 and F_3 are indicated to have zero similarity [$s_1(1, 3) = 0$] whereas in Fig. V-17 we see that F_1 and F_3 are related via $(D_1 - F_2 - D_2)$. To show such *transitive*-type similarities, we calculate a second order similarity function:

$$S_2 = S_1 \times S_1 \qquad (14)$$

where the matrices are multiplied in the usual fashion except minimum operations replace products and maximum operations replace sums. For example the $(1, 3)$ component of S_2 is computed as follows:

$$\begin{aligned}s_2(1, 3) &= s_1(1, 1) \times s_1(1, 3) + s_1(1, 2) \times s_1(2, 3) \\ &\quad + s_1(1, 3) \times s_1(3, 3) + s_1(1, 4) \times s_1(4, 3) \\ &= \max\{\min[s_1(1, 1), s_1(1, 3)], \min[s_1(1, 2), s_1(2, 3)] \\ &\quad \min[s_1(1, 3), s_1(3, 3)], \min[s_1(1, 4), s_1(4, 3)]\}\end{aligned}$$

Using values of $s_1(i, j)$ in Fig. V-18(b), we obtain:

$$s_2(1, 3) = \max\{\min[1, 0], \min[0.4, 0.67], \min[0, 1],$$
$$\min[0, 1], \min[0, 0.4]\}$$
$$= \max\{0, 0.4, 0, 0, 0\} = 0.4$$

The results of completing the calculation for all functions are given in Fig. V-18(c). In general Eq. (14) relates functions via their strongest mutual "friend," which can have important effects if the first order similarity is not as strong.

Carrying the above argument a step further, we see that F_1 and F_4 still appear to be unrelated in the S_2 matrix, Fig. V-18(c), whereas Fig. V-17 shows that F_1 is related to F_4 via $(D_1 - F_2 - D_2 - F_3 - D_3)$. Consequently we calculate a third order similarity:

$$S_3 = S_1 \times S_2 = S_1 \times S_1 \times S_1 \quad (15)$$

where products and sums have the same meanings as in Eq. (14). Figure V-18(d) shows the results of calculating S_3 for the example of Fig. V-17.

In our prsent example S_3 provides transitivity of similarity; i.e., since F_1 and F_2 are related and F_2 and F_3 are related and F_3 and F_4 are related, we obtain a nonzero similarity for F_1 and F_4. In general, if S_1 is an Nth order matrix a totally transitive similarity relationship is calculated by [32]:

$$S = S_1^{N-1} \quad (16)$$

where sums and products have the same meaning as in Eq. (14), and S_1 must be reflexive ($s_1(i, i) = 1$) and symmetric ($s_1(i, j) = s_1(j, i)$).

The similarity relation S is useful for forming partitions because it ranks the similarity of functions. In the example of Fig. V-18(d) we see that the similarity of F_2 and F_3 is 0.67, whereas the similarities of F_1 and F_2 to each other and to F_3 and F_4 is only 0.4. The interpretation is that there is stronger reason to maintain F_2 and F_3 in the same partition than F_1 and F_4; F_1 and F_4 may be separated more readily. In doing allocation we would consider putting all functions together, splitting off F_1 and/or F_4, and then splitting F_2 and F_3 as a last alternative.

The similarity matrix is more formally used to construct partitions by applying the rule that all functions with $S \geq S_T$, where S_T is a given threshold, must be placed in the same partition. By varying the threshold S_T we obtain a family of candidate partitionings, as diagrammed in Fig. V-19. In concept this partitioning algorithm may be made responsive to all other criteria in Fig. V-16. For example, we may bring maintainability into play by making S_1 be proportional to the probability that F_i must be changed, given that F_j must be changed.

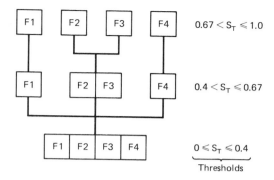

FIGURE V.19 Partitioning Alternatives as a Function of Similarity Thresholds for Similarity Matrix in Figure V.18(d)

As another example of this partitioning approach we have worked out the data accessing similarities of functions in the digital compensator example, Fig. V-10, given access frequencies listed in Fig. V-14. Figures V-20(a) and (b) are the first order and total similarities calculated according to Eqs. (14) and (16), respectively. The partitioning alternatives are presented as a function of similarity threshold in Fig. V-21.

Figure V-21 represents the results of applying our partitioning approach to the main example of this chapter. We see that the strongest partitions are (F_4, F_6) at $S = 0.65$ and (F_3, F_4, F_6, F_7) at $S = 0.45$. In this example, partitions are formed by splitting functions off a central structure. For more complex problems, we would get a more general hierarchy of partitions. The partitions may be mapped onto any number of computers (plus replications if desired). The mapping is performed by the allocation activity.

Partitioning on the basis of data accessing commonality reduces interpartition communications. This is desirable when partitions are implemented on different modules, because it reduces intermodule communications requirements. Data accessing commonality does not directly attack the problem of reducing people-costs, a primary goal of requirements engineering. As noted earlier, people-costs are largely a function of the ease of analysis and change, or maintainability, of the system. An approach to calculating a more maintainability-sensitive partitioning is described next.

We consider the situation where components of the external system (or our knowledge of them) changes. We desire a partitioning which simplifies associated changes to the computing system. First the relationships of internal change requirements to external changes are examined. In the compensator example, four external aspects are considered likely to change: actuators, the plant, sensors, and the modeling of noise characteristics in the physical process. The assumed change relationships between computing functions and external components are shown in Fig. V-22. The relative values in Fig. V-22 were determined from knowledge that F_3, F_6, and F_7 model the plant, sensors,

	F1	F2	F3	F4	F5	F6	F7	F8
F1	1	0.36	0	0	0	0	0	0
F2	0.36	1	0.38	0	0	0.33	0	0
F3	0	0.38	1	0.45	0.20	0.43	0.45	0
F4	0	0	0.45	1	0.38	0.65	0.29	0.36
F5	0	0	0.20	0.38	1	0.33	0	−0
F6	0	0.33	0.43	0.65	0.33	1	0	0
F7	0	0	0.45	0.29	0	0	1	0.36
F8	0	0	0	0.36	−0	0	0.36	1

(a) First order similarity — S_1

	F1	F2	F3	F4	F5	F6	F7	F8
F1	1	0.36	0.36	0.36	0.36	0.36	0.36	0.36
F2	0.36	1	0.38	0.38	0.38	0.38	0.38	0.36
F3	0.36	0.38	1	0.45	0.38	0.45	0.45	0.36
F4	0.36	0.38	0.45	1	0.38	0.65	0.45	0.36
F5	0.36	0.38	0.38	0.38	1	0.38	0.38	0.36
F6	0.36	0.38	0.45	0.65	0.38	1	0.45	0.36
F7	0.36	0.38	0.45	0.45	0.38	0.45	1	0.36
F8	0.36	0.36	0.36	0.36	0.36	0.36	0.36	1

(b) Total similarity — S

FIGURE V.20 Similarity Results for Compensator Example

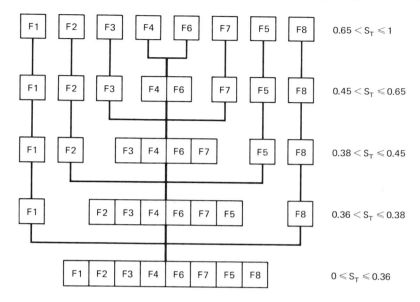

FIGURE V.21 Family of Partitions as Function of Similarity Threshold for Compensator Example

and actuators, respectively; that F_1 and F_8 interface with the sensors and actuators respectively; and from details of how F_2, F_4, and F_5 are derived. In general an in depth study is required to construct such a changeability matrix.

In addition to the direct change relationships shown in Fig. V-22, functional entities may force changes in one another. The influence of one function on another is assumed to be represented by the data access matrix, Fig. V-14. Consequently we form a total change-influence matrix by juxtaposition of the (scaled) data access matrix, Fig. V-14, and the external changeability matrix, Fig. V-22. (The data access matrix was scaled by dividing all entries by the largest number of accesses (1500); the determination of the most appropriate scale factor is a matter of more detailed analysis.) Similarity relationships were computed for the 8 × 13 matrix formed from Fig. V-14 and V-22.

The results of computing similarities based on the change-influence matrix are shown in Fig. V-23. We note that the groupings in Fig. V-23 conflict somewhat with Fig. V-21. Mainly F_6 is grouped with (F_1, F_2) and F_5 is with F_4 in Fig. V-23. In Fig. V-21, F_6 teams with F_4 strongly while F_5 is not highly similar to any other function. As we leave this example, we note that more detailed examination of using similarity to directly lower people-costs is needed.

The partitioning approach as described does not partition data entities. If a data entity is accessed only by functions within a single partition, then the

	Actuators	Plant	Sensors	Noise model
F1			0.9	
F2			0.7	0.4
F3		0.9		
F4	0.5	0.5		0.4
F5	0.5	0.5		
F6			0.9	
F7	0.9			
F8	0.9			

FIGURE V.22 Example Changeability Matrix (Assumed Relationships between Necessary Changes in Functional Entities and Changes in the External System)

data entity obviously should be assigned to that partition. The assignment of data entities which are accessed by more than one partition requires further examination. This is done in the synthesis activity.

2.3 Allocation

Candidate resources are next selected. Usually there are economic and maintainability advantages to using identical computers. In some cases customized computers may be desired for particular functions. For instance in the compensator example we might select a special computer for the plant model ($F3$), which we know (from other sources) requires relatively high throughput and performs predominantly matrix calculations. In fact, we might choose to realize $F3$ with a hardwired module if the basic plant model is unlikely to change.

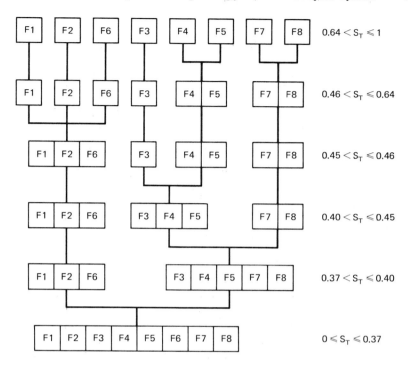

FIGURE V.23 Family of Partitions Taking into Account Change Influences

The allocation problem is applicable to solution by mathematical programming methods [33]. In particular it is naturally an integer programming problem, which is usually treated as a zero–one problem [34]. Problems of this type admit exact solutions when the number of variables is not too large (say less than 10) and the criteria and constraints are linear functions of the solution variables. The solution is then determined by explicit enumeration and evaluation of alternatives. In other cases special algorithms are required to efficiently eliminate unpromising alternatives without explicit evaluation. The branch and bound algorithm is a well known technique for this so-called implicit enumeration and evaluation.

The allocation algorithm outlined below can map either partitions or functions onto a selected number of resources (computer modules). Mapping partitions is recommended for most problems because this provides fewer alternatives. For our small example, however, we discuss mapping functions onto modules and use the similarity function to guide the mappings.

To set up the branch and board algorithm, we assign a zero–one variable to each alternative allocation decision as follows:

Let $X_{ij} = 1$ if function F_i is assigned to computer module CM_j else $X_{ij} = 0$

We then (in concept) form a decision tree as shown in Fig. V-24.† If there are M functions ($i = 1, \ldots, M$) and N modules ($j = 1, \ldots, N$), then the tree contains ($M + 1$) levels of nodes with N branches emanating from each node. At the root of the tree, all $X_{ij} = 0$. Each lower node corresponds to a decision of $X_{ij} = 1$ for a particular (i, j). For instance, following branch (k) from the root node corresponds to selecting $X_{1k} = 1$, and all other $X_{1j} = 0$. Once at the node corresponding to the decision $X_{1k} = 1$, our next decision is to assign function F_2 to a specific module, i.e., let $X_{2j} = 1$ for a specific value of (j). This takes us along the associated branch to a lower node. We continue this sequence of decisions until we reach a leaf node. The path traced out then corresponds to a complete allocation of all functions to specific modules.

Associated with each branch in the decision tree is a *cost*. For example, assigning F_1 to CM_j after assigning F_k to CM_ℓ produces a communication cost which depends on the amount of interaction between F_i and F_k and distance between CM_j and CM_ℓ. Consequently each path from root to leaf has a cost by which it can be evaluated and compared to other paths. Furthermore some paths are not feasible because they violate a constraint. For example, at least

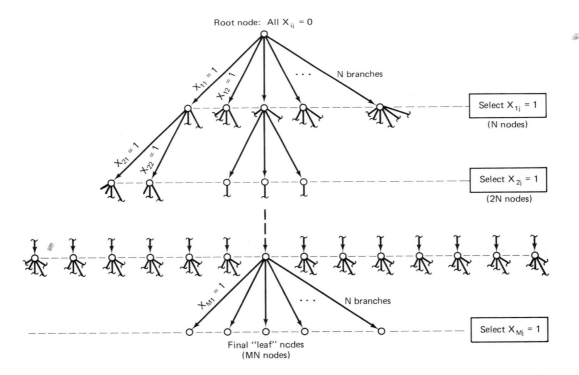

FIGURE V.24 Decision Tree Concept of Allocation

†This approach is not unique. Previously cited references describe other set-up methods.

one F must be assigned to each CM. Each F_i assigned to resource CM_j takes a portion of the resource's instruction rate and storage capacities, and eventually no new functions can be assigned.

Branch and bound algorithms search for very good (maybe optimal) solution paths by first finding a good feasible path and then backtracking to seek out improvements. The process is efficient if many potential paths can be eliminated high in the tree, before explicit evaluation. In application, therefore, the algorithm usually reorders decisions to place highest costs near the tree root and applies certain rules (or heuristics) to determine which branches are unlikely to produce good solutions. The algorithm may stop before evaluating all feasible paths, thereby yielding a suboptimal solution.

Aside from the mechanics of the branch and bound algorithm, the heart of the solution technique is the cost and constraint functions (or models). We describe some important models below:

1. Every function must be assigned to exactly one module:

$$\sum_{j=1}^{N} X_{ij} = 1 \text{ for all } i \quad (17)$$

(If we want redundant copies of a function F_5, we assign new indices—say F_9 and F_{10} are copies of F_5. Then we would constrain F_5, F_9, and F_{10} so that they cannot be assigned to the same module; see 3 below.)

2. Every module must have at least one function assigned:

$$\sum_{i=1}^{M} X_{ij} > 0 \text{ for all } j \quad (18)$$

3. Separation—F_i and F_k cannot be assigned to the same module:

$$X_{ij} + X_{kj} \leq 1 \text{ for all } j \quad (19)$$

4. Collection—F_i and F_k must be assigned to the same module:

$$X_{ij} - X_{kj} = 0 \text{ for all } j \quad (20)$$

5. No module's instruction rate (or storage) capacity can be exceeded:

$$\sum_{i=1}^{M} c_i X_{ij} - C_j \leq 0 \text{ for all } j \quad (21)$$

where c_i is the instruction rate (or storage) requirement of F_i, and C_j is the capacity of module (j).

6. The probability of losing a function F_i must be less than PL_i, given that copies of F_i are assigned to modules CM_k, $k \in K$. Let PF_k be the probability of module CM_k failing during the specified time interval.

Then the probability of all modules containing copies of F_i failing is:

$$PFF_i = \prod_k PF_k \qquad (22)$$

We require $PFF_i < PL_i$. $\qquad (23)$

7. Maximize similarity—given similarity numbers s_{ij} from the partitioning activity, then the minimum similarity of functions in module CM_k is:

$$S_k = \min_{i,j}[s_{ij} X_{ik} X_{jk}], \qquad (24)$$

and the minimum similarity over all modules would be:

$$S = \min_k S_k \qquad (25)$$

The we seek paths which maximize S.

8. Minimize module-to-module communications—let $d_{k\ell}$ be a measure of distance between module CM_k and module CM_ℓ, $d_{k\ell} = 0$ if $k = \ell$, and CV_{ij} be a measure of communication volume between function F_i and function F_j then minimize:

$$Z = \sum_{i,j,k,\ell} d_{k\ell} CV_{ij} X_{ik} X_{j\ell} \qquad (26)$$

To use these models with a branch and bound algorithm, we read in values of capacities (c_i and C_j), similarities (s_{ij}), distances ($d_{k\ell}$), communication volumes (CV_{ij}), failure probabilities (PF_k), and requirements for functional separation and collection. Then we would form solutions to maximize ($S - Z$) as calculated from Eqs. (25) and (26), subject to the constraints of Eqs. (17) to (23).

We note that the modeling equations above do not attack people-costs for testing, maintenance, etc., except by means of the similarity function. Work on such models has been reported [35, 36] but it has not been applied to requirements engineering for modular computer systems. A very simple way of indirectly reducing people-costs is to reduce allowable resource capacities C_j below actual capacities to help obtain smaller functional units and more room for growth.

Example results of performing allocation for the compensator problem (Fig. V-10) are shown in Fig. V-25. Data entities D_1 and D_3 are allocated to computer modules 1 and 2, respectively, because they are accessed only by functional entities within the modules. The other data entities remain unassigned at this point because they are accessed by multiple modules. The synthesis activity treats these unassigned data entities.

2.4 Synthesis

The goal of the synthesis activity is to represent intermodule communications requirements. The requirements are manifested by additional functional and

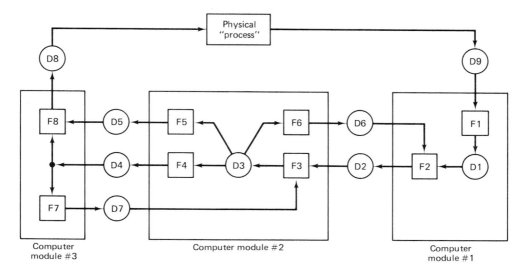

FIGURE V.25 Example Allocation of Functions to Computer Modules

data entities within communicating modules and communication paths between modules. These requirements must be explicitly accommodated in lower level design.

Before allocation, all functional and data entities are located conceptually within one computer module. After allocation some data entities must be shared by functional entities in two or more modules. Since the functional entities are not conceived to be changed by allocation, the need to pass data from one module to another creates new functional and data entities for message preparation, sending/receiving, buffering, and possibly acknowledgments. If different modules are both updating (reading and writing) a data entity, additional steps are needed to maintain data consistency.

The first step of our synthesis approach is to assume that all shared data is replicated in all modules accessing it. We represent the additional communications requirements with new data and functional entities, which we call logical buffers and logical switches. Figure V-26 is an example of the introduction of the new entities. We emphasize that Fig. V-26 is only a framework for defining requirements and that some entities may not explicitly appear in the final design.

The next step is to size the new components. During analysis we determined the data rate of D_i in number of instances per time interval ($R_d = N_d/T_d$), processing rate requirements for F_i and F_j (R_{pi} and R_{pj}), and permissible input to output delays for F_i and F_j (T_i and T_j). We size the logical switches in the same manner as illustrated for the application functions in the analysis activity: (1) describe the message preparation/reception in pseudo-

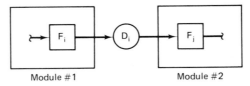

(a) Allocated functions with shared data

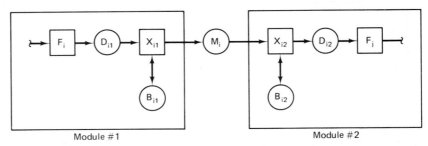

(b) Addition of logical switches (X_{i1} and X_{i2}),
logical buffers (B_{i1} and B_{i2}), and mesage M_i.
Data (D_i) is replicated (D_{i1} and D_{i2})

FIGURE V.26 Initial Representation of Communications Requirements

code; (2) estimate instructions per line of pseudocode; and (3) add up the total instructions. The sustained instruction rate for X_{i1} would then be computed (in analogy to Eq. (6)) as:

$$R_{xi1} = (I_{0xi}/T_d + I_{1xi}R_d) + R_{x0} \qquad (27)$$

where I_{0xi} and I_{1xi} are analogous to I_0 and I_1 in Eq. (1) and R_{x0} is additional control overhead associated with accessing D_{i1}, preparing the message in buffer B_{i1}, and transmitting the message M_{i1}. (Acknowledgments would increase requirements accordingly.) R_{xi2} also is calculated using Eq. (27). The processing rate requirements on the modules now must be increased by R_{xi1} and R_{xi2}. Calculation of increased storage requirements due to B_{i1} and B_{i2} is considered to be straightforward.

Where delay requirements are stated, previously apportioned delays must be reapportioned across the logical switches and the message link. The module overhead for setting up use of the link is represented in Eq. (27) by R_{x0}. The remaining link delay is primarily due to network switching and propagation delays. If this network delay is estimated (or budgeted)† to be T_n, then the reapportioned delay for F_i is

---

†The network delay T_n depends on length of message, network topology, type of switching, protocols, etc., which are largely unspecified at this point. It is appropriate to budget T_n to be less than $(T_i + T_j)/2$ in the absence of other data.

$$T'_i = \frac{R_{pi}}{R_{pi} + R_{xi1}} (T_i - T_n/2) \qquad (28)$$

and the apportioned delay for X_{i1} is

$$T_{xi1} = \frac{R_{xi1}}{R_{pi} + R_{xi1}} (T_i - T_n/2) \qquad (29)$$

where the apportionment follows the concept of Eqs. (9) and (10) and the network delay has been distributed equally between modules.

Additional processing and storage overheads and delays will be encountered when multiple modules must update the same data entity. Consistency of updating is gained by enforcing mutually exclusive access to the data, as with monitors [37], or by allowing replicated data entities and providing a means of arbitrating conflicting update requests [38]. The monitor approach is simpler and easier to size during requirements engineering. The same sizing approach as described earlier can be applied to determine monitor overhead requirements. Delays caused by processes waiting to access data are much more difficult to assess accurately. Simulation of system operation with monitors may be used for this evaluation.

3. CONCLUDING REMARKS

An approach to requirements engineering, or high-level design, of modular computer systems was described in this chapter. Requirements engineering provides specifications of the numbers and characteristics of computing modules to be employed, the logical interconnection topology (message paths between modules), the applications software modules to be mapped onto given computing modules, and requirements for intermodule control and communication to satisfy a given application or class of applications. The approach yields functional and performance requirements for each module, but it does not define the internal structures of modules. We suggest (without proof) that the same approach can be used to define each module's internal elements and interelement network.

The approach contains four major activities: analysis, partitioning, allocation, and synthesis. In analysis we apply conventional structured design principles to develop a functional computing structure for the given application. We do not assume a hierarchical control structure as in conventional designs, however, because hierarchical control usually constrains concurrent operations of the modules. Also we develop requirements not only for the applications software, as in conventional design, but for the overall module architecture and control system.

Partitioning and allocation activities yield the module architecture and map requirements onto the modules. These activities are much more complex than any analogous steps in conventional, nondistributed system design. Our approaches for partitioning and allocation are based on resource assignment techniques used in operations research and general systems design. The main issues are definition of models to relate partitioning/allocation decisions to requirements and derivation of algorithms to handle the very complex decisions necessary. We present a number of models in current use and describe two types of assignment algorithms, fuzzy clustering and branch and bound.

The task of synthesis is to explicitly define functional and performance requirements for intermodule communications. These requirements are important to design of the physical intermodule network and internal design of the modules.

The guiding motivation of requirements engineering is reduction of people-costs. This motivation differs from usual concerns with hardware costs. Typically at lower levels, the criteria for good modular system design are stated to be:

1. Minimum number of different types of elements used to construct modules: Costs of LSI modular architectures are thereby reduced, assuming that control and communication logic are not made overly complex.
2. Minimum number of logical interconnections between modules and elements: Communications complexity and delays are strongly proportional to the number of logical interconnections, i.e., the number of elements which must be connected.
3. Maximum nonsynchronous concurrency of elements: Adequate potential processing capacity can thereby be achieved with slower, cheaper modules.

We note that both the people-cost and hardware cost reduction criteria yield the same guidelines: Develop modules with low interactions and high potential concurrency.

Finally, we note the need for more complete and refined computing aids and languages for modular system design.

REFERENCES

[1] ALFORD, MACK W., "A Requirements Engineering Methodology for Real-Time Processing Requirements," *IEEE Trans. on Software Engineering*, **SE3**, *1*, January, 1977, pp. 60–69.

[2] RAMAMOORTHY, C. V., and H. H. So, "Software Requirements and Specifications: Status and Perspectives," in *Tutorial: Software Methodology*, RAMAMOORTHY, C. V. and RAYMOND T. YEH, IEEE Comp. Soc., EHO 142-0, 1978, pp. 43–164; also reprinted in MARIANI and PALMER, *Tutorial: Distributed System Design*.

[3] YOURDON EDWARD, and LARRY L. CONSTANTINE, *Structured Design: Fundamentals of a Discipline of Computer Program and Systems Design*, Englewood Cliffs, N.J.: Prentice-Hall, Inc., 1979.

[4] PALMER DAVID F., and W. MICHAEL DENNY, "Distributed Data Processing Requirements Engineering: High Level DDP Design," *Proc. COMPSAC/78*, IEEE Comp. Soc., 78CH1338-3C, 1978, pp. 352–357; also reprinted in MARIANI and PALMER, *Tutorial: Distributed System Design*.

[5] MARIANI MICHAEL P., and DAVID F. PALMER, *Tutorial: Distributed System Design*, Long Beach, Cal.: IEEE Computer Society, 1979, IEEE Catalog No. EHO 151-1.

[6] SPILLERS, W. R., ed., *Basic Questions of Design Theory*, New York, N.Y., North-Holland Co., 1974.

[7] JONES, J. C. *Design Methods: Seeds of Human Futures*, New York, N.Y., Wiley-Interscience, 1970.

[8] PETERS LAWRENCE J., and LEONARD L. TRIPP, "A Model of Software Engineering," *Proc. Third Intl. Conf. on Software Engineering*, IEEE Computer Soc. 78CH1317-7C, 1978, pp. 63–70; also reprinted in MARIANI and PALMER, *Tutorial: Distributed System Design*.

[9] BOEHM, BARRY W., "Software Engineering," *IEEE Trans. on Computers*, **C25**, *12*, December, 1976, pp. 1226–1241.

[10] BROOKS, FREDERICK P., Jr., *The Mythical Man-Month: Essays on Software Engineering* (Reading, Mass. Addison-Wesley, 1975).

[11] ALFORD, M. W., "Software Requirements Engineering Methodology (SREM) at the Age of Two," *Proc. COMPSAC/78*, IEEE Comp. Soc., 78CH1338-3C, 1978, pp. 332–339.

[12] TEICHROEW DANIEL, and ERNEST A. HERSHEY III, "PSL/PSA: A Computer-Aided Technique for Structured Documentation and Analysis of Information Processing Systems," *IEEE Trans. on Software Engineering*, **SE3**, *1*, January 1977, pp. 41–48.

[13] PARNAS, D. L., *Use of Abstract Interfaces in the Development of Software for Embedded Computer Systems*, NRL Report 8047, Washington, D.C., June 3, 1977 (AD043369).

[14] DIJKSTRA, EDSGER W., *A Discipline of Programming*, Englewood Cliffs, New Jersey: Prentice-Hall, Inc., 1976, Chap. 4.

[15] YOURDON and CONSTANTINE, *Structured Design*, p. 45.

[16] ATHANS, MICHAEL, "The Role and Use of the Stochastic Linear-Quadratic-Gaussian Problem in Control System Design," *IEEE Trans. on Automatic Control*, **AC16**, *6*, December 1971, pp. 529–552.

[17] SCHNEIDER, G. MICHAEL, STEVEN W. WEINGART, and DAVID M. PERLMAN, *An Introduction to Programming and Problem Solving with PASCAL*, New York, New York: John Wiley & Sons, 1978, Chap. 2.

[18] HOLT, R. C., et al, *Structured Concurrent Programming with Operating System Applications*, Reading, Mass. Addison-Wesley, 1978.

[19] BRINCH-HANSEN, PER, "Concurrent Programming Concepts," *Computing Surveys*, **5**, *4*, December 1973, pp. 223–245.

[20] DODSON, E. N., et al, *Advanced Cost Estimating and Synthesis Techniques for Avionics Data Processing Software and Hardware*, General Research Corporation Final Report No. CR-1-701, December 1976.

[21] LAMPORT, LESLIE, "Time, Clocks, and the Ordering of Events in a Distributed System," *Com. of ACM*, **21**, *7*, July 1978, pp. 558–565.

[22] CHANDY K. M. and JAYADEO MISRA, "Distributed Simulation: A Case Study in Design and Verification of Distributed Programs," *IEEE Trans. on Software Engineering*, **SE5**, *5*, September, 1979, pp. 440–452.

[23] ABRAMOWITZ MILTON, and IRENE A. SEGUN (Eds.), *Handbook of Mathematical Functions*, New York, N.Y., Dover Publications, Inc., 1965 Table 24.4, p. 835.

[24] YOURDON and CONSTANTINE, *Structured Design*, Chap. 14.

[25] BOND ALBERT F., and PETER C. BELFORD, "Approach to a Distributed Data Processing Architecture Methodology," *Proc. Sixth Texas Conf. on Computing Systems*, University of Texas, Austin, Tex., November, 1977 pp. 1B-13–1B-35.

[26] BUCKLES B. P., and D. M. HARDIN, "Partitioning and Allocation of Logical Resources in a Distributed Computing Environment," in M. P. MARIANI and D. F. PALMER, *Tutorial: Distributed System Design*, IEEE Computer Soc., EH0151-1, 1979, pp. 247–276.

[27] GYLYS V. B., and J. A. EDWARDS, "Optimal Partitioning of Workload for Distributed Systems," *Proc. COMPCON/76 Fall*, IEEE Computer Society, 76CH1115-5C, 1976, pp. 353–357; also reprinted in MARIANI and PALMER, *Tutorial: Distributed System Design*.

[28] MORGAN HOWARD L., and K. DAN LEVIN, "Optimal Programs and Data Locations in Computer Networks," *Com. ACM*, **20**, *5*, May 1977, pp. 315–322; also reprinted in MARIANI and PALMER, *Tutorial: Distributed System Design*.

[29] LEVIN K. DAN, and HOWARD LEE MORGAN, "Optimizing Distributed Data Bases—A Framework for Research," *AFIPS Conf. Proc.*, **44**, 1975, pp. 473–478; also reprinted in MARIANI and PALMER, *Tutorial: Distributed System Design*.

[30] BOORSTYN, ROBERT R., and HOWARD FRANK, "Large Scale Network Topological Optimization," *IEEE Trans. on Communications*, **COM25**, *1*, January 1977, pp. 37–55; also reprinted in MARIANI and PALMER, *Tutorial: Distributed System Design*.

[31] STONE, HAROLD S., "Multiprocessing Scheduling with the Aid of Network Flow Algorithms," *IEEE Trans. on Software Engineering*, **SE3**, *1*, January 1977, pp. 85–93; also reprinted in MARIANI and PALMER, *Tutorial: Distributed System Design*.

[32] NEGOITA, C. V., and D. A. RALESCU, *Applications of Fuzzy Sets to Systems Analysis*, New York, New York: John Wiley & Sons, Inc., 1975, Chap. 7.

[33] MCMILLAN, CLAUDE, Jr., *Mathematical Programming: An Introduction to the Design and Application of Optimal Decision Machines*, New York, N.Y., John Wiley & Sons, Inc., 1970.

[34] TAHA, HAMDY A., *Integer Programming: Theory, Applications, and Computations*, New York, N.Y., Academic Press, Inc., 1975.

[35] UHRIG, J. L., "Life-Cycle Evaluation of System Partitioning," *Proc. COMPSAC/77*, IEEE Computer Society No. 77CH1291-4C, November 1977, pp. 2–8.

[36] UHRIG, J. L., "Mathematical Programming Approaches to System Partitioning," *IEEE Trans. on Systems, Man, and Cybernetics*, **SMC8**, *7*, July 1978, pp. 540–548.

[37] BRINCH-HANSEN, PER, *The Architecture of Concurrent Programs*, Englewood Cliffs, N.J.: Prentice-Hall, Inc., 1977.

[38] THOMAS, ROBERT H., " A Solution to the Concurrency Control Problem for Multiple Copy Data Bases," *Proc. COMPCON 78 Spring*, IEEE Computer Society.

Chapter VI

Design and Diagnosis of Reconfigurable Modular Digital Systems

Stephen Y. H. Su

and

Yu-I Hsieh

PREVIEW

With the advances of integrated circuit technology into VLSI (very large scale integration) in which a million or more components are being fabricated into a single IC (integrated circuit) chip, three problems have become more important than ever:

1. The ability to design a digital system that stays operational in spite of the failure of software or hardware components.
2. The ability to provide a way to evaluate the system's reliability.
3. The ability to test digital systems.

This chapter deals with these three important topics. The chapter is self-contained so that readers with no background or little background in fault-tolerant computing can also understand and benefit from it.

Computers are being used in various areas of military applications, such as controlling spacecraft, missiles, and ships. Failures of a single component may result in a disaster, such as the loss of the lives of astronauts or billions of dollars of research equipment. Digital instruments are being used to monitor hospital patients. The failure of these instruments may cost the lives of patients. Due to the wide usage of computers in these and many other important tasks, fault-tolerance becomes more important than ever.

Since the 1960s, the powerful fault-tolerant computing systems have been developed for two major applications: to satisfy the need of highly reliable computers in the various aerospace projects, and for ground-based real time control applications. It is often impossible to perform manual maintenance on a complex on-board aerospace computer. They must perform properly throughout entire missions. On the other hand, the real time control systems and/or telephone switching systems can allow manual repairing.

This chapter provides the readers with the basic ideas used in various fault-tolerant computers. We summarize and explain those ideas, concepts, and techniques used in various fault-tolerant computing systems such as Bell No. 1 ESS, SIFT, etc., instead of introducing a specific system in detail. The reason we do this is that we feel that it is more important to provide readers with the fundamental concepts, to help readers to understand the existing techniques, and point out the possible future trends in fault-tolerant computing.

The chapter contains a comprehensive coverage of the new look in the fault-tolerant architectures as well as many basic concepts and definitions in such important fields such as hardware/software fault-tolerant structures, software fault-tolerant operating systems, detection, location, treatment, recovery, and reconfiguration as well as system diagnosis.

The chapter is organized into five sections. Section 1 gives the fundamentals and background in the diagnosis and fault-tolerant design of digital systems. Due to the rapid reduction of the cost of hardware, fault-tolerance using redundancy techniques becomes more feasible than ever. Section 2 presents various hardware redundancy schemes. It also briefly covers software fault-tolerance. It is important to be able to know how reliable a system is. Section 3 presents and compares the reliability and reliability-related measures in various fault-tolerant systems. The area of system level diagnosis needs more research. Section 4 provides a theoretical basis for attacking problems in this area. In the final section, we point out some unsolved problems in the area of fault-tolerant computing systems.

1. INTRODUCTION

Increasing interest and attention have been given to the areas of fault-tolerant design—the design of digital systems which can stay operational in spite of the failure of some of their components (software or hardware)—and diagnosis of digital systems. This chapter treats these two topics. The chapter is organized into five sections. The first section presents the fundamentals, basic concepts, and background on the diagnosis and fault-tolerant design of digital systems.

Section 2 deals with schemes for using redundant hardware and software to provide fault-tolerance. The following hardware redundancy schemes will be discussed: static, dynamic, hybrid, NMR (N-modular redundancy)/-bipurge, self-purging, TMR/spares, and hardware redundancy reconfiguration [1]. Section 2 also discusses some software fault-tolerant techniques [2] such as error detection, error treatment, damage assessment, and error recovery including forward and backward recovery.

In Section 3, a Markov model is utilized for finding the reliabilities of hardware redundancy fault-tolerant systems as well as graceful degradation systems. Reliability related measures such as mean time before failure (MTBF), availability, and hazard functions for these systems are also discussed. Some extension of the Markov model is made for the analysis of computation in fault-tolerant computer systems. Computation related measures such as computation reliability, mean computation before failure, and computation availability with several examples are given.

Section 4 gives the analyses of system diagnosis by the use of Preparata's graph theoretical model. Several diagnostic measures such as t-fault diagnosability with or without repair, one-step t/s *diagnosability*, t_i diagnosability, concurrent one-step t-fault diagnosability, fault-tolerant graphs, and reconfiguration trees are introduced and analyzed by examples.

In Section 5, we point out some unsolved problems in the area of fault-tolerant modular digital systems.

A digital system can be viewed as a set of computing modules together with their interrelationships. In a digital system, a *failure* is defined as an *event* in which the system does not perform its service in the manner specified (i.e., the system violates its specifications). An *error* is defined as an item of information which when processed by the normal algorithms of the system, will produce a failure, but the failure will disappear whenever the error is removed by error recovery algorithms. A *fault* is the physical or algorithmic cause of an error [3]. Obviously, the failure of components may produce some wrong information (i.e., error) in the system, so failure is a physical cause of this error and is called a physical fault of the system. Thus from the system point of view, a component failure is a fault. In summary, the relation among component failure, faults, errors, and system failures can be expressed by component failure → fault → error → system failure where "→" denotes "causes." To show the differences among faults, errors, and failures, the following two examples are given: one for hardware and another for software.

Example 1. In Fig. VI-1, breakdown damage on a transistor inside gate $G1$ may produce an error in which $G1$ has the value of logic 1 permanently. We say that the damage in the transistor is a *fault* which causes a stuck-at-1 *error*. If the *error* propagates to the output f, then the *error* causes

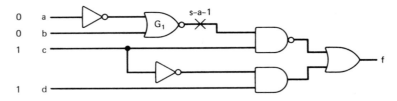

FIGURE VI.1 A combinatorial circuit with a s-a-l fault

system *failure*. In order to detect this error, we have to find the failure, produced by this error, at the output site of this circuit. If we apply the input pattern $a = b = 0$ and $c = d = 1$, then the output value is logical 0 when the error exists, and 1 when it does not exist. Hence a failure has been found, and the input pattern ($a = b = 0$, $c = d = 1$) is called a *test* for this error. ∎

Example 2. Let us consider a program P containing a subprogram Q implementing a square root algorithm for real numbers. If there is a wrong statement in Q, i.e., a design fault, then when Q accepts an input data, it will generate a wrong output, i.e., an *error* will be produced by this design *fault*. This error may be detected by a dedicated algorithm (e.g., an acceptance test routine). If this error can be removed by a recovery algorithm (e.g., rollback and reconfiguration) or be masked by an algorithm (e.g., software redundancy), then there is no *failure*; otherwise, the error in Q will be propagated to the output of program P and cause program *failure*. ∎

Figure VI-2 shows that faults can be classified into two types: physical faults and algorithmic faults. Hardware component faults can be classified by their *duration* (permanent or solid faults versus transient faults [4]), their *extents* (local faults versus distributed faults) or their *values* (determinate faults versus indeterminate faults). One important determinate fault is called the *logical lead fault* or *stuck-type fault*. It causes the input and/or output leads leads of each gate to become fixed at a constant value, either logical one (in this case, the signal at the site of fault is said to be stuck-at-one (denotated by s-a-1) or stuck-at-zero (denoted by s-a-0). A *bridging fault* [5, 6] means that two or more lines are shorted (wired) together. Two types of bridging faults are considered in the literature: the AND-type and OR-type meaning that two or more lines are short-circuited to form AND and OR logical operations, respectively. Bridging faults may be either solid or transient. Software errors can be caused by either physical faults (hardware component faults) or algorithmic faults. According to the different software development phases, algorithmic faults for software include *requirement faults, specification faults, design faults, construction faults,* and *verification faults*. An error is called a *functional error* if it affects the functional transformation relation of a computing module. A *control error* can influence the interrelationships among the computing modules (e.g., causes information to flow

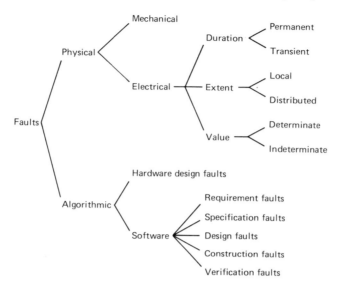

FIGURE VI.2 Classification of faults

into a wrong module). A *data error* affects the system states, i.e., contents of memories and registers. It is easier to detect functional error than control and data errors. For a functional error, a *test* of the error is input data which produce different output values from the normal output values [7]. Avizienis [8] defined a fault-tolerant compuuter system as a system possessing the following three attributes:

1. It consists of a set of hardware and software components.
2. It is initially free from both hardware and software design errors.
3. It executes programs correctly in the presence of faults.

Fault-tolerant computing is defined as the ability to execute a specified algorithm regardless of hardware and/or software failures. Recent tutorial and research papers on fault-tolerant computing can be found in [9, 10, and 11]. An overview on fault-tolerance in multiprocessor systems is given in [12].

In general, a reconfiguration can be *software-controlled,* (i.e., one that is effected by control codes), *hardware-controlled,* (i.e., one that is controlled entirely by hardware devices (e.g., self-purging systems)), or *manually controlled* (i.e., one that is controlled by human operators). By writing control codes to (micro) instructions, a programmer may establish new interconnections among computer devices (registers, counters, adders, flip-flops, logical circuits). Furthermore, in all interconnection networks, new data paths can be established via writing new control codes to connecting elements (switches). A fault-tolerant reconfiguration is also either software-controlled, hardware-controlled, or manually controlled. Software (hardware) control is understood as writing control codes (using hardware devices) to isolate faulty circuits (modules) from other circuits in the system. Manual control means physical disconnection of faulty modules from the system. The objective of reconfiguration is to cause a system to return to previously fault-free state. Also, reconfiguration need not repair damage but prevent further damage.

2. REDUNDANCY TECHNIQUES FOR FAULT-TOLERANT DESIGN

Fault-tolerance can be achieved by means of protective redundancy which is introduced into a computer system in the following forms:

1. Hardware redundancy (utilization of additional hardware)
2. Software redundancy (utilization of additional software)
3. Time redundancy (repetition of machine operations)

In the following sections, we shall discuss each one of these three redundancy techniques.

2.1 Hardware Redundancy

For hardware redundancy, replication of components can be done at various levels. Circuit level redundancy is achieved by replicating circuit elements such as diodes, transistors, resistors, etc., and interconnecting them in such a way as to *mask errors* caused by the faulty elements. For instance, Fig. VI-3 is a quadded diode that can tolerate any single diode failure, such as $D1$ open, $D2$ short, or $D4$ short.

Successful application of this approach for building fault-tolerant computers has been reported for the on-board computer for NASA's Orbital Astronomical Observatory [13]. Gate level redundancy is achieved by replicating elementary gates, and interconnected together to give the desired error correcting capability. In an LSI (large scale integration) chip, several components or gates are likely to become faulty at the same time. Therefore, chip or module level redundancy is important. In the following subsections, we will examine various hardware redundancy schemes.

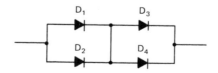

FIGURE VI.3 Quadded diode

2.1.1 Static Redundancy

A static redundancy scheme uses identical modules feeding a voter. The output takes a majority vote to provide the correct output when the majority number of modules are fault-free. A static redundant system with N modules is called NMR. Figures VI-4 (a) and (b) show triple modular redundancy (TMR) systems with nonredundant voter and redundant voters, respectively. The voters are being used in the following four types of structures:

1. Unidirectional single voter (Fig. VI-4 (a)).
2. Unidirectional redundant voters (Fig. VI-4 (b))
3. Bidirectional single voter (Fig. VI-5)
4. Bidirectional redundant voters.

Example 3. The C.vmp (a *voted multip*rocessor system) is a triplicated NMOS LSI-11 microprocessor with voting at the bus level [14]. Three

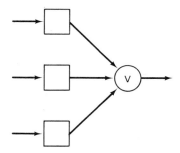

(a) With non-redundant voters

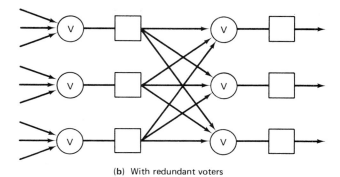

(b) With redundant voters

FIGURE VI.4 Von Neumann's multiplexing scheme

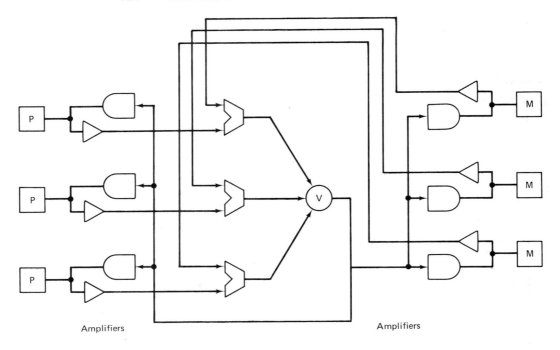

FIGURE VI.5 C.vmp voter circuit

processor-memory pairs, each pair connected via a voter circuit as shown in Fig. VI-5. Any disagreements among the memories will not propagate to the processors and vice versa. A straightforward modification of Fig. VI-5 can produce bidirectional triplicated voters; you are encouraged to draw the detail circuit. Voting is done in parallel on a bit-by-bit basis. A computer can have a failure on a certain bit in one bus, and proper operation will continue, provided that the other two busses have the correct information for that bit. ∎

Example 4. C.vmp can be viewed as a TMR CPU-voter-memory configuration. Alternatively, the triplicated undirectional voters can be built in a microprocesor system with the so-called TMR CPU-memory-voter configuration [15]. For example, Fig. VI-6 shows a microcomputer system using TMR CPU-memory-voter redundancy. Since the outputs of memory modules go through voters before feeding into the data inputs of the microprocessors, any error generated from one memory module is masked. If those registers in the microprocessors have incorrect contents due to transient (intermittent) faults, their contents will be corrected when they are loaded with new data from memory. Similarly, if the contents of memory modules are incorrect due to transient faults, they will be corrected when new data is loaded into memory modules from the microprocessors. In order to guarantee that the

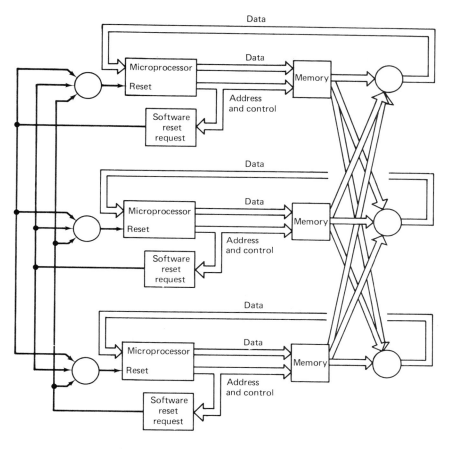

FIGURE VI.6 TMR with CPU-memory-voter

microprocessors in the same states and the memory modules have the same contents, a resynchronization subroutine can be invoked and executed. A software reset request device generates the interrupt signal and provides the starting address of the resynchronization subroutines.

The advantages of static redundancy are as follows:

1. It is easy to implement, hence the design cost is low.
2. It can mask errors instantaneously allowing programs to execute or devices to perform their functions without interruption.
3. There is no need for an error detection procedure.
4. It is easy to provide for occasional resynchronization.

The disadvantages of static redundancy are as follows:

1. All units need to be powered.
2. Replication might cause excessive circuit loading. This means that it is necessary to maintain excessive fanout capabilities of logical components in order to maintain adequate forms of signals in all logical circuits due to the replication of active modules. This might require introduction of additional active components (amplifiers) in addition to those that perform logical functions.
3. Modules are not isolated from each other since they physically connect to the voter. Faulty modules are not detected. ∎

2.1.2 Dynamic redundancy

A *dynamic redundant system* [16] consists of *one* active module with several spares, a fault detection device to detect the failure of the active module, and a switching network for switching out the faulty module and switching in the spare. In a *general dynamic redundant system* [16] more than one active module is used to perform parallel computation.

The dynamic redundancy has several advantages:

1. Only the active module needs to be powered.
2. The system survives until all spares and the active module are faulty.
3. There is good isolation among the active module and the spares.

The main disadvantages of a dynamic redundant system are as follows:

1. Error detection and recovery mechanisms are complex and cause delay.
2. The system is down if the error detecting device fails.

Permanent fault-tolerance is implemented by three distinct actions: error detection, error location, and reconfiguration.

2.1.2.1 Error detection

Error detection is the process of finding that something is wrong in the system but the error location is unknown. Error location is the process of pinpointing the exact site of failure.

Because of the requirement for high reliability and data security, fault-tolerant computer systems usually employ extensive hardware, software, and firmware checking mechanisms. It is important to detect an error as soon as it occurs. This allows the system to limit the extent of damage and makes the subsequent recover easier. *Concurrent detection* [8] of errors means that errors can be detected immediately when they occur. At the present time, concurrent detection of errors can only be achieved with hardware mechanisms such as parity checks of words stored in the memory (the simplest error detection code within a computer) or *framing checks* on communication links (among computers).

In data transmission, the specific format for each data transmission block is called a *frame*. The actual data frame which is transmitted over the communicated link is built up from the *text* by adding a sequence of header fields and finally trial fields. The frame shown in Fig. VI-7 is a SDLC frame (Synchronous Data Link Control protocol) which is implemented on IBM SNA (System Network Architecture) [17]. The *flag fields* indicate the beginning and the end of each frame which also is called the synchronization field of the frame. The *address field* is used to designate the receiver station(s). (One address may designate more than one station in the case of broadcasting communication.) The *control field* indicates whether the frame transmitted is a sequenced information frame, a nonsequenced information frame, or a supervisory frame. All sequenced information frames are transmitted and received sequentially. Nonsequenced information frames, each with a sequence number, are transmitted randomly and are assembled by the receiver according to these sequence numbers. The supervisory frames contain no information and are used to designate ready or busy conditions or for special purposes. The *text* has variable length and is inserted between the control field and the frame check sequence (FCS) field. The *frame check sequence field* (called block check field) serves as an error checking field. The transmitter makes a mathematical transformation (e.g., cyclic redundancy checking, check sum, or other methods) of all bits in a given frame, except the FCS and flag fields, and inserts the result of the transformation into the FCS field. The receiver makes the same computation (just like those the transmitter has done) and compares the computed value with the received value found in the FCS field. If these two values do not match, the receiver rejects the frame since it contains errors. This kind of checking is called *frame checking* on communication links. The use of error detection and error correction codes is a special form of mathematical transformation in frame check.

Software checking of CPU and memory can be accomplished by running diagnostic programs periodically, although without a *concurrent error detection capability* since some delays must exist between the occurrence of errors and the detection of them. On systems such as the PRIME system [18], diagnostic sequences are implemented in microcodes. This technique is called *microdiagnosis*.

Flag	Address	Control	Text	Sequence frame check	Flag
0 1 1 1 1 1 1 0	8 bits	8 bits	variable length	16 bits	0 1 1 1 1 1 1 0

FIGURE VI.7 Synchronous data link control (SDLC) frame

2.1.2.2 Error location

When an error is detected, the system has to locate the error so that it can either repair the damage or reconfigure to eliminate the cause of trouble. This is generally known as *fault diagnosis* or *fault isolation*. The level of diagnosis depends on the system design as well as maintainance policy. For dynamic redundant systems, it is sufficient to locate the fault to the level of a replaceable unit so that subsequent reconfiguration and replacement can be achieved.

2.1.2.3 Reconfiguration and recovery

After a fault has been detected and located to the level of a replaceable module, the system must initiate actions in order to eliminate further disturbances. Probably the first operational computer with full self-repair provisions was the STAR computer, built in 1971 [19]. An example of a system with extensive dynamic redundancy with *software and human support* is the No. 1 ESS system (*E*lectronic *S*witching *S*ystem) [20]. The spare modules do not participate in the computing process and hence are passive modules. Based on the experimental results reported in [21], by keeping the spares in a powered-off state, lower failure rates can be achieved, even though additional circuitry would be needed to turn on the spares under system control. The reconfiguraion and recovery include the reloading of lost programs and data from backup files, the repairing of the damaged data base, and the rescheduling of disrupted tasks.

2.1.3 Hybrid redundancy [22]

In the hybrid redundancy, the N modular redundancy (NMR) core is used with a set of spares. The spares may be kept powered off or may be powered. When one of the N modules fails, it is switched out and one of the spares in switched in. This process continues until all spares fail. The hybrid redundancy scheme has the advantages of both static and dynamic redundancy schemes, but the switching network is complex. Figure VI-8 shows the hybrid redundancy with an NMR core and standby spares. The design of the switch network depends on the types of the voter, i.e., threshold voter or majority voter. A *threshold voter* with n inputs and a threshold value of t has an output of logic 1 if and only if the number of 1's in input lines is greater than or equal to t. A *majority voter* with n inputs (n must be an odd integer) has an output of logic 1 if and only if the number of 1's in input lines is greater than or equal to $(n + 1)/2$. When $t = (n + 1)/2$, a threshold voter becomes a majority voter.

Example 5. Figure VI-9 is a hybrid redundant system with a TMR core and two standby spares. The *disagreement detector* is implemented by five exclusive-OR gates. An output of logic 1 means disagreement and an

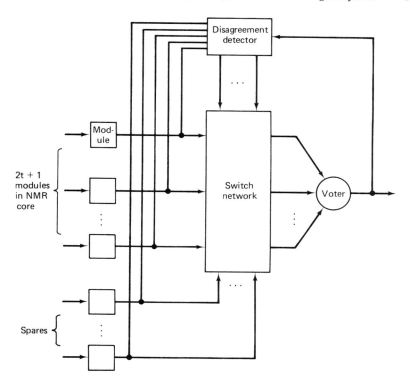

FIGURE VI.8 Hybrid redundancy with an NMR core and standby spares

output of logic 0 means agreement. The voter is a majority gate with three inputs.

For simplicity, only the functions of the switching network are described. Interested readers are referred to [23]. The switching *network* consists of three parts:

1. *Conditional Flip-Flops (C-FF)* The output C_i has a logic value 1 (0) means that the module i is functional (faulty).
2. *Cell Array (Cell 1 to Cell 5)* All cells in cell array are used to determine the first three fault-free modules.
3. *Interconnection Logic* The ith module of the first three fault-free modules determined by cell array is connected to the ith voter input through the interconnection logic where $1 \le i \le 3$. ∎

2.1.4 N-Modular Redundancy/Bipurge

In the NMR/bipurge or (NMR-simplex) system, the initial system configuration consists of an NMR core along with a switching unit. When

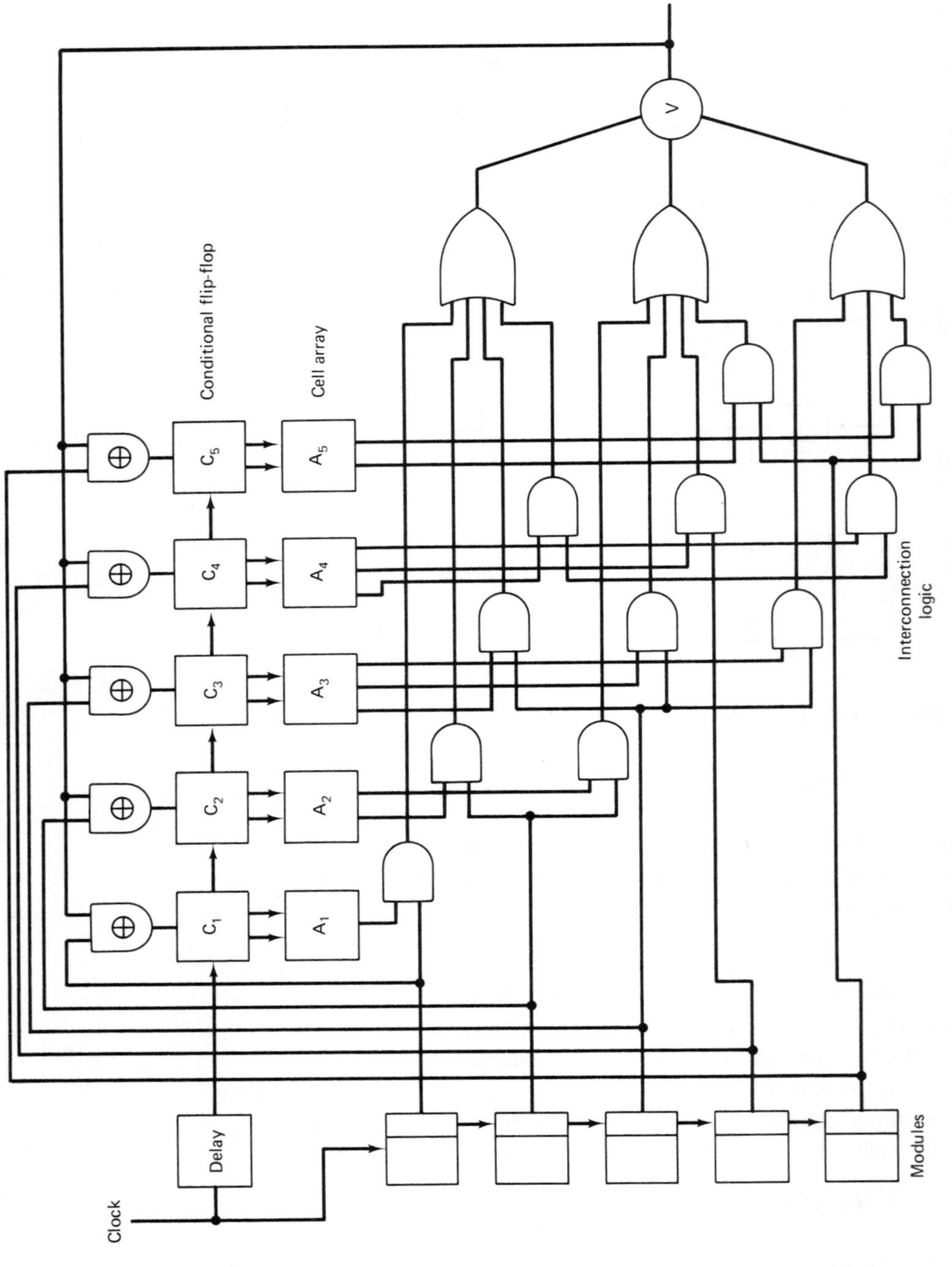

FIGURE VI.9 Hybrid redundant system

failure in a module is detected by the system (through comparing module outputs with the voter output), the failed module is switched out along with a good module to keep the number of active modules odd (because a majority voter is used). Thus the system actually degrades from a NMR to a $(N - 2)$MR, finally reaching a simplex mode of operation. This purging action can be achieved by suitable voter design.

Figure VI-10 gives a design [24] for a TMR/bipurge system $(N = 3)$. The module outputs A_i $(i = 0, 1, 2)$ in the TMR core are fed through OR gates to the majority voter.

$$V_{out} = \overline{Q_0}\overline{Q_1} (A_0 + Q_1)(A_1 + Q_2) + \overline{Q_1}\overline{Q_2} (A_1 + Q_2)(A_2 + Q_0) \\ + \overline{Q_2}\overline{Q_0} (A_2 + Q_0)(A_0 + Q_1)$$

In the fault-free state all three JK flip-flops are reset ($Q_i = 0$), then V_{out} is

$$V_{out} = A_0 A_1 + A_1 A_2 + A_2 A_0$$

However, an error on an output A_i will cause a disagreement signal which sets the ith flip-flop ($Q_i = 1$). Now assume $i = 1$, then $Q_1 = 1$, $Q_0 = Q_2 = 0$ and hence $V_{out} = A_2$. This means that the majority voter has purged out the faulty module (1) and the fault-free module (0). In general case, it purges out the faulty module i and the fault-free module $(i - 1) \mod 3$.

Similarly, for a 5MR/bipurge system,

$$V_{out} = \Sigma \overline{Q_i}\overline{Q_j}\overline{Q_k} (A_i + Q_j)(A_j + Q_k)(A_k + Q_{k+1 \,(mod\, 5)})$$
↑
all possible combinations of $i, j, k \in \{0, 1, \ldots, 4\}$

The verification of the above equation is left as an exercise for the reader.

2.1.5 N-Modular Redundancy-Unipurge [25, 26]

The NMR/bipurge technique can tolerate only the failure of $(N - 1)/2$ modules, if module failures occur sequentially one at a time. This is because a good module is purged out together with a faulty module. The NMR/unipurge scheme (also referred to as the selfpurging redundancy scheme) [25, 26], on the other hand, purges out only the failed module(s). Figure VI-11 is a self-purging system with P modules and a voter with threshold M. The voter output is 1 if and only if the weighted sum of its inputs is equal to or greater than its threshold M. Thus a 0 on one of the voter inputs does not influence the voter output as long as at least M modules are fault-free.

The E.S.i. unit in Fig. VI-11 is called the *Elementary Switch* unit for selfpurging systems [26]. Each module is associated with an elementary switch. The elementary switches allow the output signals of good modules to go to the threshold gate so that they can participate in voting, while preventing the failed modules from participating in voting by forcing 0's to the threshold

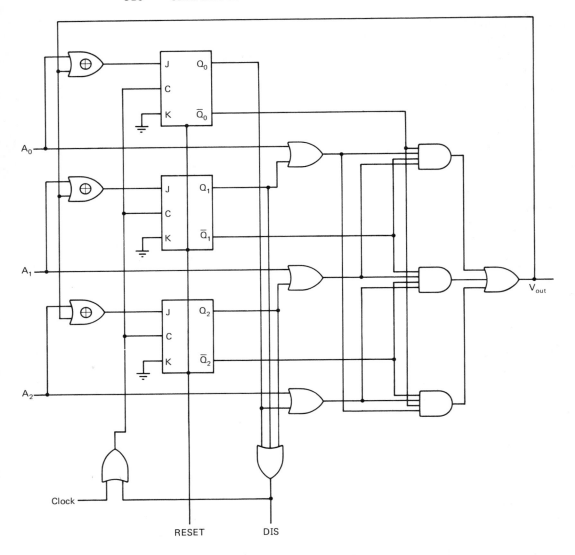

FIGURE VI.10 A voter design for a TMR/bipurge system

gate. Elementary switches are very simple. They need only detect the *first* disagreement between the module output and the voter output, and force the faulty module output to a logic 0. An elementary switch can be realized as shown in Fig. VI-12 with an exclusive -OR gate, a flip-flop, and an AND gate.

The elementary switch with its module can be constructed within a single unit called *modified module*. The initialization signal set the *asynchronous flip-flop* to an output value 1 ($Q = 1$) to allow the module output to propagate

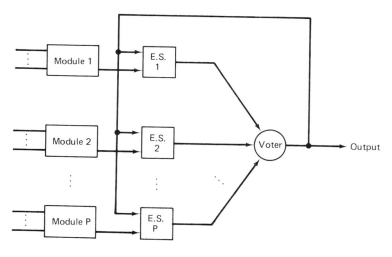

FIGURE VI.11 Self-purging system with P modules and a voter threshold of M

to the threshold gate. When there is a disagreement between the voter ouput and the module output, then the exclusive-OR gate (disagreement detector) resets the flip-flop ($Q = 0$), producing a logic value of 0 at the voter. For transient (intermittent) errors, the *retry signal* is used so that the module can be reused if the error disappears. A counter can be added in each elementary switch to keep track of the number of disagreements N_d within a specified period of time T_s. If the number of disagreements is greater than N_d within T_s, the module is declared faulty and is purged out. In general, the voter threshold value M must be greater than or equal to 2 and less than $P - 1$.

2.1.6 TMR/Spares [27]

In a TMR/spares system, when one of three active modules fails, the system will degrade into a *duplex mode* of operation with the voter degrading

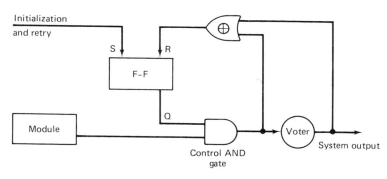

FIGURE VI.12 Elementary switch for self-purging systems

into a comparator. When a disagreement of the two remaining good modules is detected, a diagnostic program or error detection device will identify the failed unit and switch it out, and the system reverts to a *simplex mode* of operation.

2.1.7 A Scheme for Multiple Fault-Tolerance [28]

Su and DuCasse [28] present a scheme and its realization which will automatically reconfigure a five-modular redundancy system into a TMR (triple modular redundancy) when two modules fail simultaneously. When only one module fails, the system will become a TMR with a spare. The spare module will replace the faulty module in the TMR. The failure of a third module will be tolerated. This scheme is superior to the 5MR system since the scheme can tolerate either a double fault followed by a single fault or three single faults occurring one at a time which is a situation that cannot be tolerated by the 5MR. The scheme is better than the hybrid redundancy with a TMR core and two spares because the latter can only tolerate one single fault in the TMR core at any instant. (Here the effects of compensating failures (a stuck-at-0 compensates for a stuck-at-1 as far as the voter is concerned) are not considered.) This scheme can easily be generalized for handling a multi-valued five modular system.

2.1.7.1 Reconfiguration of 5MR

A maximum of two failures can be tolerated in a five modular redundancy system. Using the technique described in this section, it is possible to tolerate the failure of at least three modules.

Consider a five module system with binary variables $x_1, x_2, \ldots, x_5$ as the outputs for the modules, where $x_i \in \{0, 1\}$ for all i. If the x_i's feed into a majority gate, then the output of the majority gate, z, can be expressed as

$$z = M(x_1, x_2, \ldots, x_5) = x_1 (x_2 (x_3 + x_4 + x_5) + x_3 (x_4 + x_5) + x_4 x_5) + x_2 (x_3 (x_4 + x_5) + x_4 x_5) + x_3 x_4 x_5 \tag{1}$$

Substituting $x_1 = 0$ and $x_2 = 1$ into the above equation, we obtain

$$z = M(0, 1, x_3, x_4, x_5) = x_4 x_5 + x_3 x_4 + x_3 x_5 = M(x_3, x_4, x_5) \tag{2}$$

where the term $x_3 x_4 x_5$ has been removed by absorption. From the above observation, since the majority function is symmetric, we see that a TMR is easily obtained from a 5MR by replacing any variable by 0 and any other variable by 1. For example, if module number 1 is faulty, by letting x_1 be a 0 and x_3 be a 1, we can obtain a TMR with inputs x_2, x_4, x_5.

Based on this idea, we can design a logic network for automatically locating single or double faults and reconfiguring a 5MR to a TMR.

The block diagram for the automatic reconfigurable system is shown in

Fig. VI-13. Block A consists of five "equivalence detectors." Each has two inputs, x_i and z, and one output, g_i, where $i = 1, 2, \ldots, 5$. The output $g_i = 1$ if the ith module is faulty, i.e., x_i has a logic value different from z. In this case, Block B inhibits the signal x_i and prevents x_i from transferring to α_i. Thus α_i becomes stuck at 0. This removes the faulty module from active participation in the voting. The lead $\alpha_{i+1 \pmod 5}$ is forced to become temporarily stuck at 1. Thus, we obtain a spare. The remaining α_i's are unaffected by the flip-flop, allow the x_i to transfer to α_i. If both the ith and kth modules fail either simultaneously or sequentially, Block B causes $\alpha_i = 0$, $\alpha_k = 1$ for $k = i + 1$ or $i + 2$ and $\alpha_i = 1$, $\alpha_k = 0$ for $k = i - 1$ or $i - 2$ and again a perfect TMR system results. Suppose x_i fails first, then $g_i = 1$ yielding $\alpha_i = 0$ and $\alpha_{i+1} = 1$. If x_k fails next, where $k = i - 1$, then $\alpha_k = 0$ and α_{i+1} changes from 1 to x_{i+1}.

2.1.7.2 Design of Block A

Figure VI-14 shows the logic diagram for Block A. Initially all RS flip-flops are reset. If module i is faulty, then Z and X_i are different from each other and hence $S_i = 1$ sets flip-flop i when clock pulse p occurs. Therefore, $g_i = 1$.

2.1.7.3 Design of Block B

In this subsection, we present an informal approach to the design of network B by deriving the α's in terms of the x's and g's. The formal approach is available from the first author. If all modules are fault-free, then $g_i = 0$ for all i and network B should cause $\alpha_i = x_i$ for all i—a direct transmission from x_i's to α_i's. The function α_i is equal to the disjunct of $x_i \overline{g_i}$ and terms which are "zeroes" when all $g_i = 0$. This determines the behavior

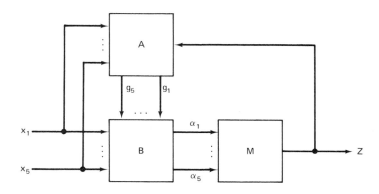

FIGURE VI.13 Block diagram for the automatic reconfigurable system

520 CHAPTER VI

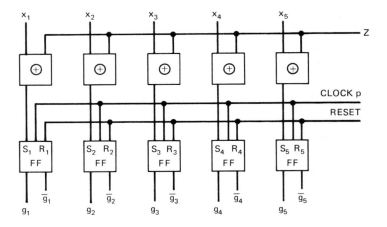

FIGURE VI.14 Logic diagram for Block A

of α_i when faults are present. The system with single and double faults is considered below.

Since the majority function is a symmetrical function, the five-module majority function can be expressed by the following equation:

$$M(x_1, x_2, x_3, x_4, x_5) = \sum_{j=1}^{5} x_j \, x_{j-1 \text{ (mod 5)}} \, x_{j-2 \text{ (mod 5)}}$$

$$+ \sum_{j=1}^{5} x_j \, x_{j-1 \text{ (mod 5)}} \, x_{j-3 \text{ (mod 5)}}$$

For example, all double faults affect one of the module pairs $(x_i, x_{i-1 \text{ (mod 5)}})$ or $(x_i, x_{i-2 \text{ (mod 5)}})$ since $(i-3) - 2 = i \pmod 5$.

For ease of understanding, we shall design Block B by informally deriving the equation for the system with a single failure. Then we shall derive the equation for the system with double failures and show that the same equation can be used for the single failure case. In each case, the equation is constructed informally by asking the question: "What terms are needed for α_i in order to satisfy each case?" Foraml derivation of the equations can be obtained from the first author.

System with single failures. For example, if the module M_1 is faulty, then network B should cause $\alpha_1 = 0$ and $\alpha_2 = 1$ and $\alpha_i = x_i$ for $i = 3, 4, 5$ so that the 5MR becomes a TMR. If module m_5 is faulty, then $\alpha_5 = 0$ and $\alpha_1 = 1$. Since the majority function is a symmetric function, network B can be synthesized using a ring structure type of logic network. Therefore, α_i can be considered the same as $\alpha_{5+i \text{ (mod 5)}}$, that is, the addition is to be taken as a modulo five addition, $\alpha_{i+5 \text{ (mod 5)}} = \alpha_i$. Figure VI-13 can then be shown in more detail as Fig. VI-15. Block A in Fig. VI-14 is the logic network inside

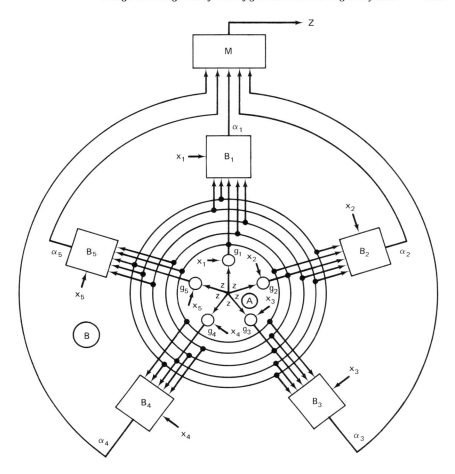

FIGURE VI.15 Detailed diagram for the system in Fig. VI.13

the five smallest circles in Fig. VI-13. Blocks B_1 to B_5 in Fig. VI-15 constitute Block B of Fig. VI-13. The center point of the circles in Fig. VI-15 represents the output of the majority gate, z.

In general, if the ith module is the only faulty module, then $g_i = 1$ and $g_j = 0$ for $j \neq i$. We want network B to satisfy three requirements: (1) $\alpha_i = 0$, (2) $\alpha_{i+1 \,(\mathrm{mod}\,5)} = 1$, and (3) $\alpha_k = x_k$ for all $k \neq i, i+1$. We now derive the equation for α_i as a function of the g_i's and x_i's. Recall that g_i is the only one out of five g_i's whose value is logic 1. Thus if g_i is an implicant of α_{i+1} (i.e., g_{i-1} is an implicant of α_i), then requirement (2) $\alpha_{i+1 \,(\mathrm{mod}\,5)} = 1$ will be satisfied. The requirements (1) and (3) will be satisfied if α_i contains the product term $x_i \bar{g_i}$. Therefore, the following equation satisfies all three requirements:

$$\alpha_i = x_i\, g_i + \overline{g}_{i-1} \qquad (3)$$

System with double failures. Due to the ring structure of the reconfiguration network as shown in Fig. VI-15, any two "nonadjacent" faulty modules in the set $\{m_1, m_2, m_3, m_4, m_5\}$ can be considered as separated by a "distance" of 2, i.e., if one faulty module is m_i, then the other one can be considered as m_{i-2}. Suppose modules number 1 and number 4 are faulty, then we can consider $i = 1$ and $i - 2 \pmod 5 = -1 \pmod 5 = 4$. Therefore, if there is a double fault among the modules, we need only consider two cases.

Case 1. Non-Adjacent Double Failures The $(i - 2)$th and ith modules are faulty. The term $g_i\, g_{I-2}$ appearing in a disjunction representing α_i will guarantee that $\alpha_i = 1$.

Case 2. Adjacent Double Failures If module i and module $i - 1$ fail, the term $g_{i-1} \cdot \overline{\sum_{k \neq i-1, i} g_k}$ will ensure $\alpha_i = 1$.

If neither module $i - 1$ nor i fails, the term $x_i \overline{g_i}$ will guarantee (in the absence of terms other than those described above) that $\alpha_i = x_i$. The following representation for α_i will then assure the desired performance of the reconfiguration network:

$$\alpha_i = g_{i-1} \overline{\left(\sum_{k \neq i-1, i} g_k \right)} + x_i \overline{g_i} + g_i\, g_{i-2} \quad (4)$$

The logic diagram for realizing α_i (each of the five square blocks B_i in Fig. VI-15) is shown in Fig. VI-16.

Example 6. To verify the above equation, let us consider the case where both m_1 and m_3 are faulty. Substituting $g_1 = g_3 = 1$ and $g_2 = g_4 = g_5 = 0$ into Eq. (4), we obtain

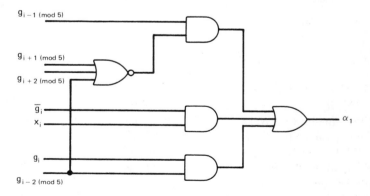

FIGURE VI.16 Logic diagram of each B_i in Fig. VI.15

$$\alpha_1 = g_5 \overline{(g_2 + g_3 + g_4)} + x_1 \overline{g_4} = 0 + 0 + 0 = 0$$

and $\alpha_3 = g_2 \overline{(g_4 + g_5 + g_1)} + x_3 \overline{g_3} + g_3 g_1 = 0 + x_3 \cdot 0 + 1 \cdot 1 = 1$

Let us now verify that Eq. (4) works for a 5MR with a single faulty module. For a single fault, one and only one g, say g_i, is equal to 1, while the other g's are 0's. In this case $g_k = 0$ for all $k \neq i - 1, i$, and in Eq. (4) the term

$$\overline{\sum_{k \neq i-1, i} g_k} = 1 \qquad (5)$$

Also,

$$g_i g_{i-2} = 0 \qquad (6)$$

Substituting Eqs. (5) and (6) into Eq. (4), we obtain Eq. (3). Therefore, Eq. (4) is valid for either single or double faults. ∎

Example 7. In a five-input system, if module 1 fails, then $g_1 = 1$ and $g_2 = g_3 = g_4 = g_5 = 0$. From Eq. (4), $\alpha_1 = 0$, $\alpha_2 = 1$, $\alpha_3 = x_3$, $\alpha_4 = x_4$, and $\alpha_5 = x_5$. If module 4 fails next, then substituting $g_1 = g_4 = 1$ and $g_2 = g_3 = g_5 = 0$ into Eq. (4), we obtain $\alpha_1 = 1$, $\alpha_2 = x_2$, $\alpha_3 = x_3$, $\alpha_4 = 0$ and $\alpha_5 = x_5$. Therefore, the system reconfigures correctly with the three fault-free modules 2, 3, and 5, forming a TMR system.

Suppose the third module now fails. The system will change the majority gate in the TMR to an OR or an AND gate. This is shown below.

1. If module 3 fails, then substituting $g_3 = g_4 = g_1 = 1$ and $g_2 = g_5 = 0$ into Eq. (4), we obtain $\alpha_1 = 1$, $\alpha_2 = x_2$, $\alpha_3 = 1$, $\alpha_4 = 0$, and $\alpha_5 = x_5$. Hence $z = x_2 + x_5$. The system will tolerate a fourth module failure as long as x_2 or x_5 has a stuck-at-0 failure.
2. If module 2 or 5 fails, by the same procedure as in (1), we can see that the system will be reconfigured from a majority gate into an AND gate. Thus a fourth failure can be tolerated as long as it is stuck-at-1 fault. ∎

2.1.7.4 Advantages of the scheme

From the system described above, we see that the proposed fault-tolerant digital system allows the system to reconfigure from 5MR to TMR following the failure of the first module. Initially this might strike the reader as being absurd since a majority gate with five inputs will tolerate the malfunction of not only one, but even two participating modules. The reconfiguration is accomplished in such a way, however, as to hold effectively in reserve, or make into a spare, the fault-free module that is removed from active participation in the subsequent vote at the same time that the faulty unit suppressed. This is done by forcing the output of the faulty module, say, the ith module, to be stuck at 0 and the output of the module next to it, that is, the $(i + 1)$th module to be stuck-at-1. Thus, the five-input NMR becomes a TMR with the $(i + 1)$th module as a spare. If a second module, that is, one of the modules in the TMR should then develop a fault, it is seen that the system frees the $(i + 1)$th module and employs it in place of the newly found malfunctioning unit. In this way, then, the network effectively employs a spare unit when in actuality it had no spares at all. Thus, after two modules have failed sequentially, we are left with a perfect TMR system.

This system can tolerate more failures than 5MR and hybrid redundancy systems. This is shown in Table VI-1. For example, in the presence of a double fault, the system can reconfigure a 5MR to a perfect TMR system. This means that the third module failure can be tolerated. Furthermore, the TMR will be reconfigured into an AND gate or an OR gate in the presence of the third failure.

The second advantage is that the circuit realization of the proposed scheme is simpler than the schemes presented in the existing literature. From Figs. VI-14 and VI-16, we see that only five gates (or 6 NAND gates) and one flip-flop are required to implement the switch.

A third advantage is that the structure of the switching network for reconfiguration is highly modular. For example, both Blocks A and B contain five identical components. This means that the testing of these components should be easier.

2.1.7.5 Disadvantage of the proposed scheme

Although the approach taken in this scheme for a five-module system can be used for deriving equations for an N-module system where $N \geq 7$ and N is odd, the complexity of Block B grows with N. Such a drawback is not serious since NMR's with $N \geq 7$ are seldom used in a practical environment.

2.2 Software Fault-Tolerance [29]

2.2.1 Software Redundant Architecture

The problem of fault-tolerant software structure can be formulated as the development of reliable software with unreliable components. This can be solved by the application of software redundancy techniques. The errors can be generated by either hardware faults or imperfect software faults; different techniques to detect these errors must be used. Software redundancy uses additional software which includes additional programs, program segments, set of instructions and microprogram steps which provide either fault detection or recovery in fault-tolerant computer systems. Software redundancy is implemented by using the following techniques:

TABLE VI-1

Faults	Effect on Proposed Scheme	Effect on 5MR	Effect on Hybrid Redundancy with TMR core
1	Reconfigure to TMR	None	None
2 simultaneously	Reconfigure to TMR	None	Failure
2 in sequence	Reconfigure to TMR	None	None
2 simultaneously, followed by 1	OR gate or AND gate, OK	Failure	Failure
3 in sequence	OR gate or AND gate, OK	Failure	None
1 followed by 2	Failure	Failure	Failure
3 simultaneously	Failure	Failure	Failure

1. Replicate those critical programs and data in different storage locations
2. Diagnostic programs at various program and microprogram levels
3. Special programs which implement program restarts and interface with the operating system

Example 8. One approach to the software redundancy is the triple modular software redundancy (static redundancy) scheme as shown in Fig. VI-17(b). Three processing modules perform the same data transformation function on a single computer. A voting module selects the result on the basis of a majority decision. Furthermore, the module also identifies the minority modules as the faulty modules. The error detection by voting module will be carried out to the module level only. Thus an error within the module is not detected. The overhead of this approach is considerable. At least three times the amount of resources and storage are needed during run time. The errors may be generated by hardware transient faults, in which case the three processing modules may be the same or not. Alternatively, errors may also be generated by software design errors, in which case the three processing modules must be developed independently. That means that the system development effort has to be triplicated too. ■

Example 9. The dynamic redundancy scheme can also be used in software redundancy by using the standby spare module and audit module as shown in Fig. VI-17(c). This scheme involves a smaller overhead in contrast to the above example. If a processing module fails, it is detected by the audit module and immediately replaced by a standby spare module which performs the same function. In order to facilitate the error checking of the audit module, redundant information should be stored in the state vector of the software system. It is then possible to test the state vector for consistency. If an error has been detected by the audit module, then the error message is

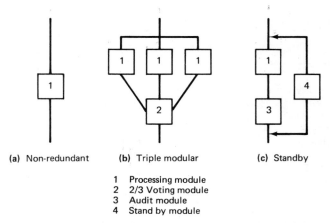

FIGURE VI.17 Block diagram of different redundancy schemes

written on an audit file and a spare module is activated. To prevent the same design error in the spare module and the processing module, different algorithms should be used. ∎

2.2.2 Software Error Diagnosis

Software error diagnosis consists of three steps: error detection, error treatment, and damage assessment.

2.2.2.1 Error detection

The *software error* detection scheme can be implemented using the following modules:

1. *Comparison module* in the dual redundancy scheme
2. *Majority voting module* in the software triple modular redundancy scheme
3. *Audit module* in the software dynamic redundancy scheme. The val-validation routine used in the audit modules can be either the *acceptance tests* which compare the desired I/O relations with those of the active module or *dynamic assertions* that are *executable predicates* (logical and relational expressions for checking purposes only) added before or after each statement in the processing module, generating a true exit (which indicates the correct execution of the statement) or a fail exit (which indicates the incorrect execution of the statement and replaces it by a spare or alternate statement).
4. *Watch-dog timer* (WDT) that can be reset at any time, generates an interrupt signal on reaching the present time limit, and detects dead functional units or tight (infinite) looping software.†

†See Chapter IV dedicated to profound treatment of infinite loops in programs and microprograms. (Ed.)

In addition to functional checks on software, a somewhat different kind of check called a *reversal check* is sometimes used. This scheme processes the results of the system activity in order to determine the corresponding inputs and checks them with the actual inputs. This scheme can only be used for those software whose input-to-output relation is either one-to-one or one-to-many but not many-to-one.

2.2.2.2 Error treatment

Once an error is detected and located at a module level, then two *error treatments* can be considered: one is the *replacement strategy* that replaces the designated faulty component by a standby spare; the other is the *graceful degradation strategy* which causes some or all of the responsibilities of the faulty software module to be taken over by other fault-free active modules. Graceful degradation involves some degree of performance and/or functional degradation but increases the reliability of the system. The error treatment can be designed to be either *manual* (manual software replacement or reconfiguration), *dynamic* (external stimuli issued manually to cause the system to reorganize its future activity), or *spontaneous* (spontaneous replacement and reconfiguration are carried out automatically by the system itself, i.e., self repair).

2.2.2.3 Damage assessment

The approach to *damage assessment* (to determine the extent of the damage caused by the error) can be designed by making use of *atomic actions*. An action is *atomic* if the process or processes performing it does not communicate with other processes while the action is being performed. Figure VI-18 shows the communication relationships in a multiprocessor environment, denoted by the dot lines, while the solid lines indicate the range of atomic actions in the system. Obviously the damage assessment can be determined by the closest atomic action in which the error was detected. For example, atomic actions 12 to 14 are for single processes only. Action 11 deals with processes 1 and 2 since there is a communication between them as shown in the dotted line. Atomic action 10 deals with all three processes. Damage in process 2 at point α may introduce errors in all three processes.

2.2.3 Software Error Recovery

Error recovery is the scheme for dealing with the damage caused by a detected error. There are two recovery schemes in the literature: backward and forward error recovery.

The *backward error recovery scheme* backs up one or more of the processes of a system to a previous state which is hoped to be error-free, before attempting to continue further operation of the system or subsystem.

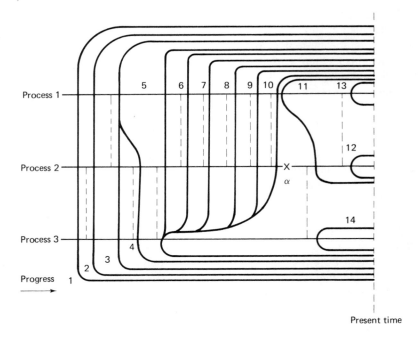

FIGURE VI.18 Atomic actions to the present

This scheme depends on the provision of *recovery points* (RP) and the state vector, which contains the states of a process or a program and can be recorded and later re-executed. The most famous technique for obtaining such recovery points is the *recovery cache-type mechanism* which records the original states of just those resources that will be modified for re-execution later if an error is detected. The backward error recovery scheme can be designed by assigning a recovery point for a process each time it enters an atomic action. Then the states at the recovery point are saved and retained for all processes involved in the given atomic action until all have reached the end of that atomic action.

Example 10. In Fig. VI-18, if an error is detected at point α in process 2, then processes 1, 2, and 3 could be backed up to the entry points defined by atomic action (labeled by integer 10). The recovery points need not be assigned at each entry point of atomic action. For example, in process 1 in Fig. VI-18 we can assign only one recovery point at atomic action 5 instead of assigning six recovery points, one at each entry of atomic actions 5 to 10. Such a reduction in the number of recovery points will reduce the overhead but increases the length of retry or alternate program segment in process. ■

Example 11. Figure VI-19 corresponds to the same processes as drawn in Fig. VI-18. In Fig. VI-19 the left brackets represent the recovery

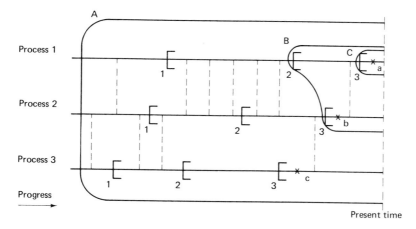

FIGURE VI.19 Recovery points assignment and domino effect

points assigned. If an error was detected at point a in process 1, only one recovery action needs to be done to back up to RP3 (recovery point labeled by integer 3). If an error was detected at point b in process 2, the process 2 has to back up to RP3 and process 1 has to back up to RP2 rather than RP3, because the communication takes place between RP3 for process 2 and RP3 for process 1. In some situations, for instance, if an error at point c in process 3 is detected, the process 3 has to back up to RP3 and process 2 has to back up to RP2. This causes process 1 to back up to RP1, and this again causes process 2 to back up to RP1, and then process 3 has to back up to RP1. Repeat this process, we have found that the error at point c in process 3 causes all processes 1, 2, and 3 to back up to the starting (entry) points of the action A. This kind of phenomenon is called the *domino effect* [30] of recovery. ∎

The *recovery line* is the minimal set of recovery points in all processes which are either in error or affected by the error. In the above example, the recovery lines for errors at point a, b, and c are those atomic actions C, B, and A, respectively, drawn in Fig. VI-19. Thus, the basic design concept for the backward error recovery scheme is to find out the recovery line for each error. Of course, for each error, only one recovery line can be found.

The *forward error recovery scheme* [31] does not attempt to repair the damage caused by the detected error but takes the following actions:

1. Report the error to the user.
2. Take the system default action or user defined action (because all errors in forward error recovery scheme are predetermined, (e.g., divided by zero)) so that the program can continue its execution. The *exception handling* is a language facility provided by programming language designers to handle the forward error recovery scheme. The

PL/I language provided such facilities for exception handling. This involves enabling and disabling the error conditions for exceptions, software detection of these exception conditions, and exception handling defined by the users or by the system. The readers are recommended to refer to any PL/I programming language reference manual or [31].

2.2.4 Time Redundancy [8]

Two major aspects of time redundancy are the repeated execution of instructions and the restart of programs after an error detection. Time redundancy can be found either in the identification and correction of those errors caused by transient faults or in program restarts after a hardware reconfiguration. This is accomplished by the rollback of single instructions, segments of programs or the entire program. To reexecute the affected program at a previously established "rollback" point or checkpoint is a common way to recover from a damaging transient. Program rollback (or checkpoint restart) requires saving the status of the program, i.e., its state vector, at each rollback point. Thus it is imperative to make an intelligent tradeoff between the two check points. In addition, the checkpoint information requires protection because its integrity is a prerequisite to a successful rollback. The repeated execution of instructions, for the purpose of detecting errors generated by transient faults, slows down the computation, so it is usually only applicable to a limit set of programs or I/O operations.

3. RELIABILITY MODELS OF RECONFIGURABLE FAULT-TOLERANT SYSTEMS

In a system which has no spares, or which has exhausted all its spares during previous recovery from failures, additional failure in the active configuration may or may not result in a system crash. Systems with the capability of *graceful degradation*, i.e., the ability to reconfigure after a hardware failure to operate with a subset of the total resources and still provide acceptable (though degraded) performance, will tolerate additional failures after the spares are exhausted. The *graceful degradation strategies* arrange for some or all of the responsibilities of the failed components to be taken over by other components already in use by the system. Reconfiguration always involves some degree of performance and/or functional degradation. Reconfigurable fault-tolerant systems can be categorized into two types:

1. *Identical-task system without intercommunications:* All active modules run the same job, and there is no intercommunication among them except resynchronization, switching, and voting. Therefore,

this kind of system either masks the errors or switches out the failed module with or without replacement. If a faulty module is replaced by a spare, resynchronization between the new active module with the previous active modules is required. All hardware redundancy systems belong to this category.

2. *Different-task system with intercommunications:* All active modules run different jobs and intercommunicate with each other by some switching networks. When a module fails, the system has to switch out the failed module, but due to the communications among good modules and the failed module, the good modules may be contaminated by the faulty module. Therefore, it is necessary to rollback to a previous good state in which no failure has occurred. Otherwise, the system fails due to unrecovered error. This kind of system consists of gracefully degradable systems, self repairing systems, and general multiprocessor systems.

In this section, we shall discuss each type of reconfigurable fault-tolerant systems.

Ng and Avizienis [32] have developed a model which is applicable to achieve fault-tolerance in closed (self repairable by standby spares or non-repairable) computer systems. They have assumed that a module is a physical unit that does not contain *internal redundancies* and cannot tolerate internal failures. Redundant structure is developed by using multiple units of these modules. They defined the following parameters to model a closed fault-tolerant computer system (The symbol "$\triangleq$" means that a parameter is defined by the string following this symbol.):

1. $N \triangleq$ Initial number of modules in the active configuration.
2. $s \triangleq$ Number of spare modules.
3. $D \triangleq$ Number of additional failures (in the absence of spares) that the active configuration can tolerate before the system fails (degree of degradation).
4. $Y \triangleq (Y[1], Y[2], \ldots, Y[D])$ is an integer vector that defines the number of modules in successive active configurations following successive failures. $Y[i]$, where $i = 1, 2, \ldots, D$, is the number of modules in the active configuration (in the absence of spares) after the system successfully recovers from the ith failure.
5. $\lambda \triangleq$ Failure rate (number of failures per unit time) of a powered module. Note that the active modules are always powered.
6. $\mu \triangleq$ Failure rate of a spare module which can be either powered ($\mu = \lambda$) or unpowered ($\mu < \lambda$).
7. $C_a \triangleq$ Prob {System recovers from active module failures when the system has a full active configuration | fault occurs}.

8. $C_d \triangleq$ Prob {System recovers from spare module failures | fault occurs}.
9. **CY** $\triangleq$ $(CY[1], CY[2], \ldots, CY[D])$ is a coverage vector where $CY[i]$ = Prob {Subsystem recovers from the ith active module failure that will cause a degradation in the active configuration | fault occurs}.

At any given time, each module is in one of three possible states:

1. It is in the failed state.
2. It is a good spare.
3. It is an active module currently participating in the computing process.

It is assumed that the failed module is isolated from the system.

3.1 System Reliability of Fault-Tolerant Computing Systems

In the following reliability evaluation, we shall adopt the model developed by Ng and Avizienis by using the vector notation $(N, D, s, C_a, C_d, \lambda, \mu, \mathbf{Y}, \mathbf{CY})$ to denote the set of parameters for a redundant system. Table VI-2 shows the defining parameters and references for some existing redundant systems. It includes a general dynamic system $_cR_s^N$ with N active modules process tasks tasks in parallel, s spares, and the probability of successful recovery for both active and spare modules is $C = C_a = C_d$. The reliability calculation of a system by the use of this model assumes that in the redundant system the

TABLE VI-2

DEFINING PARAMETERS FOR SOME EXISTING RELIABILITY MODELS
($\mathbf{CY}$ = 11 ... 1 in all cases)

System	N	D	s	C_a	C_d	λ	μ	Y	Reference
Simplex	1	0	0	1	1	λ	λ		33
Static									
TMR	3	1	0	1	1	λ	λ	2	33
TMR/	3	1	0	1	1	λ	λ	1	33
Bipurge									
NMR	$2n+1$	n	0	1	1	λ	λ	$2n, \ldots, n+1$	34
NMR/	$2n+1$	n	0	1	1	λ	λ	$2n-1, \ldots, 3, 1$	24
Bipurge									
Dynamic									
$_cR_s^N$	N	0	s	C	C	λ	μ		16
Hybrid									
$R(N, s)$	$2n+1$	n	s	1	1	λ	μ	$2n, \ldots, n+1$	22
$R_{\text{TMR/spares}}$	3	2	s	1	1	λ	μ	2, 1	27

probabilities of transitions among various states are *time-independent (homogeneous Markov chain)*.

In reliability theory, let the random variable t_F denote the time at which the system fails. Then the *system reliability* $R(t)$ is the probability that the system is operational during the time interval between 0 and t, i.e., $R(t) = \text{Pr}\{t_F > t\}$. The *cumulative distribution function* (CDF) for the errors is defined as $F(t) = \text{Pr}\{t_F \leq t\}$. Therefore, the relationship between $R(t)$ and $F(t)$ is obvious; $R(t) = 1 - F(t)$. The *probability density function* (PDF) for the interarrival time distribution between errors is then defined as

$$f(t) = \frac{d}{dt} F(t) = -\frac{dR(t)}{dt}$$

The hazard function, $Z(t)$, of the system is defined to be the ratio of $f(t)$ to $R(t)$; i.e., $Z(t) = f(t)/R(t) = -(dR(t)/dt)/R(t)$. If the hazard function $Z(t)$ is a constant, i.e., independent of t, then this constant is called the *error rate* of the system, denoted by λ.

Example 12. For hardware components, the PDF, $f(t)$, is often assumed to be an exponential distribution function and is given by $f(t) = e^{-\lambda t}$, where λ is a positive constant and $t > 0$. In this case, the following important functions are derived:

1. CDF $F(t) = 1 - e^{-\lambda t}$
2. Reliability $R(t) = e^{-\lambda t}$
3. Hazard function $Z(t) = \lambda$ ■

Example 13. It is not reasonable to assume a constant error rate for software components. Based on some experimental results [35], as errors occur the program is stopped and the error is identified, corrected, and the modality (the number of design errors) is reduced. Therefore, a *decreasing error rate* model seems to be appropriate. In [35] they defined a hazard function for a software component with a decreasing error rate as $Z(t) = \alpha \beta t^{\beta - 1}$, where $t > 0$ and α and β are positive constants, $1 > \beta > 0$. Since $Z(t) = -(dR(t)/dt)/R(t)$ we obtain

$$\frac{d}{dt} R(t) + Z(t) R(t) = 0$$

Solving this equation, we obtain the following functions:

1. Reliability $R(t) = \exp(-\alpha t^\beta)$
2. CDF $F(t) = 1 - R(t) = 1 - \exp(-\alpha t^\beta)$
3. PDF $f(t) = \alpha \beta t^{\beta - 1} \exp(-\alpha t^\beta)$

This last function is called the Weibull distribution function. ■

In the calculation of system reliability, we assume that the switching circuits and voters among the active modules are highly reliable (otherwise, the use of switching circuits decreases the reliability). The errors, generated by permanent faults or uncovered errors generated by transient faults, produce system failure. For simplicity, we assume that all active modules are identical and only single module failure can occur at any instant of time.

In Table VI-3 we summarize the system reliabilities for various existing reliability models. The system reliabilities for TMR/bipurge and TMR/unipurge systems are not given because they are a special case of NMR/bipurge and NMR-unipurge systems. The reliability expression of self-repairing degradable systems is not given here since it is very complicated. A reference is given instead. The symbol $R_s(t)$ in Table VI-3 denotes the system reliability while the symbol $R(t)$ means the reliability of each module in that system.

TABLE VI-3
SYSTEM RELIABILITIES FOR SOME EXISTING RELIABILITY MODELS

System	System Reliability $R_s(t)$	Figure	Reference
TMR	$3R^2(t) - 2R^3(t)$	VI-20	33
NMR	$\sum_{k=n+1}^{N} \left[\sum_{\substack{m=n+1 \\ m \neq k}}^{N} \frac{m}{m-k} \right] \cdot R^k(t)$	VI-21	34
	$= \sum_{k=n+1}^{N} \binom{N}{k} R^k(t)[1 - R(t)]^{N-k}$, where $n = (N-1)/2$	VI-21	34
NMR/bipurge	$\sum_{k=0}^{n} \left[\prod_{\substack{i=0 \\ i \neq k}}^{n} \frac{2i+1}{2i-2k} \right] \cdot e^{-(2k+1)\lambda t}$	VI-22	24
NMR/unipurge (selfpurge)	$\sum_{k=M}^{N} \binom{N}{k} R^k(t)(1 - R(t))^{N-k}$, where M = threshold	VI-23	26
Hybrid	$\sum_{i=n+1}^{N} A_i e^{-i\lambda t} + \sum_{j=1}^{s} B_j e^{-(N\lambda + j\mu)t}$ where $A_i = \left(\prod_{\substack{k=n+1 \\ k \neq i}}^{N} \frac{k}{k-i} \right) \cdot \left(\prod_{l=1}^{s} \frac{N\lambda + l\mu}{(N-i) + l\mu} \right)$ $B_j = \left(\prod_{k=n+1}^{N} \frac{k\lambda}{(k-N)\lambda - j\mu} \right) \left(\prod_{\substack{l=1 \\ l \neq j}}^{s} \frac{N\lambda + l\mu}{(l-j)\mu} \right)$	VI-24	22

TABLE VI-3 *(cont)*

SYSTEM RELIABILITIES FOR SOME EXISTING RELIABILITY MODELS

System	System Reliability $R_s(t)$	Figure	Reference
TMR/spares	$3e^{-\lambda t} - 3e^{-2\lambda t} + e^{-3\lambda t}$	VI-25	27
Graceful degradation without spares	$\sum_{k=0}^{D} \sum_{j=0}^{k} A_{j,k}\, e^{-(N-k+j)\lambda t}$ where $D \geq k \geq 1$ $A_{j,k} = \dfrac{1}{N-k} \cdot \prod_{i=1}^{k} CY[i] \cdot \left[\prod_{\substack{m=0 \\ m \neq j}}^{k} \dfrac{(N-k+m)}{m-j} \right] \cdot (N-k+j)$	VI-26	32
Self repairing	$e^{-N\lambda t} \sum_{j=0}^{s} A_{s,j}\, e^{-j\mu t}$ where $A_{0,0} = 1$ $A_{m,i} = \dfrac{(NC_a\lambda + mC_d\mu)\, A_{m-1,i} + (1 - C_d)\mu \sum_{k=i}^{m-1} A_{k,i}}{(m-i)\mu}$ $A_{m,m} = 1 - \sum_{i=0}^{m-1} A_{m,i}$ $s \geq m \geq i$	VI-27 and VI-28	32
Self repairing degradation	See reference.	VI-29	32

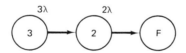

FIGURE VI.20 Reliability model for TMR systems

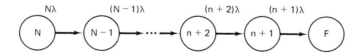

FIGURE VI.21 Markov reliability model for NMR systems

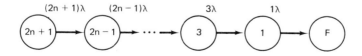

FIGURE VI.22 Markov reliability model for NMR/bipurge system

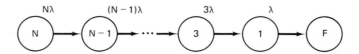

FIGURE VI.23 Markov reliability model for self-purging system

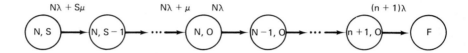

FIGURE VI.24 Markov reliability model for hybrid redundant system

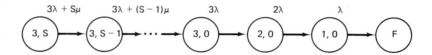

FIGURE VI.25 Markov reliability model for TMR/spares system

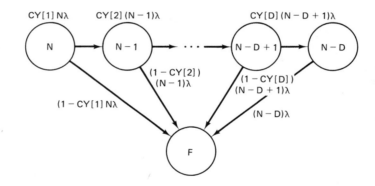

FIGURE VI.26 Markov reliability model of gracefully degradable systems

3.2 Other Reliability Related Measures

Some commonly used reliability measures [36] are given below.

1. *System reliability*

   $R(t)$ = Prob {the system operates correctly during the time interval between 0 and t}, i.e.,

   $$R(t) = \int_t^\infty f(z)\, dz$$

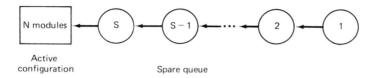

FIGURE VI.27 A self repairing system

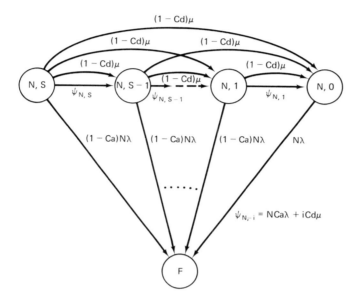

FIGURE VI.28 Model of a self repairing system

where $f(z)$ is the failure function which is the probability density function for the interarrival time distribution betwen errors.

2. *Mean time before failure*, MTBF, is the expected value of t_F, the time at which the system fails, i.e.,

$$\text{MTBF} = E[t_F] = \int_0^\infty t f(t)\, dt$$
$$= -\int_0^\infty t\, dR(t) \quad \left[\frac{d}{dt} R(t) = -f(t)\right]$$
$$= \int_0^\infty R(t)\, dt \quad [R(\infty) = 0]$$

Example 14. For a constant error rate λ, i.e., $R(t) = e^{-\lambda t}$

$$\text{MTBF} = \int_0^\infty R(t)\, dt$$

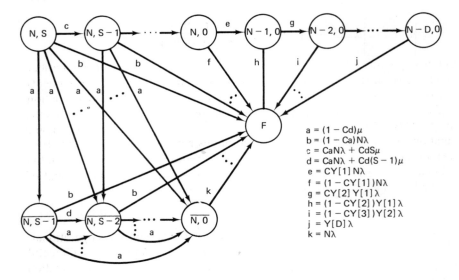

FIGURE VI.29 Markov reliability model for closed systems

$$= \int_0^\infty e^{-t}\, dt$$

$$= \frac{1}{\lambda}$$

Similarly, in a decreasing error rate, i.e., $R(t) = e^{-\lambda t^\beta}$ and $1 > \beta > 0$, then the MTBF is given by

$$\text{MTBF} = \int_0^\infty e^{-\lambda t^\beta}\, dt$$

$$= \int_0^\infty \frac{z^{1/\beta - 1}\, e^{-z}}{\beta \lambda^{1/\beta}}\, dz$$

where $z = \lambda t^\beta$

$$\text{MTBF} = \frac{\Gamma\left(\dfrac{1}{\beta}\right)}{\beta \lambda^{1/\beta}} \quad \text{where } \Gamma(\alpha) = \int_0^\infty t^{\alpha-1}\, e^{-t}\, dt \quad \blacksquare$$

3. The *mission time* is the duration in which the system reliability remains above a specific value.
4. The *system availability* is the ratio of the time that the system is operational (uptime), to the total time (the sum of the usable time and the downtime for maintenance and repair).

$$\text{Availability} = \frac{\text{Uptime}}{\text{Uptime} + \text{downtime}}$$

In this expression, the *downtime* is considered as the product of the number of maintenance/repair actions (requiring shutdown of a system or component) and the average time for each such action. The average time for each action is usually designated as the *mean time to repair* (MTTR). Hence, Availability = MTBF/(MTBF + MTTR). Let T_i denote the amount of available computation (e.g, computation time or the number of instructions executed) in system state i. Let t_i be the actual time the system spends in state i. Then the *computation capacity* in state i, denoted α_i, is defined to be the amount of available computation per unit time in system state i.

$$\alpha_i = T_i / t_i$$

Active states are defined as those states in which the system stays operational. For example, in Fig. VI-26, all states except state F are active states. Graceful degradation systems have various active states that differ in computation capacities to execute tasks.

Several reliability measures are given below.

1. The *computation reliability*, $R(t, T)$, is the probability that the system will correctly execute a task of length T starting at time t.
2. The *mean computation before failure*, MCBF is the expected amount of computation available on the system before failure.
3. The *computation threshold*, t_T, is the time at which the computation reliability reaches a specific value for a task of length T.
4. The *computation availability*, $A_c(t)$, is the expected value of the computation capacity of the system at time t.

Example 15. The graceful degradable system shown in Fig. VI-26 has the reliability given in the Table VI-3. For simplicity in Fig. VI-26 we assume that $CY[1] = CY[2] = \ldots = CY[N - D + 1] = C$ and that each module has the same computation capacity, α. Then state i has computation capacity $\alpha_i = i\alpha$. The differential equations describing this system are

$$\frac{d}{dt} P_N(t) = -N\lambda P_N(t)$$

$$\frac{d}{dt} P_{N-k}(t) = (N - k + 1)\lambda C P_{N-k+1}(t) - (N - k)\lambda P_{N-k}(t)$$

$$\frac{d}{dt} P_F(t) = (N - D)\lambda P_{N-D}(t) + (1 - C)\lambda \sum_{i=0}^{D-1} (N - i) P_{N-i}(t)$$

where $D \geq k \geq 1$.

In state $N - k$, the task of length T is related to t by $T = (N - k)\alpha t$, or $dT = (N - k)\alpha \, dt$. In considering the computation capacity in the system, the differential equations describing the system are now changed and given by

$$\frac{d}{dT} P_N(T) = -\frac{\lambda}{\alpha} P_N(T) \tag{7}$$

$$\frac{d}{dt} P_{N-k}(T) = \frac{\lambda C}{\alpha} P_{N-k+1}(T) - \frac{\lambda}{\alpha} P_{N-k}(T) \tag{8}$$

$$\frac{d}{dT} P_F(t) = \frac{\lambda}{\alpha} P_{N-D}(T) + \frac{\lambda(1-C)}{\alpha} \sum_{i=0}^{D-1} P_{N-i}(T) \tag{9}$$

Figure VI-26 is now changed to Fig. VI-30. Applying the Laplace transforms to (7) and (8), we obtain:

$$P_N(s) = \frac{1}{s + \dfrac{\lambda}{\alpha}} \qquad [P_N(0) = 1]$$

$$P_{N-k}(s) = \frac{\dfrac{\lambda C}{\alpha}}{s + \dfrac{\lambda}{\alpha}} P_{N-k+1}(s) \qquad [P_{N-k}(0) = 0]$$

$$= \left(\frac{\lambda C}{\alpha}\right)^k \cdot \frac{1}{\left(s + \dfrac{\lambda}{\alpha}\right)^{k+1}}$$

So,

$$P_N(T) = e^{-(\lambda/\alpha) T} \tag{10}$$

$$P_{N-k}(T) = \left(\frac{\lambda C T}{\alpha}\right)^k \cdot \frac{1}{k!} \cdot e^{-\lambda T/\alpha} \tag{11}$$

where $D \geq k \geq 1$. ∎

The *capacity distribution of the system*, $C_N(T)$, can be defined and calculated as

$$C_N(T) = \Pr \{\text{system executes a task of length } T \mid \text{system in state } N \text{ at the beginning of the computation}\}$$

$$= \sum_{i=0}^{D} P_{N-i}(T)$$

$$= e^{-\lambda T/\alpha} \sum_{i=0}^{D} \frac{1}{i!} \left(\frac{\lambda C T}{\alpha}\right)^i \tag{12}$$

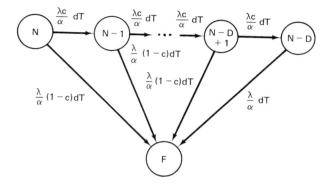

FIGURE VI.30 Markov chain of the graceful degradation system

The *mean computation before failure* for this system is

$$\text{MCBF} = \int_0^\infty C_n(t)\, dt$$

$$= \frac{\alpha}{\lambda} \sum_{i=0}^{D} \frac{C^i}{i!} \int_0^\infty z^i e^{-z}\, dz$$

where $z = (\lambda T/\alpha)$

$$= \frac{\alpha}{\lambda} \sum_{i=0}^{D} C^i \qquad (13)$$

since

$$\int_0^\infty z^i e^{-z}\, dz = \Gamma(i+1) = i!$$

Similarly, the *state capacity distribution*, C_i, of the system in state i is defined to be

$C_i(T) = $ Pr {system executes a task of length T | state i is the beginning of the computation}

For this system, C_{N-k} can be easily obtained by replacing D by $D - k$ in equation (12).

The *comuptation reliability*, $R(t, T)$, of a system is defined to be

$$R(t, T) = \sum_{i \epsilon F} C_i(T) P_i(t) \qquad (14)$$

where $P_i(t)$ is the state probability in state i.

The *computation reliability* of the system is

$$R(t, T) = \sum_{k=0}^{D} \text{Pr \{system executes a task of length } T \mid \text{system in}$$

state $N - k$ at t} · Pr {system in state $N - k$ at $t \mid$ system in state N at $t = 0$}

$$= \sum_{k=0}^{D} C_{n-k}(T) \cdot P_{N-k}(t) \tag{15}$$

Setting $T = 0$ in the computation reliability, we obtain

$$R(t, 0) = \sum_{k=0}^{D} P_{N-k}(t)$$

= Reliability of the graceful degradation system

On the contrary, if we set $t = 0$, then

$$R(0, T) = C_N(T)$$

= The capacity distribution of the system

The *computational availability*, $A_c(t)$, of the system can be derived as follows:

$$A_c(t) = \sum_{i=N-D}^{N} \alpha_i P_i(t)$$
$$= \sum_{k=0}^{D} (N - k) \alpha P_{N-k}(t)$$

Two other measures of run time performance—system throughput and response time unrelated to reliability—are briefly introduced here. The throughput is defined as the number of jobs in a job stream divided by the total time to process the stream for batch systems and is expressed in the number of jobs per minute. The *response time* for job i is the time difference ($Q_i - A_i$) between job i completion time (Q_i) and job i arrival time (A_i). The *mean response time* (MRT) is defined as

$$\text{MRT} = \frac{1}{n} \sum_{i=1}^{n} (Q_i - A_i)$$

3.3 Diagnostic Performance Measure

The terms "reconfigurable system," "dynamic system," "restructurable system," "adaptive system," and "reorganizable system" are used interchangeably in the literature [37]. In all types of parallel systems (array, pipeline, multicomputer, or multiprocessor), reconfiguration can be used to increase the system performance. In the previous subsection, we have discussed the relations among reliability and computation capacity in various reconfigurable systems. The next question is how the diagnosis and reconfiguration capabilities can be incorporated into a system.

The *diagnostic performance* of a system is defined by Saheban and Friedman [37] to be the function $N_s(X_a)$ where X_a is the number of accumulated faulty modules in the system and N_s is the maximum number of simultaneous faults which a system can diagnose (or mask) correctly. Thus $N_s(X_a)$

can be interpreted as the maximum number of simultaneous faults which the system can diagnose correctly when X_a faults occur. Obviously, N_s is a function of X_a and this function can be plotted in a diagram which is usually called the *fault-tolerant graph* (FTG). The FTG's of NMR systems and NMR/unipurge systems with a threshold value M are shown in Figs. VI-31 and VI-32, respectively. Figure VI-31 is a diagnostic performance of the system with no diagnostic reconfiguration. An NMR/unipurge system with threshold value M is equivalent to a hybrid system with PMR core ($P = 2M - 1$) and N-P spares. Thus, an NMR/unipurge system with threshold value M can tolerate $N - M$ faults (i.e., maximum value of X_a is $N - M$). The maximum number of simultaneous faults which the NMR/unipurge system can tolerate correctly before the $N - (2M - 1)$ spares are

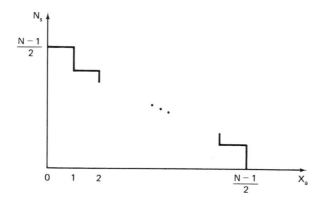

FIGURE VI.31 FTG of an NMR system

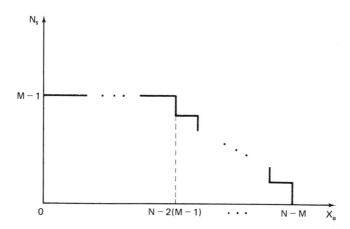

FIGURE VI.32 FTG of a self-purging system

exhausted is $M - 1$. So, when the spares are exhausted, the next fault [the $(N - 2(M - 1))^{th}$ fault] behaves like that in an NMR system.

In Fig. VI-32, we have observed that the step length L (greater than 1) of the first step is equal to the number of spares plus one, i.e., $L = N - P + 1 = N - 2(M - 1)$. For the selfpurging system, L can be increased by reducing M. However, this will reduce N_s. Alternatively, L can be increased by increasing N.

A reconfigurable multiprocessor system without spares has been considered by Saheban and Friedman [38]. The reconfiguration scheme is based on improving one of the following two performances: computational and diagnostic performance. A general architecture for multiprocessor system is shown in Fig. VI-33. There are four main parts in this model:

1. *Processor modules* (PM) which include all active facilities of a system.
2. *Computational interconnection network* (CIN) which consists of interconnection networks for computation.
3. *Diagnostic interconnection network* (DIN) which consists of interconnection network for diagnosis.
4. *Reconfigurable network control unit* (RNCU) which is reponsible for fault diagnosis and reconfiguration.

In Fig. VI-33, if the state of CIN can be changed in order to improve the computational performance, then the system is *computationally reconfigurable*. The computational performance of a fault-tolerant system with respect to an algorithm A executed by this system is a set of parameters C_j's related to the computation of A (e.g., computation time, mean computation before failure, computation reliability, etc.). In Example 15, we discussed the com-

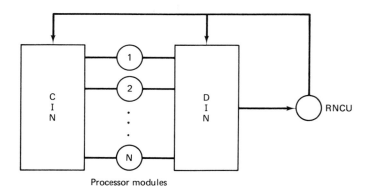

FIGURE VI.33 General architecture of fault-tolerant multiprocessor systems

putation reliability, computation availability, and mean computation before failure for graceful degradation systems. Examples of graceful degradation systems are C-mmp (Multi-miniprocessor) [14], PRIME [18], and the bitonic sorting network [39]. An example of a computationally undegradable system is the SERF (*S*ubelement *R*edundant *F*ault-*T*olerant) Computer [40].

A system is *diagnostically reconfigurable* if the state of DIN can be changed to improve the diagnostic performance of the system. The switching circuit in the selfpurging system or the hybrid redundancy can be considered as a type of implementation of DIN.

4. DIAGNOSIS OF MODULAR SYSTEMS

4.1 t-Fault Diagnosable Systems for Permanent Faults

Preparata, Metze, and Chien [41] proposed a graph theoretical model for the system diagnosis of digital systems. In their model, a system S consists of n computing modules denoted by $m_0, m_1, m_2, \ldots, m_{n-1}$. Each module is tested by a subset of other modules.

The model represents the system S by a *directed graph* (diagraph) $G(V, E)$ in which V is the set of *nodes* (each module is represented by a node in G) and E is the set of *arcs*. In this model, the arc from node i to node j represents a test link, l_{ij}, from module m_i to m_j. A system S is clled *t-fault diagnosable* if, given the response pattern of the system S, all permanently faulty modules in S can be *identified* provided that the number of faulty modules does not exceed t.

A test link l_{ij} exists if and only if module m_i tests module m_j. When this test is performed, a test outcome t_{ij} can be obtained. Let $t_{ij} = 0(1)$ if m_i judges m_j to be fault-free (faulty). The set of all test outcomes $\{t_{ij}\}$ in the system forms a *response pattern*. On the other hand, suppose the status (faulty or fault-free) of each node is known, then we can also denote the expected test outcome of a test link l_{ij}, by a_{ij} as follows:

$$a_{ij} = \begin{cases} 0 \text{ if both } m_i \text{ and } m_j \text{ are fault-free} \\ 1 \text{ if } m_i \text{ is fault-free and } m_j \text{ is faulty} \\ X \text{ (don't know) if } m_i \text{ is faulty} \end{cases}$$

The set of *all* a_{ij}'s in a system forms a *fault condition*. Let us call a possible status (each node being faulty or fault-free) of a system a *fault pattern*. There are 2^n fault patterns for an n-node system.

Assuming that no more than t faulty modules can occur in the system under test, based on the response patterns obtained, two cases can occur.

1. All faulty modules have been identified. In this case, repair can be done in one step and the system is said to be *one step t-fault diagnosable* (or t-fault diagnosable without repair).
2. Only a subset of faulty modules have been located. In this case, these faulty modules are replaced first. Then another diagnosis procedure is initiated to find the remaining faulty modules. For a system with no more than t faulty modules, up to t such sequential steps may be necessary to identify all faulty modules. Such a system is said to be *sequentially t-fault diagnosable* (or t-fault diagnosable with repair).

For each fault pattern, there is a corresponding fault condition. Let FP_i and FC_i be the ith pattern and its corresponding fault condition, respectively. Our purpose is to find two subsets of the set FC (which consists of all fault conditions for all possible fault pattens): one is a subset, FCA, of FC such that that all fault conditions in FCA are *mutually distinguishable* (i.e., if FC_i, FC_j ϵ FCA, $FC_i \cap FC_j = \phi$). The other is a subset, FCB, of FC such that all *indistinguishable* fault conditions in FCB have common faulty modules (i.e, if $FC_1, \ldots, FC_K \epsilon FCB$ and $FC_1 \cap \ldots \cap FC_k \neq \phi$ then $FP_1 \cap \ldots \cap FP_k \neq \phi$). If we can construct a subset FCA (FCB) such that for each FC_i ϵ FCA (ϵ FCB), the corresponding FP_i consists of at most t faulty modules, then the system is t-fault diagnosable without (with) repair. Obviously, if a system is t-fault diagnosable without (with) repair, then it is p-fault diagnosable without (with) repair for all $1 \leq p \leq t$. Generally when a system is called t-fault diagnosable, t always means the maximal number of faulty modules which can be located.

We first introduce a tabular approach to find out whether a system is t-fault diagnosable. Let $G(V, E)$ be the diagraph of a system, S with n modules (n nodes) and q test links (q arcs), we can construct a *fault condition table* as follows:

Case 1: If there are no faulty nodes in G, then the fault condition contains all zeroes:

$$FC_0 = \{a_{ij} = 0 \mid a_{ij} \epsilon \mathrm{E}\}$$

Case 2: If there are exactly k faulty nodes in G (where $n \geq k \geq 1$), then there are $\binom{n}{k}$ fault conditions.

Let FC_k^j be the jth fault condition for exactly k faulty nodes (where $\binom{n}{k} \geq j \geq 1$). The complete table has 2^n entries $\left(\sum_{i=0}^{n} \binom{n}{i} = 2^n \right)$ and each entry corresponds to a fault condition FC_k^j which is a string of length q consisting of 0, 1, and X. This table is called a *fault condition table,* and a table containing a subset of its entries is called a *partial fault condition table*. The system S is *t-fault diagnosable* if and only if any two different fault conditions, FC^h and FC^g, in the partial fault condition table $\{FC_0\} \cup \{FC_k^j \mid k \leq t$ and $\binom{n}{k} \geq j \geq 1\}$ are distinguishable, i.e., $FC^h \cap FC^g = \phi$ for all $\binom{n}{k} \geq h, g \geq 1$. This is due to the fact that, if $FC^h \cap FC^k \neq \phi$, then there exists a response pattern in $FC^h \cap FC^k$ that may not be diagnosed correctly.

Example 16. Suppose after testing a system, we have found a partial fault condition table for Fig. VI-34 excluding the dotted arcs as given in Table VI-4, where: Let $FC_k^j = a_{01}\ a_{02}\ a_{12}\ a_{13}\ a_{23}\ a_{24}\ a_{34}\ a_{30}\ a_{40}\ a_{41}$, we want to find t such that the system is t-fault diagnosable.

Obviously, any two different fault conditions in $Q = \{FC_0\} \cup \{FC_1^i \mid 5 \geq i \geq 1\} \cup \{FC_2^j \mid 10 \geq j \geq 1\}$ are distinguishable since the intersection of any pair in Q is empty. However, $\{FC_2^1, FC_3^1\}$ is indistinguishable since $FC_2^1 \cap FC_3^1 = XXXX\,000111 \neq \phi$ (empty). Therefore, it is a 2-fault diagnosable system.

For brevity, we shall not give the proofs for the theorems. Interested readers are encouraged to study the cited references for these proofs.

Theorem 4.1 If we assume that $G(V, E)$ is the digraph of a system S of n modules with the property that no two modules test each other in S,

TABLE VI-4

Row Number	FC_k^j		= Fault conditions									Faulty modules (m_i)
r	j	k	a_{01}	a_{02}	a_{12}	a_{13}	a_{23}	a_{24}	a_{34}	a_{30}	a_{40}	a_{41}	i
1	0	0	0	0	0	0	0	0	0	0	0	0	None
2	1	1	X	X	0	0	0	0	0	1	1	0	0
3	2	1	1	0	X	X	0	0	0	0	0	1	1
4	3	1	0	1	1	0	X	X	0	0	0	0	2
5	4	1	0	0	0	1	1	0	X	X	0	0	3
6	5	1	0	0	0	0	0	1	1	0	X	X	4
7	1	2	X	X	X	X	0	0	0	1	1	1	0, 1
8	2	2	X	X	1	0	X	X	0	1	1	0	0, 2
9	3	2	X	X	0	1	1	0	X	X	1	0	0, 3
10	4	2	X	X	0	0	0	1	1	1	X	X	0, 4
11	5	2	1	1	X	X	X	X	0	0	0	1	1, 2
12	6	2	1	0	X	X	1	0	X	X	0	1	1, 3
13	7	2	1	0	X	X	0	1	1	0	X	X	1, 4
14	8	2	0	1	1	1	X	X	X	X	0	0	2, 3
15	9	2	0	1	1	0	X	X	1	0	X	X	2, 4
16	10	2	0	0	0	1	1	1	X	X	X	X	3, 4
17	1	3	X	X	X	X	X	X	0	1	1	1	0, 1, 2

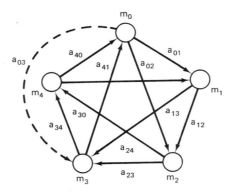

FIGURE VI.34 Diagraph G of system S

then S is *t-fault diagnosable* only if each module is tested by at least t other modules in S, i.e., the *indegree* (the number of incoming arcs) of each module has to be at least t [42].

Let us consider a system S of five modules represented by the digraph G in Fig. VI-34 excluding the dotted arc. From this figure we can observe that no two modules test each other and each module is tested by two other modules. Therefore this is a 2-fault diagnosable system. A sequence of nodes and arcs in $G(V, E)$, e.g., $m_1 a_{12} m_2 a_{23} m_3 \ldots m_{k-1} a_{(k-1)k} m_k$, is called a *dipath* from m_1 to m_k, and m_k is said to be *reachable* from m_1. A digraph $G(V, E)$ is said to be *strongly connected* if any pair of nodes in G are *mutually reachable*. The *connectivity* $k(G)$ of a digraph G is the minimum number of nodes whose removal from G yields a digraph that is not strongly connected or is trivial (i.e., consisting of only one node). For example, in Fig. VI-34, $m_0 a_{01} m_1 a_{12} m_2 a_{24} m_4$ is a dipath from m_0 to m_4; the connectivity of this digraph is $k(G) = 4$. Let $G(V, E)$ be the digraph of a system, S, of n modules. If $k(G) \geq t$ and $n \geq 2t + 1$, then S is t-fault diagnosable [42]. This is a sufficient condition for t-fault diagnosability. For example, in Fig. VI-34, $n = 5$ implies that $t \leq 2$. With the fact that $k(G) = 4 \geq 2$, we again find that the system is 2-fault diagnosable.

Theorem 4.2 Let $G(V, E)$ be the digraph of a system S of n modules. For each module $m_i \in V$, the set of modules testable by module m_i is $T(m_i) = \{m_j \mid a_{ij} \in E\}$. $T(X) = \left\{ \bigcup_{m \in X} T(m) \right\} - X$, where $X \subset V$, denotes the set of modules testable by a set of modules, X, but excluding the modules in X. Then the system S is t-fault diagnosable if and only if

1. $n \geq 2t + 1$
2. $d_{in}(m) \geq t$ for every module $m \in V$, where $d_{in}(m)$ is the *indegree* of module m.

3. for each integer p with $t > p > 0$, and each X with $|X|$ $= n - 2t + p$, then $|T(X)| > p$ where $|X|$ and $|T(X)|$ denote the numbers of modules in X (and $T(X)$) respectively [42].

Example 17. Let us consider the five-module system S shown in Fig. VI-34, including the dotted arc and excluding the arc a_{13}. If S is t-fault diagnosable, then t has to satisfy the following three conditions:

1. $n \geq 2t + 1$. In this case, $n = 5$ implies $t \leq 2$.
2. $d_{in}(m_i) = 2$ for $i = 0, 1, \ldots, 4$ which also implies $t \leq 2$.
3. To satisfy the third condition in Theorem 4.2, since from 1 and 2, the maximum value of t is 2 and thus $p < 2$, there are only two cases to consider. *Case 1*: $p = 1, t = 2$, then $n - 2t + p = 2$. So X can be any set $\{m_i, m_j\}$ where $4 \geq i, j \geq 0$ and $i \neq j$. Readers can check that for all such X, $|T(X)| \geq 2 > p$. *Case 2*: $p = 0, t = 2$; here $n + p - 2t = 1$. The number of modules testable by module m_0 is $|T(m_o)| = 3$. Similarly, $|T(m_1)| = 1$, $|T(m_2)| = 2$, $|T(m_3)| = 2$, and $|T(m_4)| = 2$. This implies all $|T(m_i)| \geq 1 > p$.

The results of 1, 2, and 3 imply that $t = 2$, and the system S represented by the digraph shown in Fig. VI-34 with a_{03} and without a_{13} is a 2-fault diagnosable system even though two nodes, m_0 and m_3 test each other. This can also be verified by constructing a partial fault condition table to cover all fault conditions for 1 and 2 faults and one fault condition for 3 faults. ■

A system S of n modules is a $D_{\delta A}$ design if and only if, in the digraph $G(V, E)$ of S, an arc a_{ij} exists from module m_i to module m_j such that $j = (i + \delta k) \mod n$ where δ is an integer less than n, and A and k are integers such that $n > A \geq k \geq 1$.

A system S of n modules of $D_{\delta A}$ design can be represented by a digraph $G(V, E)$, if each node has A *outgoing arcs* and each arc is separated by a "distance" δ (i.e., a_{ij_k} exists if and only if $j_k = (i + \delta \cdot k) \mod n$ and $A \geq k \geq 1$. Hence, $(j_{k+1} - j_k) \mod n = \delta$. Every module can test A other modules separated by a distance $\delta k (A \geq k \geq 1)$.

In a system of $D_{\delta A}$ design with n modules, the necessary and sufficient conditions that no two modules test each other is that n and δ are relatively prime. (Readers are encouraged to prove this condition as an exercise.) With the aid of Theorem 4.1, the following theorem is obtained.

Theorem 4.3 A system S of $D_{\delta A}$ design with n modules is t-fault diagnosable whenever n and δ are relatively prime and $t = A$. [41].

For example, the system in Fig. VI-34 is a D_{12} design and is 2-fault diagnosable since $n = 5, t = A = 2, \delta = 1$ and if the ith and jth module are tested by another module, then $(j - i) \mod n = 1 = \delta$. A *single loop sys-*

550   CHAPTER VI

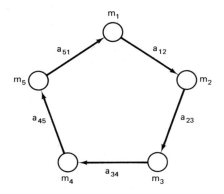

FIGURE VI.35   A single loop system of five modules

tem, as shown in Fig. VI-35, is a D_{11} design (i.e., $\delta = A = 1$). It is 1-fault diagnosable without repair due to Theorem 4.2. As a convention, we shall always order test outcomes according to the direction of arcs as shown in the next example.

Example 18.   A single loop system S of n modules is shown in Fig. VI-35. The partial fault condition table is as follows:

Let $FC_k^j = (a_{12}\ a_{23}\ a_{34}\ a_{45}\ a_{51})$

Assume the number of faulty modules does not exceed 2. The table below (excluding the last row) contains the following row pairs of indistin-

TABLE VI-5

Row Number	FC_k^j		= Fault conditions				Faulty Modules (m_i)
r	$j\ k$	$(a_{12}$	a_{23}	a_{34}	a_{45}	$a_{51})$	i
1	0 0	0	0	0	0	0	None
2	1 1	X	0	0	0	1	1
3	2 1	1	X	0	0	0	2
4	3 1	0	1	X	0	0	3
5	4 1	0	0	1	X	0	4
6	5 1	0	0	0	1	X	5
7	1 2	X	X	0	0	1	1, 2
8	2 2	X	1	X	0	1	1, 3
9	3 2	X	0	1	X	1	1, 4
10	4 2	X	0	0	1	X	1, 5
11	5 2	1	X	X	0	0	2, 3
12	6 2	1	X	1	X	0	2, 4
13	7 2	1	X	0	1	X	2, 5
14	8 2	0	1	X	X	0	3, 4
15	9 2	0	1	X	1	X	3, 5
16	10 2	0	0	1	X	X	4, 5
17	1 3	X	X	X	0	1	1, 2, 3

guishable fault conditions: (1, 6), (2, 6), (2,10), (3, 13), (4, 15), (5, 9), (6, 7), (8, 15), (9, 12), (10, 11), and (13, 14). Each pair indicate a common faulty module. We thus can identify at least one faulty module from a response pattern. For example, suppose the response pattern is (01010) which belongs to both FC_2^8 and FC_2^9. We can identify the common faulty module m_3. After m_3 has been repaired, we can continue to test the other faulty module (either m_4 or m_5). By definition, the system is 2-fault diagnosable with repair.

Now, let's assume the number of faulty modules may be 3. From the above table, we can see that FC_2^{10} and FC_3^1 are indistinguishable, however no common faulty module exists. For example, the response pattern (00101) cannot be used to identify any faulty module without the possibility of misidentifying a fault-free module. The system is not 3-fault diagnosable with repair. ■

The analysis shown in the example can readily be extended to other systems. We construct a partial fault condition table in increasing order of the number of faulty modules. Groups of indistinguishable fault conditions are identified. If they indicate no common faulty modules, the construction is terminated, and the maximum allowable number of faulty modules can be determined.

In general, we have the following theorem [43].

Theorem 4.4 A single loop system of n modules is sequentially t-fault diagnosable if and only if $n \geq m(t - m + 2) + 1$ where $m = \lfloor \frac{t}{2} \rfloor + 1$ and $[x]$ is the greatest integer less than or equal to X. If, given t, we choose the minimum number of arcs (also the minimum number of nodes) $n_0 = m(t - m + 2) + 1$, then the single loop system n_0 modules is an *optimal* (with respect to the minimum number of arcs in its diagraph) t-fault diagnosable system with repair.

4.2 t/s (t-out-of-s) Diagnosability [41]

A new diagnostic measure called *t/s (t-out-of-s) diagnosability* is introduced in [44]. A system S is *one step t/s diagnosable* if and only if, given one response pattern, the system with no more than t faulty modules can be repaired by replacing at most s modules. A one-step t/s diagnosable system replaces all suspected faulty modules, therefore, s is always greater than or equal to t. For example, a single loop system of five modules (Fig. VI-35) has been shown to be 2-fault diagnosable with repair. If we try to repair all possible faulty modules in one step for the system with at most two faulty modules, then we have to find out all suspected faulty modules. In Example 18 the table (excluding the last row) shows that a response pattern can be related to at most two fault conditions, and in term related to at most three suspected faulty modules. For example, (01110) can identify three suspected faulty modules (m_3, m_4, and m_5) via FC_2^8 and FC_2^9. In order to repair in one step, we need to replace at most three spare modules, hence, the system in Example 18 is a one-step 2/3 diagnosable system.

The following theorem has been proved in Theorem 1 of [44].

Theorem 4.5 A single loop system of n modules is one-step t/s diagnosable if and only if $n \geq s + 2$ and $s = m(t - m + 2) - 1$ where $m = \lfloor \frac{t}{2} \rfloor + 1$.

For example, a single loop system of seven modules is a one-step 3/5 diagnosable system. It consists of seven active modules. When it is diagnosed, no more than three simultaneously faulty modules are allowed, and it can be repaired by replacing no more than five modules. Obviously, a single loop system is one-step t/s diagnosable if it is t-fault diagnosable with repair. If, in Example 18, the response pattern (00101) occurs, it may belong to the fault conditions shown in rows 9, 16, or 17. There are no faulty modules common to all three fault conditions but all possible faulty modules determined by the response pattern are m_1, m_2, m_3, m_4, and m_5. This means we have to repair the whole system in one step and all five modules need to be replaced.

In the previous discussions, most of the analyses for various diagnosabilities are based on partial fault condition tables, and implies the use of three tables in the diagnostic procedures. When a system is large, this is sometimes prohibited by the complexity of analysis and the size of the tables. In following, we describe diagnosis based on strucutral analysis of response patterns.

Let $G(V, E)$ be the digraph of a system S. A set C, of nodes in G forms a *node chain* if there exists a dipath from one node to another node via a sequence of arcs where each node in C appears exactly once. A set L, of nodes in G forms a *node loop* if there exists a closed loop composed of a sequence of arcs where each node in C appears exactly once. The *length* of a node chain (node loop) is the number of nodes in the chain (loop).

Theorem 4.6 [43] In the digraph $G(V, E)$ representing a system S with no more than t faulty modules, if a subgraph G' of G contains k faulty nodes ($t \geq k \geq 0$) then a response pattern $00 \underbrace{\ldots}_{t-k} 0$ corresponding to a node chain C of $t - k + 1$ nodes in $G - G'$ (i.e., in G but not in G') indicates that the last node in C is fault-free.

The statement is still true if G' is empty and $k = 0$. This is due to the fact that there can exist at most $t - k$ faulty nodes in C. Suppose the last node was faulty, the one testing it has also to be faulty. Iteratively applying this argument, we find that all $t - k + 1$ modules in C are faulty, contradicting the previously stated fact that there can exist at most $t - k$ faulty nodes in C.

Theorem 4.7 [43] In a node loop L, if the response pattern of L is $R = 00 \ldots 0$, then nodes in L are either all faulty or all fault-free.

This is true because if we assume that any node in L is fault-free (faulty), then proceeding forwards (backwards) along L we will find all nodes fault-free (faulty).

Theorem 4.8 [43] In a single loop system S of n modules, a set of k faulty modules can produce a response pattern containing at most $2k$ 1's or at most k occurrences of the subsequence $\{01\}$ ($\{10\}$).

For example, for the system shown in Fig. VI-35, suppose modules m_2 and m_4 are faulty ($k = 2$). Then fault condition ($a_{12}a_{23}a_{34}a_{45}a_{51} = (1X1X0)$ means that the response pattern for a test may be one of the following four patterns: $\{(10100), (10110), (11100), (11110)\}$. We see that the response pattern (11110) has four 1's ($2k = 4$), and the response pattern (10110) has 2 occurrences of $\{01\}$ or $\{10\}$ subsequence. Since it is a single loop system, the last 0 and the first 1 in the response pattern form a 01 sequence.

Suppose we want to determine all suspected faulty modules and replace them in one step, as we did for one-step t/s diagnosability. The following diagnostic procedure can be applied to a single loop system with at most t faulty modules:

Step 1 Let R be the given response pattern and p be the number of 1's in R, then p may be $p = 0$, $p = 2t$, or $2t > p > 0$.

1. If $p = 0$, the system is fault-free.
2. If $p = 2t$, then flag alternate 1's by X as shown: 11 is flagged as 1^X1, 1111 as 1^X11^X1, and so on.

From Theorem 4.8; each X indicates an actual faulty module.

3. If $2t > p > 0$, then every subsequence $\{01\}$ in R indicates that there exists at least one faulty module (($m_i \xrightarrow{0} M_j \xrightarrow{1} m_k$) if m_j is fault-free, then m_k must be faulty. If m_j is faulty, then m_k may be either fault-free

fault-free or faulty), thus we flag 1 by X. If additional 1's follow $\{01\}$, flag alternate 1's by X as shown:

010 is flagged as $\underbrace{0_\uparrow 1^X_\uparrow 0}$, 0110 as $\underbrace{0_\uparrow 1^X_\uparrow 0}$, 01110 as $\underbrace{0_\uparrow 1^X_\uparrow 1 1^X_\uparrow 0}$ and so on

at least one module is faulty

The number of X's represents the greatest lower bound on the number of faulty modules.

Step 2 Apply Theorem 4.6 to the response pattern to identify all good modules, repeat this process until no good module can be identified, then mark all good modules with a check mark $\sqrt{}$ and the remaining modules are either actually or possibly faulty modules and repairing them in one step.

Example 19. Consider a single loop system of 32 modules. From Theorem 4.4, it is 9-fault diagnosable with repair; from Theorem 4.5, it is also one-step 9/29 diagnosable. Assume the response pattern R is given by $R = 0000000000\ 1\ 000000000\ 111101\ 000000$.

Step 1 $t = 9$, $s = 29$ and $p = 6$ imply that $2t > p > 0$. R is flagged according to Step 1 and is given by

$$R = \underbrace{0000000000}_{b}\ 1^X \underbrace{000000000}_{a}\ 1^X 1 1^X 1\ 0 1^X \underbrace{000000}_{b}$$

thus $k = 4$, $t - k = 5$.

Step 2 Apply Theorem 4.6 to node chain $a(b)$, all nodes in chain a(b) are good except the first five nodes.

$$R = \underbrace{\sqrt{}0\sqrt{}0\sqrt{}0\sqrt{}0\sqrt{}0\sqrt{}0\sqrt{}0\sqrt{}0\sqrt{}0\sqrt{}0\sqrt{}}_{b}\ 1^X \underbrace{000000\sqrt{}0\sqrt{}0\sqrt{}0\sqrt{}}_{a}\ 1^X 1 1^X 1 0 1^X \underbrace{000000\sqrt{}}_{b}$$

There are 15 totally good modules, and 17 suspected faulty modules to be replaced. Because at most nine faulty modules exist, at least eight fault-free modules have been replaced.

The above example clearly shows great overhead can occur in a one-step diagnosis. Instead, let us identify those definitely faulty modules by analyzing a response pattern. Repair those modules only and then test the system again, just as we did for t-fault diagnosability with repair. The analysis is pretty much like the one we just described, except for modifications on the last step: the modules to be replaced are those judged as faulty by good modules (i.e., the X in any $0\ 1^X$). In the above example

$$R = \vee 0 \vee 0 \vee 0 \vee 0 \vee 0 \vee 0 \vee 0 \vee 0 \vee 0 \vee 0 \vee 1 X_{000000} \vee 0 \vee 0 \vee 0 \vee 1 X_{11} X_{101} X_{000000} \vee$$
$$\uparrow \phantom{X_{000000} \vee 0 \vee 0 \vee 0 \vee 1}\uparrow \phantom{X_{11}}\uparrow$$

only the two modules pointed by arrows are definitely faulty, and have to be repaired before the next round of tests. The reason that the last module marked X is not definitely faulty is given below. Since 101 are the three symbols before the last X, we have $m_0 \xrightarrow{1} m_1 \xrightarrow{0} m_2 \xrightarrow{1} m_3$, where m_3, has been marked by X. This means that m_2 "thinks" that m_3 is faulty. This may happen because m_2 itself is bad. This is possible since m_1 may be bad and bad m_1 will "think" that m_2 is good (as shown by the 0 between m_1 and m_2). The following theorems have been proved in [43].

Theorem 4.9 *A system S of n modules of D_{1A} design with $A > \left[\frac{t}{2}\right]$ is t-fault diagnosable with repair if and only if $n \geq 2t + 1$.*

Theorem 4.10 *A system S of n modules of D_{1A} design with $A > \left[\frac{t}{2}\right]$ is is one-step t/s diagnosable if and only if $n \geq 2t + 1$ and $s = 2t - A$.*

Example 20. Figure VI-36 is the digraph $G(V, E)$ of a system S of seven seven modules of D_{12} design. There are 14 arcs (i.e., nA arcs) in G. From Theorem 4.9, the system is a 3-fault diagnosable system with repair, from

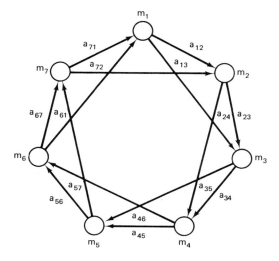

FIGURE VI.36 A D_{12} system of seven modules

CHAPTER VI

Theorem 4.10 it is a one-step 3/4 diagnosable system, and from Theorem 4.3 it is a 2-fault diagnosable sytem without repair. The readers can prove that all fault conditions are mutually distinguishable provided that the number of faulty modules does not exceed 2. If we assume that three modules m_1, m_3, and m_5 are faulty, the fault condition is given by R_3.

	a_{12}	a_{13}	a_{23}	a_{24}	a_{34}	a_{35}	a_{45}	a_{46}	a_{56}	a_{57}	a_{67}	a_{61}	a_{71}	a_{72}
$R_3 =$	X	X	1	0	X	X	1	0	X	X	0	1	1	0
$R_2 =$	1	0	1	0	1	0	1	0	1	0	0	1	1	0
$R_1 =$	1	1	1	0	1	1	1	0	1	1	0	1	1	0
$R =$	0	1	1	0	1	0	1	0	0	1	1	1	0	0

If a response pattern R_2 is obtained after a test, we have found a *node chain*, C, of the set $\{m_1, m_3, m_5, m_7, m_2, m_4, m_6\}$ with response $R_C = (a_{13}\ a_{35}\ a_{57}\ a_{72}\ a_{24}\ a_{46}) = (000^{\vee}0^{\vee}0^{\vee}0^{\vee})$. For $t = 3$, applying Theorem 4.6 to R_C, we find that modules m_1, m_2, m_4, and m_6 are definitely fault-free. Because $a_{71} = a_{61} = 1$ implies that m_1 is faulty, $a_{23} = 1$ and $a_{45} = 1$ imply that m_3 and m_5 are faulty, respectively. Hence, all three faulty modules can be repaired in one step, requiring three spares. If the response pattern is R_1 instead of R_2, we can find a node loop L of the set $\{m_2, m_4, m_6, m_7, m_2\}$ with response $R_L = (a_{24}\ a_{46}\ a_{67}\ a_{72}) = (0000)$. Applying theorem 4.7 to R_L we see that all nodes in node loop L are fault-free. Because $a_{61} = a_{71} = 1$ implies that m_1 is faulty, $a_{56} = a_{57} = a_{45} = 1$ and $a_{23} = a_{34} = 1$ imply that m_5 and m_3 are faulty, respectively. If the response pattern is R instead of R_1 and R_2, then the node chain D of the set $\{m_7, m_2, m_4, m_6\}$ has the response $R_D = (a_{72}\ a_{24}\ a_{46}) = (000^{\vee})$. Applying Theorem 4.6 to R_D we find the m_6 is fault-free ($t = 3$). Since $a_{61} = a_{67} = 1$ implies that m_1 and m_7 are faulty, there are in node chain D at most two faulty modules, so m_4 is fault-free. Hence, $a_{45} = 1$ implies m_5 to be faulty, and again node chain D now has at most one faulty module, so m_2 is fault-free. Furthermore, $a_{23} = 1$ implies that m_3 is faulty. We begin this diagnosis by assuming at most $t(=3)$ faulty modules, but the diagnostic procedure identifies four faulty modules (m_1, m_3, m_5, and m_7) which violates the assumption. If we assume that all nodes are faulty in node chain D, then $a_{56} = 0$ implies that m_5 is faulty. Furthermore, $a_{35} = 0$ and $a_{12} = 0$ imply that m_1 and m_3 are faulty, respectively. Now all seven modules are faulty, and the analysis for response pattern R implies that the system is not diagnosable if the number of faulty modules exceeds three. ∎

Example 21. Figure VI-37(a) is a seven module system S of D_{22} design. Figure VI-37 (b) is the rearrangement of (i.e., isomorphic to) Fig. VI-37(a).

From Theorems 4.9 to 4.10, the system is a 3-fault diagnosable system with sequential repair or a one step 3/4 diagnosable system. ∎

Design and Diagnosis of Reconfigurable Modular Digital Systems

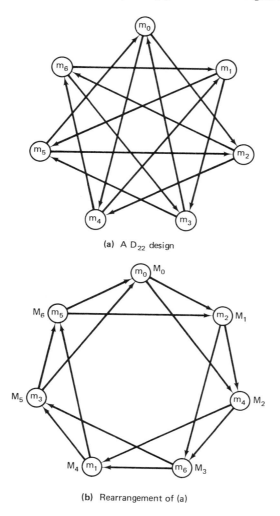

(a) A D_{22} design

(b) Rearrangement of (a)

FIGURE VI.37 A seven-module D_{22} design and its equivalent D_{12} design

In general, if a system S of n modules is a $D_{\delta A}$ design and δ and n are relatively prime, then the system S can be reconfigured to be an isomorphic D_{1A} system. Therefore, Theorems 4.9 and 4.10 for D_{1A} systems can also be applied for $D_{\delta A}$ systems if δ and n are relatively prime.

The diagnostic procedure for a t-fault diagnosable system S of n modules ules of D_{1A} design with repair is similar to that of a one-step t/s diagnosable D_{1A} system. In D_{1A} design with repair only those identified faulty modules are replaced rather than replacing all but those identified as fault-free modules.

4.3 t_i-fault Diagnosable Systems for Intermittent Faults

In addition to the diagnosability measures introduced for permanent faults, there exists an intermittent fault diagnosability. A system S is t_i-*fault diagnosable* if no more than t_i modules in S are intermittently faulty. A fault-free module will never be diagnosed as faulty and the diagnosis at any time is at worst *incomplete* but never *incorrect*. The incompleteness means that certain intermittent faults are not detected since the fault is inactive when the test is applied.

Example 22. Figure VI-34 without the dotted line shows a system of five modules of D_{12} design that is 2-fault diagnosable. We will find out whether it is 2_i-fault-diagnosable or not. If a module m_i is intermittently faulty, then $a_{ij} = a_{ki} = X$ for all j and k, where X denotes unknown. Now construct the *intermittently fault condition table* using the following correspondence.

a_{ij}	Intermittently Faulty Modules
0	None
X	m_i
D	m_j
X	m_i, m_j

where D means that $a_{ij} = 0$ if the intermittently faulty module m_j is not detected by the diagnostic test, but it will eventually become 1 if we repetitively apply the diagnostic test to m_j (i.e., when D is replaced by 1 in the intermittently fault condition table, then every entry is a fault condition provided that the intermittently faulty modules become permanently faulty). For the system shown in Fig. VI-34, the partial fault condition table is given below.

ACow10Number $FC_r =$	(a_{01}	a_{02}	a_{12}	a_{13}	a_{23}	a_{24}	a_{34}	a_{30}	a_{40}	a_{41})	Intermittently Faulty Modules (m_i)
r											i
0	0	0	0	0	0	0	0	0	0	0	none
1	X	X	0	0	0	0	0	D	D	0	0
2	D	0	X	X	0	0	0	0	0	D	1
3	0	D	D	0	X	X	0	0	0	0	2
4	0	0	0	D	D	0	X	X	0	0	3
5	0	0	0	0	0	D	D	0	X	X	4
6	X	X	X	X	0	0	0	D	D	D	0, 1
7	0	0	0	D	D	D	X	X	X	X	3, 4

If response pattern (1000000110) is obtained during testing, then from the above table we can see that this fault condition belongs to FC_1, and FC_6 and the fault is either in module 0 or in both module 0 and 1. If at most, one intermittently faulty module can exist in the system, then the response pattern can identify one intermittently faulty module m_0. We can observe that in the intermittent fault condition table, all entries are mutually distinguishable when some or all D's in each entry become 1, provided that the number of intermittently faulty modules never exceeds one. Therefore, we have shown that this system is 1_i-fault diagnosable without repair. Let us see whether the system is diagnosable for two intermittently faulty modules. Suppose we obtain the response pattern (0001000111), which belongs to FC_6 and FC_7, we cannot determine whether the modules m_0 and m_1 or m_3 and m_4 are intermittently faulty. Therefore the system is not 2_i-fault diagnosable without repair. ∎

The proof of the following theorem can be obtained in [45].

Theorem 4.11 A system S is t_i-fault diagnosable without repair if and only if in the digraph $G(V, E)$ of system S, given any two subsets S_1 and S_2 of G, and $0 < |S_1| \leq t_i$, $0 < |S_2| \leq t_i$ and $S_1 \cap S_2 = \phi$, the set R of the remaining modules of G (i.e., $R = G - (S_1 \cup S_2)$) has the property that both S_1 and S_2 receive at least one arc from R.

Example 23. Let us show that the system in Fig. VI-34 is 1_i-fault diagnosable by using the above theorem.

Case 1 Let $S_1 = \{m_i\}$, $S_2 = \{m_j\}$ where $i \neq j$ and $4 \geq i, j \geq 0$, then $R = \{m_k \mid k \neq i, k \neq j, 4 \geq k \geq 0\}$. If $j = (i + 1)$ mod 5 or $j = (i + 2)$ mod 5, then there are two arcs connected to m_i and one arc to m_j from R. If $j = (i - 1)$ mod 5 or $j = (i - 2)$ mod 5, then there are two arcs connected to m_j and one arc to m_i from R. Therefore the system is 1_i-fault diagnosable.

Case 2 Let $S_1 = \{m_0\}$, $S_2 = \{m_3, m_4\}$ and $R = \{m_1, m_2\}$, then there is no arc from R to S_1. Therefore, the system cannot diagnose two intermittent faults.

4.4 Diagnostic Reconfiguration

In general, if a system S is diagnostically reconfigurable, then once the detected faulty modules are located, two steps must be taken:

1. Purge out (or switch out) the faulty modules.
2. Determine a new diagnostic configuration (through switching networks) between the remaining modules such that some diagnostic

560 CHAPTER VI

procedures may still be applicable and the effect of the faulty modules on the diagnostic performance of the system S is minimized.

Example 24. An n-module system S of D_{1A} design has nA test links, and is A-fault diagnosable. The system shown in Fig. VI-36 is a D_{12} design of seven modules ($t = 2$). Once one or two faulty modules are located by the t-fault diagnostic procedure, it can be reconfigured to be either a D_{12} design of six modules or a D_{12} design of five modules in order to tolerate failures. Repeating this process, we obtain a *diagnostic reconfiguration tree* of this system based on t-fault diagnosability. Let $D_{1A}(n, nA)$ represent the digraph of a system of n modules of D_{1A} design with nA arcs. Then the reconfiguration tion tree of the system $D_{12}(7, 14)$ is given in Fig. VI-38.

In Fig. VI-38 we neglect all similar subtrees by giving only one of them. If our reconfiguration scheme is identified by a path given by $D_{12}(7, 14) \to D_{12}(6, 12) \to D_{12}(5, 10) \to D_{11}(4, 4) \to D_{11}(3, 3) \to DS$, the diagnostic performance of this reconfiguration scheme is given by Fig. VI-39(a) and the fault tolerant graph (FTG) of $D_{12}(7, 14)$ design without reconfiguration is given by Fig. VI-39(b).

Let $G(V, E)$ be the digraph of a system S of n modules and let each module in G has A outgoing arcs, then the system S is defined as a *one-step*

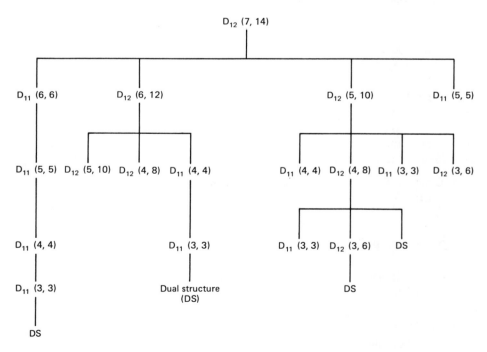

FIGURE VI.38 One reconfiguration tree of $D_{12}(7, 14)$

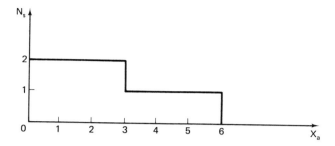

(a) FTG of reconfigurable D_{12} (7, 14) system

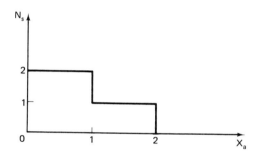

(b) FTG of D_{12} (7, 14) system without reconfiguration

FIGURE VI.39 Diagnostic performance of D_{12} (7, 14) system with and without reconfiguration

$(X_0, X_1, \ldots, X_{A-1})$ *diagnostically reconfigurable* system if at each step of diagnosis the system can exactly diagnose $N_S \leq A - i$ $(i = 0, 1, \ldots, A - 1)$ simultaneous faults as long as the number of accumulated faults, X_a, does not exceed $\left(\sum_{j=0}^{i} X_j\right) - 1$. The FTG of the system is shown in Fig. VI-40. For example, the system shown in Fig. VI-39 is not a one-step [3, 3] diagnostically reconfigurable system. A system S of n modules of D_{1A} design is said to be a reconfigurable $D_{1[X_0, X_1, \ldots, X_{A-1}]}$ design if the system can be reconfigured into $D_{1(A-i)}$ design $(A - 1 \geq i \geq 0)$ as long as the number of accumulated faults, X_a, does not exceed $\left(\sum_{j=0}^{i} X_j\right) - 1$.

The reconfiguration tree for the reconfigurable $D_{1[X_0, \ldots, X_{A-1}]}$ design of a A_{1A} system is shown as $D_{1A}(n, nA) \rightarrow D_{1(A-1)}(n_1, n(A-1)) \rightarrow \ldots \rightarrow D_{11}(1, 0)$ where $n > n_1 > \ldots > 1$.

For example, the path $D_{12}(5, 10) \rightarrow D_{11}(3, 3)$ in Fig. VI-38 is a reconfigurable $D_{1[2, 1]}$ design.

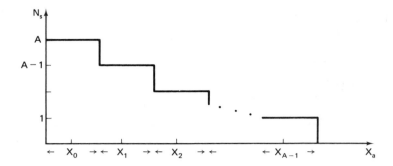

FIGURE VI.40 FTG of a one-step $[X_0, X_1, \ldots, X_{A-1}]$ diagnostically reconfigurable system

Obviously, if a system S of D_{1A} design is reconfigurable $D_{1[X_0,\ldots, X_{A-1}]}$ design, then it is also one-step $[X_0,\ldots, X_{A-1}]$ diagnostically reconfigurable. ∎

4.5 Concurrent Computation and Diagnosis

A selfdiagnosing system S is represented by a digraph $G(V, E)$ in which the set of nodes V corresponds to replaceable modules in S and the set of arcs E represents the set of testing connections among the modules of system S. Each module in system S may be in one of two states; i.e., a busy state and an idle state. These are defined as follows:

1. The *busy* state is when the module has been assigned to a certain computation.
2. The *idle* state is when the module has not been assigned to a certain computation.

All idle modules may be assigned to the diagnostic part of the system. Since only idle modules can participate in system diagnosis, the subgraph consists of *diagnostic nodes* (representing idle modules) and is obtained by removing all the computation nodes (representing busy modules) and their incoming and outgoing arcs from $G(V, E)$. Let $G_D(V', E')$ with $V' \subset V$ and $E' \subset E$ be the diagnostic subgraph which consists of all diagnostic nodes, then the digraph $G_D(V', E')$ has to maintain diagnostic capabilities in the system. The main purpose of designing *concurrent computation and diagnosis* is to maximize the number of busy modules (maximize the degree of parallelism), while maintaining the maximal diagnostic capability in the rest of the system.

A system S is *concurrently one-step diagnosable* if it is possible to identify in one step each node in the diagnostic subgraph to be faulty or fault-free without interrupting any computation nodes.

Example 25. The system in Fig. VI-41 is a system of D_{13} design which is 3-fault diagnosable. Suppose one module m_i is busy, then the diagnostic subgraph $G_D(V_1, E_1)$ has nine nodes and 24 arcs, and the indegree of each node m_j in $G_D(V_1, E_1)$ is 3 if $j \neq (i + k)$ mod 10 where $3 \geq k \geq 1$, or 2 if $j = (i + k)$ mod 10 where $3 \geq k \geq 1$. Therefore the diagnostic subgraph $G_D(V_1, E_1)$ is 2-fault diagnosable (Theorem 4.1). If it has two busy modules m_i and m_j, then the diagnostic subgraph $G_D(V_2, E_2)$ has two structures.

Case 1 If $j = (i + k)$ mod 10 and $2 \geq k \geq 1$, then each node has at least one incoming arc. This implies that the system representing $G_D(V_2, E_2)$ is 1-fault diagnosable

Case 2 If $j \neq (i + k)$ mod 10 and $2 \geq k \geq 1$, then each module has at least two incoming arcs. This implies that the system representing $G_D(V_2, E_2)$ is 2-fault diagnosable. If there are three busy modules, then the diagnostic subgraph $G_D(V_3, E_3)$ may be undiagnosable (e.g., m_i, $m_{(i+1) \bmod 10}$ and $m_{(i+2) \bmod 10}$ are busy), 1-fault diagnosable (e.g., m_i, $m_{(i+1) \bmod 10}$ and $m_{(i+3) \bmod 10}$ are busy), or 2-fault diagnosable (e.g., m_i, $m_{(i+3) \bmod 10}$ and $m_{(i-3) \bmod 10}$ are busy). If there are four busy modules then the diagnostic subgraph $G_D(V_4, E_4)$ may be undiagnosable, or 1-fault diagnosble. If there are five busy modules, then the diagnostic subgraph $G_D(V_5, E_5)$ may be undiagnosable or 1-fault diagnosable. If six modules are busy, then the diagnostic subgraph is either undiagnosable or 1-fault diagnosable. More than six busy modules

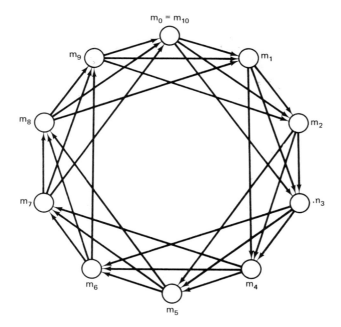

FIGURE VI.41 A D_{13} (10, 30) system

make the diagnostic subgraph undiagnosable. If the reconfiguration scheme is determined as assigning the busy modules according to the sequence defined by $m_0 \rightarrow m_5 \rightarrow m_1 \rightarrow m_6 \rightarrow m_3 \rightarrow m_8 \rightarrow m_2 \rightarrow m_6 \rightarrow m_7 \rightarrow m_9$, then the computation fault-tolerant graph (CFTG) of this diagnostic performance is shown in Fig. VI-42. The CFTG is a function, $N_S(B_A)$, which defines the maximal number of simultaneous faults diagnosable correctly by a system as a function of accumulated busy modules. The CFTG is a performance measure of concurrent computation and diagnosis. ∎

Theorem 4.12 [46] In a system of n modules of D_{1A} design with $n \geq 2A + 1$, let m be an integer such that $(m + 1)A \geq n \geq mA$, and $i = lA$ where $m \geq l \geq 0$ and l is an integer. Suppose we assign all modules $m_{j,\ (i,\ k)}$ to be diagnostic modules where $j(i, k) = (i + k)$ mod n and $A \geq t - 1 \geq k \geq 0$, and the remaining modules to be computational modules. Then the diagnostic subgraph is a *concurrently one-step t-fault diagnosable system*.

For example in Fig. VI-41, if we build up the diagnostic subgraph which consists of m_0, m_3, m_6, and m_9 and related arcs among them, then the diagnostic subgraph is a 1-fault diagnosable system with D_{11} design. Let the number of computational modules and diagnostic modules be denoted by N_C and N_D, respectively. Then Theorem 4.12 gives three cases of N_C and N_D as follows:

Case 1

$$n = mA$$
$$N_C = m(A - t)$$
$$N_D = mt$$

Case 2

$$mA + (t - 1) \geq n > mA$$

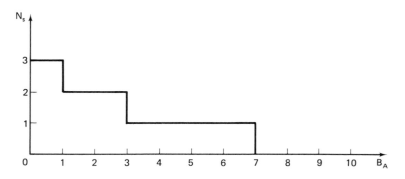

FIGURE VI.42 CFTG of the reconfiguration scheme in Example 25

$$N_C = m(A - t)$$
$$N_D = n - m(A - t)$$

Case 3
$$(m + 1)A \geq n > mA + (t - 1)$$
$$N_C = n - (m + 1)t$$
$$N_D = (m + 1)t$$

Example 26. A system S of seven modules of D_{14} design is shown in Fig. VI-43. We want to find out the maximal number of computational modules in the system which can maintain the concurrently one-step 1-fault diagnosability. In order to maintain the concurrently one-step 1-fault diagnosability, the diagnostic subgraph must contain at least a D_{11} system of three modules. One such system can be constructed to contain three modules m_0, m_4, and m_6 and four arcs a_{04}, a_{40}, a_{46}, and a_{60} which has proper subgraph D_{11} consisting of m_0, m_4, m_6, a_{04}, a_{46}, and a_{60}. Hence the maximal number of computational modules is four. ∎

5. SOME UNSOLVED PROBLEMS

A procedure to provide fault-tolerance is shown in Fig. VI-44. First the error is detected, then error diagnostics identify the site of error. If the error is permanent, the reconfiguration system is reconfigured to exclude the faulty module from the system. If the faulty module is replaced by a fault-free

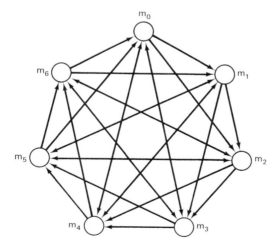

FIGURE VI.43 A D_{14} design of seven modules

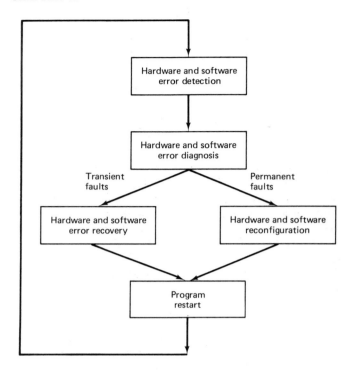

FIGURE VI.44 A procedure for fault-tolerance

module, the system is fault-tolerant. If the faulty module is not replaced, the system is gracefully degraded. For intermittent errors, a recovery procedure is activated. After either reconfiguration or recovery procedures, the program restart procedure is activated so that the program can continue its operation. The whole procedure is then repeated. Error detection, diagnosis, and reconfiguration involves both the hardware and the software. The recovery procedure deals only with the software. The whole process improves the reliability of the system. We can see that error diagnosis is a necessary but not a sufficient step for increasing the system reliability. Thus diagnostic performance can be considered as a part of reliability performance of a digital system.

In general, a fault-tolerant system design should be based on some measures of system peformance. Reliabilty can be improved through the use of redundancies and/or effective diagnostic procedures. For a permanent fault, the system switches out each faulty module. If it is a transient fault, the system runs recovery software to cause a program to rerun at a predetermined rollback point. A main method for improving system reliability is to design a complete self-repairing redundant system with permanent fault-tolerance and transient faults recovery capabilities. Reconfiguration among multiple computers and/or redundant systems can enhance the reliability performance by increasing the diagnostic performance (e.g., diagnostic or computational fault-tolerant graph) by the use of diagnostic interconnection networks (DIN). Alternatively, the system can increase its reliability performance (e.g., reliability meaure, MTBF, and availability, etc.) by using the interconnection networks among redundant modules and active modules or the computational performance (e.g., computation reliability, computation availability, and mean computation before failure, etc.) by the means of computational interconnection networks (CIN). Therefore, the design of a reconfigurable fault-tolerant multiprocessor system based on the tradeoffs among these measures of system performances is an important topic for research in the near future. In other words, the area of concurrent computation and diagnosis in multiprocessor systems (either redundant or nonredundant) needs more research.

In diagnostic performance analysis, Preparata's model is not perfect since a fault-free module may fail to detect some faults in the module under test. To cope with such an incomplete fault detection the following probabilities can be defined. Let t_{ij} be the test outcome when module m_i tests module m_j, then the probabilities of the test outcomes for incomplete fault detection under different situations are given below. [47].

$\Pr\{t_{ij} = 0 \mid m_i \text{ and } m_j \text{ are both fault-free}\} = 1$

$\Pr\{t_{ij} = 1 \mid m_i \text{ is fault-free and } m_j \text{ is faulty}\} = P_i$

$\Pr\{t_{ij} = 0 \mid m_i \text{ is fault-free and } m_j \text{ is faulty}\} = 1\text{-}P_i$

$\Pr\{t_{ij} = 1 \mid m_i \text{ is faulty and } m_j \text{ is fault-free}\} = r_i$

$\Pr\{t_{ij} = 0 \mid m_i \text{ is faulty and } m_j \text{ is fault-free}\} = 1 - r_i$

$\Pr\{t_{ij} = 1 \mid m_i \text{ and } m_j \text{ are both faulty}\} = s_i$

$\Pr\{t_{ij} = 0 \mid m_i \text{ are } m_j \text{ are both faulty}\} = 1 - s_i$

If $P_i = 1$ then it corresponds to the model proposed by Preparata, Metze, and Chien [41]. The probabilistic model of incomplete fault detection can analyze the situation when there is a probability that certain faults are not detected. Therefore, the development and analysis of such a model are worthwhile tasks.

For the measures of reliabilty and computation performances, we have introduced the redundant systems, gracefully degradable systems with or without spares, etc. It is imperative to analyze the case in which a system utilizes both redundant techniques and reconfiguration tehcniques (i.e., gracefully degradable systems). For example, a system S of five processing modules can have the following structures:

1. 5MR
2. TMR/spares

3. Dynamic redundancy (unipurge or bipurge)
4. Hybrid redundancy (TMR core + two spares)
5. Gracefully degradable structure with degradation of degree 5
6. Selfrepairing structure (e.g, two active core + three spares)
7. Gracefully degradable selfrepairing structure (e.g., three active modules + two spares)
8. Gracefully degradable TMR structure. This means that the three parallel processing units (TMR and other two active modules) have degradation capability.

The problem is: Given the specific values on the measures of reliability and computational performances, find the optimal design of a multi/computer system by specifying a cost function that includes the measures of performances.† Current research should address this type of topic or try to answer the question: Is it worthwhile to introduce redundant structure into the graceful degradation system?

Concurrent fault detection techniques are the basis for fault recovery. Approaches for developing efficient recovery algorithms (either hardware or software or both) to reach this objective determine directly the efficiency of recovery processes. A fault-tolerant distributed computing system poses two major problems: One is the performance measure of communications networks, the other is the fault-tolerant communication protocols. The fault-tolerant operating system in each computer and the network operating system among the computers become an important research area.

†More information on optimal design of a multi-computer system can be found in Chap. III and Chapter V, Sec. 2. (Ed.)

REFERENCES

[1] SU, S. Y. H. and R. J. SPILLMAN, "An Overview of Fault-Tolerant Digital System Architecture," *Proceedings of 1977 National Computer Conference,* June 1977, pp. 19–26.

[2] RANDELL, B., "System Structure for Software Fault-Tolerance," *IEEE Transactions on Software Engineering,* June 1975, pp. 220–232.

[3] MELLIAR-SMITH, P. M. and B. RANDELL, "Software Reliability: The Role of Programmed Exception Handling," *Proceedings ACM Conference on Language Design for Reliable Software,* Sigplan Notices 12, 3, March 1977, pp. 95–100.

[4] MALAIYA, Y. K. and S. Y. H. SU, "A Survey of Methods of Intermittent Fault Analysis," *Proceedings of 1979 National Computer Conference,* pp. 577–586.

[5] KARPOVSKY, M. and S. Y. H. SU, "Detection and Location of Input and Feedback Bridging Faults Among Input and Output Lines," *IEEE Trans. on Computers,* June 1980. pp. 523–527.

[6] ———, "Detecting Bridging and Stuck-at Faults at Input and Output Pins of Standard Digital Components," *Proceedings of the 17th Design Automation Conference,* Minneapolis, Minnesota, June 1980, pp. 494–505.

[7] ANDERSON, K. R. and H. A. PERLEINS (Guest Editors), "Hardware Test Technology," Special issue of *Computer*, October 1979.

[8] AVIZIENIS, A. A., "Design of Fault-Tolerant Computers," *AFIPS Conference Proceedings,* **31,** Fall Joint Computer Conference, 1967, pp. 733–743.

[9] PRADHAN, P. K., Guest editor, "Fault-Tolerant Computing," Special issue of *Computer,* March 1980.

[10] STIFFLER, J. (Guest Editor), "Fault-Tolerant Computing," Special issue of *Computer,* June 1980.

[11] REDDY, S. M. (Guest editor), "Fault-Tolerant Computing," Special issue of *Computer,* June 1978.

[12] MALAIYA, Y. K. and S. Y. H. SU, "Fault-Tolerance in Multiprocessor Systems," *Proceedings of 1980 International Conference on Circuits and Computers.* Port Chester, N.Y., sponsored by IEEE, October 1–3, 1980. pp. 710–716.

[13] KUEHN, R. E., "Computer Redundancy: Design, Performance and Future," *IEEE Trans. on Reliability,* February 1969, pp. 3–11.

[14] SIEWIOREK, P. P., V. KINI, H. MASHBURN, S. MCCONNEL, and M. TSAO, "A Case Study of C-mmp, Cm* and C.vmp: Part I: Experiences with Fault-Tolerance in Multiprocessor Systems, Part II: Predicting and Calibrating Reliability of Multiprocessor Systems," *Proc. of IEEE,* October 1978, pp. 1178–1220.

[15] WAKERLY, J. F., "Transient Failures in Triple Modular Redundancy Systems with Sequential Modules," *IEEE Trans. on Computers,* May 1975, pp. 570–572.

[16] BOURICIUS, W. G., "Reliability Modeling Techniques for Self-Repairing Computing Systems," *Proceedings 24th National Conference of the ACM,* August 1969, pp. 295–309.

[17] "Telecommunications," *IBM Sys J,* **18,** 2, 1979, pp. 189–347.

[18] BORGERSON, B. R., "A Fail-Softly System for Time-Sharing Use," *Digest of the 1972 International Symposium on Fault-Tolerant Computing,* Boston, Massachusetts, June 1972, pp. 89–93.

[19] AVIZIENIS, A. A., "The STAR Computer: An Investigation of the Theory and Practice of Fault-Tolerant Computer Design," *IEEE Trans. on Computers,* **C20,** *11,* November 1971, pp. 1322–1331.

[20] DOWNING, R. W., "No. 1 ESS Maintenance Plan," *Bell System Technical Journal,* **43,** September 1964, pp. 1961–2020.

[21] TAYLOR, D. S., "Unpowered to Powered Failure Rate Ratio: A Key Reliability Parameter," *IEEE Trans. on Reliability,* April 1974, pp. 33–36.

[22] MATHUR, F. P. and A. AVIZIENIS, "Reliability Analysis and Architecture of a Hybrid-Redundant Digital System: Generalized TMR with Self-Repair," *AFIPS Conference Proceedings,* **36,** Spring Joint Computer Conference, 1970, pp. 376–384.

[23] SIEWIOREK, P. P. AND E. J. MCCLUSKEY, "An Interactive Cell Switch Design for Hybrid Redundancy," *IEEE Trans. on Computers,* March 1973, pp. 290–297.

[24] MATHUR, F. P. and P. T. DESOUSA, "Reliability Modeling and Analysis of General Modular Redundant Systems," *IEEE Trans. on Reliability,* December 1975, pp. 296–299.

[25] CHANDY, K. M., C. V. RAMAMOORTHY, and A. COWAN, "A Framework for Hardware-Software Tradeoffs in the Design of Fault-Tolerant Computers," *AFIPS Conference Proceedings,* **41,** 1972, pp. 55–63.

[26] LOSQ, T., "A Highly Efficient Redundancy Scheme: Self-Purging Redundancy," *IEEE Trans. on Computers,* June 1976, pp. 569–578.

[27] TAYLOR, D. S., "A Reliability and Comparative Analysis of Two Standby System Configurations," *IEEE Trans. on Reliability,* April 1973, pp. 13–19.

[28] SU, S. Y. H. and E. DUCASSE, "A Hardware Redundancy Reconfiguration Scheme for Tolerating Multiple Module Failures," *IEEE Trans. on Computers,* March 1980, pp. 254–257.

[29] HECHT, H., "Fault-Tolerant Software for a Fault-Tolerant Computer," *Software Engineering Techniques,* Onlin Conference Limited, Oxbridge, England, 1976, pp. 235–348.

[30] KIM, K. H., "Error Detection, Reconfiguration and Recovery in Distributed Processing Systems," *Proceedings, First International Conference on Distributed Computing Systems,* October 1979, pp. 284–295.

[31] GOODENOUGH, J. B., "Exception Handling: Issues and a Proposed Notation," *Communications of ACM* **18,** *12,* 1975, pp. 683–696.

[32] NG, Y. W. and A. AVIZIENIS, "A Reliability Model for Gracefully Degrading and Repairable Fault-Tolerant Systems," *Digest of the 1977 International Symposium on Fault-Tolerant Computing,* 1977, pp. 22–28.

[33] BOURICIUS, W. G., "Reliability Modeling for Fault-Tolerant Computers," *IEEE Trans. on Computers,* **C20,** *11,* November 1971, pp. 1306–1311.

[34] MATHUR, F. P. and P. T. DESOUSA, "Reliability Models of NMR Systems," *IEEE Trans. on Reliability,* **R24,** June 1975, pp. 108–112.

[35] MCCONNEL, S. R., D. P. SIERWIOREK, and M. M. TSAO, "The Measurement and Analysis of Transient Errors in Digital Computer Systems," *Digest of the 1979 International Conference on Fault-Tolerant Computing,* June 1979, pp. 67–70.

[36] BEAUDRY, M. D., "Performance Related Reliability Measures for Computing Systems," *Proceedings, Fault-Tolerant Computing Symposium,* 1977, pp. 16–21.

[37] SAHEBAN, F. and A. D. FRIEDMAN, "A Survey and Methodology of Reconfigurable Multimodule Systems," *Proceedings, Computers Software and Applications Conference,* 1978, pp. 790–796.

[38] ———, "Diagnostic and Computational Reconfiguration in Multiprocessor Systems," *Proceedings of the ACM78 Annual Conference,* 1978, pp. 68–78.

[39] BATCHER, K. E., "Sorting Networks and Their Applications," *AFIPS Conference Proceedings,* **32,** Spring Joint Computer Conference, 1968, pp. 307–314.

[40] STIFFLER, J., "The SERF Fault-Tolerant Computer, Part I: Conceptual Design, and Part II: Implementation and Reliability Analysis," *Digest of the 1973 International Conference on Fault-Tolerant Computing,* 1973, pp. 23–32.

[41] PREPARATA, F. P., G. METZE, and R. T. CHIEN, "On the Connection Assignment Problem of Diagnosable Systems," *IEEE Trans. on Computers,* December 1967, pp. 848–854.

[42] HAKIMI, S. L. and A. T. AMIN, "Characterization of Connection Assignment of Diagnosable Systems," *IEEE Trans. on Computers,* January 1974, pp. 86–88.

[43] KARUANITHI, S. and A. D. FRIEDMAN, "Analysis of Digital Systems using a new Measure of System Diagnosis," *IEEE Trans. on Computers,* Feb. 1979, pp. 121–133.

[44] ———, "System Diagnosis with t/s Diagnosability," *Digest of the 1977 International Symposium of Fault-Tolerant Computing,* June 1977, pp. 65–71.

[45] MALLELLA, S. and G. M. MASSON, "Diagnosable Systems for Intermittent Faults," *IEEE Trans. on Computers,* June 1978, pp. 560–566.

[46] SAHEBAN, F., L. SIMONCINI and A. D. FRIEDMAN, "Concurrent Computation and Diagnosis in Multiprocesor Systems," *Digest of the 1979 International Symposium on Fault-Tolerant Computings,* 1979, pp. 149–156.

[47] SIMONCINI, L. and A. D. FRIEDMAN, "Incomplete Fault Coverage in Modular Multiprocessor Systems," *The ACM Annual Conference,* 1978, pp. 210–215.

Glossary of Definitions Used in the Book

(Roman numerals indicate chapters of the book, provided the definition is used in one chapter only.)

Accumulator Registers. The accumulator registers hold address information and active data for the instruction. It is popular to use these functions to assign a register in the processor that sends its content to an input terminal of the adder and receives a result produced by the adder.

Acquiring the Pulse Trains. This is the task of examining all incoming pulses which do not belong to any established pulse train and sorting them into new pulse trains, if possible. (II)

Adaptation System. This is a system for dynamic architectures in which the software that performs preprocessing of algorithms and determines the best architectures for their execution. Also it assigns the processor and memory resources for dynamic architectures and includes two subsystems—*assignment* and *monitor*. The first system assigns the hardware resource to programs and constructs a flow chart of architectural transitions. The monitor subsystem supervises correct execution of the flow chart of architectural transitions. (III)

Address Computers. Depending on the number of addresses stored in the address field of an instruction, von Neumann computers were divided into:

1. Three address computers in which each instruction stored three addresses, two for operands and one for computational results.
2. Two address computers in which each instruction stored only two addresses (either for two operands or for one operand and the result).
3. One address computer in which each instruction stored only one address, either of an operand or of the result. (I)

Address Field of Frame. The address field of frame is used to designate the receiver station(s). (One address may designate more than one station in the case of broadcasting communication.) (VI)

Allocation. Allocation is the assignment of programs and data arrays to processor and memory resources. (*see* Resource assignment)

Application of Structural Termination to Simple-loops. The structural termination analysis of simple-loops can be applied to programs whose control graphs contain any number of simple loops; structural termination is determined separately for each of them. (IV)

Architectural Switch to a New State. DC-group transition from one state to another is performed by a special $N_d \to N_f$ instruction which may be executed by any computer functioning in the state N_d. This instruction initiates the V monitor, which first checks the priority of the calling program and the availability of the equipment requested. If the program is denied execution of the $N_d \to N_f$ transition, the instruction stops, otherwise monitor checks the availability of the requested equipment. If the requested computer elements are free they fetch new meanings for the variable control codes necessary for the next state, N_f. Otherwise $N_d \to N_f$ waits until the requested computer elements stop execution in the predecessor state, N_d. The last stage of an architectural switch is establishment of new transfer modes in the connecting elements, MSE. The task of the V monitor is to identify which connecting elements should switch and which transfer mode each of them should assume. (III)

Array Adaptation. This means the system capability to change the number of concurrent arrays, the dimension of a data vector computed in each array, and the word sizes of processors working in each array. For each array, its resources may be redistributed among its processors, so that a data vector computed in each array may contain different size data items. Therefore, the architecture is capable of performing array computations with variable precision, a variable dimension of data vectors, and a variable number of data vectors.

Array System. An array system is generally understood to be a collection of N processors, $P_1, \ldots, P_N$, handling the same instruction issued by a single control unit. Each processor, P_i, is equipped with a local memory, M_i, used by P_i for storing both its operands and the computational results it obtains. An array composed of N processors may concurrently execute N identical operations with one program instruction. Each instruction handles a data vector made of operands handled by $P_1, \ldots, P_N$, respectively. An array of N processors thus implements data parallelism but no instruction parallelism.

Assessment of Supernetworks. One-stage and multistage Supernetworks are two examples of a classical tradeoff. A high rate of information processing in a one-stage network is traded for a less complex realization in the multistage network. For those applications for which speed is a major concern it is expedient to partition the entire network into several one-stage subnetworks in order to reduce the complexity of the resource. Each one-stage subnetwork may then function as a more complex node of the Supernetwork, and all such nodes are interconnected together via one of the interconnection networks described in the literature. (I)

Associative Array Control Unit. The associative array control unit exercises direct control over the associative memory arrays. (II)

Associative Memory. The associative memory is an extension of the search memory and its operation is based on the same search-by-content function implemented in the search memory. The associative memory contains additional logic at each cell which permits it to perform work-parallel logical, arithmetic, input, and output operations. (II)

Associative Parallel I/O Control Unit. The associative parallel I/O control unit controls the data transfers performed by the parallel I/O flip network. (II)

Associative Parallel I/O Flip Network. The associative parallel I/O flip network handles parallel data transfers among the four STARAN associative memory arrays, and also the parallel input and output of data to the arrays. The flip network permits the data to be permuted during these transfer operations. (II)

Associative Processor. An associative processor is a system that consists of an associative memory, together with a processor for controlling the operation of the associative memory and for interfacing with an operator. The controller is generally a sequential processor with a special interface for the associative memory and with some corresponding special instructions for implementing the control and monitor functions for the memory. (II)

Backward Error Recovery Scheme. A scheme that backs up one or more of the processes of a system to a previous state which is hoped to be error-free, before attempting to continue further operation of the system or subsystem. This scheme depends on the provision of recovery points (RP) and a state vector, which are the states of a process or a program that can be recorded and later reinstated. The backward error recovery scheme can be designed by assigning a recovery point for a process each time it enters an atomic action, then the states at the recovery point are saved and retained for all processes involved in the given atomic action until all have reached the end of that atomic action. (VI)

Basic Concepts of von Neumann Architecture are:

1. Principle of addresses
2. Fixed instruction microprograms (I)

Basic Features of PS Exchange are as follows:

1. In any $C_i(k)$ computer one $16 \cdot k$-bit word is stored in a single parallel cell of k memory elements, all with the same address. Consequently, a data array containing d $16 \cdot k$ bit words will take d parallel cells for its storage in different size computers. When a $16 \cdot k$ bit word is accessed, each of k computer elements generates the same address, resulting in a concurrent fetch of all k 16-bit bytes comprising the data word.

2. Since the minimal size computer is equivalent to one computer element (CE), the instruction size has been made coincident with that of one CE. The instruction sequences are stored in consecutive cells of one memory element (ME), and complex programs may be stored in several ME's. The instruction fetched from one ME would then be sent concurrently to all modules of the $C_i(k)$ computer. Therefore, the same program occupies the same number of cells in any size computer. (III)

Basic Principles of Microprogrammed Architecture are:

1. *Reconfiguration codes:* the microinstruction must store special reconfiguration codes that select various microoperation sequences in the instruction microprogram.
2. *Variable microprogram:* a programmer may change the instruction microprogram by selecting more task-oriented microoperation sequences.
3. Separate *control memory:* the microinstructions are stored in a separate memory called the control memory that is not used for keeping data words, with the exception of some addresses and constants.

These three principles follow each other since the variable microprogram property is implemented by reconfiguration codes that must be stored in the microinstruction field. This leads to a significant increase in the microinstruction size, since, in addition to opcode and/or address, it must store several reconfiguration codes. The width of the control memory should thus match the microinstruction size, which is as a rule much larger than the data word size.

If, on the other hand, the control memory is shared for the storage of both microinstructions and data arrays, each cell storing a data word will contain a number of unused bits. To increase the speed of computation, control memories must be very fast and thus very expensive. It therefore becomes prohibitively expensive to use the same memory for both microinstruction and data storage.

Bit-parallel Operation is a process in which all bits of a data work are processed or transferred simultaneously.

Bit-serial Operation is a process in which a data word is transferred or processed one bit at a time. (II)

Bit Size of the Reconfiguration Field. The bit size of the reconfiguration field of the microinstruction is generally the sum of the bit sizes of all the reconfiguration zones that can be specified in the computer; i.e., it is determined by the number of microoperation groups that can be selected by the programmer.

Block Oriented Application. Block oriented applications are those that deal with a number of similar "objects" in the data base or in the outside world, generally a varying number, and on all of which the same general process is performed. (II)

Boolean Data Dependence means that existence of lack of dependence can be derived from a given variable. (IV)

Branch Address Field is the field of a (micro-) (macro-) instruction that stores a jump address used for conditional or unconditional branching.

Bridging Fault. A bridging fault means that two or more lines are shorted (wired) together. Two types of bridging faults are considered in the literature: the AND-type and OR-type bridging faults which mean that two or more lines are short-circuited to form AND and OR logical operations, respectively. Bridging faults may be either solid or transient. (VI)

Busy State means that the module has been assigned to certain computation. (VI)

"Candidate" (Pulse). An input pulse which is a candidate for a new radar pulse train, but is not yet a member of an established pulse train. See Sec. 4.3. (II)

Carrier Frequency. In radar applications, the frequency of the signal which is pulse-modulated by a radar transmitter to produce a sequence of radar pulses. See Sec. 4.2. (II)

Cell. A PE for an associative processor. Also called a *word* when its contents, rather than its hardware are being discussed. (II)

Chaining Channel. The inter-cell communication network of the ALAP. See Sec. 3.2.3.2. (II)

Changeability of a Computer System is its ability to be changed or added to or to accept different I/O requirements with relatively low cost and scheduling leadtime.

Class of Microinstructions that Perform Addressing. The microinstruction format includes three fields: opcode, reconfiguration, and address. The reconfiguration field is composed of zones that establish connections between the register for data memory and processor registers. Other zones of the reconfiguration field establish connections between registers of the control unit that store base addresses, B, indices, I, and relative addresses, RA, with the memory address register that must receive the effective memory address, $E = B + I + RA$.

Class of Microinstructions that Perform Conditional Branches. The microinstruction format includes three fields: opcode, reconfiguration, and address. The opcode field shows what type of conditional test must be performed; the reconfiguration field establishes the registers connected with the adder inputs; and the address field stores the address to a location in the control memory to be jumped to if the conditional test is true (or false).

Class of Microinstructions that Perform Processor Operations. The microinstruction format that represents this class usually includes the opcode field and a reconfiguration field composed of zones that show reconfigurations between registers and the logical circuits in the processor.

Closing Predicate is a predicate that performs a conditional test, the truth or falsehood of which specifies the exit from the loop. (IV)

Collection of Functions, F_i and F_k. Two functions belonging to the same block (partition). (V)

Communication Network. A network of communication channels which permits the processors in a multiprocessor system to transfer data and flag states (usually simultaneously) among one another. (II)

Compare Register. The compare register contains the data item that is the object of the search. (II)

576 GLOSSARY

Completion Signals. Used to signal the end of some instruction intervals which last longer than one clock period (instruction fetch, I/O operation, etc.). These signals are always produced only in one CE and sent to all other CE's through pin "p" and special logic. All modules of the given computer have their "p" pins connected via connecting elements. (III)

Complex Conditional Transition in the Microprogram. If microprogram node a has more than two successors, i.e., it is followed by nodes $b, c, \ldots, k$ in the microprogram graph, the respective transition $a \to b, a \to c, a \to d, \ldots, a \to k$ is called a *complex conditional transition*. For this case node a executes several conditional test microoperations so that each arrow of the transition is distinguished from others by a unique combination of conditional signals. If node a has three successors b, c, d, then it must generate at least two values of conditional signals y and z so that transition from a to each successor is distinguished by a unique combination of y and z. For instance, $a \xrightarrow{yz} b$ is activated by yz; $a \xrightarrow{\overline{y}z} c$ is activated by $\overline{y}z$ and $a \xrightarrow{y} d$ is activated by $\overline{y} = \overline{y}z \vee \overline{yz}$. In general, if node a has k successors in the microprogram graph, then it must generate at least $\log k$ conditional signals. This will allow assigning each arrow leaving a with a unique combination of true and false values of the conditional signals generated in node a. (I)

Complex Series-parallel Control-graphs. By complex series-parallel Control-graphs we denote those Control-graphs in which at least one loop is not a simple one, i.e., at least one loop encloses another loop or else one loop encloses two or more alternative paths. (IV)

Computational Performance. The computational performance of a system with respect to an algorithm A executed by this system is a set of parameters C_j related to the computation of A.

Computer Architecture Assembled from the Universal Module (UM). Obtaining a universal module type for implementing all functional units in the resource sharply reduces the initial investments related to a system's fabrication. Indeed, since the entire architecture is assembled from copies of the same chip type with only pin-to-pin connections between modules, the assembly costs are reduced. Inasmuch as each UM executes only those microoperations which are activated inside, the only control signals that are transferred between modules are a set of completion signals sent in a time shared mode through a small number of pins. Since the modular control principle provides that each microoperation require no pins for its implementation, the computational throughput of a UM no longer depends on its pin count but only on its component count. Because each year the component count per LSI module grows exponentially whereas its pin count grows at best linearly, the independence of module throughput from pin count sharply increases the gate-to-pin ratio in a module. This leads to increased utilization of the chip space. (III)

Concurrent Computation and Diagnosis are designated to maximize the number of busy modules (maximize the degree of parallelism), while maintaining the maximal diagnostic capability in the rest of the system. (VI)

Concurrent Detection of Errors means that all errors can be detected immediately when they occur. (VI)

Concurrently One-step Diagnosable System. A system S is concurrently one-step diagnosable if it is possible, in one step, to identify each node in the diagnostic subgraph to be faulty or fault-free without interrupting any computation nodes. (VI)

Conditional Branch Instruction. Every computer usually implements six signed and unsigned tests over words A and B: $A > B, A \leq B, A \geq B, A < B, A = B, A \neq B$, and and each test is usually implemented in a single conditional branch instruction. (I)

Conditional Test Microoperations. In a microprogram graph, multiple microoperation sequences usually originate from a node performing one or several conditional test microoperations. This node will have several outgoing arrows connecting it with its successors. Selection of a successor node is determined by the value of the signal, listed by the arrow, that causes the transition. All transitions in the microprogram graph are either unconditional or conditional. (I)

Content Search (Masked). The content search is masked, i.e., restricted to the contents of some selected fields of the memory locations rather than the entire contents of these locations.

Contradiction Between One-staged and Multistaged Interconnection Networks Used in a Supersystem. An organization with one-staged connections between different functional units of a Supersystem leads to enormous complexity of the interconnection busses. On the other hand, the use of multistaged connections leads to significant communication delays which reduce throughput of the Supersystem.

Contribution of Parallel Systems to the Computations. The contribution of parallel systems to the computations is that they allow implementation of instruction parallelism and data parallelism in addition to the microoperation parallelism that was implemented in original von Neumann and microprogrammed computers.

Control Circuits are designated to cause a word to transfer between two memory elements: register–register; register–memory cell; memory cell–register. (I)

Control Error. A control error can influence the interrelationships among the computing modules (e.g., causes information to flow into a wrong module). (VI)

Control Field of Frame. The control field of frame indicates whether the frame transmitted is sequenced information frame, nonsequenced information frame, or supervisory frame. All sequenced information frames are transmitted and received sequentially. Nonsequenced information frames, each with a sequenced number, are transmitted randomly and are assembled by the receiver according to these sequenced numbers. The supervisory frames contain no information and are used to designate ready or busy conditions or for special purposes. (VI)

Control Memory in a Microprogrammed Computer is a memory that stores microinstructions with reconfiguration codes.

Control Memory for Parallel Systems. The control memory contains the system's programs, plus common data items and common working storage.

Correlation Registers contain information during program execution which enable them to determine whether or not they will receive and store new data. (II)

Cost of LSI Implementation for a Complex Parallel System. The existing experience with LSI implementation of complex parallel systems contests the indiscriminate use of the thesis that LSI technology allows one to obtain arbitrarily cheap hardware for complex systems. This hardware may become cheap only if the architecture can be assembled from LSI modules selected from a minimal number of module types and have mostly pin-to-pin connections among modules, i.e., the number of external circuits connecting LSI modules should be restricted.

Thus, in evaluating different LSI-based architectures one has to consider:

1. The number of different module types used in assembling the system and
2. The overall percentage of circuits mapped onto LSI modules in logical portion of the system.

Cycle of Macroinstruction. The time to execute a macroinstruction microprogram.

Cycle in V-graph. A looplike path in the V-graph—that is, a path whose initial and final nodes coincide—is called a "cycle." (IV)

Damage Assessment. A damage assessment approach (to determine the extent of the damage caused by the error) can be designed by making use of *atomic actions*. An action is *atomic* if the process or processes performing it does not communicate with other processes while the action is being performed. (VI)

Data Access Graph. This is a graph having two types of nodes: data and functions. An arrow from a data node to a function node represents read-only type access of the data by the function. Write-only accesses are represented by an arrow from a function node to data node. (V)

Data Access Matrix. The matrix whose rows are numbered with dedicated functions (programs) F_j and columns with data D_i. The entry (F_j, D_i) shows the number of accesses of data D_i by function F_j. Each access can be read, write, or read/write. (V)

Data Error. A data error affects the system states, i.e., contents of memory and registers.

Data Driven Process. A process which is brought to execution by the arrival of appropriate input data. (V)

Data Entity. An abstract representation of a data component (data set, file, record, etc.) that the processing is to manipulate. They are data nodes on data access graph. (V)

Data Parallelism means the capability of a computer system to process several data words at a time with a single instruction.

Data Vector. A collection of words handled in parallel by one instruction is called a data vector.

DC Group Hardware Resource. The DC group hardware resource is characterized by an F parameter (expressed in bits) which shows the maximal size of the computer which can be formed in one DC group. By fixing h, the number of bits processed by one CE, one gets $n = F/h$, where n is the number of computer elements. (III)

DC-group with Minimal Complexity. A DC-group whose interconnection bus—that connects memory and processor resources—contains the minimal number of connecting elements. For DC-group with n computer elements CE, this number is $n - 1$, since $(n - 1)$ connecting elements must be used to separate n CE, each of which functions as an independent computer. (III)

DC-group with Minimal Delay. This dynamic architecture is characterized by the minimal communication times between any two modules of the resource (PE–PE, PE–ME) which are independent of the module locations. This allows obtaining permanent and minimal times for accessing instructions and $h \cdot k$-bit data words for any variable size computer which may be formed in a DC-group. In addition, this dynamic architecture implements simple parallel between computer communications which can also be established during permanent and minimal time. (III)

Decomposition. Separation of design items or entities into component parts and interconnections. (V)

Dedicated Architecture. A dedicated architecture is an architecture that takes into account the computational specificities of some algorithm or class of algorithms. Such architectures are provided with special-purpose execution and control circuits. Special-purpose execution circuits are a set of dedicated microoperations each of which speeds up execution of the algorithm. Special-purpose control circuits are hardware realizations of a collection of dedicated instructions, each of which implements complex sequences of operations encountered in the algorithm. Such instructions may execute complex expressions such as $(a + b)^2 - (a - b)^2 - k \div a$, or $(a + b - c) \times (a - b - c)$, etc. Each of these instructions is equivalent to several conventional instructions, so their use reduces the number of memory access operations required for fetching instructions and data.

Delay of the Component. Any computer component—flip-flop, and a logical element—requires time to switch; i.e., to generate an output after it receives an input(s). This time is called the *delay* of the component. It is a standard component parameter that can be found in any circuit catalog.

Delays of Reconfigurable Interconnections. Any adaptable architecture is characterized by delays introduced by reconfigurable interconnections. If one assumes that a dedicated connection, $A \rightarrow B$, between modules A and B introduces no delays, then a reconfigurable interconnection between modules A and B, C, or D in which either interconnection $A \rightarrow B$, or $A \rightarrow C$ or $A \rightarrow D$ is activated selectively requires additional reconfiguration logic which delay specifies the delay of reconfigurable interconnections.

$D_{\delta A}$ Design. A system S of n modules of $D_{\delta A}$ design can be represented by a digraph $G(\overline{V}, E)$, each node has A *outgoing arcs* and each arc is separated by a "distance" δ (i.e., a_{ij} exists if and only if $j_k = (i + \delta \cdot k) \bmod n$ and $A \geq k \geq 1$). Hence, $(j_{k+1} - j_k) \bmod n = \delta$). Every module can test A other modules separated by a distance $\delta k (A \geq k \geq 1)$. In a system of $D_{\delta A}$ design with n modules, the necessary and sufficient conditions that no two modules test each other is that n and δ are relatively prime. (VI)

D_{11} Design. A *single loop system* is a D_{11} design (i.e., $\delta = A - 1$). It is 1-fault diagnosable without repair. (VI)

Diagnostic Nodes. All idle modules may be assigned to the diagnostic part of the system. Since only idle modules can participate in system diagnosis, the subgraph consisting of diagnostic nodes (representing idle modules) is obtained by removing all the computation nodes (representing busy modules) and their incoming and outgoing arcs from $G(V, E)$. (VI)

Diagnostically Reconfigurable System. A system is diagnostically reconfigurable if the state of the diagnostic interconnection network can be changed to improve the diagnostic performance of the system. The switching circuit in the selfpurging system or the hybrid redundancy can be considered as an example of an implementation of diagnostic interconnection network. (VI)

Digital Compensator. A digital feedback controller for maintaining a physical plant at a given output. (V)

Digital System. A digital system can be viewed as a set of computing modules together with their interrelationships. (VI)

Direct Accessing Similarity of Two Functions, F_i and F_j. The ratio, $S_i(i,j)$, of the sum of the total minimal number of data accesses specified for the functions to their total average number of data accesses. (V)

Division of Each Microinstruction into Instructions $I1$ and $I2$ leads to a division of the entire control memory into the two levels, $M1$ and $M2$; $I1$ instructions are stored in the $M1$ level, and $I2$ instructions are stored in the $M2$ level. Such a division into two levels may lead to a real savings in the number of bits only if the size of $I1$ instructions exceeds a certain critical bit size. (I)

Domination of a Closing Predicate in the Program Loop. A computed variable *dominates* a closing predicate if the value of this variable specifies the truth or falsehood of the condition tested by this predicate. (IV)

Drawbacks of Pipeline Architectures. Limited applicability is the most severe drawback of pipeline architectures.

Dynamic Architecture. A dynamic architecture permits software controlled reconfiguration of the available resources that form the system into new computers with different sizes; i.e., it creates an additional parallelism on the basis of existing equipment. It is specified by a set of architectural states each of which is distinguished by the number and sizes of computers that operate concurrently.

Dynamic Reconfiguration from one state to another is supervised by the monitor system. During dynamic reconfiguration the following actions have to be performed.

1. *Task synchronization:* For each state-successor, at least one new computer is formed from the resources sacrificed by other computer(s) that had been functioning in the predecessor state. Before giving up their resources, all such computers must finish their executions and inform the V monitor of this occasion. It then follows that the V monitor has to establish the moment when the required resource is free and the DC-group is ready for the next architectural transition. Thus arises the need to organize task synchronization.

2. *Priority analysis:* DC-group computers may compute independent programs which make conflicting calls for the next transition. The V monitor must resolve these conflicts on the basis of priorities assigned to the programs. Therefore, one must organize a priority analysis during which the V monitor allows or denies an architectural switch requested by a program.

3. *Access of variable control codes:* For different architectural states, one has to write new meanings of the variable control codes to all UM's belonging to a newly formed computer. These codes may acquire different meanings for UM's of the same computer as well as for concurrent computers existing in the same state. All the variable control codes for each state form an array which has to be stored in a memory. To speed up an architectural switch, the storage of this array should be organized in a way that would minimize the time of searching for any given item in the array.

4. *Architectural switch to a new state:* Includes establishing the appropriate transfer modes in the connecting elements. (III)

Dynamic Redundant System. A dynamic redundant system consists of *one* active module with several spares, a fault detection device to detect the failure of the active module, and a switching network for switching out the faulty module and switching in the spare. (VI)

Dynamic Software Redundancy Scheme. The dynamic software redundancy scheme means use of standby spare module(s) and one audit module. If a processing module fails, it is detected by the audit module and immediately replaced by a standby spare module that performs the same function. (VI)

Emitter. A device that emits radar pulses. (II)

Error. An error is defined as an item of information which when processed by the normal algorithms of the system will produce a failure. This failure will disappear whenever the error is removed by error recovery algorithms. (VI)

Error Detection. The process of finding that system functioning is wrong, but the error location is unknown.

Error Location. The process of pinpointing the exact site of failure. It is understood as the system capability to locate the error so that it can either repair the damage or reconfigure to eliminate the cause of trouble. This is generally known as *fault diagnosis* or *fault isolation*. The level of diagnosis depends on the system design as well as maintenance policy. For dynamic redundant systems, it is sufficient to locate the fault to the level of a replaceable unit so that subsequent reconfiguration and replacement can be achieved. (VI)

Error Recovery. The scheme for dealing with the damage caused by a detected error. (VI)

Error Treatments. Two error treatments are possible: one is the *replacement strategy* that replaces the designated faulty component by standby spares, the other is the *graceful degradation strategy* that causes some or all of the responsibilities of the faulty software module to be taken over by other fault-free active modules. The error treatment can be designed to be either *manual* (manual software replacement or reconfiguration), *dynamic* (external stimuli issued manually to cause the system to reorganize its future activity), or *spontaneous* (spontaneous replacement and reconfiguration are carried out automatically by the system itself, i.e., selfrepair). (VI)

Estrin Program Graph Model. The Estrin model is derived from the program as follows: each node α represents an instruction sequence S; control-flow from instruction sequence S_α to instruction sequence S_β is represented by an oriented edge going from node α to node β. A simple way to derive an Estrin program graph model consists of associating an entry point with (a) each labelled statement, (b) each statement immediately following a branching instruction, and (c) each statement reached as a consequence of a branching. *Exit points* are then either *branching points* or else are defined by default with respect to *entry points*, since each instruction sequence is a group of consecutive statements starting with an entry point and ending with the (nearest) exit point. Edges connect nodes in the model whenever control-flow connects the corresponding instruction sequences. (IV)

Evaluation of the Predicate. The evaluation of the predicate will in general depend from a set of variables, $v_1, v_2, \ldots$ that will be said to "dominate" the predicate itself. Consider for instance the very simple case

if (v_1) then e_1 else e_2.

The predicate is simply based upon evaluation of a boolean variable v_1: if $v_1 = \mathit{true}$ the predicate is true and control is passed to statement e_1, otherwise the predicate is false and e_2 is executed. v_1 is then the variable dominating the predicate. (IV)

Evaluation of Dynamic Architectures. Each dynamic architecture may be evaluated from two viewpoints: *adaptation* and *performance*. The adaptation aspect looks at how well a particular dynamic architecture is adapted to an algorithm to be executed. This aspect of dynamic architecture is characterized by such parameters as

1. Adaptation to program streams: gives the percentage of the resource used to compute a given set of concurrent programs;
2. Adaptation to program structures: examines execution speed-up arising from the program being computed by a particular instruction set;
3. Array adaptation: shows the percentage of redundant equipment created during array (vector) computations;
4. Bit size adaptation: measures the delay in the execution of the entire task (program) because of selection of a permanent computer size for executing all of its instructions; etc.

The performance aspect characterizes each dynamic architecture from the viewpoint of time of computation and of complexity of the respective hardware resource.

Performance of dynamic architectures is affected by the following major factors:

1. The time for architectural reconfiguration;
2. Delays introduced into signal propagation by reconfigurable busses;
3. Time of between-computer communications;
4. Modular expansion of existing systems with new hardware components;
5. Cost of realization. (III)

Exception Handling. Exception handling is a language facility provided by programming language designers to handle forward error recovery schemes. The PL/I language provides such facilities for exception handling. They involve enabling and disabling the error conditions for exceptions, software detection of these exception conditions, and exception handling defined by the users or by the system. (VI)

Execution Circuit. An execution circuit performs a functional transformation of the word while it is being transferred between memory elements. All execution circuits can be implemented with the three primitive logical components: AND, OR, and NOT. There are other primitive sets such as (AND, NOT), (OR, NOT), etc.

Failure. In a digital system, a failure is defined as an event in which the system does not perform its service in the manner specified (i.e., the system violates its specifications).

Fault. A fault is the physical or algorithmic cause of an error.

Fault Diagnosable System. Assuming that no more than t faulty modules can occur in the system under test, based on the response patterns obtained, two cases can occur:

1. All faulty modules have been identified. In this case, repair can be done in one step and the system is said to be *one-step t-fault diagnosable* (or t-fault diagnosable without repair).
2. Only a subset of faulty modules has been located. In this case, these faulty modules are replaced first. Then another diagnosis procedure is initiated to find the remaining faulty modules. For a system with no more than t faulty modules, up to t such sequential steps may be necessary to identify all faulty modules. Such a system is said to be *sequentially t-fault diagnosable* (or t-fault diagnosable with repair). (VI)

Fault-Tolerance may be achieved by means of protective redundancy which is introduced into a computer system in the following forms: (1) hardware redundancy (additional hardware); (2) software redundancy (additional software); and (3) time redundancy (repetition of machine operations). A complete procedure of providing fault-tolerance consists of the following steps: First the error is detected, then error diagnostics identify the site of error. If the error is permanent, the system is reconfigured to exclude the faulty module from the system. If the faulty module is replaced by a fault-free module, the system retains its original throughput. If the faulty module is not replaced, the system is gracefully degraded. For intermittently caused errors, a recovery procedure is activated. After either reconfiguration or recovery procedures, the program restart procedure is activated so that the program can continue its operation. The whole procedure is then repeated. Error detection, diagnosis, and reconfiguration involves both the hardware and the software. The recovery procedure deals only with the software. The whole process improves the reliability of the system. We can see that error diagnosis is a necessary but not a sufficient step for increasing the system reliability. Thus diagnostic performance can be considered as a part of reliability performance of a digital system. (VI)

Fault-Tolerant Computer System. A fault-tolerant computer system is a system possessing the three following attributes: (1) it consists of a set of hardware and software components; (2) it is initially free from both hardware and software design errors; and (3) it executes programs correctly in the presence of faults. (VI)

Fault-Tolerant Computing. The ability to execute a specified algorithm regardless of hardware and/or software failures.

Fault-Tolerant Design. The design of digital systems that can stay operational in spite of the failure of some of its components (software or hardware).

Fault-Tolerant Software Structure. The problem of fault-tolerant software structure can be formulated as the development of reliable software with unreliable components. This can be solved by the application of software redundancy techniques. The errors can be generated by either hardware faults or imperfect software faults. Software redundancy uses additional software which includes additional programs, program segments, set of instructions, and microprogram steps which provide either fault detection or recovery in fault-tolerant computer systems. (VI)

Firmware. A collection of microprograms.

Fixed Microprogram of von Neumann Architecture. The partition of the instruction into two zones—opcode and address—has led to the major characteristic of von Neumann architectures: Each instruction activates a fixed microprogram that can not change its sequence of microoperations. This means that a programmer could not alter any sequence of executing microoperations by writing a new code to the instruction field. Consequently, each instruction microprogram is implemented permanently and can not be modified via software as could be done later in microprogrammed computers. (I)

Flag (Conditiona) Signals. Used to control branching within microprograms. (*See also* Complex and Simple conditional transitions.)

Flag-Shift Instruction. The ALAP instruction which sets the cell states in preparation for a word-cycle instruction. See Sec. 3.2.3.4. (II)

Flag Fields. The flag fields indicate the beginning and the end of each frame (also called synchronization field of the frame). (VI)

Formal Correctness is a set of techniques aimed to state that each instruction of a program can be reached from the program entry point, that from any instruction the end can be reached, that all variables are defined prior to usage, etc. Formal correctness refers in fact to syntactic properties of the program, but it gives no insight into the semantics of the program, nor does it point out any problem areas excepting obvious errors such as:

- references to undefined variables,
- statements which cannot be reached from any point in the program,
- no branching to undefined labels, etc.

On the graph model, "formal correctness" means that given any edge there is at least one oriented path from START node to STOP node traversing it. (IV)

Forward Error Recovery Scheme. This scheme does not attempt to repair the damage caused by a detected error but takes the following actions: (1) reports the error to the user; (2) takes the system default action or user defined action (because all errors in forward error recovery scheme are predetermined so that the program can continue its execution). (VI)

Frame. In data transmission, a frame is a specific format for each data transmission block. The actual data frame which is transmitted over the communication link is built up from the text by adding a sequence of header fields and finally trail fields. (VI)

Frame Check Sequence Field (FCS). The frame check sequence field (or called block check field) serves as an error checking field. The transmitter makes a mathematical transformation (e.g., cyclic redundancy checking, check sum, or other methods, etc.) of all bits in a given frame except the FCS and flag fields and inserts the result of the transformation into the FCS field. The receiver makes the same computation (just like those the transmitter has done) and compares the computed value with the received value found in the FCS field. If these two values do not match, the receiver rejects the frame, since it contains errors. This kind of checking is called *frame checking* on communication links. The use of error detection and error correction codes is a special form of mathematical transformation in frame check. (VI)

Free Redundant Resource. A system resource that is not needed by a current program and may be used for other computations. (III)

Functional Error. An error is called a functional error if it affects the functional transformation relation of a computing module. (VI)

Functional Entities. Function nodes on a data access graph. They represent a function (program) that must be computed by a system. (V)

Functional Requirements. Requirements on functions and transformations to be performed by a computer system. (V)

Functional Simulator. The functional simulator is a program that causes the host machine to behave as though it were the object machine. It accepts as input a binary load module put together by the linkage editor and simulates the instruction-by-instruction execution of the program.

Functions of Monitor for Dynamic Architecture. Since for each architectural transition, at least one computer in the present state must give up its resource to form another computer in the next state, any architectural transition occurs at a moment when all computers that sacrifice their resources finish their tasks. This moment is identified by the V monitor.

Another function of the V monitor is to resolve conflicts between concurrent programs. Indeed, two programs may request transitions into two different states, causing conflicts in the system. To resolve them, the V monitor specifies the priority of a requesting program and allows or denies its call for the next transition. If the program request is allowed, the V monitor allows execution of the architectural switch instruction. It fetches special control codes to be written to the computer elements of all newly formed computers, and writes new transition modes to the connecting elements that integrate or separate these computers. These codes activate a new pattern of interconnections specific to the next architectural state. (III)

Fuzzy Clustering. A partitioning approach based on applying thresholds to fuzzy relations (specifically similarity relations). (V)

Graceful Degradation Strategies. The graceful degradation strategies arrange for some or all of the responsibilities of the failed components to be taken over by other components already in use by the system. This always involves some degree of performance and/or functional degradation.

Grouping of Functions in Accordance with Common Data Accessing Pattern. Partitioning of a set of functions into blocks such that each block includes functions with approximately similar patterns of data accesses. (V)

Hardware Redundancy. A replication of components at various levels. Circuit level redundancy is achieved by replicating circuit elements such as diodes, transistors, resistors, etc., and interconnecting them in such a way as to *mask errors* caused by the fault elements. Chip or module level redundancy is achieved by replicating system chips or modules. (VI)

Head Flag. A flip-flop associated with the ALAP flag register which permits the interchange and logical manipulation of the flag states. See Secs. 3.2.3.4 and 4.12. (II)

Hybrid Redundancy. In hybrid redundancy, the N-modular-redundancy (NMR) core is used with a set of spares. The spares may be kept power-off or may be powered. When one of the N modules fails, it is switched out and one of the spares is switched in. This process continues until all spares fail. The hybrid redundancy scheme has the advantages of both static and dynamic redundancy schemes, but the switching network is complex. (VI)

Idle State. A module that has not been assigned to a computation is in the idle state.

IFU. The interface unit of the ALAP system. See Sec. 3.2.3.1. (II)

Image Processing. The processing of digitized image data by a computer to obtain information about the image or to change the representation of the image. (II)

Immediate Operands (Constant) are usually stored in the instruction field.

Impact of LSI Technology on Computation. The impact of LSI technology on computation in general is twofold: First it has led to the cost-effective computerization of small processes, performed by microcomputers and microprocessors; i.e., it has shifted the lower bound of computerization downward. Second, by allowing the construction of complex modular computer systems that adapt their architectures to the problems being computed, LSI technology lights the way towards the computation of supercomplex parallel processes; i.e., it is shifting the upper bound of computerization upward.

Impact of LSI Technology on Computer Organization. In the area of computer organization, LSI technology has permitted new types of LSI modular architecture to be created that can adapt to the needs of an executing algorithm by reconfiguring module interconnections. Such architectures are characterized by the set of architectural states they can assume and software-controlled reconfiguration of the available system hardware resource from one architectural state to another. Each state is characterized by such things as the number and sizes of concurrently operating computers, the instruction set activated, and the type of architecture employed (array, pipeline, multiprocessor, multicomputer).

Implicit Enumeration. In evaluating decision alternatives, a technique for discarding unlikely solutions without explicit listing and evaluation. (V)

Incomplete Diagnosis. A system S is t_i-*fault diagnosable* if no more than t_i modules in S are intermittently faulty; then a fault-free module will never be diagnosed as faulty and the diagnosis at any time is at worst *incomplete* but never *incorrect*. The incompleteness means that certain intermittent faults are not detected since the fault is inactive when the test is applied. (VI)

Increase in System Throughput Caused by LSI Technology. LSI technology has greatly influenced the emergence of the following techniques as powerful methods for increasing throughput:

1. Architectural adaptation.
2. Enhanced computational parallelism. (I, III)

Initial Data Set of a Loop. A set of all constants and initial values of variables that participate in a loop computation.

Input/Output Control Unit. The input/output control unit controls all system input and output for most applications.

Input-Output Relationship Graph. In the input-output relationship graph, *a strong edge* represents a functional transform; *a weak edge* represents an assignment statement with no transform. (IV)

Instruction and Arithmetic Pipelines. The existing pipeline systems are divided into two categories—*instruction* and *arithmetic*.

Instruction pipeline systems fragment the memory access process into phases aimed at fetching the instruction and its operands from the memory. For most instruction pipelines, the entire processor operation assigned to the instruction is treated as a single phase.

Arithmetic pipelines fragment the arithmetic operations assigned to an instruction into phases. These can be either complex expressions containing several arithmetic operations or single operations such as multiplication, division, and floating-point operations that are represented by iterative computational algorithms. (I)

Instruction Buffer. The instruction buffer (PLA) holds the current instruction and pending instruction, both of which are received from the PE memories via the control unit bus.

Instruction Cycle. The time required to execute an instruction microprogram.

Instruction Fields. A von Neumann architecture requires that two fields be assigned to each instruction, *the opcode* and *an address*. In a microprogrammed computer the additional necessity of having a third field storing reconfiguration codes appeared. This field is called the *reconfiguration field*. The reconfiguration field is partitioned into several zones, each assigned to one group of reconfiguration microoperations; i.e., those activated by reconfiguration codes. By writing a code into this zone, a programmer can select one or several microoperations from the group.

Instruction Interval. The time needed for executing one set of parallel microoperations assigned to instruction microprogram. (III)

Instruction "Jump $ME_i \to ME_j$ (A_d)" concludes the task stored in ME_i and initiates fetches of the next task from ME_j, where A_d is its base address. It performs the following actions.

1. Since only one PE generates instruction addresses for the respective ME, PE_i stops and PE_j initiates generation of instruction addresses.
2. PE_j receives the base address A_d of a program stored in ME_j.
3. Connecting elements within the $C_i(k)$ computer are switched to new transfer modes, since program instructions will henceforth be coming from a different memory element ME_j but must still be distributed to all modules in the computer. (III)

Instruction Microprogram. Since a microoperation identifies an elementary processing operation, one machine instruction is usually assigned a sequence of several microoperations. This sequence is called the *instruction microprogram*.

Instruction Parallelism. The capability of a computer system to execute several instructions at a time.

Instruction Sequence. An instruction sequence as a program segment is characterized by one entry point, one exit point, and, at most, one branching, coinciding with the exit point.

Instruction Type Code. Belongs to the category of opcodes. This code distinguishes among different instruction formats and allows a further reduction in the overall instruction size since it permits the assignment of the same bits to new reconfiguration zones in each format.

Integer Programming. The set of mathematical programming approaches applicable when the decision variables can take only integer values (V)

Invariable Codes for Dynamic Architecture. The invariable codes are as follows:

1. *Function code q* determines the function performed by a given LSI module. (For all modules forming the processor, $q = 10$; for I/O, $q = 01$; for connecting elements, $q = 00$; for the monitor, $q = 11$.)
2. *Position code i* shows the position of the given CE among n other CE's. Its size is $\log_2 n$ bits where n is the overall number of CE's.
3. *Fetch instruction code m_I*, specifies the speed of the ME contained in the given CE.
4. *Slice code g* is used only when a functional element is assembled from more than one LSI module. When a 16-bit byte is received by a PE or GE, this code forces the more and less significant 8 bits to be processed by the more and less significant modules respectively. (III)

Justification for the Use of Parallel Hardware. Sufficient justification for the use of parallel hardware is a requirement that the amount of processing on each item or each set of data be sufficient to justify the time required to load the data.

Length of a Node Chain. The length of a node chain (node loop) is the number of nodes in the chain (loop). (VI)

Level of Abstraction. The level at which the concept of a system or component is represented; at lower levels of abstraction more details are presented. (V)

Life Cycle. The total life of a system or component from initial concept through operation and maintenance.

Linkage Editor. The linkage editor is a program which combines a number of relocatable object modules generated by the assembler into a single object program ready for loading and execution on the object computer.

Local Data Buffer. The local data buffer holds common operands received from the PEs via the control unit bus.

Logical Buffers. An abstract representation of storage requirements for communications. (V)

Logical Circuits in a computer may be either control circuits or execution circuits.

Logical Clock. A variable assigned to a process in order to simulate concurrency of execution. It is incremented by estimated processing and waiting time increments. (V)

Logical Network or Topology. A graph with nodes corresponding to processing facilities or processors and edges representing message flow requirements; physical message flow may have a much different network. (V)

Logical Switches. An abstract representation of switching requirements for communications. (V)

Loop Counter. The counter that counts the number of iterations for a (micro-) program loop.

Loop Enclosing an Instruction Sequence S. A loop enclosing an instruction sequence S corresponds to the possiblity of having at runtime a number $1 \leq n < \infty$ of iterations of S, and thus can be represented by $S \cup SS \cup SSS \cup \ldots S*$ which means closure over S. (IV)

LSI Chip and LSI Wafer. The term "chip" or module type is used for a circuit module cut from a large silicon wafer on which many such identical modules are fabricated. The wafer may be 2 in. or more in diameter. The ALAP memory is fabricated from full wafers, which each contain many ALAP cells, and which are not cut up into individual chips after manufacture. (II)

Machine Instructions. Codes that are executable by a computer.

Main Memory. A memory in a computer that stores instructions and data words that are in active use by the processor.

Maintaining the Pulse Trains. Once a sequence of pulses has been identified as belonging to a single pulse train, the signal-sorter must maintain that pulse train by assigning every subsequent matching pulse to it. (II)

Majority Voter. A majority voter with n inputs (n must be an odd integer) has an output logic 1 if and only if the number of 1's in input lines is greater than or equal to $n + \frac{1}{2}$. Therefore, the majority voter is a special case of the threshold voter when $t = n + \frac{1}{2}$. (VI)

Major Drawback of Architectural Dedication. The major drawback of architectural dedication is that it lowers the system's applicability. If a new algorithm appears or the old algorithm is modified, the system may not be capable of completing it in the necessary time interval since all its dedicated instructions take into account the computational specificities of the original algorithms for which the system was designed.

Major Drawback of the Use of the Maximal Concurrency Present in Algorithms. The major drawback of this technique is that it may lead to excessive complexity of the system and low utilization of the equipment in it. This is because the number of independent computers in a system is determined by the algorithm having the largest number of parallel information streams, while on the other hand, the bit size of each computer is determined by the maximal bit size of computed results that can be encountered in all the algorithms processed by this computer. It follows that in computing other less complex algorithms requiring fewer instruction and data streams or smaller word sizes, a portion of the system resource becomes redundant. (III)

Major Problems of Supersystem Design. The designers of Supersystems are faced with the following major problems:

1. Obtaining maximal throughputs while the system maintains a specified level of reliability, and
2. Maintaining the adequacy of the Supersystem for solving a class of complex algorithms during specified time intervals, while these algorithms may change or even new algorithms may be added to the class. (III)

Mask Register. The mask register determines by its content the field or fields within the memory cells at which the content search will be made. (II, VI)

(Macro-) Instruction Cycle. The time needed to execute a (macro-) instruction microprogram.

Mathematical Programming. A family of approaches for selecting decision variables which optimize a given objective function subject to given constraint functions. (V)

Memory Element and Logical Circuit. For its computations, a computer uses two kinds of information: data and instructions. Both come packaged in information words, or simply words. Each word in the computer is either stored in a *memory element* or transformed by *logical circuits*. Thus all computer components are divided into two categories: memory elements and logical circuits. (I)

Memory Elements. There are two types of memory elements: registers and memory cells. The difference between them is that registers may be connected to each other for data transfer while memory cells are always disconnected from each other. Each cell is connected to one or to several registers. (I)

Memory-to-Memory Organization of the Processor. An architecture in which each instruction has its operands fetched from the memory and the result is sent back to the memory. (I)

Meta-Assemblers. Assemblers with the capability to assemble programs for computers other than the host computer, under the direction of a user-supplied set of assembly procedures. (II)

Microassemblers. Software that translates microprograms written in an assembly language to the encoded microinstructions. (I)

Microcommand. A signal sent by the control unit to activate the transfer of a word from a source register to destination register is a microcommand.

Microinstruction. The instruction of a microprogrammed computer that stores reconfiguration codes.

Microinstruction Cycle. The time needed to execute one microinstruction.

Microoperation. The most elementary processing done in a computer consists of executing a *microoperation* during a word transfer between registers. To execute a microoperation, a data path between registers *must* contain the control circuit and may or may not contain an execution circuit. An execution circuit implements a Boolean function that transforms input data word(s) stored in one or several source registers into an output data word broadcast to one or several destination registers. Since a microoperation necessarily includes one control circuit, it is activated by the microcommand that opens this control circuit. (I)

Microoperation Delay. Since every microoperation is a circuit assembled from logical components, it introduces a delay, called the *microoperation delay*, in the generation of an output word broadcast to the destination register which is a function of input word(s) stored in source registers. To find microoperation delay one has to find the delay of the longest path of consecutive logical components the signal must pass through before it reaches the destination register. Therefore, to find the microoperation delay, one has to find all the logical paths in its circuit, find the delay of each path, PD, and then select the maximal delay: $MID = \max(PD_i)$. Microoperation delay can be expressed in nanoseconds (or microseconds) or in the number of primitive logical components (AND, OR, NOT) in the logical path with maximal delay. For the latter case assume that each component's delay is t_d, then a logical path containing k components is delayed by $k \cdot t_d$. (I)

Microoperation Description of the Architecture. An approach in which any architecture is described by the set of elementary actions (microoperations) it realizes and all computations are reduced to the activation of sequences of these actions. (I)

Microoperation Hardware Diagram, in addition to listing all registers and logical circuits, shows which microoperations are executed during word transmission between registers. To obtain its microoperation analog, a conventional hardware diagram must have all the microoperations that it executes added to it. (I)

Microoperation Set. Since each program is executed in a computer as a sequence of machine instructions and each instruction is interpreted in the computer as a sequence of microoperations, the microoperation set may be used to characterize any architecture regardless of its type, hardware complexity, or technological realization. The microoperation set is a qualitative characteristic of a computer's architecture that is independent of its technology of fabrication, the bit size of its processor(s), the size of primary memory, the number of I/O devices, etc. These, on the other hand, are quantative characteristics which specify the speed of information processing, the sizes of information arrays that can be stored in computer memories, the form in which information is received or sent to I/O devices, etc. The microoperation set of an average computer contains 120–150 different microoperations. Their typical distribution among different functional units is as follows: the processor accounts for 50–60 microoperations; the control unit implements 30–50 microoperations; I/O devices realize 30–40 microoperations; and the memory 10–20 microoperations. (I)

Microprogram. Each machine instruction activates a unique microprogram, which is a sequence of sets of microoperations that realize the instruction. If the instruction set needs to be supplemented with a new machine instruction, the computer must already contain all the microoperations to be activated by the microprogram of this instruction. Thus the microprogram is a sequence of microoperations that correspond to one instruction for a von Neumann computer and one macroinstruction for microprogrammed computer.

Microprogram (Microinstruction) Counter. The counter that stores a current address of microinstruction in the control memory. (I)

Microprogram Graph. A microprogram is a directed graph in which each node denotes one or several microoperations denoted by identifying numbers that are executed in parallel. Arrows show the sequencing of nodes, and are marked with the signals that activate transitions from one node to its successor. (I)

Microprogram Sequencer. A memory in a microprogrammed computer that stores sequence of addresses for microinstructions. (I)

Minimal Delay End-Around Carry Bus for Dynamic Architectures. A minimal delay end-around carry bus minimizes delays of the reconfiguration logic introduced into the carry-propagation path. (III)

Minimal Delay Memory-Processor Bus. The minimal delay memory-processor bus provides high flexibility for data exchanges between concurrent computers. It can make the following between-computer information exchanges possible:

1. "X processor–Y memory": For this exchange the X computer accesses the primary memory of the Y computer. This access may be performed without interrupting the operation of the processor in the Y computer.
2. "X processor–Y processor": Data exchange between the processors of computers X and Y.
3. "X memory–Y memory": Data exchange between primary memories of computers X and Y. (III)

Minimal Interconnection End-Around Carry Bus for Dynamic Architectures. The minimal interconnection end-around carry bus minimizes the number of between module interconnections between all PE's of the resource. (III)

Minimally Required Time of an Operation. A minimally required time of an operation is the delay of the longest microoperation that is included into this operation. For instance, binary addition includes two microoperations: carry propagation and modulo 2 addition. Of those two the longest is carry propagation microoperation. Therefore the minimally required time of the binary addition is the delay introduced by carry propagation microoperation. (I, III)

Mixed Architecture. Coresidence of any combination of multicomputer, multiprocessor array, and pipeline architectures. (III)

Mode of Operation of Triple Modular Redundant System. When one of three active modules fails, the triple modular redundant system will degrade into a *duplex mode* of operation with the voter degrading into a comparator. When a disagreement of the two remaining good modules is detected, a diagnostic program or error detection device will identify the failed unit and switch it out and the system reverts to a *simplex mode* of operation. (VI)

Modes of Operation for Dynamic Architecture. In a system with dynamic architecture one may distinguish two modes of computations.

- **Mode 1.** An algorithm is computed by a single architectural state that is established before computation.
- **Mode 2.** An algorithm is computed by a sequence of architectural states. The algorithm is partitioned into several tasks for this case, and for each task the architectural state is found that minimizes its computation time.

Complex real-time algorithms require mode 2 in most cases, because as a rule, these algorithms are characterized by a changeable number of information streams and changeable processing requirements. It follows that such algorithms require the architecture to reconfigure from one state to another during computation. (III)

Mode Register. This register is set from within the PE. The instruction control information received by the PE for all instructions contains a specification as to which mode or modes the PE must be set in order that the PE execute the instruction.

Modified Module. The elementary switch with its logic can be constructed within a single unit called a modified module.

Modular Control Organization. The modular control organization satisfies two basic requirements of dynamic architecture—it is able to generate variable time intervals for computer processes and provide independence of the computer control organization, regardless of the number of control units it contains. Contrary to synchronous and asynchronous control organizations that assume that control over all computer devices is performed by either a unique control unit (synchronous control) or by one central and several local control units (asynchronous control), the modular control organization eliminates the separate control unit in the architecture. Instead, every LSI module is provided with a local *modular control device* (MCD), and the functions of the control unit of any computer are performed by concurrent operations of all the MCD's contained in this computer. (III)

Modular Computer Systems are distinguished by the following characteristics:

1. Their architectures are modular.
2. They support distributed processing.
3. They provide for software-controlled reconfiguration of between-module interconnections.

Modular Expansion of Complex Parallel System. The integration of new resource units into an existing system. This source of throughput increase is presently acquiring an ever greater importance as the hardware resource becomes cheaper. However, since each attachment of an additional hardware resource is accompanied by adding both new interconnections and new complexities to the system, the modular expansion of any existing system is limited by a combination of such factors as

1. Technological constraints in the number of admissible interconnections.
2. Increase in delays introduced by the reconfiguration logic.
3. Ability of the system to maintain the specified level of reliability. (III)

Monitor. A process that controls the accessing of memory and processor resources shared by multiple processes. A monitor enforces mutual exclusion of accessing on the basis of priorities of processes and provides wait and restart-type signals.

Multicomputer System. A multicomputer system is a parallel system assembled from N computers in such a way that any two computers may communicate with each other via an interconnection bus connected with their I/O devices. A multicomputer system assembled from N computers may compute N instruction sequences concurrently, where each instruction may handle a pair of operands at a time. A multicomputer system thus implements instruction parallelism but not data parallelism. (I)

Multiphased Microinstructions. In a microprogrammed computer one microinstruction does not activate a complex sequential microprogram as was the case with von Neumann architectures. To increase the amount of computation assigned to one microinstruction it can be multiphased, however; i.e., the microcommands it generates may be activated during several clock periods. Thus, the microinstructions in microprogrammed computers may be one-phased, two-phased, three-phased, etc. (I)

Multiprocessor System. A multiprocessor system is assembled from the functional units (processors, memories, I/Os) that are used in a multicomputer system. However, the interconnection bus of the multiprocessor system allows activation of a direct communication path between any two functional units contained in the system. A multiprocessor system can therefore provide direct data exchanges between any processor and any memory, or any two processors, or any two memories, etc. For instance, if processor P_2 needs a block of data words stored in the memory M_N, then the interconnection bus reconfigures into a direct path between P_2 and M_N, allowing such an exchange directly rather than by using two buffer memories as is done in multicomputer systems. In general, multiprocessor systems are effective for executing complex algorithms having a high interaction between the concurrent instruction sequences computed in different processors; i.e., one sequence needs blocks of data words computed by another sequence. As with multicomputers, multiprocessors implement instruction parallelism but not data parallelism since a system containing N processors and N memories may compute N instructions at a time where each instruction can handle not more than two operands at a time. (I)

Multistage Network. A collection of connecting elements (ME's) that interconnect a processor resource made of microcomputers or a processor-memory resource assembled from PE's and ME's. Every pair of microcomputers (PE–ME) could communicate through several consecutively connected connecting elements (MSE's). This reduces the overall number of MSE's in the network, but introduces longer delays into each data transfer. The number of levels in the interconnection bus of a multistage network generally increases logarithmically with each increase in the number of microcomputers (or PE's and ME's); i.e., a network of n microcomputers will have about $\log_2 n$ levels and each data transfer between a pair of microcomputers will pass through $\log_2 n$ connecting elements. But this seriously reduces the speed of information processing in the network and is inadmissible in many applications. (I)

New Sources of Throughput Increase. The new sources of augmenting throughput in a Supersystem are:

1. Adaptation of hardware resources to instruction and data parallelism.
2. Reconfiguration of hardware resources into different types of architectures: array, pipeline, multicomputer, and multiprocessor. (III)

N-Modular Redundancy/Bipurge (or N-Modular Redundancy/Simplex) System. In the N-modular redundancy/bipurge (or N-modular redundancy/simplex) system, the initial system configuration consists of an N-modular redundancy core along with a switching unit. When failure in a module is detected by the system (through comparing module outputs with the voter output), the faulty module is switched out along with a good module to keep the number of active modules odd (because a majority voter is used). Thus the system actually degrades from an N-modular redundancy to an $(N-2)$ modular redundancy, finally reaching a simplex mode of operation. (VI)

Nonfree Redundant Resource. A hardware resource inside one computer, array, or pipeline that is not needed by a current program(s). However it can not be released for other computations. (III)

Nonterminating Loops. Loops for which there is at least one data set leading to nontermination. (IV)

Off-Chip Delay. The time delay in transferring a signal between chips or wafers.

One Clock Period. As a rule, a microoperation is executed during one clock period, T_0, which is the time between the fronts of two consecutive synchronization pulses, τ. The meaning of T_0 depends on the ordering of the control and execution circuits in the microoperation circuit. (I)

One Computer Element (CE). Each CE includes a 16-bit processor element (PE), a 16-bit wide memory element (ME), and a 16-bit I/O element (GE), equipped with a small memory (M(GE)), having the same width as PE. Generally the word size h of one CE depends on the current restrictions of LSI technology (chip size and pin count) and may be 4, 8, 12, 16 bits, etc. To increase h, each CE may have its processor and I/O elements assembled from several LSI modules. (III)

One Microinstruction Cycle (IC). In a microprogrammed computer with two level memory, one microinstruction cycle is the time to execute (address) instruction I2 and instruction I1. This time IC is thus divided into three phases, $F1$, $F2$, and $F3$. $F1$ and $F2$ are fetch phases that fetch $I2$ and $I1$, respectively, and $F3$ is a computational phase that executes $I1$. (I)

One-stage Networks. In this type of network, each microcomputer (processor) is connected with each of the other microcomputers (memory) through a dedicated connecting element, MSE, having h input and h output pins. Thus a network containing n microcomputers, will have $n \cdot (n - 1)$ connecting elements. (I)

One-step t/s Diagnosable System. A system S is one-step t/s diagnosable if and only if, given one response pattern, the system with no more than t faulty modules can be repaired by replacing at most s modules. A one-step t/s diagnosable system replaces all suspected faulty modules; therefore, s is always greater than or equal to t. (VI)

One-step $[X_0, X_1, \ldots, X_{A-1}]$ Diagnostically Reconfigurable System. Let $G(V, E)$ be the digraph of a system S of n modules and let each module in G have A outgoing arcs, then the system S is defined as a one-step $[X_0, X_1, \ldots, X_{A-1}]$ diagnostically reconfigurable system if at each step of diagnosis the system can exactly diagnose $N_S \leq A - i$ ($i = 0, 1, \ldots, A - 1$) simultaneous faults as long as the number of accumulated faults, X_a, does not exceed $\sum_{j=0}^{i} X_j - 1$. (VI)

Operational Requirement. A requirement on quality of operation as opposed to function or speed; e.g., reliability, changeability, and maintainability are operational requirements. (V)

Opfield. A field of a microinstruction that stores the opcode and instruction type codes. (I)

Optimization of Adaptations for Arrays. For arrays, the architecture must be capable of partitioning its resources into a variable number of concurrent arrays. Within each array it must be able to change the dimension of each data vector as well as the word size of a processor that computes one component of this vector. (I)

Optimization of Adaptations for Multiprocessors and Multicomputers. For multiprocessors, optimization means reducing the time spent by any pair of resource units involved in communication; i.e., one must construct fast reconfigurable paths that transmit data words in parallel between any pair of resource units. For multicomputers, the architecture must provide fast parallel exchanges both between a pair of computers and between any pair of functional units from two different computers. Thus, the objectives for optimizing multiprocessors and multicomputers are similar and consist of creating fast reconfigurable busses that support parallel word exchanges between any pair of functional units, either from the same or from different computers. (I)

Overlapped Execution of Phases. An overlapped execution of phases in a microprogrammed computer with two level memory means that in any given time interval the computer is executing three consecutive microinstructions in three phases, $F3$, $F2$, $F1$, of their instruction cycles. This means that for every triplet of microinstructions the second microinstruction begins executing its instruction cycle one phase delay behind the first one, and the third microinstruction begins its instruction cycle one phase behind the second and two phase delays behind the first microinstruction. For instance, if a third microinstruction is in phase $F1$ then the second microinstruction is in phase $F2$ and the first is in phase $F3$, etc. (I)

Parallel Buffer. The parallel buffer receives and sends data in parallel to and from all PEs in the array, or to and from all PEs in specified rows and columns. (II)

Parallel Instructions are executed by some or all PEs of the array simultaneously, depending on their internal states. (II)

Parallel Network Processor. The term "parallel network processor" refers to a reconfigurable parallel array processor in which:

1. The processing elements themselves are full-fledged sequential processors with local memories for storing data words.
2. The communication network connects each PE to several other PEs in a regular fashion; and
3. The PEs execute essentially the same program on different sets of data simultaneously, under central control. Each data set is stored in a local memory of one PE. (II)

Parallel Reconfigurable Data Path. Switching of propagation modes of connecting elements in the interconnection network such that the 16 k-bit processor and the 16 k-bit memory may have parallel exchanges with $16 \cdot k$-bit words. (III)

Parallel System. A parallel system generally means a system that implements either instruction parallelism, or data parallelism, or both types of parallelism, instruction and data, in a single system. (I)

Parallel-Write Operation. This operation is that of modifying selected bit positions at all cells in the associative memory, or at a selected subset of the cells. The bit positions to be modified are determined by the contents of the mask register. The contents to be inserted into those bit positions are the settings of the corresponding bit positions in the compare register. That is, the resulting setting at each selected bit position will be the same for all selected cells. The cells to be modified are selected by the settings of the match flip-flops. Parallel-write operations give the associative memory its capability to process data internally. This capability includes word-parallel arithmetic and logical operations as well as file searching and modification. (II)

Partitioning of Functions onto Computers. Partitioning of a set of functions (programs) into blocks (partitions) and assigning each block to a computer. (V)

Partitioning of the Microoperations into Groups, each of which is activated by one reconfiguration zone, is, as a rule, performed on the following basis: two microoperations belong to the same group if they have a common source or destination register or a common logical circuit. (I)

Path Types in V-graph. In the V-graph, a path is strong if it contains at least one strong edge, otherwise it is weak. A weak path is a sequence of weak edges; for any node on such a sequence, the path is the only one entering it, so that ultimately it is also the only path from the initial to the final node. (IV)

Performance Gains for Supersystem with Dynamic Architecture. A Supersystem with dynamic architecture may realize additional performance gains on the same resource by taking advantage of the following factors:

1. By reconfiguring resource into minimal size computers, it may maximize the number of programs computed by the same resource.
2. By switching the resources into different types of architecture—array, pipeline, multicomputer, or multiprocessor—it may speed up respective computations. This allows the available resources to be permanently involved in computations, even those that require dedicated subsystems such as array and pipelines.

Therefore, the use of dynamic architectures in Supersystems leads to the realization of new sources of throughput increases heretofore unused in traditional parallel systems. (III)

Performance Improvement of Dynamic Architectures. For a dynamic architecture, performance improvement may be achieved if one perfects the organizations of reconfigurable interconnections among modules and reduces the number both of different module types and of the logical circuits that interconnect LSI modules. (III)

Performance Requirement. A requirement of how well a function must be performed; e.g., performance requirements specify speed, delay, and computational accuracy. (V)

Permutation Function Implemented by a Flip STARAN Network. The permutation function implemented by a flip STARAN network with 2^n input lines is specified by the value of an integer that is input to the n control lines of the flip network. (II)

Phase Duration. The duration of one phase equals that of the longest microoperation it executes. If all microoperations assigned to a phase last one clock period, then the phase takes one clock period. If the phase executes longer microoperations such as carry propagation or memory access, it may last several clock periods. (I)

Phase Termination. Termination of a microinstruction phase may be made asynchronously by a completion signal that signifies the end of the operation or by a count of the number of clock periods that the longest microoperation requires. (I)

Physical Fault of the System. The failure of components that produce error in the system is called a physical fault of the system. (VI)

Pipeline Stage. A pipeline stage is a separate resource unit assigned to the execution of one phase. It can be a processor or a dedicated resource unit (multiplier, divider, adder, etc.).

Pipeline State of Dynamic Architecture. A pipeline state is specified by the number of concurrent pipelines and the number of stages in each pipeline. (III)

Pipeline System. A pipeline system is a parallel system containing N pipelines. Since each pipeline computes one instruction sequence in the overlapped mode, a pipeline system may computer N instruction sequences concurrently and thus implement instruction parallelism. No data parallelism is implemented since each instruction handles no more than two operands at a time.

Pipelining. In general pipelining requires partitioning of the instruction microprogram into several phases and allows for the overlapped execution of consecutive phases assigned to consecutive instructions. If the microprogram is partitioned into K phases, $F_1, F_2, \ldots, F_K$, then the pipeline that executes it contains k stages, $S_1, S_2, \ldots, S_K$, so that S_i executes phase F_i of the microprogram, etc. (I, II)

Predicate Associated with a Type-o Node. The predicate associated with a type-o node may be seen as originating the *union* of the symbols associated with the outgoing edges. For instance, "sequence *a or* sequence *b* is executed" is represented as "symbol *a* or symbol *b* is accepted," i.e., "$a \cup b$ is accepted." (IV)

Principle of Addresses of von Neumann Architecture. The principle of addresses of von Neumann architecture organizes computations in a way such that each instruction fetches operands from memory and/or sends the computational result back to memory. As a result, it is necessary for each instruction to store data addresses; i.e., to have clearly defined address fields. The operation to be performed on the data words fetched from memory is written to its opcode field. Each instruction is accordingly partitioned into the two zones—opcode and address. (I)

Principle of Sequentiality of Computations for von Neumann and Microprogrammed Computers. *Sequentiality of computations* means that each of the traditional architectures may execute only *one* operation at a time in the processor over *one* pair of data words.

This leads to the following restrictions:

1. In each type of computer (von Neumann and microprogrammed) program instructions are computed sequentially.

If the program provides for some computational concurrency, in which several instruction sequences may be executed concurrently, neither of these architectures can implement instruction *parallelism*; i.e., execute several instructions at a time. The only parallelism they allow is *microoperation parallelism* in which each type of computer may execute several microoperations at a time.

2. Each instruction may handle only two operands at a time.

If an array of operands needs to be handled, they can be computed sequentially using indexed addressing. This means that the addresses of the next pair of operands are computed in the control unit as sums of the addresses of the current pair of operands and an index increment. (I)

Priority Analysis is performed by a special control program stored in the V monitor. When a DC-group computes several independent programs, they may have conflicting calls on the next DC-group transition, i.e., two programs may ask for transitions to the N_k and N_e states respectively where $N_k \neq N_e$. Possible conflicting calls are resolved by assigning to each task of any given program a priority code. (III)

Problems of Arithmetic Pipelines. The major problems with arithmetic pipelines is the time overhead introduced by the disparity between the pipeline(s) and the algorithm being executed. As a result, pipeline systems tend to become dedicated to certain types of computation and pipeline systems usually have limited applicability. To broaden the range of their cost effective application, arithmetic pipelines offer various software controllable reconfigurations of the available hardware resource. The general idea is to reconfigure the resource, via software to reduce the dissimilarity between the pipeline and the sequences of arithmetic operations assigned to various instructions. (I)

Problems of Instruction Pipelines. The main problems faced in constructing an instruction pipeline are those of time overheads caused by conditional branches and of variations in the number of operand addresses and of the addressing procedures used in instructions.

Processor Code. Shows how many clock periods a processor dependent operation requires to execute in a given size computer. (III)

Processor-Dependent Operation. The operation in which operation time depends on the size of the processor. Each processor-dependent operation includes carry-propagation microoperation in which propagation time depends on the adder size. Examples of processor-dependent operations are: addition, subtraction, conditional branches on $\geq, \leq, <, >$, multiplication, and division. The last two operations include addition and subtraction in the operation algorithms. (III)

Processor-Independent Operation. The operation in which operation time is permanent and does not depend on the size of the processor. Thus a processor-independent operation does not include carry-propagation microoperation in its algorithm. Examples of processor-independent operations are: Boolean, conditional branches on $=, \neq$, parallel m-bit shifts where $m \geq 1$, etc. (III)

Program Computation in a System with Dynamic Architecture. In a multicomputer system with dynamic architecture the same program may be computed by a sequence of different size computers. Such computation may achieve a performance improvement due to the maximization in the number of programs computed by the same resource. (I, III)

Program Selection Code (PSC) shows the position i of the memory element ME_i that stores the currently executed program segment. For instance, if PSC = 2, instructions are fetched from ME_2, etc. This code is written to all PE's of the computer by a special "memory jump" instruction that organizes the fetching of the next program segment from a new ME. (III)

Program (Main Memory) Counter. The counter that stores a current (macro-) instruction address in the main memory.

Program Universality for dynamic architectures means the following:

1. Instructions should store no codes nor constants which change meaning depending on the computer's word size. This is achieved with the modular control

organization that provides that all control codes be stored in the MCD of all UM's.

2. None of the addresses contained in instructions should undergo any changes. This condition is satisfied if the same instruction array or data array contains, respectively, the same number of cells when the program is run on different size computers and if the array is specified by the same base address in both bases. Permanent size, s, in either case may be maintained if one uses a new memory allocation technique (PS exchanged).

3. All DC-group computers should be equipped with a unique instruction format and unique instruction set. (III)

Program Verification or Proof of Correctness consists of demonstrating the consistency between (1) a computer program, and (2) specifications or assertions describing what the program is supposed to do. (IV)

Psuedocode is a combination of natural language and structured computer code. It is useful for describing processing requirements before actual coding, and is a member of a family of the languages that have no specific syntactic rules.

Pulse Carrier Frequence. *See* Carrier frequency. (II)

Pulse Deinterleaving. *See* Signal-sorting. (II)

Pulse Train. A sequence of radar pulses output from a single radar emitter. See Section 4.6. (II)

Quality of an Architectural Adaptation. The quality of an architectural adaptation means how precise the match between the dynamically created architecture and the algorithm is sustained. (I)

Random-Access Memory. The random-access memory accepts the address of the desired memory location as input, and outputs the contents of that location in response. Also the time of accessing a random access memory is permanent and independent of the address of the location.

Reconfigurable Fault-Tolerant Systems can be categorized into two types:

1. *Identical-task system without intercommunications:* All active modules run the same job, and there is no intercommunication among them except resynchronization, switching, and voting. Therefore, this kind of system either masks the errors or switches out the failed module with or without replacement. If a faulty module is replaced by a spare, the resynchronization between the new active module with the previous active modules is required. All hardware redundancy systems belong to this category.

2. *Different-task system with intercommunications:* All active modules run different jobs and intercommunicate with each other by some switching networks. When a module fails, the system has to switch out the failed module, but due to the communications among good modules and the failed module, the good modules may be contaminated by the faulty module. Therefore, it is necessary to program rollback to a previous good state in which no failure has occurred. Otherwise, the system fails due to unrecovered error. (VI)

Reconfigurable Instruction Path. Switching of propagation modes of connecting elements in the interconnection network such that the instruction sorted in one memory element, ME, may be received by all processor elements PE that are included into one $16 \cdot k$ bit computer. (III)

Reconfigurable Parallel Array System. A parallel array system in which the dimentionality of the data vector being processed and the length of the vector elements can be varied under software control. (I, II)

Reconfiguration Codes specify what combination of registers needed by the instruction could be connected with each other and the adder terminals. (I)

Reconfiguration Field. A field of a microinstruction that stores reconfiguration codes.

Recovery Cache-type Mechanism. The recovery cache-type mechanism records the original states of just those resources that must be modified for reinstatement later if an error is detected. (VI)

Recovery Line. The recovery line is the minimal set of recovery points in all processes that are either in error or affected by the error. (VI)

Reduction in the Number of Addresses Assigned to One Instruction may lead to savings both in the number of memory accesses required and in the amount of memory needed to store the intermediate results of expressions that contain more than one processor operation executed sequentially (such as $(A^2 - B)/C + D$ or $(A - B)^2 > K$). If an algorithm contains expressions containing not more than one arithmetic operation, and the result of this operation is not used in the next instruction (one instruction executes $A + B$, the next one $C \div D$, etc.), then computation of such expressions with a two address or one address instruction leads to an increase both in the number of accesses and in the amount of memory needed when compared to a three address instruction. (I)

Register-to-Register Transfer. Each register-to-register transfer may be conceived of as a mircooperation with no execution circuit. The data word merely passes through the control circuit. (I)

Relation Among Component Failure. The relation among component failure, faults, errors, and system failures can be expressed by the following sequence: component failure $\rightarrow$ fault $\rightarrow$ error $\rightarrow$ system failure where "$\rightarrow$" denotes "causes." (VI)

Relations. For structural termination techniques extended to microprograms, relations can be:

1. *Simple transfers* of an information unit from a "source" variable to an "object" one; these correspond to simple assignments in programs, and will be represented in the graph by weak edges;
2. *Transfers through functional units* ("functional transfers"), involving some processing of the source variable(s); they correspond to "proper functions" in programs, and will be represented in the graph by strong edges.

    It remains to consider "*read*" transfers. While from an architectural point of view they do not differ from a register-to-register transfer, if we refer to the focal point of structural termination—that is, after all, a continued data dependence—it can be said that an external read, in microprograms just as in real-time programs, leads to an a priori nondeterminable value of the variables thus read. It seems therefore reasonable, in this context, to interpret a read as a functional transfer. (IV)

Requirements Engineering. An initial phase of design in which the application requirements are analyzed and a preliminary processing solution is structured. It specifies the numbers and characteristics of computing modules to be assigned for execution, the logical interconnection topology between computing modules, the applications software modules to be mapped onto given computing modules, and requirements for intermodule control to satisfy a given class of applications. It gives functional and performance requirements for each module, but it does not define the internal structures of modules. (V)

Requirements to Control Organization of Dynamic Architecture. The control organization of dynamic architectures must take into account *common* LSI technological restrictions—high cost of a module type and restricted pin count and *specific* restrictions that originate from dynamic partitioning of the resources into a variable number of concurrent computers. These requirements are:

1. Each formation of a new computer must be accompanied by establishment of its control unit via software. This control unit must generate variable operation times for processor and memory access operations.
2. The control organization of a DC-group computer must function independently of the number of control units it contains. (III)

Resource Assignment for Dynamic Architectures. To realize the capability of dynamic architectures to increase throughput of the available resources, the following problem has to be solved. For each user program one has to find a sequence of minimal size computers which may execute it. Next, the available hardware resources of the multicomputer system have to be assigned among user programs, each of which is executed by a sequence of the minimal size computers found earlier. The hardware resource assignment is then reduced to finding a flow chart of architectural states which gives the maximal concurrency in execution of the given set of user programs. (III)

Resource Assignment (Allocation). Assignment of programs (functions) to the processor and memory resources, where both types of resources are determined by requirement of programs on processor and memory resources. Resource assignment (allocation) can be *static* and *dynamic*; whereby *static assignment* assigns programs to the same processor and memory resources that do not change during computation, while *dynamic assignment* assigns the same programs to a sequence of memory and processor resources so that each time a program(s) is computed by the next processor and memory resource of a sequence.

Response Pattern. The set of all test outcomes, $\{t_{ij}\}$, in the system forms a response pattern. Suppse that the status (being faulty or fault-free) of each node is known, then we can also denote the expected test outcome of a test link, l_{ij}, by a_{ij} as follows:

$$a_{ij} = \begin{cases} 0 \text{ if both } m_i \text{ and } m_j \text{ are fault free} \\ 1 \text{ if } m_i \text{ is fault-free and } m_j \text{ is faulty} \\ X \text{ (don't know) if } m_i \text{ is faulty} \end{cases} \text{(VI)}$$

Restrictions of LSI Technology on Commercial Production of Complex Computers. LSI technology poses new restrictions on commercial production of complex computers: To create a pilot computer, one has to develop a set of module types that will be used in its assembly. However, the design of such a set requires enormous investments that can be recovered only through mass production. Since a powerful computer system can hardly be produced in volume, it is easy to see why the industry lags in manufacturing LSI modular computers capable of emulating existing general-purpose mainframes. (I)

Retained State Data. The local data that a history sensitive process must retain over a sequence of executions. (V, VI)

Reversal Check. A reversal check processes the results of the system activity in order to determine the corresponding inputs and check them with the actual inputs. This scheme can only be used for those software whose input-to-output relation is either one-to-one or one-to-many but not many-to-one. (VI)

Rollback. The rollback of a single instruction, a segment of programs, or the entire program means reexecution of the affected program at a previously established rollback point or checkpoint. This is a common way to recover from a damaging transient. Program rollback (or checkpoint restart) requires saving the status of the program, i.e., its state vector, at each rollback point. (VI)

"Round-trip" Signal. An output signal from a chip or wafer that causes inputs to that wafer from other wafers during the same clock cycle. See Sec. 4.11. (II)

Routing Logic controls the source or destination of data and control information passed between the PE and other PEs and common buses.

Sami and Stefanelli Program Scheme (otherwise called C-graph or control graph) is a program graph in which:

Each node is substituted with an oriented edge termed the "instruction edge" and oriented from entry to exit of the corresponding instruction sequence.

Each edge is kept as a "control edge"; control edges are always oriented from exit of an instruction edge to entry of a (possibly different) instruction edge.

All series connections can be subsequently simplified through merging into a single edge (merging of an instruction edge and a control edge results into an instruction edge). In the resulting graph edges represent at once instruction sequences *and* control flow; nodes simply denote "singular points" in the control flow, that is either merging points or branching points. This type of graph has proved to lend itself very easily to various preprocessing actions useful for such operating system policies as task allocation, hierarchical memory management, etc.; for example, it allows straightforward computation of "probable activities" of instruction sequences both a priori and at runtime. (IV)

Search Memory. A computer memory which can be addressed by the content of the words, rather than (or in addition to) hardware address. See Sec. 3.2.1.1. The search memory accepts the desired contents of the memory location or locations as input, and outputs flag settings indicating the memory location or locations having those contents. (II)

Second Order Similarity of Two Functions, F_i and F_j. Specified as S_2 $(i, j) = S_1(i) \times S_1(j)$, where $S_1(i)$ is ith row of the first order similarity matrix S_1, and $S_1(j)$ is jth column of the same matrix; multiplication of matrices is understood in conventional sense except minimum operations replace products and maximum operations replace sums. (V)

Select Instructions of Array Architecture are executed upon the PE's activity register, its stack, and the tag register to determine which PEs will take part in subsequent parallel arithmetic instructions. (II)

Selection of the Size of One Reconfiguration Zone. Since one zone is assigned to one group of microoperations, its size depends on two factors: the number, N, of microoperations in the group, and the encoding technique used. There are two encoding techniques: *logarithmic* encoding and *linear* encoding. Logarithmic encoding assigns each microoperation from the group with a p-bit code, where $p = \log_2 N$ bits; i.e., such an assignment leads to the minimal size of each reconfiguration zone. Linear encoding assigns each microoperation in a group to one bit. Namely, for a group containing N microoperations, the size of reconfiguration zone is N bits. Linear encoding maximizes the size of the zone and leads to a very large microinstruction size. As a result, the width of the control memory also increases significantly. However, all microoperations from the group can be executed concurrently. (I)

Self-Loops. For the program graph model, existence of a self-loop would coincide with an error situation, that is, the availability of an instruction sequence without one exit path towards the program STOP. (IV)

Separation of Functions, F_i and F_k. Two functions belonging to different blocks (partitions). (V)

Sequencing of Two Instruction Sequences. The sequencing of two instruction sequences is representable by means of the *catenation* of the corresponding symbols: "execution of instruction sequence a is followed by execution of sequence b" is represented as "symbol a is followed by symbol b." (IV)

Sequential Controller. The sequential controller provides system interfacing with the system peripheral devices and with the operator. (II)

Sequential Processor. A computer processor that executes a single instruction at a time on a single operand or a pair of operands. (I, II)

Sequential Program. A program executed on a conventional single processor. (II, IV)

Serial Instructions provide branch control of the program, based both on the results of its own sequential arithmetic, Boolean processing, and on the results of parallel PE activity. (II)

Set of Array Architectural States. Since for a given processor resource there exist different partitions into various arrays, the array dynamic architecture may be characterized by a set of array architectural states, which differ from each other along one or several attributes:

1. The number of concurrent arrays,
2. Dimensions of data vectors computed in each array,
3. Word sizes of processors working in each array. (III)

Signal Paths for Dynamic Architecture Delayed Due to Reconfiguration. One may distinguish the following signal paths that are delayed due to the reconfiguration logic:

1. *Instruction path:* In dynamic architectures the available resource may be partitioned into a variable number of concurrent computers. The instruction bus must

separate the instruction streams computed by independent computers. It follows that the instruction path is reconfigurable, and the reconfiguration logic adds additional delay to the instruction fetch time;

2. *Data exchange path:* Since each $h \cdot k$-bit computer must maintain a parallel data exchange with $h \cdot k$-bit words, and several computers of different sizes may be formed in a single state, the data bus must separate concurrent data streams processed by independent computers in one state. Thus the data path is also reconfigurable, and additional delays introduced by the reconfiguration logic are added to the data fetch times;

3. *End-around carry path:* Since the available processor resource, containing n h-bit processor modules, may be arranged into $h \cdot k$-bit processors where k can range from 1 to n, a dynamic architecture must form a reconfigurable path for the end-around carries. This means that in addition to the usual delays that characterize a static end-around-carry path, there arises an additional delay due to reconfiguration in this path;

4. *Paths for equality signals:* The equality signal produced during the comparison of two $h \cdot k$-bit operands participates in several conditional branch operations, and must be routed to the most significant processor module of the respective $h \cdot k$-bit computer. Since each processor module in a dynamic architecture may be the most significant one in some $h \cdot k$-bit computer, the path for equality signals must also be reconfigurable. Additional delay arises while propagating the equality signal(s) through the reconfigurable equality path due to reconfiguration logic. (III)

Signal-sorting. The process aimed at separating a mixed stream of radar pulses, received from a number of emitters into the separate pulse streams, one stream for each emitter. Also called *pulse-deinterleaving*. See Sec. 4.2.2 (II)

Similarity Relation. A computational relationship with range [0, 1] which indicates the similarity of functions (programs) with respect to certain characteristics. (V)

Simple Conditional Transition in a Microprogram. Transition $a \xrightarrow{y} b$, $a \xrightarrow{\bar{y}} c$ is called a simple conditional transition if node a has two succeeding nodes b and c where b is selected when $y = 1$ and c is selected when $\bar{y} = 1$ where y and $\bar{y}$ are the true and false values of signal y that marks this transition. Thus a simple conditional transition is activated by a single conditional signal y that may assume two values: a true value ($y = 1$) and a false value ($\bar{y} = 1$). (I)

Simple Control Circuit. A simple control circuit is understood to be a collection of two-input AND gates. Each such gate has one of its inputs connected to the information output (0 or 1) of one bit of the source register that stores the data word, and a second input (control input) connected to the control unit. The output of this gate is connected to the information input (0 or 1) of one bit of the destination register that is to receive this word. To activate this word transfer, the control unit sends a signal through the control line that connects all the control inputs of the AND gates together, and thus causes a word to be transferred from the source register to the destination register. (I)

Simple Loop. A simple loop is defined as a Control-graph (or, better, a segment of a Control-graph) in which one loop only exists, and there is only one possible path inside the loop. (IV)

Simulator Program. A program executing on one computer which simulates the instruction-by-instruction execution of a program written on (usually) another computer, using a machine-language program for the object computer as input.

Software Checking of CPU and Memory can be accomplished by running diagnostic programs periodically, but with no *concurrent error detection capability* because some delays exist between the occurrence of errors and the detection of them.

Software Control of Interconnections may proceed in the following directions in reconfigurable architectures:

1. It may be used to change the sizes of processors in array systems, giving the system the ability to compute a larger number of data items at the same time, thus enhancing data parallelism.
2. Reconfiguration allows any processor to be connected with any memory unit or any other processor in a multiprocessing system. Thus all external communications between different processors by way of slow I/O devices may be replaced by fast and direct processor–memory, processor–processor, or memory–memory exchanges.
3. In reconfigurable pipeline systems, reconfiguration allows changes to be made in the sequence of operational units (adders, multipliers, subtractors, dividers) connected into the pipeline in order to match the sequence of operations encountered in the program. The consequences of such a match is the minimization of the number of dummy time intervals caused by the disparity between program and pipeline structures.
4. Software control of interconnections between various computer nodes in a multicomputer system allows different topological configurations in the network to be established—such as star, closely connected graph, hierarchical pyramid, binary tree—depending on the structure of the algorithms being computed.
5. Since all faulty modules detected by diagnostic tests may be isolated via software and replaced by spares, software-controlled reconfiguration may lead to enhanced system reliability. (I)

Software-controlled Microprograms are those in which some sequences of activated microoperations could be selected by the programmer when he or she writes new reconfiguration codes to the instruction fields.

Software Error Detection Scheme. A software error detection scheme can be implemented using the following modules:

1. *Comparison module* in a dual redundancy scheme.
2. *Majority voting module* in a software triple modular redundancy scheme.
3. *Audit module* in a software dynamic redundancy scheme. The validation routine used in the audit modules can be either the *acceptance tests* which compare the desired I/O relations with those of the active module or *dynamic assertions* which are *executable predicates* (logical and relational expressions for checking purposes only) added before or after each statement in the processing module, generating a predicate true exit (which indicates the correct execution of the statement) or a predicate fail exit (which indicates the incorrect execution of the statement and replaces it by a spare or alternate statement).

4. *Watch-dog timer* (WDT) which can be reset at any time and generates an interrupt interrupt signal on reaching the preset time limit, can detect dead functional units or tight (infinite) looping software. (VI)

Software Error Diagnosis consists of three steps: (1) error detection, (2) error treatment, and (3) damage assessment. (VI)

Software Errors can be caused by either physical faults (hardware component faults) or algorithmic faults. According to the different software development phases, algorithmic faults for software include *requirement faults, specification faults, design faults, construction faults,* and *verification faults*. (VI)

Software Modules are groups of subfunctions which are to be considered as a unit. (V)

Software Redundancy is implemented using the following techniques: (1) replication of critical programs and data in different storage locations; (2) diagnostic programs at various program and microprogram levels; (3) special programs which implement program restarts and interface with the operating system. (VI)

Source Instructions are lines of high level language. (V)

Sources in the V-graph are nodes into which no edge enters. Accordingly, a source in the V-graph represents a variable whose value is never modified. (V)

Speed-up in Adaptable Array Computations. To speed-up array computations, each array must contain $16 \cdot k$-bit processors which are not necessarily adjacent. Then one instruction may handle a data vector whose components (data words) are processed by arbitrary $16 \cdot k$-bit processors formed in the resource, i.e., there will be no limitation on where data items handled by one array are stored.

On the other hand, each $16 \cdot k$-bit processor of the array which contains k processor elements and handles $16 \cdot k$-bit words, must be assembled only from adjacent PE's. Indeed, to perform fast computations, a reconfigurable processor bus must be dedicated. Otherwise, if it shares interconnections with the memory-processor bus and uses multi-staged interconnection network, the bus introduces significant delays, since in propagating from one PE to the next more significant PE, carry signal must pass through $2 \log_2 n$ connecting elements where n is the number of processor elements in the resource.

However, if the processor bus is dedicated and includes not only adjacent PE's, the number of possible combinations of PE that may be contained in one $16 \cdot k$-bit processor grows exponentially. This leads to an exponential increase in the complexity of the bus. Therefore, if the processor bus is not dedicated, it becomes very slow; if it is dedicated but not assembled from adjacent PE's it only becomes very complex.

On the other hand, should each $16 \cdot k$-bit processor be assembled from adjacent PE's, one may construct very simple and fast processor busses which perform very fast propagations of carries and overflows and take minimal hardware complexity. (III)

Static and Adaptable Architectures. Computer architectures may be divided into two categories: *static* and *adaptable*. Static architectures do not adapt via software to the programs being computed, and adaptable architectures do. Adaptable architectures may, at present, be partitioned into three classes: *microprogrammable, reconfigurable,* and *dynamic*—depending on the level of reconfiguration performed. (I)

Static Hardware Redundancy Scheme. This scheme uses identical modules feeding a voter and the output takes a majority vote to provide the correct output when the majority number of modules are fault-free. (VI)

Strong and Weak Paths. On an I/O relationship graph a path from input node v_i to outpute node v_j is strong if it contains at least one strong edge, weak if otherwise. (IV)

Strong Submatrix S. A strong submatrix S of a matrix M as a submatrix over $\bar{I}$ is characterized by the following features:

1. There is at least one nonzero entry in each column.
2. There is at least one $+1$ entry, in the matrix. (IV)

Structural Nontermination of a Program Loop. A situation such that a variable that dominates the loop closing predicate after a transient number of iterations either becomes a constant or begins to change its value with a periodical law assuming a finite cyclic set of values. (IV)

Structural Termination Analysis examines structures of high level statements and determines termination property of program loops. (IV)

Structural Termination of the Program Loop. A situation when a variable that dominates the loop closing predicate assumes values that lead to exit from the loop after k iterations, where k depends on the initial data set. (IV)

Stuck-type Fault. *Logical lead fault* or stuck-type fault causes the input and/or output leads of each gate to become fixed at a constant value, either logical one (in this case, the signal at the site of the fault is said to be stuck-at-one denoted by s-a-1) or logical zero (or stuck-at-zero, denoted by s-a-0). (VI)

Submatrix of a V-matrix M over a subset $\bar{I}$ of the index set I is a matrix obtained from M by deleting all rows and columns marked with indices not belonging to $\bar{I}$. (IV)

Supersystems. The systems that possess an extreme computational power and are capable of solving problems of enormous complexity are now called Supersystems. The problems for Supersystems may be nonreal-time and real-time. (III)

Symbolic Assembler. The symbolic assembler is a program that accepts a source program module coded in a symbolic machine-level language and generates a relocatable binary object module from it.

Symbolic Execution selects a particular path in the program and validates it by assigning values to the input data and then "executing" the program path with reference to such symbolic values. When the "execution" has been completed, the final symbolic representation obtained can be analyzed to determine path correctness, giving data-independent results. (IV)

Synchronization Mechanism for Input-Output Relationship Graph. If V_S is the set of variables acting as synchronization mechanism and controlling access to a variable v_i, dependence of v_i from V_S must be stated. This is performed by considering the synchronization mechanisms as proper functions, and as such represented in the graph by strong edges going from "controlling" to "controlled" nodes. The problem is thus reduced to a simple examination of structural termination characteristics. (The predicate upon the semaphores will need to be examined.) (IV)

Synthesis is the procedure aimed at representing intermodule communication requirements. These requirements allow designing of the physical intermodule network and internal design of the modules. (V)

T-fault Diagnosible System. The graph theoretical model represents the t-faults diagnosable system S by a *directed graph* (digraph), $G(V, E)$ in which V is the set of *nodes* (each module is represented by a node in G) and E is the set of *arcs*. The arc from node i to node j represents a test link, l_{ij}, from module m_i to m_j. A system S is called t-fault diagnosable if, given the response pattern of the system S, all permanent faulty modules in S can be *identified* provided that the number of faulty modules does not exceed t. (VI)

Techniques for Describing Microprograms. There are two techniques of describing an instruction microprogram: (1) a symbolic technique using one of existing hardware description langauges, and (2) a directed graph in which the steps in the microprogram are presented as a directed graph. (I)

Test of the Error. For a functional error, a test of the error is the input data which produces different output values from the normal output values. (VI)

The (i, j) Connection is the connection for the read and write signals between address connecting elements, ASE_j (assigned to PE_i) and memory connecting element, MSE_i (assigned to ME_j). In each ASE and MSE the (i, j) connection takes only 2 pins. In order to implement the address and h-bit data paths between every pair PE–ME, the reconfigurable memory-processor bus has to have all (i, j) connections, where i, $j = 1, \ldots, n$. (III)

Three-Address Instructions may achieve a saving in both the number of memory accesses and memory size in comparison with one and two address instructions for executing of algorithms that contain simple expressions needing no more than one processor operation. (I)

Threshold Voter. A threshold voter with n inputs and a threshold value of t has an output logic 1 if and only if the number of 1's in input lines is greater than or equal to t. (VI)

Time for Architectural Reconfiguration. The time for architectural reconfiguration includes the following actions:

1. The time to search, fetch, and write all the necessary control codes to those modules of the resource that are formed into new computers in the next architectural state;
2. The time to switch the reconfigurable busses that must separate computers of the next architectural state and provide correct information exchanges between them.

Since each time an algorithm requires an architectural reconfiguration the computers affected stop, the time of architectural reconfiguration has to be added to the time of algorithm computation. It then follows that the time of reconfiguration has to be minimized in order to not offset speed advantages gained by dynamic architectures due to their adaptations to the executed algorithm. (III)

Time of Microcommand Delay. Microcommand generation begins only after a new word is written to the sequencer of the instruction register; i.e., after a synchronization pulse. Therefore, the time of microcommand delay includes time for decoding a word stored in the sequencer or instruction register. Techniques for finding the microoperation delay were introduced in Chap. I, Part I, Sec. 3. (I)

Time Redundancy. Two major aspects of time redundancy are the repeated execution of instructions and the restart of programs after an error detection. Time redundancy can be found either in the identification and correction of those errors caused by transient faults or in program restarts after a hardware reconfiguration. (VI)

Time Restriction in a Complex Real-Time Algorithm. For a complex real-time algorithm, there is a time restriction between the moment a particular set of data streams enters the system and the moment the system produces a response specified by this set. Complex real-time problems impose more severe requirements on Supersystems, because in addition to the demands for handling an enormous amount of information, they require that this information be handled during specified and, as a rule, very small time intervals. Thus real-time problems like these require Supersystems with the highest throughputs attainable by current technology. (III)

Time Tag. A time field appended to a message for timing simulation and synchronization. (V)

TOA. The time-of-arrival of a radar pulse at the radar receiver. See Section 4.2. (II)

Traditional Sources of Augmenting System Throughput. One may increase the throughput of a complex system by drawing on the following traditional sources:

1. The use of high speed components.
2. Modular expansion of the system with new equipment (adding new computers, processors, I/O units, etc.).
3. Equipping the system with a dedicated architecture that is very effective for a given class of applications.
4. Utilization of the maximal concurrency present in programs.
5. Application of special types of architecuture: pipeline, array, associative.
6. Optimization in data exchanges. (III)

Transient Interval of the Loop. An initial sequence of iterated values assumed by the variable(s) that dominates the predicate until it begins to assume typical values that lead to loop termination or nontermination. (IV)

Transition Graph of a Finite-state Recognizer Automaton. An oriented graph with one initial node, one terminal node, and such that any of its edges traversed by at least one path from initial to terminal node can formally be considered as the transition graph of a finite-state recognizer Automaton.

Triple Modular Software Redundancy. The triple modular software redundancy (static redundancy) scheme means three processing modules perform the same data transformation function in a single computer. A voting module selects the result on the basis of a majority decision. Furthermore, it also identifies the minority modules as the faulty modules. The error detection by voting module will be carried out to the module level only. Thus an error within a module is not detected. (VI)

Two Types of Busses for Dynamic Architecture. To implement a dynamic architecture one must use two types of busses:

1. Reconfigurable memory-processor bus which performs variations in the number of instruction and data streams.
2. Reconfigurable processor bus which implements variable $16 \cdot k$-bit processors.

The architectural realization of these busses affects both the speed of computations and complexity of the resources organized into a particular dynamic architecture. (III)

Type-i Nodes are nodes in which one or more arrows merge while there is only one outgoing arrow: in graph-theory terms, a type i node has indegree $i \geq 1$, outdegree $\sigma = 1$. With reference to programming practice, a type i node is the entry point of an instruction sequence which can be referenced by more than one control statement. (IV)

Type-O Nodes are nodes having only one incoming arrow while one or more arrows depart; in graph-theory terms, a type O node has indegree $i = 1$, outdegree $O \geq 1$. With reference to programming, a type O node with $O > 1$ stands for a conditional branch statement that specifies all exits from a given instruction sequence only (just as type i node can be entry node of one instruction sequence only). (IV)

Types of Conditional Tests. In a computer two types of conditional tests are distinguished—signed (arithmetic) and unsigned (logical). For a signed test, both operands A and B are signed numbers, so that each test involves comparisons of their signs and magnitudes provided signs are the same. For an unsigned test only magnitudes of A and B are compared. Accordingly, A and B are unsigned. Otherwise, if they are signed and signs are different, a complemented number (with negative sign) has to be transformed into a magnitude form before an unsigned comparison may proceed. This reduces an unsigned conditional test into a signed one with equal signs. (I)

Types of Instruction Intervals for Dynamic Architecture. The instruction cycle may contain the following types of intervals:

1. Intervals, T_p, for processor-dependent operations: $T_p = p \cdot t_0$ where p is the number of clock periods a processor-dependent operation requires to execute in a given size computer.
2. Intervals, T_0 for processor-independent operations (Boolean, equality, inequality, shift, etc.). These are of permanent duration equal to one clock period t_0.
3. Intervals, T_I, for instruction fetch: $T_I = m_I \cdot t_0$ where m_I is the number of clock periods required to fetch an instruction from a given memory element.
4. Intervals, T_E, for data-word fetch: $T_E = m_E \cdot t_0$ where m_E specifies the speed of the slowest memory element contained in the primary memory of the given computer. This originates from the fact that a $16 \cdot k$-bit word is stored in a parallel cell made of k 16-bit bytes. Thus to fetch this word, one has to take the access time of the slowest memory element. (III)

Types of Instruction in a Microprogram Computer. In a microprogrammed architecture, two types of instructions appear: *microinstructions,* each of which activates a set of parallel microoperations, and *(macro) instructions,* each of which activates a microprogram represented by a sequence of microinstructions. Microinstructions are stored in the control memory; (macro) instructions are stored in the main memory. (I)

Types of Interconnection Networks. Two types of interconnection networks are possible:

1. One-staged networks in which any pair of resource units is connected via one dedicated connecting element. If a Supersystem has n resource units it will

require about n^2 connecting elements to organize full one-staged connectedness among them. Each communication path will introduce the minimal communication delay, however, equal to that of one connecting element.

2. Multistaged networks in which any pair of PE's and ME's are connected via a sequence containing $\log_2 n$ connecting elements. This will reduce the overall number of connecting elements and the complexity of the memory-processor interconnection network; however, each communication path between a pair of PE's and ME's will have a communication delay equal to that of the $\log_2 n$ consecutive connecting elements in the path. To form a communication path between two processor elements, PE_i and PE_j in the memory-processor interconnection network will require $2 \log_2 n$ connecting elements since this communication is established as $PE_i - ME - PE_j$, and the paths $PE_i - ME$ and $ME - PE_j$ each incorporate $\log_2 n$ connecting elements. Since n may be a very large number, for a Supersystem such delays could become significant. For instance, for a Supersystem having $n = 1000$ PE's and ME's, a communication path between any pair of PE's and ME's will go through 10 connecting elements. Especially critical is the formation of variable size processors, composed of k PE's, through a path activated inside the memory processor network. Not only are all processor signals (carries, generates, and propagates) delayed by the 2 $\log_2 n$ connecting elements from one PE to the next more significant PE, but these signals must pass through m levels of gates inside each connecting element, thus leading to an overall time for $h \cdot k$-bit additional proportional to $2 \cdot m \cdot k \cdot \log_2 n \cdot t_d$ where t_d is one gate's delay. The Banyan network implementation gives $m = 4$ since the processor signals in each connecting element pass through four consecutively connected gates. (III)

Types of Parallel Systems. Four types of parallel systems may currently be distinguished: multicomputer, multiprocessor, array, and pipeline. The appearance of each of these systems was dictated by the peculiarities of the complex parallel algorithms it was designed to compute. (I)

Unbalanced System. An unbalanced system is a system in which the individual PE's can operate independently with high throughput, but whose throughput is greatly reduced when they must communicate with one another. (II)

Unconditional Transition in the Microprogram Directed Graph. Transition $a \xrightarrow{x} b$ from node a to node b is called *unconditional* if node a has only one succeeding node b. Each unconditional transition is activated by the unconditional signal that may assume only one value. For instance, all transitions in a microprogram directed graph are unconditional if they are activated only by synchronization pulse τ which signifies the end of the microoperation executed in the present clock period and activates the transition to the next microoperation executed during the next clock period. (I)

Use of High Speed Components. The use of high speed components means implementation of a system from the fastest components which are available at the moment the system construction begins. (III)

Utilization of the Maximal Concurrency Present in Algorithms. This is the technique of finding the maximal number of instruction and data streams present in a complex algorithm and assigning a separate computer (processor) to each such stream. Interest in this technique is caused by the advances in LSI technology that significantly reduce the cost and enhance the reliability of computer components. It accordingly makes feasible the design of reliable complex computer systems containing many more hardware resources than ever before. This leads to a sharp increase in the number of instruction and data streams that may be computed in parallel in a single system. (III)

V-graph. The V-graph is derived from the V-matrix as follows:

1. There is one node in the V-graph for each variable appearing in the V-matrix. Thus, refer back to Example 4, and to the related V-matrix (Fig. IV-3); in order to build the V-graph, four nodes marked v_1, v_2, v_3, and v_4 are introduced.
2. Each entry $v(j, 1) \neq 0$ in the V-matrix is represented in the V-graph by an oriented edge going from node v_i to node v_j. The edge is strong (full arrow) if the entry has value $+1$, it is weak (dotted arrow) if the entry has value -1.
3. No other edges appear in the graph. (IV)

V-matrix. The V-matrix is a matrix whose *columns* represent *input values* of the variables, while *rows* represent *output values* of the variables. (It is therefore a square matrix.) Any given element $v(j, i)$ of the V-matrix contains information on existence and nature of a path from input node v_i to output node v_j; more precisely, each such element can be assigned one—and only one—of three values $+1, -1, 0$, according to the following rules:

1. $v(j, i) = +1$ if (and only if) there is a *strong* path from input v_i to output v_j
2. $v(j, i) = -1$ if (and only if) there is a *weak* path from input v_i to output v_j
3. $v(j, i) = 0$ if (and only if) there is *no* path from input v_i to output v_j (IV)

Variable and Invariable Control Codes of Dynamic Architecture are the codes used for reconfiguration and stored in each LSI module of the resource. Variable control codes change during each reconfiguration. Invariable control codes do not, and are written to each LSI module only once during the assembly stage. (III)

Variable Codes of Dynamic Architecture. The variable codes are as follows:

1. *Processor code p* sets the time for a processor dependent operation.
2. *Data fetch code m_E* is the time for fetching a $16 \cdot k$-bit word from the primary memory of $16 \cdot k$-bit computer. It specifies the speed of the slowest ME contained in this computer.
3. *Computers size code k* specifies the number of computer elements contained in computer. This code participates in forming a variable $16 \cdot k$-bit adder provided each PE is assembled from more than one module.
4. *Significance code b* marks the most significant, least significant, and middle computer elements contained in computer. (For a 64-bit computer, containing computer elements CE_1, CE_2, CE_3, and CE_4, the most significant, CE_1, stores $b = 10$, the least significant, CE_4, stores $b = 01$, and the middle ones, CE_2 and CE_3, store $b = 00$.). This code establishes connections for a variable $16 \cdot k$-bit adder. Note: All variable codes except b are the same for all LSI modules of the computer. (III)

Variable Size Computer, $C_i(k)$. Each $C_i(k)$ computer ($k = 1, 2, \ldots, n$) integrates k computer elements and $k - 1$ connecting elements. The i shows the position of the computer's most significant CE. This computer handles $16 \cdot k$-bit words and has a primary memory $16 \cdot n$-bits wide. By changing the number of computer elements in $C_i(k)$, one obtains 16, 32, 48, ..., $16 \cdot n$-bit computers. (III)

Variables. For structural termination analysis extended to microprograms, variables have to be defined with reference to machine architecture: we state that a variable is an information unit capable of being independently accessed, processed, read, or written. In this context, an n-bit register R appears as an n-bit variable, a carry flag C is a one-bit variable, etc. Therefore, the set of variables, V, comprises the names of:

1. All registers identifiable by a unique name (this allows also for I/O ports) and definable as a collection of bits connected via identical logical circuits with other registers or logical circuits;
2. In accordance with what has been done for programs, a "dummy" representing all constants. (IV)

Vertical Microprogram. A microprogram in which each microinstruction is made up of one microcommand only. (IV)

VLSI. Very large scale integration is the fabrication of 100,000 or more gates on a single wafer.

Window. A range of values permissible for a parameter, as specified by upper and lower limits. See Sec. 4.6.1. (II)

Word-cycle Instruction. The principal ALAP instruction. See Sec. 3.2.3.4. (II)

Bibliography of Basic Topical References

I. VON NEUMANN ARCHITECTURE

BURKS, ARTHUR W., HERMAN H. GOLDSTINE, and JOHN VON NEUMANN, "Preliminary Discussion of the Logical Design of an Electronic Computing Instruction," (Pt. I, vol. 1), Report prepared for U.S. Army Ordinance Department, 1946, in A. H. TAUB (ed.), *Collected Works of John von Neumann*, vol. 5, pp. 34–79, New York, The Macmillan Company, 1963.

GOLDSTINE, H. H., and JOHN VON NEUMANN, "On the Principles of Large Scale Computing Machines," unpublished, 1946, in A. H. TAUB (ed.), *Collected Works of John von Neumann*, vol. 5, pp. 1–32, New York, The Macmillan Company, 1963.

GOLDSTINE, H. H., and JOHN VON NEUMANN, "Planning and Coding Problems for an Electronic Computing Instruction," (Pt. II, vol. 1), Report prepared for U.S. Army Ordinance Department, 1947, in A. H. TAUB (ed.), *Collected Works of John von Neumann*, vol. 5, pp. 80–151, New York, The Macmillan Company, 1963.

GOLDSTINE, H. H., and JOHN VON NEUMANN, "Planning and Coding of Problems for an Electronic Computing Instruction," (Pt. II, vol. 2), Report prepared for U.S. Army Ordinance Department, 1948, in A. H. TAUB (ed.), *Collected Works of John von Neumann*, vol. 5, pp. 152–214, New York, The Macmillan Company, 1963.

GOLDSTINE, H. H. and JOHN VON NEUMANN, "Planning and Coding of Problems for an Electronic Computing Instruction," (Pt. II, vol. 3), Report prepared for U.S. Army Ordinance Department, 1948, in A. H. TAUB (ed.), *Collected Works of John von Neumann*, vol. 5, pp. 215–235, New York, The Macmillan Company, 1963.

II. MICROPROGRAMMED COMPUTERS

WILKES, M. V., "The Best Way to Design an Automatic Calculating Machine," *Manchester University Computer Inaugural Conference,* July 1951, published by Ferranti Ltd., London.

WILKES, M. V., and J. B. STRINGER, "Microprogramming and the Design of the Control Circuits in an Electronic Digital Computer," *Proceedings Cambridge Philosophical Society,* Pt. 2, **49,** April 1952, pp. 230–238.

ROSIN, R. F., "Contemporary Concepts of Microprogramming and Evaluation," *Computing Surveys,* **1,** *4,* December 1969, pp. 197–212.

HUSSON, SAMIR S., *Microprogramming: Principles and Practices,* Englewood Cliffs, N.J., Prentice-Hall, Inc., 1970.

ABD-ALLA, A. M., and D. C. KARLGAARD, "Heuristic Synthesis of Microprogrammed Computer Architecture," *IEEE Trans. Computers,* **C23,** *8,* August 1974, pp. 802–807.

AGRAWALA, A. K., and T. G. RAUSCHER, "Microprogramming: Perspective and Status," *IEEE Trans. Computers,* **C23,** *8,* August 1974, pp. 817–837.

JONES, L. H., "A Survey of Current Work in Microprogramming," *Computer,* **8,** *8* August 1975, pp. 33–37.

AGRAWALA, A. K., and T. G. RAUSCHER, *Foundations of Microprogramming: Architecture, Software and Applications,* New York, Academic Press, 1976.

SALISBURY, A. B., *Microprogrammable Computer Architectures,* New York, Elsevier, 1976.

JONES, L. H., "Instructions Sequencing in Microprogrammed Computers," *AFIPS Conf. Proc.* NCC, **46,** 1977, pp. 91–98.

RAUSCHER, T. G. and P. M. ADAMS, "Microprogramming: A Tutorial Survey of Recent Developments," *IEEE Transactions on Computers,* **C29,** January, 1980, pp. 2–20.

III. PARALLEL PROCESSING: GENERAL QUESTIONS

GILL S., "Parallel Programming," *Computer Journal,* **1,** *1,* April 1958, pp. 2–10.

HELLERMAN, H., "On the Organization of a Multiprogramming-Multiprocessing System," *IBM Research Report RC-522,* 52 pp., Yorktown Hts., N.Y., September 1961.

BALDWIN, F. R., W. B. GIBSON, and C. B. POLAND, "A Multiprocessing Approach to a Large Computer System," *IBM System Journal,* **1,** September 1962, pp. 64–76.

CODD, E. F., "Multiprogramming," *Advances in Computers,* **3,** New York, Academic Press, 1962, pp. 78–153.

BUSSELL, B., and G. ESTRIN, "An Evaluation of the Effectiveness of Parallel Processing," *IEEE Pacific Computer Conference,* 1963, pp. 201–220.

CONWAY, M. E., "A Multiprocessor System Design," *AFIPS Conference Proceedings FJCC,* **24,** 1963, pp. 139–146.

CRITCHLOW, A. J., "Generalized Multiprocessing and Multiprogramming Systems," *AFIPS Conference Proceedings Fall Joint Computer Conference,* **24,** 1963, pp. 107–126.

SQUIRE, J. S., and S. M. POLAIS, "Programming and Design Considerations of a Highly Parallel Computer," *AFIPS Conference Proceedings Spring Joint Computer Conference,* **23,** 1963, pp. 395–400.

BLAAUW, G. A., "Multisystem Organization," *IBM System Journal,* **3,** *2,* 1964, pp. 181–195.

ANDERSON, JAMES P., "Program Structures for Parallel Processing," *Communications ACM,* **8,** *12,* December 1965, pp. 786–788.

McCullough, J. D., K. H. Speierman, and F. W. Zurcher, "Design for a Multiple User Multiprocessing System," *AFIPS Conference Proceedings Fall Joint Computer Conference*, Pt. I, **27,** 1965, pp. 611–617.

Katz, J. H., "Simulation of a Multiprocessor Computing System," *AFIPS Proceedings Spring Joint Computer Conference,* **28,** 1966, pp. 127–139.

Murtha, J. C., "Highly Parallel Information Processing Systems," *Advances in Computers,* **7,** New York, Academic Press, 1966, pp. 2–116.

Lehman, M., "A Survey of Problems and Preliminary Results Concerning Parallel Processing and Parallel Processors," *Proceedings IEEE,* **54,** *12,* December 1966, pp. 1889–1901.

Roberts, Lawrence G., "Multiple Computer Networks and Intercomputer Communication," *ACM Symposium on Operating System Principles,* Gatlinburg, Tenn, October 1–4, 1967.

Hobbs, L. C., et al, ed., *Parallel Processor Systems, Technologies and Applications,* New York, Spartan Books, 1970.

Bell, C. Gordon, *Computer Structures: Readings and Examples,* New York, McGraw-Hill, 1971.

Anderson, G. A. and E. D. Jensen, "Computer Interconnection Structures, Taxonomy, Characteristics and Examples," *ACM Computing Surveys,* **7,** *4,* December 1975, pp.197–213.

Kuck, D., "Parallel Processor Architecture—A Survey," *Proceedings of the 1975 Sagamore Computer Conference on Parallel Processing,* Long Beach, Cal, IEEE Computer Society, 1975.

Baer, J. L., "Multiprocessing Systems," *IEEE Trans. Computers,* **C25,** December 1976, pp. 1271–1277.

IV. PARALLEL SYSTEMS

1. MULTICOMPUTER SYSTEMS

Rothman, S., "R/W 40 Data Processing System," *International Conference on Information Processing and Auto-math 1959,* Ramo-Wooldridge, Division of Thompson Ramo Wooldridge, Inc., Los Angeles, Cal, June 1959.

Porter, R. E., "The RW-400—A New Polymorphic Data System," *Datamation,* **6,** *1,* January/February 1960, pp. 8–14.

West, George P., and Ralph J. Koerner, "Communications within a Polymorphic Intellectronic System," *AFIPS Proceedings West Joint Computer Conference,* **18,** 1960, pp. 225–230.

Plugge, W. R., and M. N. Perry, "American Airlines' 'SABRE' Electronic Reservations System," *AFIPS Proceedings West Joint Computer Conference,* May 1961, pp. 593–602.

Segal, R. J., and H. P. Guerber, "Four Advanced Computers—Key to Air Force Digital Data Communication System," *AFIPS Proceedings East Joint Computer Conference,* **20,** 1961, pp. 264–278.

Davies, D. W., K. A. Bartlett, R. A. Scantlebury, and P. T. Wilkinson: A Digital Communication Network for Computers Giving Rapid Response at Remote Terminals, *ACM Symposium on Operating System Principles,* Gatlinburg, Tenn, October 1–4, 1967.

2. MULTIPROCESSOR SYSTEMS

Pariser, J. J. and H. E. Maurer, "Implementation of the NASA Modular Computer with LSI Functional Characters," *AFIPS Conference Proceedings,* **35,** AFIPS Press, 1969, pp. 231–245.

Davis, R. G., and S. Zucker, "Structure of a Multiprocessor Using Microprogrammable Building

Blocks," *Proceedings of National Aerospace Electronics Conference*, Dayton, Ohio, IEEE Press, 1971, pp. 186–200.

DAVIS, R. L., S. ZUCKER, and C. M. CAMPBELL, "The Building Block Approach to Multiprocessing," in *AFIPS 1972 Spring Joint Computer Conference*, AFIPS Press, **40**, 1972, pp. 685–703.

REIGEL, E. W., D. A. FISHER, and V. FABER, "The Interpreter—A Microprogrammable Processor," *AFIPS Conference Proceedings*, AFIPS Press, **40**, 1972, pp. 705–723.

ENSLOW, P. H., JR. (ed.), *Multiprocessors and Parallel Processing*, New York, John Wiley and Sons, 1974.

ENSLOW, P. H., JR., "Multiprocessor Organization—A Survey," *ACM Computing Surveys*, **9**, *1*, March 1977, pp. 103–129.

SWAN, R. J., A BECHTOLSHEIM, K. LAI, and J. OUSTERHOUT, "The Implementation of the Cm* Multi-Microprocessor," *AFIPS Conference Proceedings*, National Computer Conference, AFIPS Press, 1977, **46**, pp. 645–655.

SWAN, R. J., S. H. FULLER, and D. P. SIEWIOREK, "Cm*—A Modular, Multi-Microprocessor," *AFIPS Conference Proceedings*, National Computer Conference, AFIPS Press, **46**, 1977, pp. 637–643.

3. ARRAY SYSTEMS

UNGER, S. H., "A Computer Oriented Toward Spatial Problems," *Proceedings IRE*, **46**, *10*, October 1958, pp. 1744–1750.

HOLLAND, JOHN, "A Universal Computer Capable of Executing an Arbitrary Number of Subprograms Simultaneously," *AFIPS Proceedings East Joint Comp. Conf.* **13**, 1959, pp. 108–113.

SLOTNICK, DANIEL L., W. CARL BORCK, and ROBERT C. MCREYNOLDS, "The SOLOMON Computer," *AFIPS Proceedings Fall Joint Computer Conference* **21**, 1962, pp. 97–107.

GREGORY, J., and R. MCREYNOLDS, "The SOLOMON Computer," *IEEE Trans. Electronic Computers*, **EC12**, *6*, pp. 774–781, December 1963.

SMITH, R. V., and D. N. SENZIG, "Computer Organization for Array Processing," *IBM Research Report RC* 1330, Yorktown Hts., N. Y., December 1963.

BARNES, GEORGE H., et al, "The ILLIAC IV Computer," *IEEE Trans. on Computers*, **C17**, *8*, August 1968, pp. 746–757.

KNAPP, MORRIS A., et al, "Applications of ILLIAC IV to Urban Defense Radar Problem," *Parallel Processor Systems, Technologies and Applications*, L. C. HOBBS, et al, ed., New York, Spartan Books, 1970.

EVENSEN, ALF J. and JAMES L. TROY, "Introduction to the Architecture of a 288-Element PEPE," *Proceedings of the 1973 Sagamore Computer Conference on Parallel Processing*, New York, IEEE, 1973.

REDDAWAY, S., "DAP-A Distributed Array Processor," *Proceedings of the First Symposium on Computer Architecture*, pp. 61–72, 1973.

BATCHER, K., "STARAN Parallel Processor System Hardware," *AFIPS Conference Proceedings*, National Computer Conference, 1974, **43**, pp. 405–410.

QKAGA, Y., H. TAJIMA, and R. MORI, "A Novel Multiprocessor Array," *Proceedings of the Second Euromicro Symposium on Microprocessing and Microprogramming*, Venice, 1976, pp. 83–90.

LIPOVSKI, G. JACK and ANAND TRIPATHI, "A Reconfigurable Varistructure Array Processor," *Proceedings of the 1977 International Conference on Parallel Processing*, Long Beach, Cal, IEEE Computer Society, 1977.

BATCHER, K., "The Multidimensional Access Memory in STARAN," *IEEE Transactions on Computers*, **C26**, February 1977, pp. 174–177.

HINER, FRANK P., III, "Tracking Array Processor," *Proceedings of the 1978 International Conference on Parallel Processing*, Long Beach, California, IEEE Computer Society, 1978.

4. PIPELINED SYSTEMS

General Questions

RAMAMOORTHY, C. F., and H. F. LI, "Pipeline Architecture," *ACM Computing Surveys*, **9**, *1*, March 1977, pp. 61–102.

IRWIN, M. J., "Reconfigurable Pipeline Systems," *Proceedings 1978 ACM Annual Conference*, **1**, pp. pp. 86–92.

Instruction Pipelines

IBBETT, R. N., and P. C. CAPON, "The Development of the MU5 Computer System," *Communications of the ACM*, **21**, *1*, January 1978, pp. 13–24.

Arithmetic Pipelines

ANDERSON, D. W., F. J. SPARACIO, and R. M. TOMASULO, "IBM System 360 Model 61, Machine Philosophy and Instruction Handling," *IBM Journal of Research and Development*, January 1967, pp. 8–24.

WATSON, W. J., "The TI ASC—A Highly Modular and Flexible Super Computer Architecture," In *AFIPS 1972 Fall Joint Computer Conference*, Montvale, N.J., AFIPS Press, 1972, pp. 221–228.

THOMASIAN, A., and A. AVIZIENIS, "A Design Study of a Shared-Resource Computer System," *Proceedings of the Third International Symposium on Computer Architecture*, 1976, pp. 105–111.

RUSSELL, R. M., "The CRAY-1 Computer System," *Communications ACM*, **21**, January 1978, pp. 63–72.

REDDI, S. S., and E. A. FEUSTEL, "A Restructurable Computer System," *IEEE Transactions on Computers*, **C27**, *1*, January 1978, pp. 1–20.

5. ASSOCIATIVE PROCESSING

LEE, C. Y., and M. C. PAULL, "A Content Addressable Distributed Logic Memory with Applications to Information Retrieval," *Proceedings IEEE*, **51**, June 1963, pp. 924–932.

FULLER, R. H., and R. M. BIRD, "An Associative Parallel Processor with Application to Picture Processing," *AFIPS Conference Proceedings*, Fall Joint Computer Conference, **27**, Fall 1965, pp. 105–116.

CANNELL, M. H., et al, *Concepts and Applications of Computerized Associative Processing, Including an Associative Processing Bibliography*. U. S. Department of Defense Communications, Document No. AD879281, December 1970.

KATZ, JESSE H., "Matrix Computations on an Associative Processor," *Parallel Processor Systems, Technologies and Applications,* L. C. HOBBS et al, ed., New York, Spartan Books, 1970.

LOVE, H. H., and D. A. SAVITT, "An Iterative-Cell Processor for the ASP Language," *Associative Information Techniques,* E. L. JACKS, ed., New York, American Elsevier, 1971.

RUDOLPH, J. A., "A Production Implementation of an Associative Array Processor—STARAN," *AFIPS Conference Proceedings,* **41,** 1972, pp. 229–241.

LOVE, H. H., JR., "An Efficient Associative Processor Using Bulk Storage," *Proceedings of the 1973 Sagamore Computer Conference on Parallel Processing,* New York, IEEE, 1973.

BERRA, P. B., and A. K. SINGHANIA, "Some Timing Figures for Inverting Large Matrices Using the STARAN Associative Processor," *Proceedings of the 1975 Sagamore Computer Conference on Parallel Processing,* Long Beach, Cal, IEEE Computer Society, 1975.

THURBER, K. J., and L. D. WALD, "Associative and Parallel Processing," *Computing Surveys,* **7,** December 1975, pp. 215–255.

BATCHER, KENNETH E., "The Flip Network in STARAN," *Proceedings of the 1976 International Conference on Parallel Processing,* Long Beach, Cal, IEEE, 1976.

LOVE, HUBERT H., JR., "Radar Data Processing on the ALAP," *Proceedings of the 1976 International Conference on Parallel Processing,* Long Beach, Cal, IEEE Computer Society, 1976.

BOULIS, ROGER L. and RUDOLF O. FAISS, "STARAN E Performance and LACIE Algorithms," *Proceedings of the 1977 International Conference on Parallel Processing,* Long Beach, Cal, IEEE Computer Society, 1977.

FINNILA, CHARLES A. and HUBERT H. LOVE, JR., "The Associative Linear Array Processor," *IEEE Trans. on Computers,* **C26,** *2,* February 1977, pp. 112–125.

LOVE, H., JR., "A Modified ALAP Cell for Parallel Text Searching," *Proceedings of the 1977 International Conference on Parallel Processing,* Long Beach, Cal, IEEE, 1977.

BATCHER, K. E., "The Massively Parallel Processor (MPP) System," *Proceedings, AIAA Computers in Aerospace Conference II,* 1979.

V. MODULAR COMPUTER SYSTEMS

1. COMPUTER ORGANIZATIONS AND TECHNOLOGICAL PROGRESS

BORGERSON, B. R., "The Viability of Multimicroprocessor Systems," *Computer,* **9,** January 1976, pp. 26–30.

KARTASHEV, S. I., and S. P. KARTASHEV, "LSI Modular Computers, Systems and Networks," *Computer,* **11,** July 1978, pp. 7–15.

VICK, C. R., "Research and Development in Computer Technology, How Do We Follow the Last Act?" Keynote Speech, *Proceedings of International Conference on Parallel Processing,* 1978, pp. 1–5.

WELCH, T., and S. S. PATIL, "An Approach to Using VLSI in Digital Systems," *Proceedings 5th Annual Symposium on Computer Architecture,* 1978, pp. 139–143.

DAVIS, A. L., "A Data Flow Evaluation System Based on the Concept of Recursive Locality," *1979 National Computer AFIPS Conference Proceedings,* **48,** AFIPS Press, 1979, pp. 1079–1088.

DURNIAK, A., "VLSI Shakes the Foundations of Computer Architecture," *Electronics,* **52,** *11,* May 1979, pp. 111–133.

MOREI, R., et al, "Microcomputer Applications in Japan," *Computer,* **12,** May 1979, pp. 64–74.

2. MODULAR COMPUTER SYSTEMS: HARDWARE DESIGN TECHNIQUES

Computer, July, 1978, Special Issue on Modular Computers and Networks, edited by S. I. KARTASHEV and S. P. KARTASHEV.

KARTASHEV, S. I., and S. P. KARTASHEV, "LSI Modular Computers, Systems and Networks," *Computer,* **11,** July 1978, pp. 7–15.

KARTASHEV, S. P., and S. I. KARTASHEV, "Synthesis of a Modular Dedicated Network Assembled from Microcomputers," *Proceedings 1978 International Conference on Parallel Processing,* IEEE Catalog No. 78CH1321-9C, August 1978, pp. 108–114.

DURNIAK, A., "VLSI Shakes the Foundations of Computer Architecture," *Electronics,* **52,** *11,* May 24, 1979, pp. 111–133.

TJADEN, G. S., and M. COHN, "Some Considerations in the Design of Mainframe Processors with Microprocessor Technology," *Computer,* **12,** *8,* August 1979, pp. 68–74.

3. MODULAR COMPUTER SYSTEMS: SOFTWARE DESIGN TECHNIQUES

ALFORD, MACH W., "A Requirements Engineering Methodology For Real-Time Processing Requirements," *IEEE Transactions on Software Engineering.* **SE3,** *1,* January 1977, pp. 60–69.

FREEMAN, P. and A. I. WASSERMAN, *Tutorial on Software Design Techniques,* IEEE Computer Society, Catalog No. 76CH1145-2C, 1977.

TEICHROW, DANIEL, and ERNEST A. HERSHEY, III, "PSL/PSA: A Computer-Aided Technique for Structured Documentation and Analysis of Information Processing Systems," *IEEE Trans. on Software Engineering,* **SE3,** *1,* 1977, pp. 41–48.

ALFORD, M. W., "Software Requirements Engineering Methodology (SREM) at the Age of Two," *Proceedings CompSac/78,* IEEE Computer Society, Catalog No. 78CH1338-3C 1978, pp. 332–339.

PALMER, DAVID F., and W. MICHAEL DENNY, "Distributed Data Processing Requirements Engineering: High Level DDP Design," *Proceedings Comp/Sac/78,* IEEE Computer Society, 1978, pp. 352–357.

RAMAMOORTHY, C. V., and H. H. SO, "Software Requirements and Specifications: Status and Perspectives," In *Tutorial: Software Methodology,* IEEE Computer Society, EHO 142-0, 1978, pp. 43–164.

MARIANI, MICHAEL P., and DAVID F. PALMER, *Tutorial: Distributed System Design,* Long Beach, Cal, IEEE Computer Society, 1979, IEEE Catalog No. EHO 151-1.

YOURDON, EDWARD and LARRY L. CONSTANTINE, *Structured Design: Fundamentals of a Discipline of Computer Program and Systems Design,* Englewood Cliffs, N.J., Prentice-Hall, Inc., 1979.

4. BIT-SLICED MODULAR COMPUTERS

General Questions

LAPIDUS, G., "MOS/LSI Launches the Low-cost Microprocessor," IEEE *Spectrum,* **9,** *11,* November 1972, pp. 33–40.

WRIGHT, D., "Microprocessor Survey," in Miniconsult Ltd., *Microcomputer Fundamentals and Applications,* London, 1975, pp. 15–50.

622 BIBLIOGRAPHY

ALEXANDRIDIS, N. A., "Bit-Sliced Microprocessors, PLA's and Microprogramming in Replacing Hardwired Logic," *Proc. International Symposium on Wired Logic versus Programmed Logic,* Lausanne, Switzerland, March 1977.

ADAMS, W. T., and S. M. SMITH, "How Bit-Sliced Familes Compare," Part I, *Electronics,* **51,** *16,* August 3, 1978, pp. 91–98; Part II, *17,* August 17, 1978, pp. 96–102; *18,* August 30, 1978, pp. 138–139.

ALEXANDRIDIS, N. A., "Bit-Sliced Microprocessor Architecture," *Computer,* **11,** *6,* June 1978, pp. 56–80.

WHITE, D. E., *Bit-Slice Design,* New York, Garland STMP Press, 1980.

Descriptions of Bit-Sliced Computers and Microprocessors

SCHULTZ, G. W., and R. M. HOLT, "MOS LSI Minicomputer Comes of Age," *AFIPS Conference Proceedings,* **41,** 1972, pp. 1069–1080. (Analysis of AMI7200 8-bit processor slice)

GPC/P Product Description, Publication No. 4200005B, Santa Clara, Cal, National Semiconductor, October 1973.

IMP-16C Application Manual, Santa Clara, Cal, National Semiconductor, January 1974.

RATTNER, J., J. C. CORNET, and M. E. HOFF, JR., "Bipolar LSI Computing Elements Usher in a New Era of Digital Design," *Electronics,* **47,** *18,* September 5, 1974, pp. 89–96. (Description of the Intel 3000 series)

REYLING, G., JR., "Considerations in Choosing a Microprogrammable Bit-Sliced Architecture," *Computer,* **7,** *7,* July 1974, pp. 26–29. (Description of National Semiconductor IMP series)

AM 2901, AM 2909 Technical Data Manual, Sunnyvale, Cal, Advanced Micro Devices, Inc., 1975.

HOFF, M. E., JR., "Designing Central Processors with Bipolar Microcomputer Components," *AFIPS Conference Proceedings,* **44,** Fall 1975, pp. 55–62. (Description of Intel 3000 series)

HORTON, R. L., G. ENGLADE, and G. MCGEE, "I²L Takes Integration a Significant Step Forward," *Electronics,* **48,** *3,* February 6, 1975, pp. 83–90. (Analysis of Texas Instruments SBP0500)

Intel 8080 Microcomputer System User's Manual, Santa Clara, Cal, Intel, September 1975.

Series 3000 Microprogramming Manual, Santa Clara, Cal, Intel, 1975.

MICK, J., "AM 2900 Bipolar Microprocessor Family," *Electronics, Eighth Annual Workshop on Microprogramming,* Chicago, Ill, September 1975, pp. 56–63.

IMP: The Modular Concept for Microprocessors from 4 Bits to 16 Bits, Santa Clara, Cal, National Semiconductor, 1975.

990 Computer Family Systems Handbook, Manual No. 945250-9701, Austin, Texas Instruments, October, 1975.

A Microprogrammed 16-Bit Computer, Advanced Micro Devices, Sunnyvale, Cal, 1976.

Macrologic Bipolar Microprocessor Databook, Mountain View, Cal, Fairchild Inc., 1976.

HOFF, M. E., JR., J. SUGG, and R. YARA, "Central Processor Designs Using Intel Series 3000 Computer Elements," *Intel Series 3000 Reference Manual,* Santa Clara, Cal, Intel, 1976, pp. 3-19 to 3-76.

Series 3000 Reference Manual, Santa Clara, Cal, Intel, 1976.

M10800 High Performance MECL LSI Processor Family, Phoenix, Motorola 1976.

SMITH, R. J., "Bit Slice Processor Converts Radar Position Coordinates," *Electronics,* **49,** *8,* April 15, 1976, pp. 136–138. (Describes the use of 5701/6701 series)

SN74S481/SN54LS/SN74LS481 4-Bit-Slice Schottky Processor Elements Data Manual, Austin, Texas Instruments, 1976.

COHEN, H. I., L. S. SLETZINGER, and J. L. ARDINI, "Hardware Designs to Simplify the Programming of a High Speed Signal Processor," *Wescon 77 Conference Record,* San Francisco, Paper No. 20/2, September 1977. (Describes the use of 10800 series)

COLEMAN, V. M. W. ECONOMIDIS, and W. J. HARMON, JR., "The Next Generation Four-bit Bipolar Microprocessor Slice—The AM2903," *Wescon 77 Conference Record,* San Francisco, Paper No. 16/4 September, 1977.

KARTASHEV, S. I., and S. P. KARTASHEV, "A Microprocessor with Modular Control as a Universal Building Block for Complex Computers," *Proceedings Third EUROMICRO Symposium,* Amsterdam, North-Holland, 1977, pp. 85–91.

LAU, S. Y., "Design High-performance Processors with Bipolar Bit Slices," *Electronic Design,* **25**, 7, March 29, 1977, pp. 86–95. (Signetics version of the 3000 series)

LOWE, E. H., "A 16-bit Minicomputer for Missile Guidance and Control Applications," *Proceedings 1977 Joint Automatic Control Conference,* San Francisco, June 1977, pp. 17–21. (Computer that uses the 2900 series)

McWILLIAMS, T. M., S. H. FULLER, and W. H. SHERWOOD, "Using LSI Processor Bit-slices to Build a PDP-11—A Case Study in Microcomputer Design," *AFIPS Conference Proceedings,* **46**, 1977, pp. 243–253. (The use of the 3000 series for PDP-11)

MICK, J. R., and R. SCHOPMEYER, "MOS Support Microprocessor Teams with Bit-Slice Prototypes for Easier Microprogram Debugging," *Electronics,* **50**, *19,* September 15, 1977, pp. 127–130. (The AMD System 29 microcomputer development system)

MC14500B Industrial Control Unit Handbook, Phoenix, Motorola 1977.

NEMEC, J., G. SIM and B. WILLIS, "A Primer on Bit-sliced Microprocessors," *Electronic Design,* **25**, *3,* February 1, 1977, pp. 52–60.

TOKORO, M., et al, "PM/II - Multiprocessor Oriented Byte-sliced LSI Processor Modules," *AFIPS Conference Proceedings,* **46**, 1977, pp. 217–225. (Describes an experimental 8-bit processor slice)

TSOLIS, S., "IMP-16 Helps Small Planes Fly a Straight Course," *Electronics,* **50**, *12,* June 9, 1977, pp. 147–150.

DAPL User's Manual, Santa Monica, Cal, Zeno Systems, 1977.

The Am2900 Family Data Book, Sunnyvale, Cal, Advanced Micro Devices, 1978.

BORGERSON, B. R., G. S. TJADEN, and M. L. HANSON, "Mainframe Implementation with Off-the-shelf LSI Modules," *Computer,* **11**, *7,* July, 1978, pp. 42–48.

HORTON, R. L., L. COOPERSMITH, and R. M. BERGLER, "Make the Most of Bit-slice Flexibility and Design High-performance Processors," *Electronic Design,* **26**, *21,* October 11, 1978, pp. 226–235. (Describes the Texas Instruments 481 series)

MSC-85 User's Manual, Santa Clara, Cal, Intel, 1978.

MRAZEK, D., "Bit-slice Parts Approach ECL Speeds with TTL Power Levels," *Electronics,* **51**, *23,* November 9, 1978, pp. 107–112. (Describes National Semiconductor's version of the 2900 series)

MC10801 Data Sheet, Phoenix, Motorola 1978.

SHAVIT, M., *An Emulation of the Am9080A,* Sunnyvale, Cal, Advanced Micro Devices, 1978. (2900-based emulator for the 8080)

BLOOD, W. R. JR., "High Density Raises Sights of ECL Design," *Electronics,* **52**, *3,* February 1, 1979, pp. 99–107. (The Motorola 10800 series)

Questions of Implementation

ALTMAN, L., "Schottky-TTL Controller Put on a Chip," *Electronics,* **47**, *5,* March 7, 1974, pp. 159–160. (The Monolithic Memories 6701 4-bit processor slice)

RALLAPALLI, K. and P. VERHOFSTADT, "MACROLOGIC - Versatile Functional Blocks for High Performance Digital Systems," *AFIPS Conference Proceedings,* **44**, 1975, pp. 67–73.

WYLAND, D. C., "Design Your Own Computer by Using Bipolar/LSI Processor Slices," *Electronic Design,* **23**, *20,* September 27, 1975, pp. 72–78. (The Monolithic Memories 6701)

KLINE, B., M. MAERZ, and P. ROSENFELD, "The In-circuit Approach to the Development of Microcomputer Based Products," *Proceedings IEEE*, **64,** June 1976, pp. 937–942.

LOUIE, G., "Disk Controller Designed with Series 3000 Computing Elements," *Intel Series 3000 Reference Manual*, Santa Clara, Cal, Intel, 1976, pp. 3–9 to 3–17.

"A Guide to the Selection of Support Components for the Series 3000 Microprocessor," *Applications Memorandum No. 1,* Sunnyvale, Cal, Signetics, circa 1976.

CUMMINGS, G. A., and G. S. MILLER, "Application of a Bipolar Microprocessor Chip Set to Control Systems," *Proceedings 1977 Joint Automatic Control Conference,* San Francisco, June 1977, pp. 40–45. (Controller that uses the 2900 series)

MCCASKILL, R., "Wring Out 4-bit μP Slices with Algorithmic Pattern Generation," *Electronic Design,* **25,** *4,* May 10, 1977, pp. 74–77.

FARLY, B., "Logic Analyzers Aren't All Alike," *Electronic Design,* **26,** *3,* February 1, 1978, pp. 70–76.

HAYES, J. P., "Component Expansion Techniques in Computer Design," *Digital Processes,* **4,** *4,* 1978, pp. 295–312.

Technologies

SHIMA, M., and F. FAGGIN, "In Switching to n-MOS Microprocessor gets a 2-microsecond Cycle Time," *Electronics,* **47,** *8,* April 18, 1974, pp. 95–100. (The Intel 8080)

CHU, P., "ECL Accelerates to New System Speeds with High-density Byte-slice Parts," *Electronics,* **52,** *16,* August 2, 1979, pp. 120–125. (Fairchild 100220 series)

Emulation

Signetics 8080 Emulator Manual, Sunnyvale, Cal, Signetics, March 1977. (An 8080 emulator based on the 3000 series)

LAU, S. Y., "Emulate your MOS Microprocessor," *Electronic Design,* **26,** *8,* April 12, 1978, pp. 74–81. (The Signetics 8080 Emulator based on the 3000 series)

Microassembly

POWERS, V. M., and J. H. HERNANDEZ, "Microprogram Assemblers for Bit-Sliced Microprocessors," *Computer,* **11,** *7,* July 1978, pp. 108–120.

5. MODULAR PARALLEL ARCHITECTURES

Reconfigurable Architectures

ESTRIN, GERALD, "Organization of Computer Systems, the Fixed Plus Variable Structure Computer," *AFIPS Proceedings WJCC,* **14,** *1960,* pp. 33–40.

ESTRIN, G., "Parallel Processing in a Restructurable Computer System," *IEEE Trans. on Electronic Computers,* **EC12,** 1963, pp. 747–755.

MILLER, R. E. and J. COCKE,, "Configurable Computers: A New Class of General Purpose Machines," In *International Symposium on Theoretical Programming,* Lecture Notes, in *Computer Science,* **5,** Berlin, Germany, Springer-Verlag, 1974.

ANDERSON, G. A. and E. D. JENSEN, "Computer Interconnection Structures, Taxonomy, Characteristics, and Examples," *ACM Computing Surveys,* **7,** *4,* December 1975, pp. 197–213.

REDDI, S. S. and E. A. FEUSTEL, "An Approach to Restructurable Computer Systems," In *Parallel Processing* (Lecture Notes in *Computer Science*), **24**, Berlin, Germany, Springer-Verlag, 1975, pp. 319–337.

PAXER, Y. and M. BOZYIGIT, "Variable Topology Multicomputer," *Proceedings Second Euromicro Symposium on Microprocessing and Microprogramming,* Venice, 1976, pp. 141–149.

WITTIE, L. D., "Efficient Message Routing in Megamicrocomputer Networks," *Third Annual Symposium on Computer Architecture,* 1976, pp. 136–140.

PEASE, M. C., "The Indirect Binary N-Cube Microprocessor Array," *IEEE Trans. on Computers,* **C26,** 5, May 1977, pp. 458–473.

SIEGEL, H. J., "Interconnection Networks for SIMD Machines," *Computer,* **12,** 6, June 1979, pp. 57–65.

SIEGEL, H. J., R. J. MCMILLEN, and P. T. MUELLER, JR., "A Survey of Interconnection Methods for Reconfigurable Parallel Processing Systems," *Proceedings Conference NCC AFIPS,* AFIPS Press, **48,** 1979, pp. 529–542.

Dynamic Architectures

KARTASHEV, S. I., and S. P. KARTASHEV, "Designing LSI Metacomputer System with Dynamic Architecture," Dynamic Computer Architecture Association, Lincoln, Nebraska, 1974.

KARTASHEV, S. I., and S. P. KARTASHEV, "Designing of LSI Metacomputer System with Dynamic Architecture Made of Microcomputers," *Proceedings Third Annual International Symposium on Mini- and Microcomputers and their Applications,* Zurich, Switzerland, 1977, pp. 88–93.

KARTASHEV, S. I., and S. P. KARTASHEV, "A Multicomputer System with Software Reconfiguration of the Architecture," *Proceedings of the Eighth International Conference on Computer Performance,* SIGMETRICS CMG VIII, Washington, D.C., 1977, pp. 271–286.

KARTASHEV, S. I., and S. P. KARTASHEV, "Dynamic Architectures: Problems and Solutions," *Computer,* **11,** July 1978, pp. 26–40.

KARTASHEV, S. I., and S. P. KARTASHEV, "Selection of the Control Organization for a Multicomputer System with Dynamic Architecture," *Proceedings Fourth Euromicro Symposium on Microprocessing and Microprogramming,* Munich, Germany, 1978, pp. 215–227.

KARTASHEV, S. I., and S. P. KARTASHEV, "Software Problems for Dynamic Architecture: Adaptive Assignment of Hardware Resources," *Proceedings Computer Software and Applications Conference (CompSac),* 1978, pp. 775–780.

KARTASHEV, S. P., and S. I. KARTASHEV, "Adaptable Pipeline System with Dynamic Architecture," *Proceedings of the 1979 International Conference on Parallel Processing,* pp. 222–230.

KARTASHEV, S. I., and S. P. KARTASHEV, "The Evolution in Dynamic Architectures," *Microprocessors and Microsystems,* **3,** 6, July 1979, pp. 249–256.

KARTASHEV, S. I., and S. P. KARTASHEV, "Multicomputer System with Dynamic Architecture," *IEEE Trans. on Computers,* **C28,** 10, October 1979, pp. 704–721.

KARTASHEV, S. P., and S. I. KARTASHEV, "Performance of Reconfigurable Busses for Dynamic Architectures," *Proceedings of the First International Conference on Distributed Computing Systems,* Huntsville, Alabama, 1979, pp. 261–273.

KARTASHEV, S. I., S. P. KARTASHEV and C. V. RAMAMOORTHY, "Adaptation Properties for Dynamic Architectures," 1979 National Computer Conference, *AFIPS Conference Proceedings,* AFIPS Press, 1979, **48,** pp. 543–556.

Data Flow Architecture and Language

ADAMS, D. A., *A Computation Model with Data Flow Sequencing,* Computer Science Department, School of Humanities and Sciences, Stanford University, Technical Report CS 117, December 1968, 130 pp.

DENNIS, J. B., "First Version of a Data Flow Procedure Language," *Lecture Notes in Computer Science,* **19,** Berlin, Germany, Springer-Verlag, 1974, pp. 362–376.

DENNIS, J. B. and D. P. MISUNAS, "A Preliminary Architecture for a Basic Data-Flow Processor," *Proceedings of the Second Annual Symposium on Computer Architecture,* Houston, January 1975, pp. 126–132.

RUMBAUGH, J. E., "A Data Flow Multiprocessor," *IEEE Trans. on Computers,* **C26,** 2, February 1977, pp. 138–146.

ARVIND, K., P. GOSTELOW, and W. PLOUFFE, *An Asynchronous Programming Language and Computing Machine,* Department of Information and Computer Science, University of California, Irvine, Report No. 114A, December 1978.

BOUGHTON, G. A., *Routing Networks in Packet Communication Architectures,* S. M. THESIS, M.I.T., Department of Electrical Engineering and Computer Science, June 1978, 93 pp.

COTE, W. F., and R. F. RICCELLI, "The Design of a Data Driven Processing Element," *Proceedings 1978 International Conference on Parallel Processing,* August 1978, pp. 173–183.

ACKERMAN, W. B., "Data Flow Languages," *Proceedings of the ACM 1979 National Conference,* New York, June 1979, pp. 1087–1095.

ACKERMAN, W. B. and J. B. DENNIS, *VAL: A Value Oriented Algorithmic Language, Preliminary Reference Manual,* Laboratory for Computer Science, M.I.T., Technical Report TR-218, June 1979, 80 pp.

ARVIND, K., and R. E. BRYANT, "Design Considerations for a Partial Differential Equation Machine," *Proceedings of Scientific Computer Information Exchange Meeting,* Livermore, Cal, September 1979, pp. 94–102.

BROCK, J. D. and L. B. MONTZ, "Translation and Optimization of Data Flow Programs," *Proceedings of the 1979 International Conference on Parallel Processing,* Bellaire, Mich, August 1979, pp. 46–54.

DAVIS, A., "A Data Flow Evaluation System Based on the Concept of Recursive Locality," *Proceedings of the ACM 1979 National Computer Conference,* New York, June 1979, pp. 1079–1086.

DEFRANCESCO, N., et al, "On the Feasibility of Nondeterministic and Interprocess Communications Constructs in Data-Flow Computing Systems," *Proceedings First European Conference on Parallel and Distributed Processing,* Toulouse, France, February 1979, pp. 93–100.

DENNIS, J. B., "The Varieties of Data Flow Computers," *Proceedings First International Conference on Distributed Computing Systems,* October 1979, pp. 430–439.

DENNIS, J. B., C. K. C. LEUNG, and D. P. MISUNAS, *A Highly Parallel Processor Using a Data Flow Machine Language,* Laboratory for Computer Science, M.I.T., CSG Memo 134–1, June 1979, 33 pp.

WATSON, I. and J. GURD, "A Prototype Data Flow Computer with Token Labelling," *Proceedings of the ACM 1979 National Computer Conference,* New York, June 1979, pp. 623–628.

VI. MODULAR COMPUTER SYSTEMS: SOFTWARE ISSUES

1. DISTRIBUTED PROCESSING FOR REAL-TIME APPLICATIONS

VICK, C. R., J. E. SCALF, and W. C. MCDONALD, "Distributed Data Processing for Real-time Applications," *Proceedings Sixth Texas Conference on Computing Systems,* 1977, pp. 174–191.

DAVIS, C. G., and C. R. VICK, "The Software Development System: Status and Evolution," *Proceedings Conference on Computer Software and Applications (CompSac),* 1978, pp. 326–331.

FITZGIBBON, H., B. BUCKLES, and J. SCALF, "Distributed Data Processing Design Evaluation Through Emulation," *Proceedings Conference on Computer Software and Applications (CompSac),* 1978, pp. 364–369.

MCDONALD, W. C., and J. M. WILLIAMS, "The Advanced Data Processing Testbed," Proceedings Conference on *Computer Software and Applications (CompSac),* 1978, pp. 346–351.

2. CONCURRENT PROGRAMMING AND SIMULATION

BRINCH-HANSEN, PER, "Concurrent Programming Concepts," *Computing Surveys,* **5,** *4,* December, 1973, pp. 223–245.

DODSON, E. N., et al, *Advanced Cost Estimating and Synthesis Techniques for Avionics Data Processing Software and Hardware,* General Research Corporation Final Report No. CR-1-701, December 1976.

BRINCH-HANSEN, PER, *The Architecture of Concurrent Programs,* Englewood Cliffs, New Jersey, Prentice-Hall, Inc., 1977.

HOLT, R. C., et al, *Structured Concurrent Programming with Operating Systems Applications,* Reading, Mass, Addison-Wesley, 1978.

LAMPORT, LESLIE, "Time, Clocks, and the Ordering of Events in a Distributed System," *Com. of ACM,* **21,** *7,* July 1978, pp. 558–565.

THOMAS, ROBERT H., "A Solution to the Concurrency Control Problem for Multiple Copy Data Bases," *Proceedings COMPCON 78 Spring,* IEEE Computer Society.

CHANDY, K. MANI, and JAYADEO MISRA, "Distributed Simulation: A Case Study in Design and Verification of Distributed Programs," *IEEE Transactions on Software Engineering,* **SE5,** *5,* September 1979, pp. 440–452.

3. PARTITIONING AND RESOURCE ALLOCATION (ASSIGNMENT)

CHU, W. W., "Optimal File Allocation in a Multiple Computing System," *IEEE Transactions on Computers,* **C18,** *10,* October 1969, pp. 885–889.

MCMILLAND, CLAUDE, JR., *Mathematical Programming: An Introduction to the Design and Application of Optimal Decision Machines,* New York, John Wiley and Sons, 1970.

LEVIN, K. DAN, and HOWARD LEE MORGAN, "Optimizing Distributed Data Bases—A Framework for Research," *AFIPS Conference Proceedings,* **44,** 1975, pp. 473–474, *also* reprinted in MARIANI and PALMER, *Tutorial: Distributed System Design.*

NEGOITA, C. V. and D. A. RALESCU, *Applications of Fuzzy Sets to Systems Analysis,* New York, John Wiley and Sons, 1975, Chap. 7.

TAHA, HAMDY, A., *Integer Programming: Theory, Applications, and Computations,* New York, Academic Press, 1975.

BALACHANDRAN, V., J. W. MCCREDIE, and V. I. MIKHAIL, "Models of the Job Allocation Problem in Computer Networks," *Digest of Papers COMPCON 76 Fall,* IEEE Press, 1976, pp. 211–214.

GYLYS, V. V., and J. A. EDWARDS, "Optimal Partitioning of Workload for Distributed Systems," *Proceedings COMPCON/76 Fall,* IEEE Computer Society, 76CH1115-5C (1976), pp. 353–357; 353–357; *also* reprinted in MARIANI and PALMER, *Tutorial: Distributed System Design.*

IGNIZIO, J. P., *Goal Programming and Extensions,* Indianapolis: D. C. Heath and Co., 1976.

BOND, ALBERT F., and PETER C. BELFORD, "An Approach to a Distributed Data Processing Architecture Methodology," *Proceedings Sixth Texas Conference on Computing Systems,* University of Texas, Austin, 1977, pp. 1B-13–1B-35.

BOORSTYN, ROBERT R., and HOWARD FRANK, "Large Scale Network Topological Optimization," *IEEE Trans. on Communications,* **COM25,** *1,* January 1977, pp. 37–55.

JENNY, C. J., "Process Partitioning in Distributed Systems," *Digest of Papers NTC '77,* 1977, pp. 31:1-1–31:1-10.

LEE, R. P. and R. R. MUNTZ, "On the Task Assignment Problem for Computer Networks," *Proceedings of the 10th Hawaii International Conference on System Sciences,* Honolulu, Hawaii, January 1977, pp. 5–9.

MORGAN, HOWARD L., and K. DAN LEVIN, "Optimal Programs and Data Locations in Computer Networks," *Com. ACM,* **20,** *5,* May 1977, pp. 315–322.

STONE, HAROLD S., "Multiprocessing Scheduling with the Aid of Network Flow Algorithms," *IEEE Transactions on Software Engineering,* **SE3,** *1,* January 1977, pp. 85–93.

UHRIG, J. L., "Life-Cycle Evaluation of System Partitioning," *Proceedings COMPSAC/77,* November 1977, pp. 2–8.

STONE, H. S. and S. H. BOKHARI, "Control of Distributed Processes," *Computer,* **11,** *7,* July 1978, pp. 97–106.

UHRIG, J. L., "Mathematical Programming Approaches to System Partitioning," *IEEE Trans. on Systems, Man, and Cybernetics,* **SM8,** *7,* 1978, pp. 540–548.

RAO, G. S., H. S. STONE, and T. C. HU, "Assignment of Tasks in a Distributed Processor System with Limited Memory," *IEEE Trans. on Computers,* **C28,** *4,* April 1979, pp. 291–299.

4. THEORETICAL MODELS IN (MICRO) PROGRAMMING

BÖHN, C., and G. JACOPINI, "Flow Diagrams, Turing Machines and Languages with only Two Formation Rules," *Commun. Ass. Comput. Mach.,* **9,** *5,* May 1966, pp. 366–371.

MARTIN, B., and G. ESTRIN, "Models of Computations and Systems: Evaulation of Vertex Probabilities in Graph Models of Computations," *Journal Ass. Comp. Mach.,* **14,** *2,* April 1967, pp. 281–299.

BAER, J. L., *Graph Models of Computations in Computer Systems,* UCLA 10P14/51-, report N.68-46, 1968.

BOOK, R., S. EVEN, S. GRIEBACH, and G. OTTO, "Ambiguity in Graphs and Expressions," *IEEE Trans. on Computers,* **C20,** *2,* February 1971, pp. 149–153.

KLEIR, R. L., and C. V. RAMAMOORTHY, "Optimization Strategies for Microprograms," *IEEE Trans. on Computers,* **C20,** *7,* July 1971, pp. 783–794.

ROSE, D. J., and R. A. WILLOUGHBY, "Sparse Matrices and Their Applications," *Proceedings, Symposium on Sparse Matrices and Their Applications,* Yorktown Heights, New York, September 1971.

RAMAMOORTHY, C. V., and M. TSUCHIYA, "A High-Level Language for Horizontal Microprogramming," *IEEE Trans. on Computers,* **C23,** *8,* August 1974, pp. 791–801.

5. PROGRAM PREPROCESSING

SAMI, M. G., R. STEFANELLI, *On the Determination of Probable Activities of Instruction Sequences in Program Schemes,* Politecnico di Milano, IEELC Internal report 72–12, 1975.

KARTASHEV, S. I., S. P. KARTASHEV, and C. V. RAMAMOORTHY, "Adaptation Properties for Dynamic Architectures," *AFIPS Conference Proceedings, National Computer Conference,* **48,** 1979, pp. 543–556.

6. (MICRO) PROGRAM VALIDATION

FLOYD, R. W., "Assigning Meanings to Programs," *Proceedings Symposium on Applied Mathematics,* **19,** 1967, pp. 19–32.

HOARE, C. A. R., "An Axiomatic Basis of Computer Programming," *Commun. Ass. Computing Mach.,* **12,** October 1969, pp. 576–580, 583.

LEMAN, G. B., W. C. CARTER, and A. BIRMAN, "Some Techniques for Microprogram Validation," *Proceedings, IFIP Conference on Information Processing,* 1974.

FAIRLEY, R., "An Experimental Program-Testing Facility," *IEEE Trans. on Software Engineering,* **SE1,** December 1975.

LONDON, R. L., "A View of Program Verification," *Proceedings 1975 International Conference on Reliable Software,* pp. 534–545.

SAMI, M. G., and R. STEFANELLI, "On Structural Termination of Programs and Microprograms," *Proceedings, First EUROMICRO Symposium on Microprocessing and Microprogramming,* Nice, France, 1975.

HOWDEN, W. E., "Symbolic Testing and the DISSECT Symbolic Evaluation System," *IEEE Transactions on Software Engineering,* **SE3,** July 1977.

CLARKE, L. A., "Testing: Achievements and Frustrations," *Proceedings Computer Software and Applications Conference,* CompSac, 1978, pp. 310–314.

LISKOV, B., "Introduction to CLU," *Proceedings Computing Systems Reliability,* Toulouse, France, September 1979.

VII. MODULAR COMPUTER SYSTEMS: FAULT-TOLERANCE

1. FAULTS

MALAIYA, Y. K., and S. Y. H. SU, "A Survey of Methods of Intermittent Fault Analysis," *Proceedings of 1979 National Computer Conference,* pp. 577–586.

MCCONNEL, S. R., D. P. SIERWIOREK, and M. M. TSAO, "The Measurement and Analysis of Transient Errors in Digital Computer Systems," *Digest of the 1979 International Conference on Fault-Tolerant Computing,* June 1979, pp. 67–70.

KARPOVSKY, M., and S. Y. H. SU, "Detection and Location of Input and Feedback Bridging Faults Among Input and Output Lines," *IEEE Trans. on Computers,* **C29,** 6, June 1980, pp. 523–527.

KARPOVSKY, and S. Y. H. SU, "Detecting Bridging and Stuck-at Faults at Input and Output Pins of Standard Digital Components," *Proceedings of the 17th Design Automation Conference,* Minneapolis, Minn, June 1980.

2. GENERAL QUESTIONS OF FAULT-TOLERANT COMPUTING

SU, S. Y. H. and R. J. SPILLMAN, "An Overview of Fault-Tolerant Digital System Architecture," *AFIPS Conference Proceedings, 1977 National Computer Conference,* v.46 June 1977, pp. 19–26.

REDDY, S. M. (guest editor), "Fault-Tolerant Computing," Special issue of *IEEE Transactions on Computers,* **C27,** 6, June 1978, pp. 481–560.

PRADHAN, P. K., (guest editor), "Fault-Tolerant Computing," Special issue of *IEEE Computer,* **13,** 3, March 1980, pp. 6–55.

STIFFLER, J. (guest editor), "Fault-Tolerant Computing," Special issue of *IEEE Transactions on Computers*, **C29**, *6*, June 1980, pp. 417–546.

3. HARDWARE REDUNDANCY

BOURICIUS, W. G., "Reliability Modeling Techniques for Self-Repairing Computing Systems," *Proceedings 24th National Conference of the ACM*, August 1969, pp. 295–309.

KUEHN, R. E., "Computer Redundancy: Design, Performance and Future," *IEEE Trans. on Reliability*, February 1969, pp. 3–11.

MATHUR, F. P. and A. AVIZIENIS, "Reliability Analysis and Architecture of a Hybrid-Redundant Digital System: Generalized TMR with Self-Repair," *AFIPS Conference Proceedings*, **36**, Spring Joint Computer Conference, 1970, pp. 376–384.

CHANDY, K. M., C. V. RAMAMOORTHY, and A. COWAN, "A Framework for Hardware-Software Tradeoffs in the Design of Fault-Tolerant Computers," *AFIPS Conference Proceedings*, Fall Joint Computer Conference, **41**, 1972, pp. 55–63.

SIEWIOREK, P. P. and E. J. MCCLUSKEY, "An Interactive Cell Switch Design for Hybrid Redundancy," *IEEE Trans. on Computers*, **C22**, *3*, March 1973, pp. 290–297.

4. FAULT-TOLERANT SYSTEMS

DOWNING, R. W., "No. 1 ESS Maintenance Plan," *Bell System Technical Journal, 43*, September 1964, pp. 1961–2020.

AVIZIENIS, A. A., "Design of Fault-tolerant Computers," *AFIPS Conference Proceedings*, **31**, Fall Joint Computer Conference, 1967, pp. 733–743.

BATCHER, K. E., "Sorting Networks and Their Applications," *AFIPS Conference Proceedings*, **32**, Spring Joint Computer Conference, 1968, pp. 307–314.

AVIZIENIS, A. A., "The STAR Computer: An Investigation of the Theory and Practice of Fault-tolerant Computer Design," *IEEE Trans. on Computers*, **C20**, *11*, November 1971, pp. 1322–1331.

BORGERSON, B. R., "A Fail-Softly System for Time-Sharing Use," *Digest of the 1972 International Symposium on Fault-Tolerant Computing*, Boston, June 1972, pp. 89–93.

RENNELS, D. A. and A. AVIZIENIS, "RMS: A Reliability Modeling System for Self-Repairing Computers," *Digest of the 1973 International Symposium on Fault-Tolerant Computing*, June 1973, pp. 131–135.

STIFFLER, J., "The SERF Fault-Tolerant Computer, Part I: Conceptual Design, and Part II: Implementation and Reliability Analysis," *Digest of the 1973 International Conference on Fault-Tolerant Computing*, 1973, pp. 23–32.

SAHEBAN, F. and A. D. FRIEDMAN, "A Survey and Methodology of Reconfigurable Multi-module Systems, *Proceedings*, CompSac '78, Computers Software and Applications Conference, 1978, pp. 790–796.

SIEWIOREK, P. P., V. KINI, H. MASHBURN, S. MCCONNEL and M. TSAO, "A Case Study of C-mmp, Cm* and C.vmp: Part I: Experiences with Fault-Tolerance in Multiprocessor Systems," *Proceedings of IEEE*, October 1978, pp. 1178–1220.

MALAIYA, Y. K., and S. Y. H. SU, "Fault-Tolerance in Multi-processor Systems," *Proceedings of 1980 International Conference on Circuits and Computers*.

TAYLOR, D. S., "A Reliability and Comparative Analysis of Two Standby System Configurations," *IEEE Trans. on Reliability*, April 1973, pp. 13–19.

TAYLOR, D. S., "Unpowered to Powered Failure Rate Ratio: A Key Reliability Parameter," *IEEE Trans. on Reliability,* April 1974, pp. 33–36.

WAKERLY, J. F., "Transient Failures in Triple Modular Redundancy Systems with Sequential Modules," *IEEE Trans. on Computers,* May 1975, pp. 570–572.

LOSQ, T., "A Highly Efficient Redundancy Scheme: Self-Purging Redundancy," *IEEE Trans. on Computers,* June 1976, pp. 569–578.

SU, S. Y. H., and E. DUCASSE, "A Hardware Redundancy Reconfiguration Scheme for Tolerating Multiple Module Failures," *IEEE Trans. on Computers,* March 1980, pp. 254–257.

5. SOFTWARE-FAULT-TOLERANCE

GOODENOUGH, J. B., "Exception Handling: Issues and a Proposed Notation," *Communications of ACM 18,* **12,** 1975, pp. 683–696.

RANDELL, B., "System Structure for Software for a Fault-Tolerant Computer," *Software System Engineering,* June 1975, pp. 220–232.

HECHT, H., "Fault-Tolerant Software for a Fault-Tolerant Computer," *Software System Engineering,* On Line, Uxbridge, 1976, pp. 235–348.

MELLIAR-SMITH, P. M. and B. RANDELL, "Software Reliability: the Role of Programmed Exception Handling," *Proceedings ACM Conference on Language Design for Reliable Software,* Sigplan Notices 12, 3, March 1977, pp. 95–100.

KIM, K. H., "Error Detection, Reconfiguration and Recovery in Distributed Processing Systems," *Proceedings,* First International Conference on Distributed Computing Systems, October 1979, pp. 284–295.

6. RELIABILITY MODELS

BOURICIUS, W. G., "Reliability Modeling for Fault-Tolerant Computers," *IEEE Trans. on Computers,* **C20,** *11,* November 1971, pp. 1306–1311.

MATHUR, F. P. and P. T. DESOUSA, "Reliability Models of NMR Systems," *IEEE Trans. on Reliability,* **R24,** June 1975, pp. 108–112.

MATHUR, F. P. and P. T. DESOUSA, "Reliability Modeling and Analysis of General Modular Redundant Systems," *IEEE Trans. on Reliability,* December 1975, pp. 269–299.

NG, Y. W., "Reliability Modeling and Analysis for Fault-Tolerant Computers," Ph.D. Thesis, Computer Science Department, UCLA, 1976, pp. 186–197.

NG, Y. W. and A. AVIZIENIS, "A Reliability Model for Gracefully Degrading and Repairable Fault-Tolerant Systems," *Digest of the 1977 International Symposium on Fault-Tolerant Computing,* 1977, pp. 22–28.

7. TESTING AND DIAGNOSIS OF MODULAR SYSTEMS

FORBES, R. E., et al, "A Self-diagnosable Computer," *AFIPS Conference Proceedings, 1965 Fall Joint Computer Conference,* **27,** pp. 1073–1086.

PREPARATA, F. P., G. METZE and R. T. CHIEN, "On the Connection Assignment Problem of Diagnosable Systems," *IEEE Trans. on Computers,* December 1967, pp. 848–854.

LEVITT, K. N., et al, "A Study of the Data Communication Problems of a Self-repairable Multiprocessor," *AFIPS Conference Proceedings, 1968 Joint Computer Conference,* **32,** pp. 515–527.

RAMAMOORTHY, C. V., and L. C. CHANG, "System Modeling and Testing Procedures for Microdiagnostics," *IEEE Trans. on Computers,* **C21,** *11,* November, 1972, pp. 1169–1183.

HAKIMI, S. L. and A. T. AMIN, "Characterization of Connection Assignment of Diagnosable Systems," *IEEE Trans. on Computers,* **C23,** *1,* January 1974, pp. 86–88.

CIOMPI, P. and L. SIMONCINI, "Design of Self-diagnosable Minicomputers Using Bit-sliced Microprocessors," *Journal of Design Automation and Fault-Tolerant Computing,* **1,** October 1977, pp. 363–375.

CARLSTEAD, R. H., and R. E. HUSTON, "Test Techniques for ECL Microprocessors," *Digest 1977 Semiconductor Test Symposium,* Cherry Hill, New Jersey, 1977, pp. 32–35.

KARUNANITHI, S. and A. D. FRIEDMAN, "System Diagnosis with t/s Diagnosability," *Digest of the 1977 International Symposium on Fault-Tolerant Computing,* June 1977, pp. 65–71.

SAHEBAN, F. and A. D. FRIEDMAN, "Diagnostic and Computational Reconfiguration in Multiprocessor Systems," *Proceedings of the ACM78 Annual Conference,* Washington D.C. 1978, pp. 68–78.

SIMONCINI, L. and A. D. FRIEDMAN, "Incomplete Fault Coverage in Modular Multiprocessor Systems," *The ACM78 Annual Conference,* Washington D.C., 1978, pp. 210–215.

ANDERSON, K. R. and H. A. PERLEINS (guest editors), "Hardware Test Technology," Spcial issue of *Computer,* **12,** *10,* October 1979, pp. 7–61.

KARUNANITHI, S. and A. D. FRIEDMAN, "Analysis of Digital Systems using a new Measure of System Diagnosis," *IEEE Trans. on Computers,* v.C-28, no 2. February 1979, pp. 121–133.

SAHEBAN, F., L. SIMONCINI, and A. D. FRIEDMAN, "Concurrent Computation and Diagnosis in Multiprocessor Systems," *Digest of the 1979 International Symposium on Fault-Tolerant Computings,* 1979, pp. 149–156.

SRIDHAR, T., and J. P. HAYES, "Testing Bit-sliced Microprocessors," *Digest Ninth International Symposium Fault-Tolerant Computing,* Madison, Wisconsin, June 1979, pp. 211–218.

Description of Systems

This book provides a comprehensive description of the industrial systems listed below. The data here is organized as follows: Topic, system name, pages in the book, comprehensive outside references.

RECONFIGURABLE ARRAY SYSTEMS

SOLOMON 1 COMPUTER, Chapter II, page XXX.
    SLOTNICK, DANIEL L. et al, "The Solomon Computer," *Proceedings, 1962 Fall Joint Computer Conference,* pp. 97–107.

ILLIAC IV, Chapter II, page XXX.
    SLOTNICK, DANIEL L. et al, "The Solomon Computer," *Proceedings, 1962 Fall Joint Computer Conference,* pp. 97–107.

PEPE, THE PARALLEL ELEMENT PROCESSING ENSEMBLE, CHAPTER II, PAGE XXX.
    VICK, C. R., "Pepe Architecture—Present and Future," 1978 National Computer Conference, *AFIPS Conference Proceedings,* **47,** AFIPS Press, 1978, pp. 981–992.
    MARIANI, M. P. and E. J. HENRY, "PEPE—A User's Viewpoint; A Powerful Real Time Adjunct," 1978 National Computer Conference, *AFIPS Conference Proceedings,* **47,** AFIPS Press, 1978, pp. 993–1002.

CRANE, B. A., M. J. GILMARTIN, P. T. RUX, and R. R. SHIVELY, "The PEPE Computer," *IEEE Compcon 72 Digest*.

WILSON, D. E., "The PEPE Support Software System," *IEEE Compcon 72 Digest*, September 1972, pp. 61–64.

MERWIN, R. E. and C. R. VICK, "An Architectural Description of A Parallel Element Processing Ensemble," International Symposium on Computer Architecture, Grenoble, France, 1973.

WELCH, H. O., "Numerical Weather Prediction in the PEPE Parallel Processor," 1977 International Conference on Parallel Processing.

BLAKELY, C. E., "PEPE Application to BMD Systems," 1977 International Conference on Parallel Processing.

EVENSEN, A. J., "PEPE Hardware and System Overview," 1977 International Conference on Parallel Processing.

ASSOCIATIVE ARRAY SYSTEMS

STARAN, Chapter II, page XXX.

TM, Goodyear Aeorspace Corporation, Akron, Ohio

RUDOLPH, J. A., "A Production Implementation of an Associative Array Processor—STARAN," *Proc. FJCC 72*, pp. 229–241.

ASSOCIATIVE LINEAR ARRAY PROCESSOR (ALAP), Chapter II, page XXX.

L. C. HOBBS, et al, ed., *Parallel Processor Systems, Technologies and Applications*, New York, Spartan Books, 1970.

LOVE, HUBERT H., "Radar Data Processing on the ALAP," *Proceedings of the 1976 International Conference on Parallel Processing*. Long Beach, Cal, IEEE Computer Society, 1976.

FINNILA, CHARLES A. and HUBERT H. LOVE, JR., "The Associative Linear Array Processor," *IEEE Transactions on Computers*, **C26**, 2, February 1977, pp. 112–125.

HOLLAND MACHINE, Chapter II, page XXX.

HOLLAND, J. H., "A Universal Computer Capable of Executing an Arbitrary Number of Sub-Programs Simultaneously," *1959 Proceedings of the Eastern Joint Computer Conference*, pp. 108–113.

INDEX

Ability to reconfigure resources, via software, into different types of architecture, 93
Abstract relations, 405, 406
Acceptance test routine, 504
Access, bit-slice access mode, 147
   sequential access capability, 139
Accessing necessary variable control codes, 308
   similarity, direct data accessing, 485
Accumulator, 163
   registers, 128
Accuracy of computations, 260
Acquiring pulse trains, 174
Active configuration, 531
Activity register, 132–34
Adaptable architectures, 73, 83
   evolution, 92
   for supersystems, 73
   past and present, 81
Adaptation, dynamic, 83
   to computational specificities, 77
   of hardware resources on instruction and data parallelism, 259
   microprogrammable adaptation, 83
   on operation sequences, 268
   preprocessing, 70
   properties exhibited by modular architectures, 93
   reconfigurable adaptation, 83
   of supersystem architecture, 383
   system, 92, 94
Adaptations, new adaptations that improve performance of available resources, 92
Adaptive assignment and task execution time, 370
Adder/multiplier unit, 128
Add-from-memory instruction, 189
Adding multiple memory arrays, 205
Addition, 157
   instruction, 361, 372
Additional failures, 531
   priority array, 310
Address, effective address for instruction, 222
   computers, 31
   connecting element, 318
   encoding and decoding logic, 139
   field of frame, 511
   generation for instructions and data words, 294
   mode information, 134
Addressing individual cells, 139
   procedure, 297
Adjustment to changeable number of information streams, 74
Advanced image-processing tasks, 238
   instruction station, 126

Advances in LSI technology, 259
ALAP array, 204
   assembler simulator, 195
   cell, 157
   clock, 215
   coding, 216
   degenerate memory cell, 209
   demonstrator system, 206
   fault-tolerant, 208
   instruction memory, 208
   instruction memory address, 214
   memory, 203, 206, 212
   memory array, 156, 204
   memory design, 236
   performance, 193
   word-cycle instruction, 208
Algorithm, complete signal-sorting algorithm, 175
   complexity, 111
   for constructing a program graph, 351
   for pulse-train acquisition, 177
Algorithmic faults, 504
Alignment in bit sizes, 369
Allocation, 463, 481
   of data arrays, 380
Analysis of computed variable to obtain its maximal bit size, 357
   of control statement IF which specifies exit from loop, 358
   techniques for user program written in high-level language, 350
Application, file-processing, 154
   of modeling behavior of physical system, 167
   radar pulse-deinterleaving, 170
   radar track-while-scan, 131, 134
   signal-sorting, 170
   of structural termination to simple-loop analysis, 432
   software, 458
Architectural adaptation, 74, 259
   new forms to algorithms, 81
   state, 1, 262
   switch instruction, 277
   transition, 277, 283, 308, 318, 336
Architecture of microprogrammed computers, 35
   modular architecture, 83
   types of reconfigurable architecture, 84
Area of applications for von Neumann computers, 32
Arithmetic, 67, 119
   associative operations, 138
   boolean instruction-execution logic, 132
   cell's arithmetic and logical unit, 142
   field-selective arithmetic, 150

   operations, 133
   parallel arithmetic and boolean operations, 149
   register, 133
   unit, 128, 132
Array, 57, 93, 265
   adaptation, 264
   of complex logic modules, 141
   computations, 264, 265
   configuration, 123
   control unit, 125
   dimensions of variables computed in graph nodes, 367
   machines, 109
   memory array module, 147
   multi-dimensional array, 147
   network, 123
   operation, 106
   organization, 124, 147
   parallel array systems, 102
   processing, 73
   processor, parallel array, 142
   reconfigurable parallel array system, 105
   reconfiguration, 61
   routing network, 124
   square-array organization, 147
   STARAN memory array, 147
   system, 61, 68, 84, 92
Assessing software reliability, 388
Assessment of supernetworks, 89
Assignment of first 12 pulses, 183
   of fuzzy clustering assignment, 482
   of hardware resources, 92
   on implicit enumeration, 482
   of system resource among programs, 373
Associative alternate chaining channel, 159
   array control unit, 144
   cells, interchanging contents of associative cells, 155
   chaining channel, 155, 157
   fault-tolerant memory design, 159
   flag-shift instructions, 158
   global state, 158
   interface unit, 155
   Linear Array Processor (ALAP), 155
   local states, 158
   memories, high-speed peripheral devices for associative memories, 142
   memory, 135, 138, 140–142
   memory, control settings for associative memory, 141
   memory wafer, 139
   output operations, 133
   parallel I/O control unit, 144
   parallel I/O flip network, 144

636 INDEX

Associative *(cont.)*
  processors, 113, 116, 135, 136, 141, 142, 190
  word-cycle instruction, 158
Asynchronous computer architecture, 79
  control organization, 287
  flip-flop, 516
  operation, 208, 212
  sequential machine, 290
Atomic actions, 527, 528
Audit file, 525
Average and peak numbers of accesses, 480
  delay, 477
  instruction-execution time, 189
  throughput, 477
Avoiding serial operations, 155

Background, general benefits and applicability of reconfigurable parallel array systems, 108
Backtracking in branch and bound algorithm, 493
Backward error recovery scheme, 527, 529
Ballistic Missile Defense algorithms, 72, 247
Bandwidth of interconnection bus, 59
Barrel switch, 128
Base address, 85, 298, 326
Basic architectural concepts for parallel systems, 2
  concepts of pipelined computing, 63
  features of PS exchange, 293
  logic equations for ladder module, 139
  modeling methodology, 399
  pipeline, 270
  principles of microprogrammed architecture, 35
  principles of von Neumann and microprogrammed architectures, 56
  von Neumann architectural concepts, 31
Beginning of path segment, 165
Bidirectional chaining operations, 202, 203
  triplicated voters, 508
Binary multiplication, 290
Binary tree network, 83
  variables, 518
Bit-by-bit synchronization of operation, 125
Bit-parallel, 117, 132, 138
  logic, 137
  operation, 101
  search memories, 136
Bit-serial arithmetic and boolean operations, 157
  channel, 156
  logic, 138
  machine, 138
  operation, 101
  requirement, 141
Bit sizes of computed variables, 361, 364, 365
Bit-slice access mode, 147
Bit-sliced microprogrammed computer, 77
Block oriented application, 112
  tasks, 238
Boolean data dependence, 393
  functions, 128
  operations, 149
Bounds, lower and upper bounds of computerization, 74
Branch and bound algorithm, 491, 493
Branching points, 400

Bridging fault, 504
Bucket technique, 116
Buffering, 495
Built-in fault tolerance, 167, 219
Bulk storage, 144
Bus, common bus, 103
Busses described in literature:
  time-shared, cross-bar switch, multiport-memory, 252
  widths of the data busses, 144
Bypassing of unneeded pipeline stages, 268
Byte exchanges between memories of two computers, 336

C-MOS logic, 203
Capacity distribution of system, 540
Carry bits, 138, 150
  propagation times, 336
Catenation, 444
Cell, 135
  array (cell 1 to cell 5), 513
Cells, non-matching cells, 136
Cell's arithmetic and logical unit, 142
  logical complexity, 138
Central control unit, 123
Centralized controllers, 472
CE resource diagram, 374
Changeable operation times in pipelines, 269
Chip, off-chip delays, 199
  on-chip delays, 199
Chip-size limitation, 80
Chip type constraint, 286
Circular node, 406
Clock pulse, 214
Closing predicate, 395, 433
Combinations of flip-permutations and shift-permutations, 153
Common data accessing patterns, 481
  data bus, 124
  data paths, 129
  LSI technological restrictions, 280
Communication bandwidth, 130
  channel considerations, 115
  cost, 492
  intercell communication techniques, 161
  inter-processor communication, 131
  paths, 279
  of processing control, 162
Communications processing, 144
  requirements, 495
Comparand register, 136
Compare logic, 135
  register, 136
Compilation time, 394
Completion signal, 284
Complexity of (micro) programs, 387
  of supersystems, 92
Computational interconnection network, 544
  phases, 63
  requirements, 72
Computed variables, 350, 360
Computer aids, 457
  architecture assembled from universal module (UM), 287
  elements, 273, 277, 278
  modules, 456, 459
  nodes, 83
  requests, 280
  size code, 285, 346
  transitions from one instruction interval to another, 284
  writing of program instructions to all modules in computer, 281
Concurrently one-step diagnosable, 564
Conditional branch instructions, 21, 24

branch microinstruction, 51
branch microprogram, 25
branching, 290
  pipeline drains due to conditional branching, 270
  test, 21, 306
  test microoperations, 12
  unconditional transitions, 12
Conditions of structural termination, 393
  which guarantee termination properties, 389
Conflicts between concurrent programs, 277
  between memory accsses, 134
  in noiland machine, 168
Connecting, mapping conecting elements on copies of same LS module type, 87
  elements, 252, 273, 279
Connection of pair of microcomputers, 86
Connectivity, 548
Consequence of modular control organization, 287
Constant error rate, 533
Constraints, 455
Construction falts, 504
  of diagram of hardware resource, 367
Context-sensitive language translation, 238
Contradiction between one-staged and multistaged interconnection networks used in supersystem, 252
  between system's throughput and its reliabiliy, 72
  encountered in design of supersystem, 248
Contribution of parallel systems to computations, 2
Control of accessing, 464
  channels, 103, 105
  circuit follows executional circuit, 9
  circuit precedes executional circuit, 8
  circuits, 4, 5
  circuits, special purpose control circuits, 253
  codes for architectural state, 279, 311, 312
  codes stored in each LSI module, 281, 282
  console, 131, 132
  error, 504
  field of frame, 511
  flow, 464
  flow of (micro) program, 393
  flow of representation, 390
  graph, 403
  of input, output, and instruction execution, 131
  instructions, 146
  lines, 86
  memory, 35, 36, 45, 77, 78, 120, 146
  organization of dynamic architecture, 280
  processor, 45
  program control distributed, 162
  settings for associative memory, 141
  signals, 28, 132, 133, 214
  states, 142
  unit, 4, 120
  unit bus, 124
Controllable data transfer, 4
Controller memory, 208
Controllers for arrays, 237
Content-addressable memory, 135, 155
Conventional or general-purpose computation, 74

INDEX 637

testing criteria, 392
Copies of module type, 75
Coresidence of several types of architectures, 270
Correct paths for equality signals via software, 300
Correlation registers, 133
Cost-effective design, 80, 86
Cost of realization, 341
Count over number of iterations, 168, 357
Coverage vector, 532
Cross-bar switch, 252
CRT terminal, 144
Cumulative distribution function (CDF), 533
Current microinstruction address, 45
Cycle of macroinstruction, 46
  in V-graph, 426

Damage assessment, 526, 527
Damaging transient, 530
Data Access graphs, 464
  access matrix, 480, 483
  acquisition statements, 405
  communication channels, 156
  dependent branching, 476
  driven processes, 474
  entities, 464
  error, 505
  fetch mode mE, 285
  flow, 92, 464
  flow analysis, 391
  and instructions, 4
  management, 144, 201
  manipulation operations, 212
  parallelism, 57
  reduction operation, 161
  replication of shared data, 495
  retrieval operations, 142
  routing node, 163
  states, 142
  storage requirements, 480
  structures, 219
  update of same data entity, 497
  vector, 57, 108, 254
  vector processed by single instruction, 74
  widths of data buses, 144
  word fetch, 281
DC group flow chart, 377
  hardware resource, 277
  with minimal complexity, 273, 336, 338
  with minimal delay, 317
  resource, 273
  transition from one state to another, 227, 273, 314
Decision tree, 492
Degenerate ALAP instruction memory, 209
Degrading pipeline performance, 70
Degree of degradation, 531
Delayed addition in processor, 65
Delay of component, 7
  requirements, 496
Delays, 128, 497, 340
Dependence on input data, 391
Description language, 1
  of universal module, 187
Design faults, 504
  languages, 457
Designing systems with dynamic architectures, 245
Destination register, 4, 36
Determinate faults, 504
Development of adaptable architectures, 81
Diagnosis of digital systems, 502

Diagnosis interconnection network, 544
  performance measure, 543
  program, 525
  reconfiguration, 543, 559
  reconfiguration tree, 560
  unit, 131
Diagram of bit sizes for all program graph nodes, 350
  of hardware resource, 373
Digital, fault-tolerant digital system, 523
Dimension of each data vector, 93
Dimensional, multi-dimensional array, 147
Dimensionality of processor array, 240
Dimensions of data arrays, 367
Direct communications between different functional units in system, 257
  process-memory exchanges, 59
  transfer of microcommands, 251
Disagreement detector, 512
Disparity between program and pipeline structures, 83, 93
Displacement, 326, 332
Distributed architectures, 388
  control arrays, 161
  control parallel processors, 116
  faults, 504
  program control distributed, 162
Distributing DC group primary memory among various user programs, 379
Dynamically reconfigurable systems, 102

Effect of LSIs on supersystem architectures, 173
Effectiveness of pipeline system, 68
Effects of LSI technology on computation and computer organization, 70
Efficient recovery algorithms, 568
Elementary cycles of execution, 46
  switch, 515
Elements, mapping connecting elements on copies of same LSI module type, 87
Elimination of time of memory accesses from time of execution, 267
Empty spaces created in primary memory because of variable word sizes, 300
Enable/disable state of processing element, 128
Encoded microcommands, 80
Encoding of microinstruction, 43
Enhanced computational parallelism, 74
Entry points, 400
Equality signal generated in k • h-bit computer, 306
Equivalence detectors, 519
Equivalent regular expression, 445
Error checking, 525
  detection, 510, 526
  detection procedure, 509
  location, 510, 512
  message, 525
  rate, 533
  recovery, 527
Errors in firmware, 387
Establishment of proper transfer modes in connecting elements, 308, 317
  via software of control unit, 280
Estrin model, 400
Evaluation of predicate, 402
Evolution of adaptable architectures, 92
  of computer design, 80
Exception handling, 529
Excessive complexity of resources, 254, 256

Exchanges of temporary results between tasks in pipelines, 69
Executable predicates, 526
Execution activity, 479
  of additional program streams on same hardware resource, 91
  of algorithms by dedicated subsystems, 262
  control logic, 132
  portion of universal module, 288
  speed-up in pipeline, 269
Exit points, 400
Explicit replication of functions, 478
Exponent correction, 29
Exponential growth of number of connecting elements, 252
Exponents AE, 15
Extents, 504
External circuits connecting LSI modules, 80
  connections to wafers, 122
  devices, 85
Extreme throughputs, 72

Fabrication cost for computer hardware, 71
Failure, 504, 522
  to compute algorithms under time restrictions, 32
  rate of powered module, 531
False (specified with jump address) program sequences, 270
Family of bit-sliced computers, 76
  of modular computers, 74
Faster execution of microprograms, 45
Fault, associative fault-tolerant memory design, 159
  condition table, 546
  diagnosable system, 546
  diagnosis, 512
  duration, 504
  free active modules, 527
  isolation, 110, 512
  isolation instructions, 159
  pattern, 545
  recovery, 568
  repair wafer, 110
  t-fault diagnosable, 545
  tolerance, 505, 506, 566
  tolerant at cell level, 161
  tolerant design, 502
  tolerant-designed hardware, 389
  tolerant digital system, 523
  tolerant graph, 543
  tolerant software structure, 524
  wafers, fault tolerance, 237
Faults, transient (intermittent), 508
Faulty elements, 110
  non-adjacent faulty modules, 522
  software module, 527
Fetch data code mE, 282
  instruction code mI, 282
Field-programmable arithmetic operations, 194
  operations, 158
Field-selective arithmetic, 150
File for established pulse train, 177
  management task, 116
  managing operations, 142
  organization-and-search scheme, 116
  processing application, 201
  retrieval applications, 238
Filling of empty spaces created in memory element, 382
Finding priority of requesting program, 308
  time of computing each task, 370

638   INDEX

Firmware, hardware, software and firmware checking mechanisms, 510
Fixed instruction microprograms, 32
   microprogram of von Neumann architecture, 32
Flag fields, 511
   shift operations, 159, 215, 216, 223, 225
Flexibility of bus, 326
Flip network, 150
   permutation flip network, 147
Floating-point arithmetic, 141
   instruction, 132
   multiplication, 18
   multiplication instruction, 25
   result, 30
Flow chart of architectural states, 92
Formal correctness, 391, 403
   verification, 392
Formation of additional computers from idle resources, 260
Forward error recovery scheme, 529
   technique, 527
Frame, 511
   check sequence field (FCS), 511
Free requested resource, 309
Full scale STARAN systems, 143
   self-repair provisions, 512
Function code, 286, 288, 308
   symmetric majority function, 518
Functional error, 504
   mapping, 481
   requirements, 455
   simulator, 221

Gains in performance, 91
General dynamic redundant system, 510
   purpose instructions, 73
   signal-processing tasks, 176
Generation of instruction addresses only in PE contained in same CE, 297
Goodyear STARAN, 143
GOTO statements, 399
Graceful degradable system without spares, 535
   degradation, 131, 530
   degradation strategy, 527, 530
Graph models, 393

Hardware faults, 524
   fabrication cost for computer hardware, 7
   software and firmware checking mechanisms, 510
Hard-wired digital logic, 170
Hierarchical memory management, 402
   pyramid network, 83
   structure chart of elaboration, 471
Highly-parallel architectures, 219
Highly reliable computations, 248
Holland ADD instruction, 166
   conflicts in Holland machine between paths connecting programs to their respective operands, 168
   NOP instruction, 166
   SET REGISTERS command, 167
   STORE instruction, 166
   total of eight instructions in Holland machine, 166
   TRANSFER ON MINUS instruction, 166, 167
Honeywell 645 data-processing system, 144
Horizontal microprogramming, 409, 453

((i, j) connection, 320
I/O-memory exchange, 59
overhead, 456
Idle state, 562
ILLIAC IV, 122, 134
Image-processing applications, 118
Images, application, processing of digitized images, 154
Implementation of separate dedicated subsystems, 271
Implicit enumeration and evaluation, 491
Improving on von Neumann's architecture, 33, 56
Increase in delays or complexities introduced by interconnection logic, 251
   in number of processor registers, 34
   in number of programs computed by same hardware, 83
   in system throughput caused by LSI technology, 74
Increases in computational throughput, 57
Incremental value, 85
Independence of control organization of number of control units in it, 281
Independent computers, 91
Indeterminate faults, 504
Indexed, 404
Indexing values, 129
Inductive assertion, 391
Infinite looping software, 526
   methodology for analyzing existence of possible infinite loops, 388
Information exchange between processor and memory of two different computers, 325
   gathering, 391
   stream with high priority, 260
Input data rates, 476
Input-output relationships, 407, 428
Instruction, arithmetic and boolean instruction-execution logic, 132
   arithmetic pipelines, 67
   buffer, 125, 128
   cycle, 281
   and data byte reception, 294
   decoding, 119, 371
   decoding logic, 163
   effective address for instruction, 222
   estimates, 476
   execution in Holland machine, 166
   fetch, 265, 281, 284, 290, 371
   fields, 36
   formats of DC group computers, 297
   Holland ADD instruction, 166
   Holland NOP instruction, 166
   Holland STOP instruction, 166
   Holland STORE instruction, 166
   Holland TRANSFER ON MINUS instruction, 166, 167
   interpreted as sequence of intervals, 290
   intervals, 281
   intervals executing processor dependent operation, 290
   "jumpME$i \rightarrow$ ME$j(Ad)$", 298
   microprogram, 11, 21, 291
   parallelism, 57
   pipeline, 68
   processing overlap, 189
   rates, 456, 480, 496
   sequencing, organization time intervals, 290
   set, 3, 456
   set adaptations, 73
   streams, parallel instruction, 92
   type code, 41
   "X memory—Y memory", 335
   "X processor—Y memory", 325, 327
   "X processor—Y processor", 332

Instructions, associative flag-shift instructions, 158
   fault-isolation instructions, 159
Integer arithmetic and logical instructions, 133
   programming, 491
Interactive system, 477
Intercell communication techniques, 161
Interconnection logic, 513
Interconnections between wafers, 110
   cost-effective organization of reconfigurable interconnections, 86
Interface functions, 464
   (square) nodes, 406
Intermittent, transient (intermittent) faults, 508
Intermittently fault condition tables, 558
Intermodule communications requirements, 487
Internal communication for DC group, 279
   flags, 157
   processing capability, 133
   select logic, 134
Internally-stored data sets, 130
Interpretive analysis of programs, 391
Interprocessor communication networks, 135
Interrupt signal, 526
Interval duration, 283
   sequencer, 283, 284, 291, 371
Intra-cell communication logic, 201
Invariantcodes, 286
Inversion of matrix, 112
Iterations, 168

Ladder, basic logic equations for ladder module, 139
Large computer architectures, 1
   delays introduced by reconfiguration logic into carry propagation times, 336
   module size, 82
   number of parallel instruction streams, 92
Latch, external latch, 139
Leaf node, 492
Least significant memory element, 297
Left transfer of connecting element, 273, 300
Length of node chain, 552
Less-than comparison, 157
Life cycles of supersystems, 73, 94
Limitations of LSI technology, 80
Linear array, 124
   encoding, 36, 38
   organization of chaining channel, 161
Linkage editor, 221
Loading on processor, 111
Local control unit instructions, 128
   end-around carry path, 305
   microcommands, 284, 292
Locational constraints, 482
Lock-step nature of system operation, 135
Logarithmic encoding, 36, 38
Logical, cell's arithmetic and logical unit, 142
   clock, 479
   communications network, 460
   device ladder, 139
   functions performed by cell, 141
   lead fault, 504
   relational expressions, 526
Longest operation in one LSI module, 281
Look-ahead, ALAP look-ahead

INDEX 639

instruction, 208
Loop enclosing an instruction sequence S, 444
  enclosing two or more parallel alternative edges, 437
Low-cost hardware, 75
  LSI modular computers, 74
Lower and upper bounds of computerization, 74
LSI, advances in LSI technology, 71
  implementation of individual functional units, 251
  module, 75
  modules with high throughputs, 2, 82, 83, 287
  technological advances, 12
  technological restrictions, 280
  technology and traditional architectures, 79
  or VLSI wafer, 110

Machine instructions, 478
  reliability, 387
Macroinstruction address, 45
Macroinstructions, 35, 40, 45
Main memory counter, 45
Maintainability of system, 487
Maintenance control, 131
  pulse-train maintenance operations, 174
Major drawback of architectural dedication, 254
Majority function, 520
  symmetric majority function, 518
  voter, 512
Malfunctioning of single PE, 134
Manipulation of very large matrices, 123
Mantissa addition, 18
  multiplication, 28
  overflow, 30
  product, 30
Mantissas, 15
Manual, 527
Mapping connecting elements on copies of same LSI module types, 87
  microprogrammed architecture onto LSI modules, 79
Masked, content search masked, 135
Masking errors, 509
  operation, 149
Mask register, 136
Match flip-flop, 136, 139
Matrix, first order similarity matrix, 485
  multiplication, 423
Maximal concurrency present in programs, 250
ME resource diagram, 374, 379
Mean computation before failure, 539, 541
  response time, 542
  time before failure, 537
  time to repair, 539
Memories, bit-parallel search memories, 136
Memory access and input output (I/O) rates, 456
  access interval, 284
  access operations, 32, 33
  accesses, 33, 63
  address modification, 128
  ALAP memory, 206
  allocation techniques, 293
  array memory, 17
  array module, 147
  associative fault-tolerant memory design, 159
  bit-slice memory-access mode, 149
  connecting element, 318, 320
  economy accomplished with two-level control memory, 52
  elements, 4
  instruction memory word, 209
  -memory exchanges, 264
  portion of sequencer, 290
  search memory implementations, 135
Merger of multicomputer and multiprocessor architectures, 271
Message streams, 479
Methods for increasing throughput, 74
Microcommand, 5, 292
Microcompilers and microassemblers, 408
Microcomputer, parallel transfer of h-bit word from one microcomputer to another, 86
Microcomputer-oriented industry, 71
Microcomputers, 74, 78
Microdiagnosis, 511
Microinstruction classes, 40
  counter, 52
  cycle, 46
  set, 41
  size, 40
Microinstructions, 35, 40, 50
Microoperation, 3, 6, 413
  approach, 2
  description of architecture, 1
  hardware diagram, 18, 25
  and microcommand, 4
  parallelism, 56
Microorder, 409
Microprocessor, simplified circuits, 79
  system, 508
Microprocessors, 74
Microprogram, 3, 25, 388, 392, 399, 404, 409
  directed graph, 25
  graph, 11
  reliability, 389
  sequence, 50
Microprogrammable, 81
  adaptation, 83
  computer, 253
Microprogrammed architecture, 57, 76, 81
  architecture with two-level memory, 49, 52
  computers, 1, 30, 33
Microprogramming, 387
Microprograms, 45
  that are more task-oriented, 77
Middle PE, 305
Million instructions per second (MIP), 134
Minimally required times of operations, 286
Minimization of component count, 80
  of dummy intervals created in pipeline, 93
  of idle resources, 260
Mission time, 538
Mixed architecture, 270
Mode flip-flops within each PE, 129
  register, 120, 128
Modeling timing and synchronization characteristics, 415
Modern progress in adaptable architectures, 83
Modes of operation for dynamic architecture, 339
Modified displacement constant, 327
  module, 516
Modular architecture, 70, 74, 259
  computers from off-the-shelf modules, 78
  computer systems, 2
  control organization, 281, 284, 320
  control organization and LSI technology, 286
  expansion of complex parallel system, 72, 250, 341, 343, 347
  increments, 76
  multi-valued 5-modular system, 518
  N-modular redundancy/oipurge, 513
  redundancy, 5-modular redundancy, 518
  triple modular redundancy, 507, 518
Module, 75
  type fabrication, 78
  types, high cost of developing new module types, 81
  ports for inserting new data, 161
  program streams, 162
Module, memory array module, 147
Moment when required resource is free, 308
Monitoring, passive monitoring system, 170
Motorola M10800, 78
Movement of upper bound of computerization upward, 71
Multi-array networks, 125
Multicomputer computations, 262
  system, 57, 59, 83, 84, 92, 102, 279
Multi-dimensional access memory, 147
  processing arrays, 167
Multi-object tracking mode, 135
Multimicrosystems, 78, 84
Multiphased microinstructions, 41
Multiple controllers, 205
  data stream, 123
  failures, 110
Multiple-array PEPE systems, 189
  systems, 192
Multiplicand, 28
Multiplier, 28
Multiport-memory, 252
Multiprocesses environment, 527
Multiprocessing system, 82
Multiprocessor, 57, 59, 263
  architecture, 26, 398
  computations, 263
  system, 59, 102, 415
Multiprogramming mode of operation, 31
Multistage networks, 88, 252
Multi-values 5-modular system, 518
Mutually distinguishable, 546
  reachable, 548

N-modular redundancy, 506, 512, 513, 514
n-to-one throughput, 239
Nested loops, 451, 475
Network, 83, 513
  for communication of data, 162
Networks assembled from h-bit microcomputers, 90
  of communication channels, 162
New adaptations that improve performance of available resources, 92
No-transfer mode of connecting element, 294
Node chain, 552, 556
  loop, 552
Nodes, 446
  square nodes representing "interface" with rest of program, 406
Non-free redundant resource, 256
Non-recursive addition, 361
  multiplication with complex changeable factors, 362
  multiplication with simple changeable factors, 362

Non-redundant voter, 507
Non-terminating loops, 393
Nonsequenced information frame, 511
Normalizations, 29, 30
n^2-to-one throughput, 239
Numbers of data accesses per unit time, 480
   of gates per wafer, 234

Objective of reconfiguration, 506
   to identify or classify sources of individual pulse streams, 170
Obsoletion of complex systems, 248
Off-chip delays, 199, 235
Off-the-shelf modules, 78, 79
On-chip delays, 199
One address computer, 31
   dimensional array, 118
   microinstruction cycle, 46
   phased, 42
   phased microinstruction cycle, 52
   stage networks, 88, 89, 252
   step t-fault diagnosable, 546
   step t/s diagnosable, 551
   step $[X_0, X_1, ..., X_{A-1}]$ diagnostically reconfigurable system, 561
Operand access, 65
Oprands, 132
Operation code, 164
   parallel-by-word operation, 155
   search-by-content operation, 136
   sequencer, 283, 284, 291
   serial-by-bit operation, 155
   speed up, 269
Operational requirements, 455
Operations, bi-directional chaining operations, 202, 203
   field-programmable, 158
   flag-shift operations, 159
   text-processing operations, 161
Operator interface, 120
Opportunities of LSI technology, 80
Optimization of adaptations for arrays, 93
Opimizing compilers, 391
Organization of instruction sequencing nd variable time intervals, 291
   of instruction storage, 293
   of pipeline system, 67
   of 16 • k-bit processor, 288
Output values, 408
Overflow-in and overflow-out pins in each UM, 288
Overlapped execution, 50

P-resource diagram, 373
Parallel access disk, 13
   arithmetic and boolean operations, 149
   array processor, 142
   buffer, 120
   -by-word operation, 155
   byte exchanges between computers, 326
   complex parallel systems, 1
   connection between processors of two computers, 332
   data streams, 478
   exchange between two processors, 331, 344
   instructions, 131, 134, 146
   interconnections between two processor modules, 331
   network processor array, 120
   network processing arrays, 223
   network processors, 116, 121, 134
   processing, 115
   processing memory contents, 161
   program streams, 72
   series-parallel programs, 389
   systems, 57
   transfer of h-bit word from one microcomputer to another, 86
   -write operation, 138
Parallelism, 56
Parameter of iterations, 357
Parity checks of words, 511
Partial fault condition table, 547
Partial product, 28, 290
Partitioning and allocation, 456
   input data among multiple arrays, 205
   maintainability-sensitive partitioning, 487
   of microinstruction set into microinstruction classes, 40
   of microoperations into groups, 36
Passive modules, 512
   monitoring system, 170
Path, 407, 425, 446
   for overflow, 304
   types in V-graph, 425
PC boards, 122
PE multi-purpose, 162
   storage register, 165
Penetration of microdevices, 71
PEPE, 134
   processing elements, 131
Percentage of circuits mapped onto LSI modules, 81
Performance criteria, 482
   degradation from sequential processing, 190
   degrading pipeline performance, 70
   of dynamic architectures, 338
   and/or functional degradation, 527
   measure of communications networks, 568
   new adaptations that improve performance of available resources, 92
   requirements, 455
   of supersystems, 93
Period length λ, 434
Peripheral hardware, 144
Permanent faults, 504
Permutation flip network, 147
PEs directly linked to one another in string fashion, 163
   local memory, 132
Phase-array radar system, 130
Phase duration, 42
   lags, 235
   overlap, 51
Phased execution of microinstructions, 41
Physical faults, 504
Pin-count restriction in LSI module, 80, 81, 87, 286
Pin-limited wafers, 141
Pin-to-pin connections, 80
Pipeline, 57, 93
   adaptation to parallel streams, 267
   architectures, 93
   with changeable operation time in stage, 269
   configuration, 205
   drains due to conditional branching, 270
   fast information exchanges between pipeline stages, 70
   races, 269
   stage, 66
   states of architecture, 267
   structures, disparity between program and pipeline, 83
   systems, 63, 67, 101
Pipelining, 65, 92, 267
Pixels, rotation and interpolation of pixels, 154

Popular mainframes, 75
Position codes, 286, 320, 326
Power LSI modular computer, 75
   parallel system, 93
Precision of computations, 260, 262
Predicate associated with type-o node, 444
   fail exit, 526
   true exit, 526
Preprocessing, 389
   actions, 402
   for reliability purposes, 389
Primary memory, 280
Principle of addresses, 31, 56
Priority analysis, 308, 309
   of calling program, 277, 314
   codes, 279
Probabilistic model of incomplete fault detection, 567
Probability density function, 533
Problem of conditional branch in pipelines, 270
Process/processor synchronization, 453
   retry or alternate program segment in process, 528
   synchronization, 479
Processes that allow cost-effective computerization, 71
   for data sets, 130
Processing, application, processing of digitized images, 154
   of image data from multiple sensors, 130
   query-processing, 108
   radar data processing, 108
   text processing, 108
Processor code, 282, 284
   data-sequential processor, 101
   dependent operations, 91, 269, 280
   direct processor-memory exchanges, 59
   independent operation, 269
   memory communication, 74
   microoperations, 292
   parallel array processor, 142
   parallelism, 453
   -processor exchange, 59, 263
   resource, 265, 374
   sequential processor, 101
   -supervisor in array, 265
Program characterized by structural termination, 395
   computation in system with dynamic architecture, 293
   control distributed, 162
   correctnness, 394, 415
   disparity between program and pipeline structures, 83
   execution, 128
   graph, 350
   graph model, 399
   loop, 357
   preprocessing, 70, 92
   residing in array, 163
   restart, 566
   retry or alternate program segment in process, 528
   rollback, 531
   segment stored in memory element, 298
   selection code, 321
   storage in several memory elements, 297
   termination, 388, 392
   universality, 293
   user code, 314
   validation dynamic analysis, 390
   written in high level language, 350
Programming methodology, 390
   reconfigurable parallel array

INDEX   641

processors, 218
systems with dynamic architecture, 348
Programs having different sequences of operations, 69
series-parallel programs, 389
Proliferation of adaptable architectures, 74
   of modular architectures, 81
   of supernetworks, 86
Proof of correctness, 391
Propagating network, 139
Propagation delays, 496
Propagations of carries and overflows, 267
Protective redundancy, 506
Pseudocode, 474-79
Pulse, acquiring pulse trains, 174
   carrier frequency, 172
   file for established pulse train, 177
   objective to identify or classify sources of individual pulse streams, 170
   streams, 172
   time-of-arrival, 172
   train, 188
Purging action, 515

Quality of architectural adaptation, 93
Query-answering, 239
   processing, 108
Question-answering systems, 108, 238
   tasks, 238

Radar data processing, 108
Random-access memory, 135
Real-time algorithms, 72, 92, 260
   control, 455
   system, 477
   text processing, 238
Receiver of data on chaining channel, 157
Reconfigurability of interconnections between modules, 82
   of PIPE, 134
Reconfigurable, 81
   adaptation, 83
   architectures, 82, 388
   array systems, 61, 135
   busses for end-around carry, 301
   characteristics, 142
   cost-effective organization of reconfigurable interconnections, 86
   existing reconfigurable systems, 90
   fault-tolerant systems, 530
   interconnections among modules, 81
   microprogrammable, reconfigurable, and dynamic properties, 83
   multiprocessor system without spares, 544
   network control unit, 544
   parallel array processors, 121
   parallel array systems, 102, 105
   parallel hardware, 169
   paths for processor signals, 301, 304, 306, 336
   pipeline systems, 83
   supernetworks, 84, 85
   types of reconfigurable architecture, 84
Recovery action, 529
   algorithm, 504
   cache-type mechanism, 528
   line, 529
   point, 528
   procedure, 566

Recursive addition, 361
   floating-point multiplication, 365
Reduced information extracted from program, 394
Reduction in number of addresses, 33
   in number of module types, 76
Redundancy, time, 530
Redundant resource, 73, 91, 254, 256
   system, 532, 567
   voters, 507
Register files, 413
Register-to-register transfer, 4, 5
Relation among component failure, 503
Reliability calculation, 532
   complex algorithms requiring extreme reliability, 72
   and computation performances, 568
   of computer components, 74
   evaluation, 532
   of software and firmware, 387
   of supersystem, 72
Reliable modular computer systems, 74
   software, 389
Relocatable object modules, 221
Repairable, computer system self-repairable by standby spares, 531
   non-repairable computer systems, 531
Repeated execution of instructions, 530
Replacement strategy, 527
Replication of critical programs and data, 525
   of modules, 80
Request for architectural transition, 309
Requests for communication, 279
Requirement for supersystems, 72
   of two memories, 77
Rerouting of data, 134
Resetting enable flags, 188
Reshuffle operands among cells, 159
Resolving conflicts among I/O elements, 279
Resource assignment for dynamic architectures, 309
Resource diagram, 318, 350
   requirements of algorithm, 254
   resources, new adaptations that improve performance of available resources, 92
Response pattern, 545
   time, 542
Restricted bandwidth of interconnection bus, 58
Result of exclusive-OR operation, 137
Resynchronization subroutine, 509
Retrieval, data-retrieval operations, 142
Reversal check, 527
Right transfer of connecting element, 273
Ring structure of fault-tolerant reconfiguration network, 522
   structure type of logic network, 520
Rollback, 530
Rotation and interpolation of pixels, 154
Rounding, 29
Routing logic, 119, 128
   network, 124

Sami and Stefanelli program scheme, 401
Search-by-content function, 138
   -by-content operation, 136
   memory, 135, 138
   operations, 108
   and write new control codes, 338
Selection logic for communication operations, 240
   of size of one reconfiguration zone, 36
Self-loops, 402

Self-purging redundancy scheme, 515
Semantically equivalent program, 399
Semantic characteristics, 394
Separation of concurrent instruction streams, 294
Sequence of macroinstructions, 45
   of nodes requiring same computer size, 350
   of operational units, 83
Sequenced information frame, 511
Sequences of best architectural states, 70, 92
   of instruction intervals, 283
Sequencing of two instruction sequences, 44, 297
Sequential architecture, 477
   pipeline processor, 192
   processing of subsets of cells, 139
   program, 399, 415
Sequentiality barrier, 57
   of computations, 56
Sequentially t-fault diagnosable, 546
Serial bit-by-bit inputs, 220
   instructions, 131
   processor, 135
Series-parallel programs, 389
   structures, 399
   structures of graph models, 393
   transition graph, 445
Series, step-by-step identification and reduction to series-parallel structure, 446
Set of architectural states, 260
   of array architectural states, 265
Setting enable flags, 188
   match flip-flops, 140
Shared bus, 294
   processing with another computer, 161
   resources, 415
Shift permutations, 150, 153
   register, 135, 157
Shuffling operands, 161
Signal, general signal-processing tasks, 176
   paths for dynamic architecture delayed due to reconfiguration, 340
   processing, 73, 173
   sorter, 170
   sorting, 192
   sorting algorithms, 173, 177
   sorting algorithms with no prior knowledge of pulse train characteristics, 177
   sorting program, 215, 216, 223
   sorting system, 170, 203
Signed, 23
Significance code, 285, 346
Significant time losses caused by frequent memory access operations, 32
Similarity, third order similarity, 486
Simple conditional transition, 12
   control circuit, 4
   loop, 424
   loop analysis, 432
Simplex mode, 518
Simplified circuits typical of microprocessor, 79
   instruction set, 78
   modular computers, 79
Simulations, 478
Single failures on wafer, 110
   fault, 518
   instruction stream, 123
   loop system, 550
   word-chain operation, 226
Slice code, 286
Software architecture design, 45
   compatibility, 78

Software *(cont.)*
  control of interconnections, 82
  controlled activation and deactivation of interconnections with other modules, 74
  controlled fault isolation, 110
  controlled microprograms, 34
  design fault, 504
  error, 504
  error detection scheme, 526
  error diagnosis, 526
  error recovery, 527
  errors, 504
  fault-tolerance, 524
  formation of new end-around carries, 303
  formation of variable size processors, 300
  hardware, software and firmware checking mechanisms, 510
  modules, 458, 459
  programming tools, 220
  reconfigurable, STARAN, software-reconfigurable array, 150
  reconfiguration of path for end-around carry, 303
  redundancy, 504, 506, 524
  reliability, 388, 390
  validation, 390
Solid faults, 504
SOLOMON, 134
Sorting out stream of intermixed pulses, 170
Source, 36
  instructions, 478
  register, 4
Sources, 427
Spare, failure rate of spare module, 531
Spares, computer system self-repairable by standby spares, 531
Special instruction, "X processor—Y memory", 325
  programming for reconfigurable pipelines, 69
  programming techniques, 32
Specified applications, 219
Speed-up in array computations via reconfiguration, 267
  of between-computer communications, 91
Square-array organization, 147
  nodes representing "interface" with rest of program, 406
Stack, 133
Standard memory units, 77
Standby spares, 512, 525
Star configuration, 85
  network, 83
STARAN, 143
  flip network, 144
  full scale STARAN systems, 143
  memory arrays, 146, 147
  software-reconfigurable array, 150
State capacity distribution, 541
  vector, 528
  vector of software system, 525
Statement referenced by Do statement, 351
Static and adaptable architectures, 81
  hardware redundancy scheme, 507
  program validation static analysis, 390
Step-by-step parallelism, 415
Step multiplication, 157
Strong cycles, 428
  edge, 406, 425
  submatrices, 438
  submatrix S, 428
  and weak paths, 407

Structural termination, 390, 393, 394, 398, 413, 438, 451, 453
  testing, 390
Structure, step-by-step identification and reduction to series-parallel structure, 446
  of statements, 394
Structured design, 458, 463
  programming, 389, 399
Stuck-at-1, 518
  -at-0, 518
  -type fault, 504
Submatrix of V-matrix, 428
Subsidiary pipeline, 270
Subtraction, 157
Supernetwork, 85, 86
  topological configurations of supernetwork (array, star, binary tree, etc.), 84
Supersystem, 72, 74, 92
  design, 94
  for complex real-time algorithms, 248
  with dynamic architecture, 260, 261
  modular expansion of existing supersystem, 72
Supersystems with adaptable architectures, 94
  adaptable architectures, for supersystems, 73
Supervisory frame, 511
Surveillance, applications, surveillance systems, 144
Switching network for reconfiguration, 524
  system architecture, 70
Switch reconfigurable busses, 338
Symbolic execution, 391
  technique, 11
Synchronization, 388, 453
  bit-by-bit synchronization of operation, 382
  mechanism for input-output relationship graph, 418
  modeling techniques for timing and synchronization problems, 399
Synchronous and asynchronous control organizations, 281, 287
  computer arhitecture, 79
Synthesis, 463
  activity, 456, 494
System, ALAP demonstrator system, 206
  default action, 529
  different-task system with intercommunications, 531
  with double failures, 522
  fault-tolerant digital system, 523
  full scale STARAN systems, 143
  Honeywell 645 data-processing system, 144
  identical-tasks system without intercommunications, 530
  maintainability of system, 487
  passive monitoring system, 170
  reliability, 72, 122, 536, 583
  self-repairing systems, 531
  signal-sorting system, 170
  with single failures, 520
  635-multi-user time-shared operating system, 144
  throughput, 135
Systems, gracefully degradable, 531

t-fault diagnosable, 545, 546, 548
  systems for permanent faults, 548 with repair, 555
Task allocation, 402

  synchronization, 308
Techniques for describing microprogram, 11
Technological advances in seventies, 93
Temporary results, 32
Tentative time for executing task, 370
Termination, conditions, 389
  of microinstruction phase, 42
  of (micro)program, 393
  of program loops, 394
  properties, 388
Testing and validation of aerodynamic design, 247
  for availablity of requested equipment, 308
Text-processing operations, 161
Three address computers, 31
  instructions, 34
Three-phased, 42
Three-valued matrices for loop behavior, 393
Threshold value M, 543
  voter, 512
Throughput, contradiction between system's throughput and its reliability, 72
Time of architectural reconfiguration, 339, 343, 346
  to execute one graph node, 371
  to execute one instruction, 371
  of exernal information exchanges, 257
  of microcommand delay, 9
  of operation in pipeline stage, 269
  overheads, 68, 264
  restriction in complex real-time algorithm, 247
  saving strategies, 128
  shared, busses described in literature: time-shared, crossbar switch, multiport-memory, 252
  tag, 249
Timing analysis of processes critical with respect to time, 175
  and control signals, 132
  in microoperation execution, 8
  modeling techniques for timing and synchronization problems, 399
  or synchronization of programs and microprograms, 415
  and synchronization problems, 399
Top-down structured design, 458
Topological configurations in network, 83
Trace capability of simulation, 222
  printout of execution, 223
Transfer mode of connecting element, 294, 315
  of word from one register to another, 4
Transform of fixed point number into floating-point form with a smaller bit size, 369
Transient faults, 504
  (intermittent) faults, 508
Transition graph of finite-state recognizer automaton, 444
Translating data access graph into language representation, 474
Trigonometric functions, 73
Triple modular redundancy, 507
  modular redundant/spares, 517
  modular redundant system, 518
  modular software redundancy, 525
Triplicated unidirectional voters, 508
True (incremental) program sequence, 270
t/s (t-out-of-s) diagnosability, 551

INDEX 643

Types of adaptations, 93
  of conditional tests, 23
  of instruction intervals, 281
  of instructions in microprogrammed computer, 35
  of interconnection networks, 252
  of parallel systems, 57
  of reconfigurable architecture, 84

UM performing function of processor element, 288
Unconditional, 12
Unconstricted, programs with unconstricted structures, 389
Unique microinstruction format, 41
Univac 1108, 78
Universal LSI module, 287, 292, 308
Universality of executed programs, 293
Unmasked character positions, 136
Unrelated applications, 167
Unsigned, 23
Update of same data entity, 497
Updating and access of variable, 420
  of monior idle equipment code, 309
  of all pulse trains, 191

Upper bound, movement of upper bound of computerization upward, 71
Use of dedicated architectures, 253
  of high speed components, 249, 250
  of microoperation approach of computer architecture, 21
User programs, 90, 92
Utilization of maximal concurrency present in algorithms, 254

V-graph, 425
V-matrices, 408
V-matrix, 408
Validation, program validation ynamic analysis, 390
  of sequential programs, 415
Variable codes, 284
  microprogram, 35
  number of concurrent arrays, 93
  size computer, $C_{.i}(k)$, 273
  time intervals, 281
  usage tables, 391
Variables referenced through addressing, 404
Variation of intervals, 282

Vertical microprogram, 413, 415
VLSI fabrication of parallel arrays, 110
Voter, triple modular redundancy CPU-memory-voter configuration, 508
Voting at bus level, 507

Wafer, number of pins on, 141
Wafers, fault tolerance, 237
  pin-limited wafers, 141
Watch-dog timer (WDT), 526
Weak cycle, 428
  edge, 407
  self-loop, 428
Word comparison, 288
  cycle operations, 190, 203, 216, 225
  cycles, 203
  instruction memory word, 209
  serial, 120
  size, 456
  size of processor, 93
  size required by program, 260
  slice access mode, 147